Joann Weller

Joann Weller

SUPERVISORS' SAFETY MANUAL

9th Edition

National Safety Council

Itasca, Illinois

Project Editor: Patricia M. Laing
Cover and interior design: Hughes Design/Communications

COPYRIGHT, WAIVER OF FIRST SALE DOCTRINE

DISCLAIMER

Library of Congress Cataloging-in-Publication Data
Supervisors' safety manual / National Safety Council.—9th Edition
        p.   cm.
    Includes bibliographical references and index.
    ISBN 0-87912-197-1
    1. Industrial safety—United States—Management—Handbooks, manuals, etc.
    I. National Safety Council.
    T55.S8156   1997
    658.3′82—dc21   97–8993
                        CIP

2M300   Product Number: 151430000

CONTENTS

PREFACE

The roles and responsibilities of first-line supervisors and team leaders have changed. You are still the direct link between management and the workforce. You must still produce quality goods and/or services. You are still responsible for quality job training, employee motivation, development of good safety attitudes, and detection of hazardous conditions and unsafe work practices. And you must do all these things in an environment that is in the midst of the "information explosion."

You need to understand the techniques and psychology of human relations. You should know the fundamentals of loss control, the applicable industry standards, and federal and local safety and health regulations. You must know all the potential hazards your workers can encounter, how to prevent them, and/or what safeguards and personal protective equipment are needed, how to use them, and how to enforce use of them. You must be able to assist in accident investigations that occur in your area of responsibility. And you should be able to recognize when your workers need to seek help for stress; problems in adjusting to workshift changes, stressful environments, or job changes; personal problems; or drug or alcohol abuse—and how to encourage them to get the help they need.

The ninth edition of the *Supervisors' Safety Manual* provides the basic information, references, and resources you will need to meet these responsibilities and duties. It has been revised to reflect changes in organization structure and management philosophies, in particular the emphasis on empowered employees and self-directed teams. All technical information has been reviewed, revised, and updated by experts from a variety of businesses and industries.

The text is written in easy-to-understand language with many bulleted lists for quick reference. The objectives at the beginning of each chapter help to focus learning and the reviews of key points at the end of chapters aid retention of knowledge. The Accident Case Studies offer readers an opportunity to apply learning and test their knowledge.

Even though much detailed information has been included, the purpose of the book is not to serve as a complete handbook. Rather, it is designed to emphasize important issues to be considered by the supervisors who are responsible for safety and health in their organizations.

A completely rewritten and updated Supervisors' Development Program is also available to support your training for your responsibilities and duties. It consists of printed facilitator guides, participant workbooks, and 15 videos with leaders' guides. *Supervisors' Safety Manual* is the sourcebook for the Supervisors' Development Program.

The National Safety Council gratefully acknowledges the expertise and assistance provided by the Technical Advisory Team in reviewing the manuscript of this 9th edition of the *Supervisors' Safety Manual*. The following individuals reviewed part or all of the manuscript:

Michael R. Chambers, Manager, Corporate Safety and Health Issues, Eastman Kodak Company

Pam Harris, Browning Ferris Industries, Inc.

Wayne C. Loomis, CSP, CIH, Manager, Operations Safety, Xerox Corporation

Allan Manzer, CSP, Senior Safety Engineer, Eastman Kodak Company

Richard J. Reber, MIS, ASP, Minnesota Safety Council

Maryellen Skan, MPH, Minnesota Safety Council

The Council also recognizes with appreciation the technical reviews provided by Council staff, including:

Dale Haskins
Ronald Koziol
Joseph Lasek
Barbara Jean LoMastro
Amber Nicholson
Roseann Solak

INTRODUCTION

As modern production processes have become more highly sophisticated, so has our understanding of how to create a safe, healthful work environment. In the past, "accident prevention" simply meant freedom from serious injuries and property loss. Today, however, "loss control" covers not only injury but occupational disease and environmental concerns along with fire and property damage control. This new understanding of safety is reflected at all levels of management from the CEO to the line supervisor or team leader. Loss control is considered so essential a function that in most companies safety performance is included as part of a supervisor's or team leader's overall job evaluation.

As a result, today's supervisor or team leader must become skilled in three aspects of loss control: (1) learn to recognize hazards, (2) learn the acceptable level of risk for department operations, and (3) learn how to control the hazards to prevent injuries, illness, and property damage. The National Safety Council's *Supervisors' Safety Manual* presents the skills a supervisor or team leader will need to perform these responsibilities successfully.

SUPERVISOR: MANAGER OF SAFETY PROBLEMS

The supervisor or team leader, as a member of the company's management team, shares responsibility for maintaining a safe, productive workplace. He or she must communicate and enforce rules and procedures, train workers, and represent the interests of both the organization and employees.

The supervisor or team leader must constantly watch over and inspect both the workplace and work procedures, keeping in mind the three E's of safety:

engineering, education, and enforcement. It is the supervisor's or team leader's job to work with safety and health professionals, designers, engineers, maintenance, and personnel staff to engineer as many hazards out of the workplace as possible, to educate employees in safe work practices and procedures, and to enforce all safety rules and policies. In this role, the supervisor or team leader acts as investigator, safety researcher, and advocate.

For example, if an employee complains of headaches while performing a degreasing procedure, it is up to the supervisor or team leader to investigate the situation. It may be that a degreasing chemical is causing the employee's headaches. If an investigation by the safety and health professional shows this is the case, there are three options: *eliminate* the hazard by substituting a different chemical, *isolate* the hazard by installing surface or point-source ventilation, or *compensate* for the hazard by issuing personal protective equipment.

The supervisor or team leader may be involved in determining which of the three options is the best solution to the problem. For example, in this case, personal protective equipment is desirable only when all other options have failed. The equipment does not remove the hazard—it only decreases the risk for the worker.

Once the best solution has been chosen, the supervisor or team leader may be involved in presenting the plan to upper management for approval of the required action. At this point, the supervisor or team leader becomes an advocate, presenting the benefits of the plan on behalf of the workers as well as the organization. *Supervisors' Safety Manual* is designed to take you step by step through the process of learning how to manage and control safety problems.

HOW THIS BOOK IS ORGANIZED

The supervisor or team leader manages safety problems by managing the people, equipment, and environment in the workplace. To assist you in these tasks, this book has been organized as follows:

Chapters 1 through 7 focus on the basic safety knowledge and skills that every supervisor or team leader must acquire: an understanding of loss control and the fundamentals of safety management, communication, human performance management, employee safety and health training, inspection, and accident investigation. These chapters orient the supervisor to his or her management responsibilities on the job.

Chapters 8 through 12 cover what a supervisor must know to meet OSHA and other regulatory standards to maintain a safe, healthful workplace: industrial hygiene, personal protective equipment, ergonomics, hazard communication, and environmental management. Understanding the worker-machine-environment relationship can help to reduce hazards, accidents, and job-related illnesses considerably.

Chapters 13 through 18 focus on the general principles for safeguarding, maintaining, and using tools and equipment safely to prevent accidents and property damage and loss. In these chapters, you will learn how to make the day-to-day production process safer for all concerned and how to work with various safety experts from within and outside the company to solve safety-related problems.

No safety program can eliminate all the risks and hazards from a job, but *Supervisors' Safety Manual* can help you come as close to this ideal as possible. The more you know about the hazards in your workplace and how to control them, the more everyone in the company will benefit.

1

SAFETY MANAGEMENT

After reading this chapter, you will be able to:

- Identify the trends in safety management
- Explain the principles of safety, health, and environmental management
- Define the terms *accident, hazard, hazard control,* and *loss control*
- Explain some of the direct and indirect costs of accidents
- Describe the approach of team-based organizations
- State some techniques to empower and motivate employees
- Explain the Continuous Improvement Model
- Apply principles and strategies for managing change

Managing safety starts from the premise that most accidents can be prevented. It includes the premise that incidents are near-accidents and that so-called accidents are not random events, but rather preventable events. They can be prevented with proper hazard identification and evaluation, management commitment and support, preventive and corrective procedures, monitoring, evaluation, and training. Team leaders and supervisors are important links in the chain of safety consciousness. They are responsible for watching over the workplace and protecting their employees from faulty equipment, carelessness, and the many other potential hazards on the job.

TRENDS IN SAFETY MANAGEMENT

Occupational accidents have been known in human activity for as long as there have been workplaces, and documented in historical records long before the construction of the Egyptian pyramids. However, the concept that accidents could be reduced through safety management is relatively new. Only in the past century have work groups paid much attention to organized or systematic safety practices.

Changing Focus

Until the beginning of the twentieth century, many owners and society at large took a fatalistic view of safety. Much of the history indicates that they saw accidents as regrettable but not preventable. As farms failed, farm workers and new immigrants swelled the city labor markets, providing plenty of workers willing to risk life and limb for a decent wage. But tragedies like the Triangle Shirtwaist Company fire on March 25, 1911, in which more than 146 workers, all women and many underage, died because of locked fire exits, led social reformers and labor unions to push for more humane workplaces.

Physical Workplace. By 1910, the reformers succeeded in getting workers' compensation laws passed. Now employers had a greatly increased financial stake in making the workplace safer, and general support for safe workplaces was increasing in society at large. In 1913, the National Safety Council was established to provide research and information on how to increase safety.

The 1920s were a time characterized by interest in scientific management. Many companies still treated their employees as interchangeable parts but also recognized that improved safety led to improved productivity. Many employers began inspecting the physical work space, keeping it clean and monitoring working conditions. In the 1930s, H. W. Heinrich wrote the first significant book on modern safety management, a text called *Industrial Accident Prevention*. Companies began to study not just physical workplaces, but human contributions to accidents. The emphasis was on safety on the factory floor at this point, since industrial accidents tended to be far more numerous and more serious than those in office settings.

Health Issues. In the 1940s, the focus of safety management was broadened to include prevention of not only accidents but also illnesses. Workers' compensation was updated to cover disability caused by work-related disease and industrial hygiene problems. Management style, however, was still in the "command-and-control" mode, a legacy of the military model.

In the 1950s management adopted the behavioral approach and turned to technology for solutions. Companies began to apply management and motivational principles in an effort to change workers' behavior. At the same time, some came to believe that technology had all the answers. Emphasis was on engineering—prevention of safety and health problems by redesigning or modifying equipment.

Environmental Issues. During the 1960s, national concerns about safety broadened to include the office environment, and government began to recognize many serious environmental risks. In the late 1960s, workers' compensation laws were also updated to include compensation for illness and disability caused by noise.

In 1970, the Williams-Steiger Occupational Safety and Health Act was passed, and the federal government established the Occupational Safety and Health Administration (OSHA) in 1971. OSHA was given the power to monitor companies' compliance with newly legislated safety and health standards, and impose heavy fines on those companies not in compliance. Safety and health efforts began to reflect employee participation in management.

For at least two decades gradual changes in safety and health management had occurred, until by the 1980s, priorities had shifted from hazard identification to hazard prevention. Next came the inception of a general movement toward proactive, rather than reactive, safety and health management. In the '90s, safety experts worked to create systems to identify and eliminate workplace hazards. The focus is now on integrating safety, health, and environmental management throughout the business strategy. Every employee, from CEO to forklift driver, is now seen as responsible for safety. Most managers today accept that good safety/health/environmental programs contribute to the health of the company's bottom line.

Changing Views of Causation

The understanding of what causes accidents has changed over the years. At one time, safety experts believed in the *domino theory*. That is, they thought a mishap or loss was caused by one specific unsafe act or condition, and that removal of that act or condition would solve the problem.

The overly simplistic domino theory has been replaced by the *theory of multiple causation*. Safety experts now try to look for all the factors that may have contributed to a given incident. Studying all the contributing causes leading to the loss helps them to identify the root or key causes.

Frequency refers to the number of times a specific incident occurs. *Severity* describes how serious the consequences are (which may be measured for injuries and illnesses by the number of affected employees). Some safety experts once concentrated on reducing only high-frequency accidents, but modern risk assessment considers both variables.

Changing Management Styles

At one time, the role of managers and supervisors was to dictate what should be done and the role of employees was to follow those orders unquestioningly. Now companies are coming to recognize that management does not have "all of the answers" and that other employees are capable of contributing far more to the success of the business. Just as employees are sharing responsibility for manufacturing and customer service and other "line" decisions, they are also participating more in safety and health decisions. This approach invests all employees in making the workplace safer.

As employees are permitted more participation in making workplace decisions, they tend to take more and more responsibility for their safety and health. Current management styles of empowerment, total quality management, and continuous improvement recognize that safety is everyone's job, and that the people most likely to spot problems on the job are those who are performing the job. It is becoming more common for decisions to be made by the people who are "closest to the action," i.e., empowerment.

Some American corporations have concluded that self-directed work teams are among the most productive ways to organize the workforce. In successful teams, each member buys into the team's and the company's goals. No matter what product or service the company provides, one of those goals is always to achieve a safe working environment. Each employee's job description should spell out accountabilities for safety and health. In some companies, senior management has committed to the safety and health policy and addressed it in the business plan.

PRINCIPLES OF SAFETY, HEALTH, AND ENVIRONMENTAL MANAGEMENT

The management of safety, health, and environmental programs has evolved over the years. Some forward-thinking companies now have programs that include written principles that guide safety and health efforts. These principles are called Key Results Areas, or KRAs.

Definitions

To make sense of these principles, it helps to define the terms used in safety, health, and environmental programs.

Accidents. At one time, one of the most important criteria for measuring safety performance was how much time was lost due to accidents. Minor injuries, property damage, and near misses didn't count. But this approach is not accident prevention; it's accident reaction, after the fact.

An *accident* is an unplanned, undesired event, not necessarily injurious or damaging, that may disrupt the completion of an activity.

A "near" accident or "near miss" is an example of an incident resulting in neither an injury nor property damage. However, a near accident has the potential to inflict injury or property damage if the cause is not corrected. About 75% of industrial injuries are forecast by near accidents or near misses. It's in a supervisor's best interest to find and eliminate the causes of these near misses to prevent them from recurring or becoming serious accidents.

In 1931, H. W. Heinrich conducted a study, which concluded that for every serious accident there are about 29 accidents that result in minor injuries, and 300 that produce no injuries. This means that managers who react only to major injuries are ignoring 99.7% of the accidents that occur. Heinrich pointed out that the same factors that cause a "near miss" incident on one occasion, could in a future incident result in major injuries. Figure 1–1 illustrates the multiplier effect of ignoring accidents that seem minor.

Hazards. *A hazard* is any condition in the workplace that, by itself or by interacting with other variables, could cause death, injuries, property damage, or other losses.

For example, employees continue to operate a portable drill, though one of them has felt the tingle of a slight electric shock while using it. The defect presents a definite hazard, which could cause a potentially harmful accident.

Hazard Control. *Hazard control* involves developing a program to recognize, evaluate, and eliminate (or at least minimize) the destructive effects of hazards in the workplace.

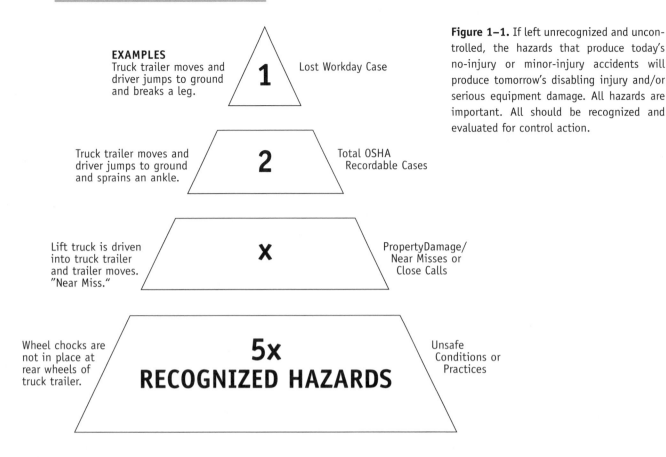

EXAMPLES

Truck trailer moves and driver jumps to ground and breaks a leg. — **1** — Lost Workday Case

Truck trailer moves and driver jumps to ground and sprains an ankle. — **2** — Total OSHA Recordable Cases

Lift truck is driven into truck trailer and trailer moves. "Near Miss." — **X** — PropertyDamage/ Near Misses or Close Calls

Wheel chocks are not in place at rear wheels of truck trailer. — **5x RECOGNIZED HAZARDS** — Unsafe Conditions or Practices

Figure 1–1. If left unrecognized and uncontrolled, the hazards that produce today's no-injury or minor-injury accidents will produce tomorrow's disabling injury and/or serious equipment damage. All hazards are important. All should be recognized and evaluated for control action.

Hazard control incorporates hazard recognition through a formalized inspection process; evaluations; safe operating procedures; safe building, machinery, equipment, and process designs; employee training; inspection and testing programs; and effective communication about hazards and their control. A hazard control program coordinates shared responsibility among departments and underscores the relationships among workers, their equipment, and the work environment.

For example, there should be a system in place for reporting the faulty portable drill. Encouraging employees to report problems can prevent injuries and save lives. All employees at all levels have a vested interest in keeping the workplace free of hazards. Personal injury, damage to equipment, and losses of time and production affect every person in the company. Employees should be praised for cooperation, and never criticized for bringing to light problems whose solution might temporarily slow down the workflow. Management must support the program and quickly remedy hazards to maintain workers' confidence.

The system can use a simple report form like that in Figure 1–2. Employees who notice a hazard can report it to supervisors or, if they wish to remain anonymous, put it in a suggestion box.

Loss Control. *Loss control* is the prevention of occupational injuries, illnesses, and accidental damage to the company's property. Losses from hazards include:

- Product liability
- Production losses
- Injury to employees
- Property damage
- Public liability
- Business interruptions

Effective loss control requires prompt and focused response to evidence of hazards—large or small. Too many supervisors shrug off minor injuries or property damage without bothering to investigate their causes and try to prevent a recurrence. The degree of loss resulting from an accident can be partly a matter of chance. It is far easier to control or eliminate the hazards that lead to accidents than to minimize the damage once an accident occurs or is in progress.

Supervisors and team leaders must look for the cause(s) regardless of the results. The hazardous condition or action that causes a near accident this time may bring about a serious injury or a fatality next time. Likewise, the hazard that produces only minor property damage this time may cause serious, expensive property damage next time. This is why loss control is now considered so essential that most companies include safety performance as part of their supervisors' overall job evaluation.

Incident (No-Injury Accident) Report Form
(This must be completed IMMEDIATELY after an accident when there is no injury.)

Exact location of incident: _____ Department:_____

Occurrence date: _____ Time: _____ Date reported:_____

Employee involved: _____ SS# _____ Employee ID:_____

Job title: _____ Employment date: _____ Time on present job: _____

1. Property damaged: _____

    Cost: $ _____   Length of downtime: _____

2. Unsafe condition at time of incident: (be specific) _____

    _____

    _____

3. Unsafe practice contributing to the incident: (be specific) _____

    _____

    _____

    _____

4. Witness(es) to incident: _____

5. Sequence of events: (detailed) _____

    _____

    _____

    _____

6. What can be done to prevent a recurrence of this incident?_____

    _____

Supervisor: _____ Department: _____ Date: _____
Immediately forward copies of this report to Department Management

Figure 1–2. This sample form can be used for reporting incidents that involve no injuries.

The Prevention Principle

The best safety, health, and environmental management system is preventive: it seeks and eliminates the *causes* of hazards. This system builds a solid foundation for preventing mishaps and achieving zero accidents.

Accountability

For the safety/health/environmental program to be most effective, all employees throughout the organization must actively support it. Holding employees accountable requires evaluating their success in identifying and eliminating workplace hazards and mishaps. Employers and employees are accountable under OSHA for maintaining safe working conditions and practices. Managers may be directly accountable through their budgets for costs of workers' compensation and insurance costs of operations under their control.

Accountability rewards employees who report near-accidents and who otherwise take responsibility for safety in their areas. It penalizes employees who refuse to take the safety issue seriously or accept responsibility for it. Accountability can also help employees who meet safety performance goals earn recognition that ranges from simple compliments to bonuses or days off.

The Human Factor

Most safety experts agree that most mishaps in the workplace involve human errors of omission or commission. A major step toward improving safety is eliminating those factors in the organizational system that contribute to human error. This includes not only worker errors but also errors by designers, engineers, purchasing agents, supervisors/team leaders, and anyone else.

As industrial designers have become more aware of safety issues, they have begun to invest in and design more provisions for safety: automatic shutoffs for equipment, foolproof operating mechanisms, and ergonomic workstations, for example. In *Turn Signals are the Facial Expressions of Automobiles*, Donald A. Norman points out that a multimillion-dollar jet plane's performance should not depend on the stability of a three-cent polystyrene cup sitting on top of a badly located but crucially important lever in the cockpit (Reading, MA: Addison-Wesley, 1992, p. 167). A custom-designed part in a well-located area may cost more, but will also provide the required control reliably.

Organizations cannot afford to wait until they can build facilities, processes, and equipment to design for safety. They must modify or replace existing ones that are hazardous in any way. Designing for safety has the effect of designing for efficiency, and even

productivity, since it reduces employee fatigue, injuries, and illnesses. Safe workplace design reflects physical and psychological compatibility between the employee and the process, methods of operation, equipment, materials, and machinery.

Integration into the Business Plan

Another principle of the safety, health, and environmental program is its newly acknowledged relevance to a company's business plan. CEOs are now recognizing the benefits of incorporating their safety and health policy into the Business Plan. Like other management processes, safety and health efforts should be planned, budgeted, measured, and evaluated.

Management should figure out what safety and health measures are needed to support every strategy and goal in the Plan. For example, one business strategy for a clothing wholesaler might be to have sales representatives increase the revenues they bring in by increasing the number of customers they visit. A safety strategy to support that might be to institute a preventive maintenance program for the vehicle fleet to reduce vehicle down-time due to operational failures or accidents.

PLANNING SAFETY, HEALTH, AND ENVIRONMENTAL MANAGEMENT SYSTEMS

Comprehensive safety, health, and environmental systems should be proactive and preventive. They should be integrated systems that involve everyone in the company, starting with a solid commitment from top management.

To create a comprehensive system, a company must first decide on its vision or mission statement. A vision statement is, by definition, general and uncomplicated—for example, "zero accidents." Once the company creates this overarching vision, its planners identify specific strategies for achieving it, and managers create the policy and procedures to define the who, what, when, where, why, and how of achieving the vision and accomplishing its mission.

Top management should be involved in the development of the safety, health, and environmental policy. It ought to cover management's intent, the scope of the policy and procedures, acceptable risks, the roles and responsibilities of various employees, the budget, and the degrees of authority different people have.

Components of Comprehensive Safety/Health/Environmental Management

Decades of research have led the National Safety Council to identify 14 components that contribute to a successful system for managing safety, health, and the environment (see *14 Elements of a Successful Safety & Health Program*, National Safety Council, 1994). All effective programs contain these elements, but the emphasis varies according to individual companies' needs.

1. Hazard recognition, evaluation, and control
2. Workplace design and engineering
3. Safety performance management
4. Regulatory compliance management
5. Occupational health
6. Information collection
7. Employee involvement
8. Motivation, behavior, and attitudes
9. Training and orientation
10. Organizational communications
11. Management and control of external exposures
12. Environmental management
13. Workforce planning and staffing
14. Assessments, audits, and evaluations

Steps toward Safety, Health, and Environmental Management

There are five steps in planning systems for safety/health/environmental management.

1. Work with senior management to shape and guide the organization's safety and health policy. Build the commitment from top management. Create a vision statement and define the safety, health, and environmental policy. Make sure the issue is addressed in the Business Plan and aligned with the business strategy.
2. Identify and communicate the safety and health rules of everyone in the organization. Spell out responsibilities. Include accountabilities in each employee's job description, and evaluate everyone, based on those accountabilities.
3. Analyze the work and the workplace continually, to identify all existing and potential hazards. Most analyses should be data driven, but some can be more imaginative (for example, what-if accident scenarios). Set priorities and focus on continuous improvement. Write down the procedures for analysis and follow them.
4. Set goals and implement actions that will remove and prevent hazards. Get everyone involved and use a continuous improvement process. The actions should include case management, workplace design, and improvement of accident investigation procedures. Use data to measure and evaluate the results.
5. Train and coach managers, supervisors, and employees. Provide thorough communication to create awareness. Provide feedback to encourage

learning; and reward people for doing a good job.

Budgeting for Safety, Health, and Environmental Management

A budget is a financial plan that shows the funds an organization will need to achieve its goals. Budgets generally describe how much money, time, and material will be spent, and on what. They also show how much money will be earned, and from what sources. Budgets are prepared annually, quarterly, monthly, and in other increments, as needed.

There are many expenses involved in running a comprehensive safety, health, and environmental effort. They include everything from accident investigation costs to eyewash fountains, from printing of newsletters to training and incentives. It is important to remember that the effort yields many savings also. A cost/benefit analysis can prove to management that the program is worth the expense. The question is not whether the program has costs, but whether it is preventive and produces benefits greater than the costs.

In order to convince management to invest in training, equipment, or material, safety professionals detail the reasons why these investments are needed. They can calculate the costs of accidents or injuries, including direct and indirect costs. Direct costs include medical, legal, insurance premium increases, injury compensation, and time and material costs to repair equipment or tools. Indirect costs include hiring and training replacements for injured workers, time lost when other employees help with, or watch, discuss, or otherwise react to the accident, machine or equipment downtime; accident investigation costs (interview and administrative time), overtime required to catch up after the accident; and the cost of supervisory time spent in handling it.

The managers then compare the purchase cost to the cost of one accident, to see how many prevented accidents it takes to pay for the expected costs for training, equipment, or material. They determine the likelihood that the purchase will prevent accidents and the number of potential preventable accidents to calculate the potential dollar cost of loss prevention in the first year. Then they use the company's profit margin to figure how much product the business must sell to make up for the loss. Usually it takes only a few prevented accidents to pay for the purchase. If the purchase would likely prevent many accidents, it effectively saves the firm money.

Total accident costs can be compared to an iceberg, as shown in Figure 1–3. Experts calculate that for every dollar of direct accident costs, there may be $4 to $10 of insured costs, $5 to $50 of uninsured ledger costs, and $1 to $3 of uninsured miscellaneous costs.

Safety pays in two ways. First, it results in fewer

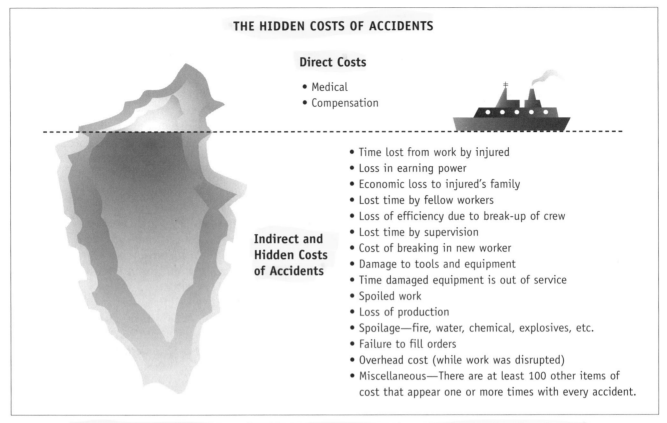

THE HIDDEN COSTS OF ACCIDENTS

Direct Costs

• Medical
• Compensation

Indirect and Hidden Costs of Accidents

• Time lost from work by injured
• Loss in earning power
• Economic loss to injured's family
• Lost time by fellow workers
• Loss of efficiency due to break-up of crew
• Lost time by supervision
• Cost of breaking in new worker
• Damage to tools and equipment
• Time damaged equipment is out of service
• Spoiled work
• Loss of production
• Spoilage—fire, water, chemical, explosives, etc.
• Failure to fill orders
• Overhead cost (while work was disrupted)
• Miscellaneous—There are at least 100 other items of cost that appear one or more times with every accident.

Figure 1–3. Like an iceberg, the hidden costs of accidents are not visible on the surface but are there just the same.

disabling injuries, lower workers' compensation costs, and lower replacement costs. Second, it frees up managers from time spent managing safety crises, so they can concentrate on quality; productivity; proactive safety, health, and environmental issues; and competitiveness in the market.

SAFETY, HEALTH, AND ENVIRONMENTAL MANAGEMENT IN TEAM-BASED ORGANIZATIONS

Traditionally, most businesses were organized as hierarchies. Decisions were handed down from managers at the top of the pyramid. Front-line workers had very little say in how they did their jobs, and no say at all in how the company was run. Managers were responsible for their departments and supervisors acted mainly as their enforcers. They assumed that if workers were given any autonomy, they would duck out of as much work as possible.

In recent years, however, there has been a major change in the attitudes and theories of American businesses. In many of them today, the managers no longer dictate orders. They share power with their employees, who tend to respond with increased productivity because they are part of the team.

Businesses today are experimenting with several approaches to organizing their workforce. One of the most effective appears to be Self-Directed Work Teams (SDWTs).

Characteristics of Team-Based Organizations

A Self-Directed Work Team is organized as a group of employees from different functions and departments who meet periodically (perhaps daily) to pool their knowledge and collaborate on creative problem solving. The team is ongoing, so its members may work together for several years. The members are multiskilled and responsible for a whole task or core process. They take ownership of their successes and failures. They constantly cross-train each other and are always learning and improving. Their range of responsibilities is always increasing, often encompassing duties that were once the province of supervisors. They are empowered to make decisions on scheduling and vacations, to set goals and performance targets, and to interact with customers, vendors, and others. Their rewards are tied to their performance.

Self-Directed Teams (also known as high-performance teams) have become popular because of their track records. Many of them have achieved higher quality, lower costs, faster service to customers, higher morale, and greater worker satisfaction than the old hierarchy-based system. Many companies say that their people are their most important resource, but many of those who have reinvented themselves into Self-Directed Teams can demonstrate that they mean it.

According to Ann and Bob Harper's *Team Barriers*, there are 10 key elements to high-performing teams:

1. Shared goals and mission
2. Climate of trust and openness
3. Open and honest communication
4. Sense of belonging
5. Respect for differences
6. Continuous learning/improvement
7. Ability to measure and self-correct
8. Interdependence
9. Consensus decision making
10. Participative leadership

Cultural Values of Teams. The Self-Directed Teams must value productivity above all else. They achieve it by being flexible, imaginative, and motivated. They believe that knowledge is power, and that power is most useful when shared rather than hoarded. This system requires company management to provide feedback to everyone, handing out productivity reports, cash-flow statements, and customer complaints. Every employee has a right to this information: after all, in order to help meet the company's goals, they must know what these goals are.

The Self-Directed Team system also values autonomy, respect, and dignity. It is assumed that people can be trusted to do their jobs without external monitoring and all employees will treat each other as responsible, intelligent adults.

Leadership Traits. One thing that the SDWT system has proven is that leadership does not depend on a degree from the right school or a formal background in corporate management. Anyone can be a leader, given enough training and enough self-confidence. In fact, more and more, leadership duties are shared in rotation among all the members of a team. Each team member is responsible for a designated area (usually operations, training, safety, scheduling, human resources, communication, or quality control) for a specific time period.

For example, one person receives training in safety leadership, and for the next several months attends all plantwide safety meetings as the team's representative. Then a co-worker is trained in safety, and takes over the team's safety leadership role. This approach gives all team members a chance to develop their leadership abilities, and also to gain experience in interaction with other teams.

Team-based organizations know that, as W. Edwards Deming said, the job of management is not supervision but leadership. Leading does not mean barking orders; but rather providing guidance, information, encouragement, and inspiration. It means motivating employees to share the organization's vision and to strive to achieve it.

Leaders act as role models. By their own behavior, they show employees how to behave and what to value. Role models live up to the highest standards of the company. Leaders are role models every day, on a continuing basis, whether they want to be or not. In addition, they also coach and teach their people both practical skills and details of the corporate culture.

Empowered Employees

"Empowered employees" are those who have the freedom to make many of their own decisions, and the training to make good ones. Their jobs are designed to provide ownership of complete task areas and responsibility for their success. Their tasks are defined so that they are responsible for a meaningful process or output, have the power to make decisions and commit organizational resources, and continually measure their own progress. They share authority, resources, and recognition for the work process and results. They receive the skills training they need to do their jobs well, which is tremendously empowering.

Empowerment tends to give workers a reason beyond a paycheck for going to work. It enriches their work life and promotes job satisfaction. Employees who are empowered do more than just stamp out machine parts or answer telephones or blindly follow a limited set of steps. They are in a position to implement the organization's vision and to understand how they personally contribute to achieving it.

For this system to work, employees must be treated with respect and honesty; management must recognize that front-line employees *are* the organization. As such, these employees want to know what's really going on. They need information on all aspects of the business, not just their own small segment. They need to know about costs, budgets, quality standards, customers, vendors/suppliers, and competitors. When top management shares all this information freely, the teams are more likely to support its actions.

To be empowered, people need to be trained and educated continuously in three areas:

- Technical training (their own skills, new skills, quality, and safety)
- Interpersonal/team training (communication, team development, conflict management, leadership skills, decision making, problem solving, and so on)
- Administrative skills training (for the tasks supervisors used to do, like scheduling, budgeting, and interviewing)

It has been demonstrated that employees who have

the power to fix problems on their own do so faster and better than when they have had to go through layers of management for approval. This leads them to take more pride in their work and counteracts the dehumanizing effects of severely limited, boring, and repetitive tasks. This system is based on the belief that empowerment becomes a way of life and is not just the management "fad of the month."

Employee Involvement in Safety

Some companies have a safety leader whose job is to get employees involved in the safety, health, and environmental program. In companies where the team system is used, this work is done through management supervisors and team leaders, who are responsible for motivating people to care about safety. Employee safety committees, contests, newsletters, and posters are some of the tools they can use. The safety expert may work as an "inside consultants" to provide direction and technical support for others.

The following steps would be helpful for anyone in the process of creating a safety, health, and environmental team.

1. Create a character vision for yourself and live it daily.
2. Respect others and accept their differences.
3. Involve others in the process of creating your safety and health vision and goals.
4. Encourage participation by inviting, and listening to, opposing points of view.
5. Involve others in solving, and making decisions related to, safety and health problems.
6. Build an atmosphere of respect, trust, and inquiry.
7. When others need help, the safety, health, and environmental professional provides coaching, resources, support, and assistance.
8. Give recognition and rewards often and consistently.

(See also Chapter 5, Promoting Safety and Health.)

Maintaining Interest and Motivation

Motivation involves moving people to action that will support the company's desired goals. In occupational safety and health, motivation increases employees' awareness, interest, and willingness to act in ways that improve their own safety and that of co-workers. Motivation aims primarily at changing behavior and attitudes. It is generally defined by three factors: (1) direction of behavior, (2) intensity of action, and (3) persistence of effort.

Motivation efforts should support the mainline safety, health, and environmental system, not take its place. Chances of success increase when the following factors are present:

- Management demonstrates its commitment at every

opportunity
- The program is energized through slogans, performance recognition, and discipline
- Workplace conditions are safe and healthful
- Tools, equipment, and workplace layout are well-designed
- Maintenance is effective
- Training and supervision are abundant and effective

Job Safety Analysis. Job safety analysis can lay the groundwork for behavior change because employees participate in identifying hazards and finding ways to eliminate them. When all workers are involved in formulating safety needs and devising a Safety Plan, each one has a greater stake in carrying out the plan. Such participation should be reinforced through an open communication system that welcomes employees' safety concerns and responds to them promptly. (For instructions on how to do job safety analyses, see Job Safety Analysis in Chapter 5, Promoting Safety and Health.)

Employee Surveys. Employee surveys help to identify safety priorities and build employee morale. They also provide a baseline for later evaluations of the effectiveness of safety initiatives. Group feedback is important. When a work group accepts safety improvement objectives, its members reinforce each other's behavior through peer pressure.

Fear. Many supervisors believe that fear can sometimes be an effective motivator. This strategy attempts to change attitudes about the risks of hazardous behaviors by first instilling fear with statistics, photographs, or other information, and then reducing the fear by providing ways to prevent the danger or lower the risk. Fear messages may work better with new employees than with seasoned workers, who are likely to want to use their own experiences to discount the message. ("For years I've checked to see if circuits were energized by quickly touching them with the back of my hand, and I've never gotten injured!") Fear messages are also more effective on employees who are not under direct supervision, since they are on their own, and may tend to take the information more seriously.

Total Quality Management Approach. One approach to motivating employees that has been effective is the Total Quality Management (TQM) model. TQM works to change employees' attitudes in order to achieve a change in behavior. For example, it strives to train employees to an awareness of the importance of safety. This awareness will lead them to be more careful about always wearing safety equipment and checking the automatic shutoffs on their tools, for example.

TQM emphasizes process, or root cause, improvements. Such process changes flow from workers' involvement and empowerment as both individuals and teams. People are far more likely to support change if both the objectives and the methods used to achieve them are based on their own recommendations, instead of being imposed by management.

Nothing less than a cultural change is called for in order to make TQM work. Management must give up trying to "manage" employees to conform to existing systems. Instead, managers must be willing to completely revamp processes and systems. They must realize that employees are in the best position to know how those processes could work better. The corporate culture must be flexible and responsive to employees' needs. It must engender trust and cooperation between labor and management.

TQM is more difficult and time-consuming than simple reinforcement/feedback motivational techniques. Its effects are, however, more wide-ranging and last much longer.

Continuous Improvement

The Continuous Improvement Model is a framework for safety presented in the National Safety Council's *Agenda 2000 Safety/Health/Environment* program. It is shown in Figure 1–4.

Continuous Improvement is a process-oriented business approach that emphasizes the contributions people make to long-range, permanent solutions to problems. It is the cornerstone of Total Quality Management. Since everyone is always working to improve something, the only constant is change. This is also one impediment to a successful TQM program. Constant change is somewhat uncomfortable because, among other things, it fosters a sense of uncertainty and insecurity. People like to establish a pattern and stick to it. Their natural resistance to change must be overcome if TQM is to be fully successful.

To apply the Continuous Improvement Model, companies must understand causes before they try to design solutions. Improvements, whether dramatic or incremental, occur regularly. The 14 components of effective safety, health, and environmental programs listed earlier in the chapter facilitate the Continuous Improvement Model.

Barriers and Challenges to Team-Based Organizations

Though the team structure has proven its effectiveness again and again, some organizations are still skeptical. There are many barriers that can either keep a company from trying teams or can sabotage the structure if it is undertaken halfheartedly. Among them are the following:

- Lack of trust
- A climate of fear

- The uncertainty of constant change
- Not involving *all* the key stakeholders
- Not involving unions (forming a partnership)
- Lack of top leadership's demonstrated commitment
- Lack of top management responsibility or role
- Not allowing enough time to change
- Resistance from any of the partners (supervisors, managers, support people, unions, or labor force)
- Inadequate training
- Systems and structures not designed to support teams
- Workers who zero in on their co-workers' imperfections
- Failure to redefine the role of leadership
- Leaders who won't let go
- Too little or too much structure
- No transition plan
- Failure to provide work security assistance
- Failure to communicate what's happening
- Treating change like a program instead of a process
- Overwhelming team members with too much responsibility before they have had adequate training
- Team members assuming too much responsibility before they have had adequate training
- A history not conducive to employee involvement
- Failure to educate everyone in the workplace about self-direction
- Failure to understand that TQM self-direction is a total cultural transformation, not just a modification
- Amount of time and effort involved

The long-term improvements that teams generate can more than offset these problems. The potential of the Self-Directed Work Teams is to boost productivity and competitiveness, enhance people's quality of life, and turn many jobs from drudgery to fulfilling work in which the worker is integrally involved.

INFLUENCING AND MANAGING CHANGE

The team-based, empowered, continuous-quality workplace is changing all the time. On the one hand, that's a good thing, because change is what enables the organization to remain viable and keep up with the competition. On the other hand, though, people tend to crave stability, and find constant change difficult. They may resist even changes for the better because the known is less threatening than the unknown.

Employees at all levels, even (especially) upper management, may fight change in several ways: logically, emotionally, and through group influence. Even if their real reason is to avoid emotional turmoil, they may be quite good at rationalizing logical reasons why the change is a mistake. When employees offer data that suggest the change may not work, top management

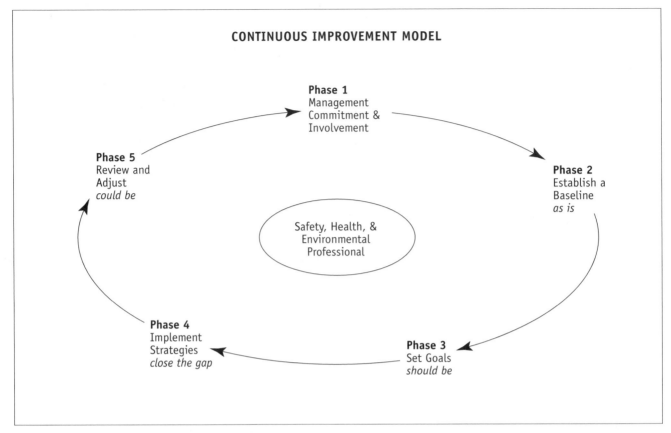

Figure 1–4. The Continuous Improvement Model.

and team leaders need to pay attention, listen to the potential problems, and the employees' ideas on how to handle them. When people's resistance is entirely emotional, it is still necessary to listen. This time the purpose is to reassure them and explain the benefits the change will bring them.

Sometimes only one or two people in the group feel strongly anti-change, but they manage to convince the whole team to resist. The key is to identify those lead resistors and then listen to their reasons. People resist change imposed by someone else, but if the change is their idea, or they have contributed to it in even a small way, they are more likely to welcome it.

Change often precedes perception. Sometimes people cannot understand why they should follow a particular procedure until they try it. Then it begins to make sense to them. Change involves breaking old habits, both individual and organizational, and replacing them with new ones. Such disruption can be wrenching, so people need an incentive to make the effort. They need to be convinced that the change will make their lives better. Without such a conviction, they will just be going through the motions.

Change has a ripple effect. When it occurs in one area of the organization, it will affect other areas. Change reaches critical mass slowly, then speeds up.

Strategies for Managing Large-Scale Change

To restructure an entire company or even a single department, top executives need a change strategy that encompasses a period of months. They should get and use input from as many people as possible before even beginning to design the change. Then they should solicit support from as many people as possible, beginning where they know they have the most support.

They need to communicate clearly the expected positive results of the change. They will need to plan and communicate the logistics, as well. Communications and public relations pieces should be designed to integrate the new, desired culture.

Executives should pace the implementation. A common mistake is to try to make everything happen at once. They should study the organization's dynamics and then use them to create change. They need to provide training and education about the change. A checklist for managers will show them how to support and enhance the process.

The Change Masters should set the example. They should model the behavior they are trying to achieve, whether it be sharing information or wearing protective gear on the factory floor. At the same time, they need to be patient. It is the nature of human behavioral change to start slowly.

Steps for Influencing Individual Change

Supervisors and team leaders can find themselves in the position of coaxing individuals to embrace change. A meeting to coax an individual to embrace a change should be well-planned, using the following five steps:

1. The first step is to plan ahead. Consider beforehand what the person's leadership style is, and whether he or she is supportive, neutral, or unsupportive of your goals. How will the change benefit this individual? Practice your meeting with a supportive colleague.
2. The second step is to begin the influencing meeting positively. Create a relaxed environment. State your reason for meeting and gain agreement on the goal of the meeting. Establish trust.
3. The third step is to discuss the change. Lay out your ideas. Then discuss the other person's concerns. Are they logical, emotional, or group influenced? Deal with those concerns respectfully.
4. The fourth step is to engage in mutual problem solving. List the areas where you agree and those where you disagree. Share important criteria and priorities. Design a solution that addresses both your priorities and the other person's.
5. The fifth and final step is to gain commitment. If the person is not yet willing to support the change, ask for a commitment to try it out briefly. If the individual won't agree to that, ask for a commitment to just think about what you've said. Think of yourself as a low-key salesperson here.

Nine Mistakes of Change Management

Here are nine common but critical *mistakes to avoid* when you try to influence change.

1. Acting without seeking input
2. Seeking input but ignoring it
3. Acting without laying the proper groundwork
4. Forgetting to see how the change might be a threat to some people
5. Forgetting to explain what's in it for them
6. Becoming impatient
7. Ignoring the benefits of small, incremental changes
8. Trying to change too much too fast
9. Failure to adequately communicate

SUMMARY OF KEY POINTS

Key points covered in this chapter include:

- The focus of safety has changed and expanded over the years to encompass not just the physical workplace but also health and environmental issues. Safety experts know now that hazards have multiple causes, and that both the frequency and the severity of accidents must be controlled.
- All near misses and minor accidents should be investigated. They are an early warning system that can help prevent serious accidents later on. The ultimate goal of any safety program is loss control.
- All employees must be accountable for maintaining safe working conditions and practices. They must remember that safety is everyone's job.
- The safety, health, and environmental policy should be integrated into the company's Business Plan and supported by top management.
- The National Safety Council has identified 14 components of a comprehensive safety, health, and environmental program. They range from hazard recognition, evaluation, and control to assessments, audits, and evaluations.
- Indirect costs add significantly to the costs of accidents. Safety experts can run a cost-benefit analysis to demonstrate to management why certain training, equipment, or material is worth the investment.
- Team-based organizations give their employees a great deal of autonomy and a corresponding amount of responsibility. Teams are becoming increasingly prevalent because they can improve quality while reducing costs.
- Empowered employees are those who have the freedom to make their own decisions and the training to make good ones. They receive constantly updated training in technical, interpersonal/team, and administrative skills.
- Companies and team leaders can motivate employees through job safety analysis, in which they participate in devising the Safety Plan, and employee surveys. Fear can sometimes be an effective motivator, depending on the audience. Total quality management (TQM) is a good motivator, but requires a cultural change in the organization.
- Continuous improvement is a process-oriented approach that emphasizes people's contributions to long-range solutions to problems.
- There are many barriers to the team-based structure, but the payoff is believed to be worth the time and effort it takes to overcome them.
- The team-based, empowered, continuous-quality-improvement workplace is changing all the time. Many people are inclined to resist those changes, but motivational techniques for implementing them can be used both throughout the organization and with individuals.

2

COMMUNICATION

After reading this chapter, you will be able to:

- Describe the elements of good communication

- Explain how two-way communication works and why feedback is important

- List and discuss the filters that interfere with good communication

- Discuss the three methods of communication: oral, written, and nonverbal

- Explain the importance of listening, and how to improve listening skills

- Describe the principles of interviewing and consulting

Communication means sharing information and ideas with others and being understood. Without understanding, there is no communication.

An essential element of any safety and health program is good communication. Safety professionals must communicate the program's contents to supervisors and team leaders, who must, in turn, effectively communicate to their teams. They must be able to effectively explain their plans for preventing accidents and creating a safer workplace.

It is important to remember that communication is a two-way street. Feedback from employees is critical. Involving employees in aspects of the safety program will increase their commitment to carrying it out. If management just hands down rules, some employees may spend more time and energy in trying to outwit them than in following them. Besides, no one knows the operation or the hazards of a particular job better than the person who accomplishes it.

Feedback is important also to give employees a voice when they disagree with the safety mechanisms already in place. They may feel that safety is not emphasized enough, or they may think a specific procedure doesn't work well. By listening to this feedback, the team leaders and supervisors will gain important insights. Listening is every bit as important a communication skill as speaking. In fact, research shows that most supervisors and managers spend 50% more time listening than speaking.

ELEMENTS OF GOOD COMMUNICATION

Communication consists of three basic elements: the sender, the message, and the receiver. These three elements can create a surprising number of problems. The method of communication that the sender chooses may be unclear, the form of the message may be unappealing, or the receiver may erect barriers that block or filter the message.

Supervisors and team leaders can benefit from using open, two-way communication and from soliciting feedback. They can take account of their listeners' possible filters and try to overcome them. They can provide reinforcement.

Open Communication

An open communication system encourages the free exchange of ideas. It's a bit like the Internet; every member of the department is hooked up to every other. The supervisor/team leader acts more as a facilitator than a disseminator of information. One message this system always sends is that employees' opinions are respected. Because of this, it's a good morale builder.

The open system is far more effective (and more realistic) than the old restricted communication system. In a closed system, supervisors insisted that all communications pass through them and discouraged employees from talking to each other. The closed system was more about control than about communication.

Feedback

To be effective, a communication system must be two-way; it must provide for feedback. Feedback assures the sender that the message has been received. It also provides a way to check on whether the message has been understood. A diagram of successful two-way communication looks like this:

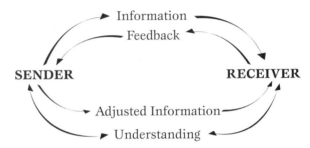

Communication improves if the receiver has a chance to ask questions and express what he or she perceives the message to be. The sender can then identify any misunderstandings and can adjust the communication to resolve them. Depending on the situation, the best type of communication to use on the job is usually one-on-one and face-to-face.

For example, Marianne Osborne is a supervisor at a fast-food restaurant. She explained to a new employee the steps to take in serving a customer. Then she followed up by saying, "Just to be sure I've covered everything, can you repeat what I said?" When he left out a step, she knew what needed repeating.

Good supervisors also solicit feedback in their written communications. An example is this sentence at the end of a memo: "Please call me by Friday to give me your view of these proposals."

Feedback improves the receiver's understanding. It also helps the sender improve future communications. A supervisor who gets the same questions several times learns what he or she needs to clarify in a message.

Open-ended, "you-based" questions can be especially productive because they can build the answerer's ego, and may also help to reassure him or her that there are no wrong answers. Examples are "How do you feel about this approach?" and "What do you like about your job?" Contrast that with "Is there enough light in this work area?" There's no wrong answer, but there's a very short answer: yes or no. Open-ended questions elicit more information and help to empower the person who answers.

Communication Needs

The organization needs frequent, informal communication with employees to update them on progress toward safety goals or to ask them to take action. It's important to be brief. Some examples of communication needs include the following:

- Company policy may require a written summary of information.
- Senior management and key staffers in other areas of the company need to be informed about the status of various safety and health programs or told about progress toward previously published goals.
- A team needs an official summary of safety meetings detailing what actions the team has taken and whether they are effective or need adjusting.
- Formal communication is needed with groups or individuals outside the organization.
- Employees need information about safety and health. The results of safety and health programs must be presented to large groups of employees and to management.
- Employees need ongoing reminders of key safety tips and general safety awareness.
- A team, a manager, or a whole department needs to illustrate the results of a safety program or progress toward goals for other employees.
- Employees often need more information than can be presented verbally in a meeting, or the information presented at meetings needs to be reinforced, or an employee who is not able to attend the meeting needs to receive the information.
- Management needs to provide ongoing reinforcement of training and communication of routine safety information.
- Managers must communicate information required by law.
- Companies may make public statements regarding safety information.

Ways to Communicate

Safety professionals, supervisors, and team leaders need to use various formats to communicate information. Some possible formats include the following:

- A *memo* is a good option. For example, a safety professional might send a weekly memo to people in his or her department to update them on the progress toward a safety goal. A memo should be brief, bulleted, and limited to one main topic.
- A *formal report* presents comprehensive and specific information in detail. A typical report includes benefits (such as reductions in workers' compensation costs) of safety and health programs, costs, and relevant statistics.

- *Meeting minutes* provide an ongoing history of events and are an important part of a safety and health program.
- *Letters* that detail relevant information about a safety and health program are often sent to other companies or community groups.
- An *informal talk* is a good way to present such information as safety tips, demonstrations (how to use personal protective equipment), or results of special safety programs. The talk can be supplemented with literature, such as employee booklets.
- A *formal presentation* is needed. When making a formal presentation, be sure to plan carefully and to use visual aids such as slides or overhead transparencies to summarize key information. A printed handout enhances retention of information.
- *Posters* can provide general safety awareness and remind employees of key safety tips. While they are a valuable addition to a safety and health communication program, posters should never be used as the only method to transmit information.
- *Charts and graphs* are especially helpful in these situations. Consider posting information (especially in chart/graph form) in a cafeteria, break room, and work area.
- *Handouts* such as safety and health booklets are important links in any program. Employees can read them at their own pace. Multilingual versions can help fill the gaps for employees whose first language is not English.
- *Newsletters* can be a good communication tool. They can include safety tips, laws and regulations, results of safety efforts (such as the number of days since a lost-time injury), and messages from management.
- *Legally required communications*, in general, must be made in the format prescribed by the specific laws and regulations.
- *Material Safety Data Sheets* and *HAZMAT materials* can be used for employees responsible for safe handling, transportation, or emergency response to incidents involving hazardous materials.
- Safety information can be included in other public communications, such as *annual and quarterly reports* or *press releases*.

Communication Filters

Communication filters are perceptions—preconceptions, biases for or against various actions, special points of view, and perspectives—through which the receiver interprets messages. Such filtering can change the message in ways the sender never intended. They can even block communication entirely. The more a sender knows about the audience, the better he or she can predict and overcome these filters.

Before supervisors begin to speak or write a message to their employees, they should consider their audience (their receivers) carefully. What are their concerns, problems, knowledge levels, age ranges, biases? Mentally putting oneself in the place of the receivers is a good way to project how they will respond to a message. The supervisor who tailors messages to the audience's needs is more likely to achieve their understanding.

Some of the most common filters through which people perceive messages are knowledge level, personal biases, moods, and physical comfort levels.

Knowledge. Effective communicators adapt their messages to the receiver's level of knowledge. When they are instructing a new employee in his or her first after-school job, they keep the language simple and to the point. By contrast, when they are discussing a new work procedure with an employee who has 20 years of experience, they can use more technical language. They may also acknowledge that experience by asking the employee's opinion about the procedure. He or she may have ideas for streamlining or otherwise improving it.

Biases. Biases can have a definite bearing on how people hear and understand communications. Biases are usually based on people's past experiences and on their attitudes, beliefs, and values. A bias may lead someone to tune out part or even all of the message. For example, consider a 60-year-old union employee listening to a supervisor who is fresh out of school, inexperienced, and perhaps of a different gender or race. This front-line worker may be inclined to assume the supervisor doesn't know anything about the topic of the communication. If the experienced worker feels this way, he or she may ignore the whole message.

These situations can be quite challenging for supervisors. An advantage of the team approach is that the group comes up with solutions instead of having them dictated by management. People often respond better to communications that show respect for their expertise and value of their knowledge.

Mood. The receiver's mood can be a serious filter to getting a message through. Listeners who have something else on their minds may not be paying attention. An employee who is worried about the knock in his or her car engine, or the persistent cough of a child at home may not absorb the message at all.

A partial solution here is to encourage feedback. Suppose the supervisor is warning an employee about the hazards of the task he or she is about to start. Asking "At this point, what are some of the problems you will be watching out for?" tests whether the receiver has heard the message. If not, the supervisor repeats it—twice, if necessary. It isn't enough to ask "Do you understand?" because most people simply reply "yes."

Mood filters are factors in both sides of the two-way communication. Supervisors must never let their own problems distract them so much that they fail to hear what their employees are saying. People are more inclined to listen to people who listen to them.

Physical Limitations. Physical limitations, such as poor hearing or vision, can inhibit communication. The handicap need not be severe to have an effect. For example, an employee might not realize or want to admit he or she needs glasses and thus may miss information posted on a bulletin board or TV monitor. Or, an employee might have difficulty hearing, but be embarrassed to ask a supervisor to repeat an instruction. Asking the employee to repeat an instruction helps ensure that the message reaches its target.

Mental Limitations. Where people have diminished mental capacity, communication can require more direct, "hands-on" communication and subsequent reinforcement.

Reinforcement

When a worker does something wrong, most supervisors may be quick to point out the mistake. When a worker does something right, however, far too many of them neglect to comment. Praising employees for a job well done is an effective way to motivate them.

Psychological studies have shown again and again that the carrot is more effective than the stick. Most people like to take pride in their work and make an effort to do it well, especially if they know they're appreciated. As long as praise is sincere, there is no such thing as too much. Supervisors should always follow the rule to praise in public and reprimand in private. Embarrassing workers by pointing out their mistakes in front of others does far more harm than good.

METHODS OF COMMUNICATION

The three methods of communication are oral, written, and nonverbal. Different methods are appropriate in different situations. Supervisors may choose to combine the methods. For example, they may explain job procedures face-to-face, pointing out the hazards and showing operators how to do the jobs safely. They may also give workers a job safety analysis form that reinforces these points.

Oral Communication

Oral communication has the advantage of being immediate and tailored to the individual listener(s). It is also less formal than written communication.

Oral communication may take the form of a "tailgate" meeting in which the team leader or supervisor discusses with several people the job they are about to start. The group members have the opportunity to ask questions during the explanation. This immediate feedback can help to ensure that everyone knows what to do. If the workers ask no questions, the group leader should ask questions to make sure they understood the explanation.

Face-to-face communications work best for day-to-day liaison, exchange of information, periodic status reports, and production meetings. Telephone communications are fine for the quick exchange of information, but they lack the warmth of face-to-face discussions. Oral communications should always be two-way. Team leaders and supervisors can use two-way oral communications to convey information and also to build relationships with workers.

Some people outline on paper what they plan to say. They may not know exactly what they want to say until they think about it enough to write it down. The process of writing clarifies their thinking, even if they do not use their notes in the meeting.

Written Communications

Messages that are formal, official, or have long-term impact should be written. This method is good for dealing with complicated or technical subjects because it gives the receiver something to refer to. Since written communications don't lend themselves to immediate feedback, it's a good idea to follow up with spoken questions to make sure the receiver understood the message.

Written communications in business are generally in the form of memos, letters, reports, or bulletin board notices. A good example is a work order request that a supervisor sends to the maintenance department, asking it to fix a problem found on a safety inspection. An oral follow-up to this communication might be a phone call asking when the job can be completed.

E-mail is a kind of hybrid. It's written communication, yet it has the immediacy and informality of talking. E-mail, faxing, and other high-tech forms have extended the breadth and reach of communications.

Nonverbal Communications

Actions really do speak louder than words. Supervisors are watched carefully by their people, so the example they set is even more important than what they say. For example, a supervisor who tells employees how important it is to wear personal protective equipment (PPE), but then doesn't wear it, is sending conflicting messages. Guess which version employees will pay more attention to?

By the same token, no matter how much pressure supervisors are under, they should strive not to pass on those pressures. Employees who see their boss take shortcuts and work in an unsafe manner are likely to do the same, and serious accidents can result.

Nonverbal communication includes body language and facial expression. A supervisor who assures workers that he wants to hear their opinions, but then rolls his eyes or drums his fingers while they're talking, is canceling out his spoken message of caring. Employees read and understand nonverbal messages, even unintentional ones, just as clearly as they do oral and written communications.

Nonverbal communication works both ways. Supervisors and team leaders also notice the body language of their people. It can provide clues to how well they've received messages and to what's really on their minds.

EFFECTIVE LISTENING

Respect
Reflect
Respond

So far this chapter has concentrated mainly on the sending portion of communication. The receiving portion is equally important. In every field, effective leaders must be skilled listeners.

Certainly first-line supervisors must know how to listen. Bob Shaffer, a supervisor at a steel mill, complained, "My people just don't listen to me!" But his employees said, "Bob says he has an open-door policy, but when I try to talk to him, he just doesn't listen. Either he's busy doing something else, or he winds up doing all the talking."

There are four distinct steps in the listening process. Breaking the process down and mastering one step at a time can turn even Bob Shaffer into a good listener.

1. *Sensing.* The first step is purely mechanical: Did the listener hear the words that were spoken? If he or she can repeat the sense of the words, this step has taken place.
2. *Interpreting.* How did the listener understand the words that were spoken? The same words can have different meanings for speaker and listener. Follow-up questions may be needed at this stage.
3. *Evaluating.* At this stage, the listener determines whether he or she agrees with the message. Evaluation can take place only after the listener understands the message.
4. *Responding.* Response may be a simple nod or shake of the head or an "I see." Before making a lengthier response, the listener should be sure the speaker has finished a particular point.

The Importance of Effective Listening Skills

How important is listening in a supervisory job? The typical manager spends about 70% of the workday communicating. Studies have classified this communication as follows:

Writing	9%
Reading	16%
Speaking	30%
Listening	45%

Our educational system does not prepare students well for jobs involving listening–even though nearly every job involves a lot of it. Every student takes classes in reading and writing. Some take communications and speech classes. But few schools offer much instruction in listening after the elementary grades. Yet this skill accounts for nearly half of a manager's communication time.

How good a listener are you? Most people rate themselves as average or below-average listeners. Tests conducted at the University of Minnesota showed that immediately after hearing a 10-minute reading, the average listener retained about 50% of the material. A few days later, the retention rate had dropped to between 25% and 30%. In this test, participants were making an effort to listen well. Average listeners retain only a 25% to 33.3% of what they hear, even when they try. Clearly, we all have considerable room for improvement.

Researchers say people can learn to double their listening ability. Like runners preparing for a marathon, however, they have to work at it. They must build up their skills gradually by working at the four steps in the listening process described earlier.

Many accidents are caused by poor listening. A worker may be distracted when the supervisor explains how to avoid the hazards of a job. Or perhaps the supervisor doesn't pay attention when the worker mentions a problem that occurred the last time such work was done. The fact is that a better understanding of a problem by both supervisors and employees can reduce accident potential.

While injury-causing accidents are the main concern, there are also the costs of rework or scrapped work caused by poor listening: letters containing errors, damaged customer relations, perhaps even lost business. Improving listening skills could improve companies' bottom lines and workers' paychecks.

Barriers to Effective Listening

There are many factors that keep people from achieving their potential as good listeners. The barriers may be words, emotions, or distractions. A look at these barriers may reveal some ways to improve in this important area.

Word Barriers. Different people have various "turn-off" words or phrases. When one of these is spoken, the listener stops paying attention to what is being said and starts focusing on what the turn-off word brings to mind. For example, the word "death" is a turn-off word for some people because it reminds them of a disturbing family tragedy. When they hear it, they may stop listening for 15 seconds—or 15 minutes. Some words that may have the potential to turn people off include layoff, panic, grievance, abortion, and taxes, as shown in Figure 2–1. Indeed, this turn-off phenomenon is one reason why people are often advised not to discuss politics or religion.

What words turn off your listening? To improve your listening ability, figure out which words turn you off and why they affect you, then work to overcome the problem. When you hear one of these words, concentrate on what else the speaker is saying to keep your mind from being distracted by thoughts that the turn-off word evokes. You are the only one who can overcome the effects of turn-off words on your listening ability.

Emotional Barriers. Emotions can also block listening. When people become angry, they concentrate on the source of the anger rather than on what is being said. They stop listening and start thinking of things to say that will support their argument and questions that will trip up the speaker.

Anger is just one emotion that can be a barrier to listening. Others include hate, fear, suspicion, jealousy, distrust, and overenthusiasm, as shown in Figure 2–2.

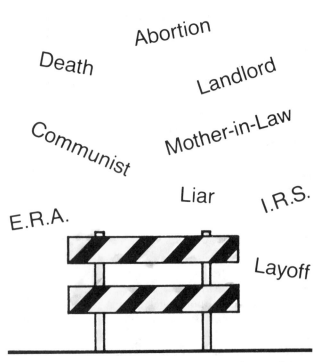

Figure 2–1. Turn-off words that recall emotional memories can interrupt your listener's concentration for 15 seconds or even 15 minutes. Avoid using negative words.

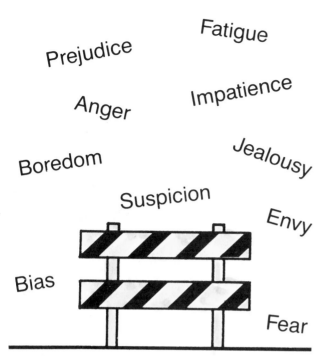

Prejudice Fatigue Anger Impatience Boredom Jealousy Suspicion Envy Bias Fear

Figure 2–2. The listener needs to control emotional barriers in order to hear and understand the entire communication.

Any one of these responses can block people from hearing what is said.

Keeping emotions under control can significantly improve listening skills. Even if you are angry at the speaker or suspicious of his or her motives, try to concentrate on what is being said. Consider how emotions affect your listening during conversations with your employees and your union steward. How about with your spouse and your children? Listening skills are useful in all aspects of life. They will help you communicate better with your family and friends as well.

A good tool for situations in which anger may cloud listening is the *heated discussion rule*. Participants must agree to its provisions in advance. If the speaker says something you disagree with, signal that you wish to speak. When she is finished speaking, she turns the discussion over to you. Before making your statement, you must first state the speaker's position and state why she feels as she does about the matter. If the speaker agrees with your statement of her position, you can then go on to state your own case. This process forces each person to listen in order to get a turn to speak.

In heated discussions, the points of disagreement are often not as serious as they appear to the combatants. When responding in a sensitive situation, begin by stating the other person's position. A good way to start is "As I understand what you have said, you feel this way. . . ." Actually listening to the other person's supporting comments instead of thinking of a rebuttal can do a lot to improve understanding. It may turn out that the two viewpoints are not really so far apart.

Distractions. Another reason for failure to listen is distractions. External noises, such as machines or equipment operating, can derail one's train of thought. Personal problems can also interfere with effective listening. Supervisors and team leaders should work to overcome the many distractions from within and without so all members of the team can pay attention. Leaders should also look for ways to reduce distractions for the team, for example, by holding meetings in a quiet spot away from the regular work area.

Keys to Improved Listening

Most people speak at about 125 to 150 words per minute. Yet most people can listen at speeds of 600 or more words per minute. That time differential might seem to ensure good listening. In fact, many people have so much downtime while trying to listen that their minds wander. That is one reason why they hear and remember only 25% to 30% of what is said.

Effective listeners use the time differential to think along with the speaker, mentally outlining the points. They evaluate the speaker's credibility and analyze whether his or her nonverbal messages agree with the verbal ones.

Making good use of the time differential can help people improve their understanding, the key to communications. Here are six general rules for better listening:

1. *Stop talking.* You can't listen while you are talking.
2. *Repeat/paraphrase.* Once the speaker is finished, repeat what you heard to verify your understanding.
3. *Clarify/probe.* If any of the message is unclear to you, ask questions until you understand it.
4. *Maintain eye contact.* Meeting the speaker's eye helps you concentrate on what is being said. It also shows that you are paying attention.
5. *Empathize.* Put yourself in the speaker's place to get a better understanding of why he or she holds the opinions expressed.
6. *Share responsibility for communication.* The receiver is just as responsible as the sender for good communication.

There are also some don'ts to remember: Don't interrupt, don't make assumptions, and don't jump to conclusions (think before you respond).

INTERVIEWING AND CONSULTING

Safety professionals often need to gather information

from employees at all levels of the organization. They may need to conduct accident investigations, plant inspections, safety audits, job safety analyses, hiring interviews, and internal consultations. Their communication skills play a large role in the success of all of these missions.

Principles of Interviewing and Consulting

1. Think of safety and health as a service function. Instead of seeing your safety and health department as an unwelcome enforcer, think of the people in your company as your internal customers (i.e., intra-company customers). Find out what they need and give it to them; be their internal consultant regarding the safety and health needs that they express.
2. Think win-win. Far from conflicting with other organizational goals, safety and health priorities can enhance them. In the long run, safety doesn't cost; it pays.
3. Build a relationship on joint problem solving. Focus on your internal customers' problems and concerns, not just your own. Working together leads to win-win solutions.
4. Build trust and respect over the long haul. Trust does not occur overnight. It takes time to build a reputation for being fair, honest, and dependable. Expect to earn people's respect and to learn to respect them in turn.
5. Find facts, not fault. Don't accuse people or look for someone to blame. Instead, show that you want to work with them to resolve the problem.
6. Listen more than you talk. You can't meet people's needs if you don't know what they are. Let your internal customers do at least 75% of the talking.

Steps for Conducting Interviews and Consultations

Once safety and health professionals master the principles of interviewing and consulting, it is time to think about the interviews with supervisors and team leaders. Here are five steps for achieving useful interview results (Figure 2–3).

1. *Prepare* for the interview or consultation. Create an objective, prepare your questions, and have a plan of action. Plan and rehearse your opening, and eliminate distractions. Make sure you allow enough time for the interview.
2. *Know* your internal customer's style and prepare for it. Factors to consider include whether the person is easy or difficult to get along with, task- or process-oriented, quick to action or reflective, and time-conscious or accommodating.

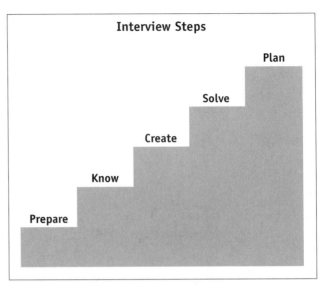

Figure 2–3. These five steps help you achieve more effective interviews.

3. *Create* a positive environment. Choose a comfortable meeting room and arrive on time. Establish a relaxed atmosphere with small talk, and maintain eye contact. Explain the reason for the consultation. Show respect for the person with whom you're working.
4. *Solve* problems jointly. Find out what is important to your customer. Exchange ideas, ask open-ended questions, and ask for suggestions. As always, listen more than you talk.
5. *Plan* the action you will take. Put your agreements in writing. Follow through on your commitments and reward your customers for following through on theirs.

A few questions are provided here that might be helpful to ask during the interview:

- What are three things our department could do to help you in your safety and health efforts?
- What are your most pressing operational problems right now?
- How can we help?
- How would your employees like to be rewarded for their safety and health efforts?

SUMMARY OF KEY POINTS

Key points covered in this chapter include:

- Good communication is essential for accident prevention. How well you send and receive messages will determine your success.

- Communication involves a sender, a receiver, and a message that is understood by both. Two-way, face-to-face communication is the best way to convey messages on the job. It allows the receiver to ask questions. This feedback lets the sender know that the message has been understood.

- Messages can be filtered by the receiver's level of knowledge, biases, or moods. It is important to tailor the message to the audience.

- Oral communication is immediate and personal. Written communication can be kept as a reference, so it is good for technical or complex messages. Nonverbal communication (both body language and facial expressions) should reinforce verbal communication.

- Listening is critical to good communication. Steps in the listening process are sensing, interpreting, evaluating, and responding to messages. People usually hear only 25% to 30% of what is said, but they can double that if they work at it.

- Listeners should overcome word and emotional barriers and avoid distractions. Six rules for better listening are to stop talking, repeat/paraphrase, clarify/probe, maintain eye contact, feel empathy, and share responsibility for communicating.

- Safety professionals draw on all their communication skills in interviewing and consulting with their internal customers. Five steps to a good interview are (1) prepare, (2) know, (3) create, (4) solve, and (5) plan.

- Treat the individual with dignity and respect.

3

HUMAN PERFORMANCE
MANAGEMENT

After reading this chapter, you will be able to:

- Discuss the importance of human relations on the job

- Describe how the high-performance team model works

- Identify what leadership styles are effective

- Apply the concepts of empowerment

- Manage challenges such as shift changes, climate extremes, clean rooms, substance abuse, and stress

- List issues of off-the-job safety

Every employee has needs beyond just earning a paycheck. People need to feel appreciated, valued, and respected. They need to know that they are contributing something important. A workplace that provides the opportunity to meet those needs will be rewarded with happier, more productive employees. Business management theory today recognizes that to get the best out of workers, managers, supervisors, and team leaders must also provide the environment that permits workers to contribute and to develop as human beings, not just as cogs in a machine. In an increasingly competitive world, few organizations can afford not to get the best their employees have to give.

To motivate people to do well, companies must meet some of their workers' social and personal needs. Most people spend about 40 hours a week at their jobs, so they naturally seek validation and self-esteem at work. Rather than seeing this fact as a burden, companies should see it as an opportunity. Improving human relations improves productivity; it's as simple as that.

THE HIGH-PERFORMANCE MODEL

High-performance, self-directed, and self-managed are different names for the same type of team. High-performance teams are fast, flexible, and driven by customer needs and a belief in values. They are led, not managed, often by people who take turns being leaders. They share power rather than hoarding it, and they respect people's dignity.

The high-performance model is an inverted pyramid, in contrast to the old-fashioned model that narrows toward the top. In the high-performance company, customers are on top. Everyone in the organization listens to the customers' definition of quality and works to exceed expectations. Work teams are designed around core processes. Every individual worker is empowered to make decisions. The teams can move quickly to meet customer and market demands, which are always changing.

High-performance leaders are participative. They know, often firsthand, what goes into every job their people do. They support their teams by coaching, counseling, training, listening, removing barriers, and providing resources. Upper management empowers the teams by passing along information about the business in general, as well as their core responsibilities in particular. Managers provide time and money for training, and reward employees for doing their jobs well. Handing over the day-to-day work to the teams frees these managers to concentrate on long-term strategic plans.

The high-performance CEO doesn't waste time protecting his or her turf. The CEO sees his or her role as leading the organization into the future. The CEO and managers alike recognize that employees can be trusted with more and more responsibility as the high-performance model permeates the company. They know that people who feel a sense of ownership will not abuse their new powers.

The high-performance organization has a "flattened" shape, composed of fewer layers of management than the "traditional" or "military" corporate management model. Work teams not only do the work, but plan it, schedule it, build in quality, and make decisions. There are as few layers as possible between them and top management.

Since the company recognizes that people are assets that can appreciate in value, it invests in them accordingly. It puts a tremendous amount of effort into providing training and learning opportunities. Its goal is to create multiskilled, flexible teams who can customize quality products at the lowest cost. It cultivates partnerships with unions, vendors, suppliers, and customers.

Flaws of the Traditional Model

The traditional hierarchical pyramid structure, composed of many workers at the bottom and a few privileged managers at the top, was based on the military command-and-control model, and has been popular since the Industrial Revolution. Decisions were all made at the top and passed downward, through many ranks, to the workers. That model functioned well enough when resources seemed limitless and competition was relatively thin, but it is not working anymore. It cannot move fast enough to accommodate the pace of change required today. It was a model meant to be used in the same way for an indefinite time. Its creators did not plan for an economy in which change must be an ongoing, eternal process for continuous improvement to become a reality.

Perhaps its most fatal flaw has been that it motivates workers through fear and control, instead of giving them opportunity to care about their work. Many jobs are too dull, fragmented in content, and isolated from context, when they could be multiskilled and personally rewarding. Individual employees' performances have often had little effect on their earnings. They are compensated for showing up, but not rewarded for giving their all.

Making the shift from the traditional hierarchy to the high-performance model can't be done with a little tinkering and rearranging. It requires nothing less than a fundamental change in the organization's outlook. The hardest part of achieving a high-performance workplace is to change the attitudes of a critical mass of people, from the CEO on down. Figure 3–1 shows the shift in state of mind that a

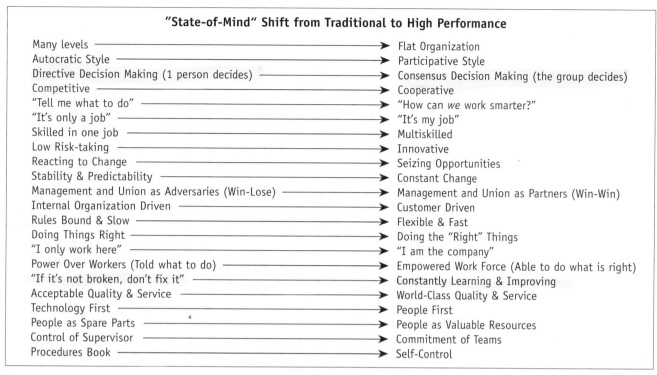

Figure 3-1. Achieving a high-performance workplace requires shifting the state-of-mind from the traditional hierarchical model to self-directed work teams. *Source*: Harper, B, and Harper, A. *Team Barriers: Actions for Overcoming the Blocks to Empowerment, Involvement, & High-Performance.* Mohegan Lake, NY: MW Corporation, 1994. Reprinted with permission.

company needs to make from the traditional to the high-performance mode.

Employee Satisfaction

It is important to remember the human part of employee relations. People respond well to being treated like human beings, with recognition of their feelings. Conversely, they respond badly to being treated like unthinking, unfeeling robots. Human beings need security, group identity, recognition, and fulfilling work.

Security. People have a basic need to feel secure. Employees want to know what is expected of them and how to do their jobs safely and efficiently. A well-trained employee will feel more self-confidence that comes from doing a job well. Being good at a job is a key component of employee satisfaction.

Group Identity. People tend to want to feel they are part of the group. Successful high-performance work teams build in a team identity. It is a positive feedback loop: The better job they do, the more pride they take in their work, and the more pride they take in their work, the better the job they do. Team leaders should try to create an atmosphere of group solidarity and friendliness. Smart leaders can make peer pressure work to their advantage by maintaining a culture that encourages safety consciousness, productivity, and a concern for quality.

Recognition. Employees who receive praise from their leaders for well-done jobs have been found to maintain higher standards of their own volition. Those who feel unappreciated are likely to do the minimum amount of work required to get by. Supervisors who try to lay down the law may be surprised at how creative employees can be in getting around that law. Workers usually meet hard-line tactics with passive resistance, work slowdowns, or equipment failures; it creates a lose-lose situation.

Although experienced workers generally know when their work is good, they are pleased when the boss knows it too. Everyone likes to receive external validation. Praise should be widespread–but always sincere. A phony compliment is worse than none at all. Criticism, whether justified or not, threatens a person's self-respect. If you have any doubts about the unproductive nature of rough criticism, consider how it feels to be on the receiving end.

Fulfilling Work. People want their jobs to provide opportunities for personal satisfaction and growth. They tend to feel a sense of accomplishment when they first learn a job, but this feeling needs to be reinforced periodically. This is one advantage of the multiskills training typical of the high-performance model. It keeps people feeling challenged and interested.

Employees are more likely to feel satisfaction with their job when they understand how it fits into the company's overall pattern. Sometimes their work is too small a part of the operation for them to see clearly what contribution they are making. Team leaders can meet this need by explaining the company's overall objectives and showing employees why their jobs are vital to the company. This is a good time to emphasize the importance of safety to the entire production process.

Most people feel good about working together to achieve a common goal. They crave a sense of purpose and belonging. It is a boost to their self-esteem when managers show that they realize that the line people are the ones who actually get the work done.

Once he or she determines that an employee can handle an assignment, the team leader should turn it over to the employee. The team leader should make sure the problem is clearly defined, the tools are available to solve it, and there is an agreed-upon completion date. Then the leader should get out of the employee's way. Employees who receive this kind of trust usually knock themselves out to do the job well. Every time that the managers turn to them for advice and solutions, they improve their skills, and grow in both knowledge and self-esteem.

Leadership

In the past, many bosses believed that workers inherently dislike work and will avoid it if possible. So management must control them, direct them, and threaten them with punishment, to coerce them into meeting the organization's goals. (This is sometimes known as Theory X, after a study conducted by Douglas McGregor.)

Many bosses have come to realize that threats of punishment are among the least effective ways to get people to work toward organizational goals. Motivating them to share those goals is the most effective way. Theory Y states that people not only accept responsibility, but actively seek it. Satisfaction results from reaching their goals.

Leadership styles range along a continuum from the autocratic to the permissive. Although the ideal style varies depending on the circumstances and the personalities involved, the most effective leadership style tends to be the participative one. Here the team leaders share power and responsibilities. Team members set objectives, develop plans, identify problems and solutions, and evaluate alternative courses of action. There is a free flow of information, not just within the team but throughout the organization.

Differences between Leaders and Managers. Leaders may be managers. In fact, one of the functions of leaders is to manage. But managing is a minor part of what leaders do. Leadership is much broader than and uniquely different from managing. Managers plan, organize, staff, direct, and control. Leaders also do these things, but not exclusively, and they do them in a different way and at a different level of purpose.

Leaders plan, but their planning is more global in terms of setting the agenda for the future. Leaders organize, but their organizational responsibilities involve mobilizing resources and building institutional arrangements that will enable the organization to achieve its long-term goals. Leaders staff, but their staffing function involves identifying and developing talent. Leaders direct and control, but they direct toward defining end purposes and control by empowering people and holding them responsible for finding their own means to those ends.

Leaders differ from managers in that they go far beyond the performance of management functions. Typically, managers focus on operating their area of assigned responsibility for efficiency, cost containment, and compliance with delivery schedules. Leaders are also interested in these things, but they go beyond performance requirements to the effectiveness of the operation in meeting the larger needs of the organization, or of society as a whole.

Managers and leaders tend to think in different ways. Managers are analytical and convergent. Leaders are intuitive and divergent. Managers make decisions and solve problems for their employees. Leaders set a direction and then empower and enable employees to make their own decisions and solve their own problems. Managers emphasize the rational and tangible. Leaders emphasize intangibles such as vision, values, and motivation. Mangers think and act for the short term. Leaders think and act for the long term. Managers accept organizational structure, policies, procedures, and methodology as they exist. Leaders constantly seek to find a better way.

Leadership Qualities. Effective leaders acknowledge publicly that most of their operations are based on ideas that have come from the team members. Contributing solutions helps individuals buy into the corporate game plan. People will work much harder to implement actions that they came up with, than actions that are imposed from outside.

In the same vein, good leaders always show respect for employees, both as individuals and as a group. They are sensitive to workers' needs. They make themselves available and sympathetic when workers do turn to them for help solving problems. But they try not to let their own problems affect the group.

Good leaders act as buffers between employees and upper management by trying not to pass on the pressures they feel to their workers. They strive to create a harmonious climate that minimizes conflict and confusion. They have enough confidence in their own abilities and judgment that they don't need to prove others wrong.

Fairness is paramount in all dealings with people. Supervisors who play favorites are playing with fire. Leaders must live up to what they say by behaving in an exemplary fashion. For example, a team leader can't insist that workers break down departmental barriers and then proceed to carry on a feud with the team leader's counterpart in the accounting department.

Measuring Leadership Effectiveness. Leaders should review their own performance regularly. This helps them nip problems in the bud before they become a concern of upper management. They measure their effectiveness by determining how well they are meeting the organization's goals and objectives. Leaders should ask themselves the following questions:

- Do accident investigations identify the root causes?
- Are these problems then corrected?
- Is the total cost of accidents increasing or decreasing?
- Is the cost of production increasing, decreasing, or remaining the same?
- Is the quality of production improving or declining?
- How about the quantity?
- Is the follow-up to safety inspections adequate?
- Are problems corrected or ignored?
- Do all employees receive thorough safety training?
- Are job safety analyses conducted?
- Is absenteeism increasing or decreasing?
- Is employee turnover increasing or decreasing?
- Are employees being treated with dignity and respect?

CHALLENGES

Sometimes leaders need to use their human relations skills to deal with employees' special problems that affect their job performance or relationships with co-workers. It's not the job of a leader to straighten out workers' private affairs or personality problems. The leader's primary concern is the safety and productivity of employees. He or she should try to determine which problems are temporary and can be handled on the job. Employees with more severe problems should be referred to the human resources department or Employee Assistance Program (EAP) for help.

Fortunately, much undesirable behavior is temporary. Like all human beings, workers sometimes experience money problems, relationship conflicts, or other crises that can affect their performance on the job. Leaders should know their team members well enough to recognize when they are behaving erratically. If the leader is supportive and approachable, employees will feel they can admit that they are not able to perform exacting or hazardous work that day. On the other hand, nonhazardous physical labor is often just what a person needs to escape temporarily from a personal problem.

Employees may react to an upsetting experience with an emotional outburst, from swearing to crying to throwing things. Team leaders can reduce the intensity of such outbursts through permissive listening. Employees' freedom to say what they want can reduce tension that might otherwise lead to an accident. It is important that leaders not judge such remarks. Employees who are on edge need to blow off steam without fearing that it will be held against them. Diligent observation and sensitive, sympathetic listening are important skills for all leaders.

Potential challenges to a smoothly running workplace include stresses caused by shiftwork, extreme climactic conditions, clean rooms, and substance abuse. Most of these problems can be ameliorated with stress management techniques.

Shiftwork and Shift Changes

In our around-the-clock society, many companies must be in constant operation. At least a quarter of the American labor force is engaged in some type of industrial or service activity that requires work around the clock. As the United States economy becomes ever more globalized, this trend may increase. Shift workers are assigned to the night shift or rotating shifts (a combination of day, afternoon, or night shifts). Such shiftwork and shift changes can create stress and pose serious health and safety problems.

Humans are physically and psychologically daytime organisms. We are used to being active during the daylight hours and resting during the hours of darkness. Our biological clock is reset daily by the rising and setting sun. This internal mechanism regulates involuntary responses like reaction time, respiration rate, digestion, and body temperature. Most of these physiological functions have cycles of high and low activity during a 24-hour period. Shiftwork schedules violate both the body's natural rhythms and the dominant social rhythms. This disruption can lead to:

- Higher risk of accidents and injuries due to slower reaction time or lower attention and motivation levels
- Negative effects on family and safety and thus on mental health
- Poorer working conditions, such as lighting, ventilation, and food services
- Chronic fatigue and sleep disorders
- Digestive problems, poor nutrition, and poor eating habits
- Potentially slower action of some prescription medications
- Extreme mood swings and/or chronic irritability
- A debilitating sense of isolation.

Some people simply cannot tolerate shiftwork. Others take it in stride for years but slowly experience increasing problems. It appears that as workers grow older, especially after the age of 50, their ability to tolerate shift changes declines. This shows up in the cluster of problems just listed, known as shift maladaption syndrome (SMS). Certain medical problems, such as insulin-dependent diabetes, seizure disorders, and bipolar brain diseases, generally disqualify an individual for shiftwork and should be carefully evaluated by a medical team.

The typical shift schedule is an 8-hour, 4-crew configuration that rotates through a 28-day calendar and averages about 42 hours of scheduled work per week. Crews may spend 5 to 7 "days" on one shift before rotating to the next time period.

Moving from one 8-hour shift to another is the equivalent of jumping one time zone. Studies of jet lag suggest that the human body can adapt better to forward-rotating time moves (that is, the person works day shift for perhaps a week, then afternoon shift, then night shift). The biological clock can adapt to schedule changes of 2 to 3 hours in a day, but an 8-hour move overwhelms the body's ability to adapt, even to a forward change.

Workdays 10 or 12 hours long are very popular in refineries and chemical plants. Workers like these schedules because they maximize the amount of time off; a typical 12-hour pattern produces 4 extra days off per month. Management likes these long shifts because they control salary costs.

These long-day schedules have disadvantages, however. Perhaps the major one is the potential for excessive exposure to noise and chemicals; standards must be modified to reflect the extended exposure period. Also, some jobs are too physically or psychologically demanding to be done for 12 hours. Workers may have trouble staying alert for so long. Workers over 50 years of age may not be able to handle extended hours. Routine overtime is difficult to schedule, and pay periods (including overtime and holiday pay) may need to be recalculated.

Worker surveys reveal that most employees are dissatisfied with their shiftwork schedules. In response, companies have tried to redesign shift schedules. Such redesign is generally harder to do effectively than it looks, but these guiding principles can help:

- All schedules are compromises among medical considerations, work preferences, and operational necessities.
- Schedule redesign is an interactive process in which workers must be included. Worker surveys are important and should be conducted in a scientific, confidential manner. No redesign will satisfy all parties equally.
- Detailed industrial hygiene and ergonomics analyses must be performed of all chemical and physical hazards if extended-hour schedules are considered.

- Any new schedule must be reevaluated after a trial period of at least three to six months.

Improving workplace conditions can minimize the effects of shiftwork. More difficult tasks should be scheduled earlier in the shift and adequate time should be allowed for all jobs. Workers should be encouraged to take breaks when they get tired. (The incidence of errors and accidents is highest between the hours of 4:00 and 6:00 a.m.) A well-lit, well-ventilated work area helps people stay awake and alert. Noise and isolation should be reduced. Cafeteria services should be available to provide nutritious meals.

Finally, employers should educate workers about shiftwork effects and stress management. Workers and their families can be informed through employee newsletters, management/union bulletins, discussion groups, and safety talks.

Extreme Climactic Conditions

Another element of the work environment that can lead to safety issues is the need to work in extreme conditions of heat or cold. These factors can impair the ability of workers to do their jobs and can also contribute to accidents.

Heat Stress. The human body works optimally at a core (internal) temperature of about 98.6 F. If the core temperature rises, the worker may suffer heat stroke or heat disorder and may have trouble thinking clearly. As the heat and humidity levels of the work environment rise above the comfort zone, the worker's energy may be diverted to keeping the body cool instead of performing the job.

Indicators of heat stress, besides a rise in body temperature, are increased heart rate and increased sweat volume. These three physiological factors are good evaluation tools for excessive exposure to heat. When employees must work in hot environments—say, construction workers at a jobsite in the summer—they should be trained to monitor their own responses before they become dangerously overheated. They should learn what first aid is appropriate for each heat-related disorder.

They should also be trained in the actions individuals can take to reduce the risk of a heat disorder. A leading practice is fluid replacement. A person can lose up to 6 quarts (about 12 pounds) of water in one day, as sweat evaporates to cool the body. Thirst is not a good indicator of how much liquid is needed; people rarely feel thirsty enough to replace as much water as they need to. They should make an effort to drink small quantities of cool water, diluted iced tea, artificially sweetened lemonade, or commercial fluid-replacement drinks as often as possible. This helps instill drinking as a habit, and the volume does not cause discomfort.

Another hygiene factor is acclimation, or the body's adaptation to prolonged daily heat stress exposures. The ability to work increases and the risk of heat disorders decreases with acclimation. Workers should know that they will be able to work better after a few days of heat exposure, but that they should not push themselves too hard at first.

Diet, lifestyle, and general health status are all heat stress hygiene factors. Employees working in extreme heat should not eat large meals during breaks, because they increase circulatory load and metabolic rate. They should maintain a healthy weight; obesity also increases the risk of heat-related disorders. A normal, healthful diet usually provides enough salt to meet heat stress needs, although drinks that have balanced electrolytes may be helpful.

Lifestyle encompasses a healthful diet, enough sleep, exercise, and little or no alcohol. Abuse of alcohol or drugs has been implicated in heatstroke. Exposure to heat stress immediately before work also increases the risk of a heat disorder at work, since the effects are cumulative.

A chronic illness such as heart, lung, kidney, or liver disease suggests a potential for lower heat tolerance. Workers suffering from any chronic disease should tell their doctors about their occupational exposure to heat stress.

Another principle of safe hygiene practice is self-determination. That is, workers must be made aware of their responsibility for monitoring their own responses to heat stress. Workers should terminate their exposure to heat stress at the first signs of heat-related disorders or extreme discomfort. Ignoring symptoms can lead to serious injury. Workers should also adjust their workload demands to their physical needs, perhaps by taking more frequent breaks or alternating demanding tasks with easier ones.

Medical surveillance includes evaluating which workers are at substantial risk for heat-related disorders, treating such disorders, and assessing the information collected from heat-related disorder incidents.

Heat-alert programs are a collection of activities undertaken in anticipation of heat stress conditions. Such conditions might include the approach of summer, a maintenance outage, special operating conditions, or a heat wave. The heat-alert committee should set the training schedule, check the operability of heat stress controls, set criteria for a heat-alert state, and establish how it will be announced.

The three major factors in heat stress are work demands, environmental conditions, and clothing requirements. The metabolic cost of doing work is the greatest contributor to heat gain. For specific jobs, heat stress and its physiological strain on workers can be controlled through engineering controls, administrative controls, and personal protection.

Engineering controls can reduce or contain a hazard. For heat stress, they are directed toward reducing physical work demands, reducing external heat gain from the air and hot surfaces, and enhancing external heat loss by increasing sweat evaporation and decreasing air temperature.

Administrative controls change the way work is performed to limit exposures or risks. For heat stress, they include acclimation; pacing, scheduling, and sharing of work; and preplanned work times, self-determination, and personal monitoring. For example, the number of workers can be increased and overtime can be restricted to avoid heat stress buildup.

Personal protection for heat stress usually involves personal cooling to absorb body heat: circulating air under the clothing (through a high-pressure air line or a portable blower), circulating cool water through tubes and channels around the body, or using ice vests. It may also include reflective clothing, but that solution should be reserved for high-radiant-heat conditions because radiant clothing reduces sweat evaporation. Workers who wear SCBAs in combination with full-body impervious suits (for example, at hazardous work sites) are especially vulnerable to heat stress. They may need extra liquids and rest breaks (see Tables 3–A and 3–B).

Modifications in work practices can help reduce heat stress. Factors to consider include the following:

- If possible, reschedule heavy or difficult tasks for cooler days or cooler parts of the day.
- Provide a cool (ideally, air-conditioned) place for break times.
- Provide a water fountain or other plentiful source of cool water near the work area.
- Remind workers to pace themselves and take breaks as often as they need to.

Cold Stress. The human body is less capable of coping with heat loss than heat gain. Safety behaviors include increasing clothing insulation, increasing activity, and seeking warm locations. Workers' subjective responses are a tipoff to cold stress in the workplace.

Two climactic factors influence the rate of heat exchange between a person and the environment: air temperature and air speed. The windchill factor, or equivalent chill temperature (ECT), takes both of these into account.

As with heat stress, general controls to fight cold stress include training, hygiene practices, and medical surveillance. Training should include how to identify and treat hypothermia. Hygiene practices center on fluid replacement with warm, sweet, noncaffeinated drinks. Self-determination is important here, too. Any worker who experiences extreme discomfort or any symptoms of cold-related disorder should stop

Table 3–A. Heat-Related Disorders, Including the Symptoms, Signs, Causes, and Steps for First Aid and Prevention

Disorder	Symptoms	Signs	Cause	First Aid	Prevention
Heat stroke	Chills Restlessness Irritability	Euphoria Red face Disorientation Hot, dry skin (usually, but not always) Erratic behavior Collapse Shivering Unconsciousness Convulsions Body temperature ≥104 F (40 C)	Excessive exposure Subnormal tolerance (genetic or acquired) Drug / alcohol abuse	Immediate, aggressive, effective cooling. Transport to hospital. Take body temperature.	Self-determination of heat stress exposure. Maintain a healthy life-style. Acclimation.
Heat exhaustion	Fatigue Weakness Blurred vision Dizziness, headache	High pulse rate Profuse sweating Low blood pressure Insecure gait Pale face Collapse Body temperature: Normal to slightly increased	Dehydration (caused by sweating, diarrhea, vomiting) Distribution of blood to the periphery Low level of acclimation Low level of fitness	Lie down flat on back in cool environment. Drink water. Loosen clothing.	Drink water or other fluids frequently. Add salt to food. Acclimation.
Dehydration	No early symptoms Fatigue / weakness Dry mouth	Loss of work capacity Increased response time	Excessive fluid loss caused by sweating, illness (vomiting or diarrhea), alcohol consumption	Fluid and salt replacement.	Drink water or other fluids frequently. Add salt to food.
Heat syncope	Blurred vision (grey-out) Fainting (brief black-out) Normal temperature	Brief fainting or near-fainting behavior	Pooling of blood in the legs and skin from prolonged static posture and heat exposure	Lie on back in cool environment. Drink water.	Flex leg muscles several times before moving. Stand or sit up slowly.
Heat cramps	Painful muscle cramps, especially in abdominal or fatigued muscles	Incapacitating pain in muscle	Electrolyte imbalance caused by prolonged sweating without adequate fluid and salt intake	Rest in cool environment. Drink salted water (0.5% salt solution). Massage muscles	If hard physical work is part of the job, workers should add extra salt to their food.
Heat rash (prickly heat)	Itching skin Reduced sweating	Skin eruptions	Prolonged, uninterrupted sweating Inadequate hygiene practices	Keep skin clean and dry. Reduce heat exposure.	Keep skin clean and periodically allow the skin to dry.

Note: Salting foods is encouraged as both treatment and prevention of some heat-related disorders. Workers on salt-restricted diets must consult their personal physicians.

work and find a place to get warm. Medical certification is suggested for workers who are routinely exposed to temperatures below minus 11 F.

Controls for specific workplaces are engineering, administrative, and personal protection. Engineering controls include providing for general or spot heating (especially hand warming), minimizing air movement, and reducing conductive heat transfer (for example, by making sure tools are insulated and chairs are not metal). They also involve redesigning equipment and processes to control systemic and local cold stress, and providing warming shelters for workers exposed to temperatures below 19 F.

Administrative controls try to reduce exposure time, allow individual control over work, and provide for mutual observation. They include setting up work-rest cycles, scheduling work at the warmest times, moving work to the warmest areas, assigning additional workers, and avoiding long periods of sedentary effort. They also encourage self-pacing and extra breaks as needed, establishing a buddy system, allowing for reductions in the productivity of workers wearing protective clothing, providing a conditioning period for new employees, and monitoring weight changes for dehydration (see Table 3–C).

Personal protection is fundamental in managing cold stress. Clothing should be selected carefully in consultation with knowledgeable vendors and with the workers themselves. Employees should be educated about the roles of clothing items and what factors might reduce their effectiveness.

Insulated clothing is the first line of defense

against cold. Insulated underwear and overalls are usually enough to maintain comfort. Moisture dramatically reduces the insulating properties of clothing. Exterior clothing should be waterproofed and interior clothing should be able to vent perspiration.

Since even mildly cold temperatures can reduce manual dexterity, gloves should be worn at temperatures below 61 F. The gloves should fit well and should not interfere significantly with grip strength. When manual dexterity is not important, mittens are preferable because they keep hands warmer than gloves do. Pants should be lapped over

boot tops to keep snow out. Boots should be waterproof and have air insole cushions and felt liners.

During severe windchill conditions, workers should wear masks or scarves. Extendible fur-lined hoods can protect the face. Workers who are using eye protection should have double-layered goggles with foam padding. Full-face respirators must have separate channels to prevent fogging and frosting of the facepiece.

Controlling exposure to heat and cold pays off by reducing the number and severity of accidents and illnesses.

Clean Rooms

Clean rooms are engineered to protect the silicon chips that are packed with integrated circuits, but these clean rooms place significant demands on workers' health, safety, and ergonomic well-being and ergonomics interests.

Clean rooms protect chips from contamination by airborne particles. This level of air purity requires engineering steps that can create hazards for the workforce. For example, ventilation hoods in clean rooms push air away from the product and toward the worker—potentially increasing the person's exposure to chemical vapors.

Most of the air entering a clean room is recirculated from core areas due to the expense and difficulty of constantly cleaning, filtering, and controlling the temperature and humidity of external air. Usually only 10% of clean-room air is replaced by outside air. This recycling increases workers' exposure to air contaminants. With the advent of vertical-laminar-flow (VLF) ventilation, air is refiltered 9 or 10 times per minute, which redistributes the air and any chemical vapors to the entire work area. VLF ventilation is used in a wide range of manufacturing and health care facilities.

The ideal chip manufacturing environment is semiarid: hot, dry, and windy. This environment is less than ideal for humans. It can lead to many health problems, including chronic skin rashes, allergies, respiratory problems, and eye irritations.

Another source of potential health problems is constant use of protective garments designed to protect the chips, not the workers. A typical person in street clothes may shed a million particles per minute sitting still. This number is multiplied tenfold during activity. Clean-room suits are designed to trap all these particles. The suits require special laundry and repair facilities. Workers have experienced contact dermatitis from the residue of the laundry chemicals.

The headgear of protective garments leads to facial acne in some workers. This problem has been attributed to the relatively hot ambient air temperatures and the constant rubbing and irritation that full facial

Table 3–B. Overview of Specific Controls for Heat Stress Provided by NIOSH in the Criteria Document for Heat Stress

Item	Actions for Consideration
Controls	
M, Body heat production of task	Reduce physical demands of the work; powered assistance for heavy tasks.
R, Radiative load	Interpose line-of-sight barrier; furnace wall insulation, metallic reflecting screen, heat reflective clothing, cover exposed parts of body.
C, Convective load	If air temperature is above 35 C (95 F); reduce air temperature, reduce air speed across skin, wear clothing.
	If air temperature is below 35 C (95 F); increase air speed across skin and reduce clothing.
E$_{Max}$, Maximum evaporative cooling by sweating	Increase by: decreasing humidity, increasing air speed Decrease clothing
Work practices	Shorten duration of each exposure; more frequent short exposures better than fewer long exposures. Schedule very hot jobs in cooler part of day when possible.
Exposure limit	Self-limiting, based on formal indoctrination of workers and supervisors on signs and symptoms of overstrain.
Recovery	Air-conditioned space nearby.
Personal protection R, C, and E$_{max}$	Cooled air, cooled fluid, or ice cooled conditioned clothing Reflective clothing or aprons
Other considerations	Determine by medical evaluation, primarily of cardiovascular status Careful break-in of unacclimatized workers Water intake at frequent intervals to prevent hypohydration Fatigue or mild illness not related to the job may temporarily contraindicate exposure (e.g., low-grade infection, diarrhea, sleepless night, alcohol ingestion)
Heat wave	Introduce heat alert program

(Reprinted from DS *NIOSH Criteria for a Recommended Standard . . . Occupational Exposure to Hot Environments—Revised Criteria 1986.* Washington, DC: U.S. Government Printing Office, 1986.)

garments can produce. Similarly, PVC gloves used in clean rooms can lead to skin problems.

There are psychological issues that clean rooms raise. Workers operate in a sterile, unchanging environment, completely isolated physically and emotionally from the outside world. Protective equipment heightens the sense of isolation and dehumanization because it makes communicating and socializing with co-workers difficult. To preserve the purity of the environment, employees are discouraged from moving around or talking to anyone. No large-scale studies have yet been done, but some workers report feeling depressed and alienated.

Substance Abuse

Employees' abuse of alcohol and drugs costs industry millions of dollars each year. It is often the cause of lowered productivity, increased absenteeism, inefficiency, high employee turnover, increased injury rates, and incidents arising from behavioral problems. Employees who have a substance abuse problem often call in sick and engage in unsafe work practices. Their conduct can be disruptive and demoralizing to co-workers and can have a ripple effect, leading co-workers to be less productive and even to quit their jobs. Supervisors and team leaders should be able to recognize these problems and take steps to deal with them.

Dependence on alcohol or other drugs is a major cause of impaired job performance, accidents at home and at work, higher insurance rates, increased absenteeism, family strife, and rising crime rates. These illnesses know no boundaries and no generation gap. People of all ages, races, and socioeconomic groups are vulnerable to their destructive effects.

In our society, some people believe that having a cocktail or beer with lunch can't hurt anything. But the effects of alcohol do not end with the meal. Supervisors can make it clear that drinking must be confined to off-duty hours. There's no excuse for any employee to indulge when it will result in impaired judgment and reaction time on the job.

Symptoms of Substance Abuse. Many companies believe the best solution to substance abuse problems is professional assistance. The employee's immediate supervisor is the first member of management likely to observe unusual employee behavior. Early

Table 3–C. Cold-Related Disorders, Including the Symptoms, Signs, Causes, and Steps for First Aid

Disorder	Symptoms	Signs	Causes	First Aid
Hypothermia	Chills Pain in extremities Fatigue or drowsiness	Euphoria Slow, weak pulse Slurred speech Collapse Shivering Unconsciousness Body temperature <95 F (35 C)	Excessive exposure Exhaustion or dehydration Subnormal tolerance (genetic or acquired) Drug/alcohol abuse	Move to warm area and remove wet clothing Modest external warming (external heat packs, blankets, etc.) Drink warm, sweet fluids if conscious Transport to hospital
Frostbite	Burning sensation at first Coldness, numbness, tingling	Skin color white or grayish yellow to reddish violet to black Blisters Response to touch depends on depth of freezing	Exposure to cold Vascular disease	Move to warm area and remove wet clothing External warming (e.g., warm water) Drink warm, sweet fluids if conscious Treat as a burn, do not rub affected area Transport to hospital
Frostnip	Possible itching or pain	Skin turns white	Exposure to cold (above freezing)	Similar to frostbite
Trench Foot	Severe pain Tingling, itching	Edema Blisters Response to touch depends on depth of freezing	Exposure to cold (above freezing) and dampness	Similar to frostbite
Chilblain	Recurrent, localized itching Painful inflammation	Swelling Severe spasms	Inadequate clothing Exposure to cold and dampness Vascular disease	Remove to warm area Consult physician
Raynaud's disorder	Fingers tingle Intermittent blanching and reddening	Fingers blanch with cold exposure	Exposure to cold and vibration Vascular disease	Remove to warm area Consult physician

Note: Hypothermia is related to systemic cold stress, and the other disorders are related to local tissue cooling.

recognition of the problem is essential to an effective program of evaluation, treatment, and rehabilitation. Awareness about alcohol and drug abuse is a basic management responsibility.

In their daily contact with employees, supervisors are expected to be familiar with appearance, behavior, and work patterns, and be alert for any changes. Such changes can result from many problems—for example, illness, fatigue, or the side effects of prescription drugs. Because behavioral problems can take many forms and have many causes, supervisors should avoid drawing hasty conclusions.

It is not the job of supervisors to diagnose; it is their job to recognize changes in an employee's behavior or job performance that might indicate a problem. Possible symptoms include sloppiness, slurred speech, and unsteady movements. Symptoms may arise suddenly or gradually. They may be episodic or continuous. In any case, they require immediate action. Ignoring the symptoms of drug abuse means ignoring a major threat to the safety of not only the employee in question but also his or her co-workers.

Substance Abuse and the Americans with Disabilities Act (ADA). Section 510 of the ADA provides that the term "individual with a disability" does not include individuals who are "currently engaging in the illegal use of drugs." Section 512 of the ADA amends the Rehabilitation Act definition of "individual with handicaps" to make a parallel exception for current use of illegal drugs under its Title V. Court decisions under Section 504 of the Rehabilitation Act had applied the statutory exemption to hold that individuals whose current illegal drug use affected their work performance or threatened safety were excluded from the class of "individuals with handicaps."

Sections 510 and 512 of the ADA expand the statutory exclusion to all current use of illegal drugs regardless of whether it can be shown to adversely affect job performance or safety. Individuals who are discriminated against on the basis of illegal drug use will not be considered an "individual with a disability" under the ADA, and they will be unable to invoke the protection of the law to challenge such discrimination.

The ADA does cover alcoholics or people addicted to the use of drugs if such conditions substantially limit their major life activities. However, it does not protect a person who currently engages in the illegal use of drugs if the employer or any other entity covered by the ADA acts on the basis of such use. A person taking drugs under the supervision of a health care professional is not excluded from the protection of the ADA. In addition, the exclusion does not apply to a person who has successfully completed a supervised drug rehabilitation program or is otherwise successfully rehabilitated and is no longer using drugs illegally. Employers may adopt reasonable policies

and procedures, including drug testing, to ensure that an individual who is participating in a rehabilitation program or has been rehabilitated is no longer engaging in the illegal use of drugs.

Drug and Alcohol Interactions. Studies show that at any given time, 10% to 20% of the United States population is taking prescription medication. Add to this the percentage of people using drugs illegally, and it becomes clear that a substantial portion of the working population is exposed to the effects of drugs.

When drugs are combined with alcohol, they produce a variety of effects that impair workers severely. Concentrations of alcohol and drugs remain in the bloodstream much longer than most users realize, and the effects may appear unexpectedly. For example, they can make vehicle operators drowsy or slow down their reflexes.

To help detect and prevent substance abuse, every time a job applicant or employee visits a doctor or nurse, some health education or counseling should be given. Occupational health units or an EAP are an effective way to provide these services.

Countermeasures. Dealing with substance abuse begins with education. Supervisors should read as much about the subject as possible. They can inform their people about the problems substance abuse can cause during a five-minute safety talk. Or they can break it down to specific, more detailed subjects, such as drinking and driving, over the course of several safety talks. The company should establish and publicize a written policy.

The second step is to identify substance abusers. Alcoholics are not always falling-down drunk. Sometimes they look sober but seem to have a hangover all the time. Drug addicts come in all ages. People who regularly take large doses of codeine or sedatives may be as dangerous as those abusing illegal drugs. However, supervisors should not become drug detectives, constantly looking for pills or joints. They should concentrate on watching for changes in employees' work behavior patterns, personal relationships, and moods.

Identifying employees with problems is much easier if a company takes a positive approach. Supervisors might emphasize that their goal is to retain employees who are substance dependent by helping them to break the dependence before they become unemployable. They should make it clear that substance abuse programs are completely confidential.

Next, get help for the workers in question. Supervisors should show employees who have a substance abuse problem that they genuinely care about them. Substance abuse is an illness and must be handled by people who are trained to deal with it. Contact local organizations that counsel addicts, as well as Alcoholics Anonymous and community groups and churches, to consult with team leaders and advise them how to proceed.

For more information on alcohol and drug abuse programs, contact:

Alcohol and Drug Problems Association of North America (ADPA)
1555 Wilson Blvd., Suite 300
Arlington, VA 22209
(703)875–8684

Alcoholics Anonymous World Services
475 Riverside Dr.
New York, NY 10163
(212)870–3400

National Institute on Drug Abuse
5600 Fishers Lane #10–05
Rockville, MD 20857
(301)443–6480   (800)843–4971
Internet: http://www.nida.nih.gov/.

Stress Management

Stress is defined as any of the factors that accelerate the rate of aging through the wear and tear of daily living. The combination of job stress and personal stress can increase the likelihood of accidents and decrease productivity. Stress management, then, is important in helping companies reduce accidents and increase productivity.

Stress is present in everyone's daily life. It can be positive or negative. Examples of positive stress include getting promoted, getting married, and buying a house. Examples of negative stress include being the subject of disciplinary action, having unrealistic deadlines or abnormally large workloads, and divorce. Sources of negative stress can be as minor as losing a favorite pen or as aggravating as an ongoing conflict with a co-worker.

Anger as a habitual response to stress creates more stress. Responses that focus on problem solving and communication are more effective. How people choose to react to stress, on and off the job, is more important than what the stressors are. No one can eliminate stress, but everyone can learn mechanisms for coping with it.

Stress has both physical and mental effects on people. Employees who master the following skills can reduce their stress levels and improve their health.

1. Specific relaxation techniques. One example is autogenic relaxation, which uses full, comfortable breathing and concentration techniques to achieve a form of self-hypnosis that relaxes mind and body. It often incorporates positive affirmations to combat the negative self-talk that so many people torture themselves with. Another example is deep muscle relaxation, a breathing technique that is easy to learn and can quickly make the breather feel more relaxed.

2. Values clarification techniques. These help people figure out what goals they want to achieve in their work and personal lives. They identify their basic values and determine ways they can express these values more fully. These techniques help employees sharpen their prioritizing and planning skills.

3. Exercise. A sound exercise program incorporates strength building, stamina, and flexibility. At least 30 minutes of cardiovascular stamina-building activity at least 3 times a week should be included.

4. Nutrition. A well-nourished body can fight off the effects of stress. Employees should eat a variety of unprocessed and little-processed foods, drink at least 6 glasses of water a day, and eat several small meals each day (including breakfast) instead of a single large one. Many people find broad-based vitamin-mineral supplements useful.

Although it is important for individuals to manage their stress, it is just as important for the organization to avoid contributing to their stress levels more than necessary. Two leading causes of stress in organizations are (1) unrealistic workloads, and (2) tight deadlines. The following steps can help reduce stress throughout the company.

- Survey and carefully consider staffing requirements of various departments. Saving the salaries of a few employees may turn out to be quite expensive in its long-term effect on productivity and morale.
- Survey departments to determine typical deadlines. Are they realistic? If deadlines cannot be altered, develop time management techniques for employees.
- Provide classes on stress management, conflict resolution, and communication. Much stress results from misunderstandings between workers.
- Survey the work environment for safety- or health-related issues that have not been addressed. Unresolved concerns in those areas can cause stress.

OFF-THE-JOB SAFETY

Part of appreciating workers is recognizing that they don't go into suspended animation when they walk out the door. Their home lives have a huge effect on their attitudes and performance at work. The company safety program should cover safety off the job too, for both financial and psychological reasons.

Most employers incur group medical insurance costs for off-the-job injuries and illnesses that are not work-related, and some employers pay wages for time off during these medical absences. Collecting and analyzing off-the-job injury data from insurance claim forms can help identify unsafe or unhealthy off-time activities that may affect the company. The effectiveness of off-the-job safety programs can be measured in terms of cost savings not only in insurance but in employee turnover, training and retraining, and absenteeism.

Off-time safety programs also send the message to employees that the company really does care about them. Safety and health newsletters and take-home posters can get whole families involved in becoming safety conscious. Inculcating the "safety on-and-off the job" mindset can improve the life of the community as a whole.

SUMMARY OF KEY POINTS

Key points covered in this chapter include:

- The high-performance team is the same as the self-directed work team. It is fast, flexible, and driven by customer needs and a belief in values. The high-performance model is "flattened," not hierarchical. Work teams not only do the work, but plan it, schedule it, build-in quality, and make decisions.
- Employee satisfaction leads to better performance. People crave security, a group identity, recognition, and fulfilling work. They want their jobs to provide opportunities for personal satisfaction and growth.
- One of the most effective leadership styles is the participative one, in which leaders share power and responsibilities. Team members set objectives, develop plans, identify problems and solutions, and evaluate courses of action. Information flows freely throughout the organization.
- Shiftwork, which is becoming ever more widespread, creates stress and poses health and safety problems. Shiftwork schedules violate both the body's natural rhythms and the dominant social rhythms. Improving workplace conditions can minimize the negative effects of shift changes on employees.

- Extremes of heat and cold can impair workers' ability to do their jobs and can contribute to accidents. Employees should be trained in heat- and cold-stress hygiene practices. Examples are drinking plenty of fluids to combat heat stress and wearing insulated clothing to protect against cold stress. Acclimation is the body's ability to adapt to prolonged daily exposures to temperature extremes.
- Clean rooms are designed to protect silicon chips from contamination during their manufacture. Clean-room engineering creates several hazards for the workforce, among them high exposure to chemical vapors and a semiarid environment that can result in chronic skin rashes, allergies, respiratory problems, and eye irritations. The protective garments worn in clean rooms can also be hazardous to workers' health. The psychological environment may lead workers to feel depressed and alienated.
- Employees' abuse of alcohol and drugs costs industry millions of dollars each year. It can cause lowered productivity, increased absenteeism, inefficiency, high turnover, increased injury rates, and incidents arising from behavioral problems. The combination of drugs and alcohol can impair workers severely. Dealing with substance abuse begins with educating supervisors and ends with getting the affected employees into an Employee Assistance Program (EAP).
- Stress is defined as any of the factors that accelerate the rate of aging through the wear and tear of daily living. The combination of job stress and personal stress can increase the number of accidents and decrease productivity. Stress has both physical and mental effects, but people can learn techniques to reduce the amount of stress in their lives and companies can take steps to do the same.
- The company safety program should cover safety off the job too, for both financial and psychological reasons. It can reduce the costs of insurance, absenteeism, and turnover. It can help convince employees that the company cares about them.

SAFETY AND
HEALTH TRAINING

After reading this chapter, you will be able to:

- Describe the goals and benefits of training
- Identify the needs of adult learners
- Develop or select an appropriate training program
- Describe the factors involved in delivering a safety training program
- Explain different methods of training
- Identify key topics for effective new-employee orientation

When a safety or health problem is being caused by a lack of the right attitude, knowledge, or skills, one solution can be appropriate training. Training focuses mainly on behavior or performance change. It can also influence attitude by inspiring confidence and the belief that improvement is possible. In a work setting, the goal of improved training is learning that leads to improved performance on the job. When safety and health are the primary issues, the goal is that workers learn how to keep themselves and others safe and healthy.

A key element in every successful organization is effective job orientation, including safety and health training. The responsibility for employee training rests with the top management. For the organization to achieve its objectives, employees must perform at a certain level. To achieve that level of performance, management needs to establish worker training policies. To achieve the organization's objectives, those responsible for safety and health—whether safety and health professionals, front-line supervisors, or team leaders—must:

- Be trained in the proper methods of leadership and supervision
- Know how to detect and eliminate (or control) hazards, investigate accidents, and handle emergency situations
- Understand how to train employees in the safest and most efficient ways to do their jobs
- Know and support all company health and safety policies and procedures

The safety and health professional must make sure that the training provided meets all regulatory training obligations and compliments/supports the company's business goals.

TRAINING AS PART OF THE SAFETY AND HEALTH PROGRAM

Utilizing effective training programs will help to ensure that employees know how to do their jobs using safe methods. Training provides the "how-to," and usually also the "what." It also provides the "why," to the extent that people need to know safe ways to complete their tasks.

Training prepares people for behavior change. In education, the focus is on information that may or may not be used on the job. In training, the focus is specifically on learning how to do the job properly, and how to apply the new information and skills on the job, in order to create and maintain a safer work environment.

Roles of the Safety Professional

Safety professionals may play many roles or have a myriad of responsibilities in a comprehensive safety and health program. They may conduct the training, coordinate its development with internal or external instructional designers, design and develop the program, purchase materials from vendors, contract with a trainer, or manage the entire training function. No matter what their role, it is important to have a clear idea of the needs and problems of the company or facility and the workers.

Benefits of Training

The benefits of safety and health training include the following:

- Fewer incidents and accidents
- Reduced costs
- Reinforcement of the organization's operational goals
- Improved performance
- Compliance with government, union, and other organizations' safety laws, rules, and guidelines

Training and Nontraining Solutions

Before exploring training in more detail, it is important to understand how training fits into an overall plan to address safety and health issues in an organization. In most situations, the success of training is improved if other nontraining solutions to needs and problems are also being implemented in the organization.

Sometimes a training program is not the solution needed in order to affect on-the-job performance. Millions of dollars are wasted every year in training to solve a problem that actually needed a different solution. Nontraining solutions (actions taken to resolve safety and health problems that do not require training) may include actions like the following:

- Workers can be instructed by a sign showing what safety goggles are needed in the work area.
- Supervisors may offer bonuses to employees with good overall safety records.
- The engineering department may redesign the workflow of a manufacturing site to protect workers who have had to walk near heavy machinery.

Analyzing problems and assessing needs must be done before making decisions about training or nontraining solutions. Management should consider all training and nontraining alternatives to safety and health challenges. For example, suppose a certain task is performed infrequently and workers often fail to remember all steps. Or perhaps it has been discovered that a task is too complex to perform from memory. In situations like these, nontraining solutions such as the ones listed here may be as effective as a training program.

- Job aids
- Task procedure instructions
- Material Safety Data Sheets (sections on protection information, special precautions, spill/leak procedures, etc.)
- Flowcharts
- Checklists
- Diagrams
- Troubleshooting guides
- Decision tables
- Reference manuals
- Help Desks or Hotlines
- Reward systems
- Improved physical work environments
- Improved work processes

Training and nontraining problems or needs and their related solutions can be categorized as (1) selection and assignment, (2) information and practice, (3) environment, and (4) motivation/incentive.

1. Selection and assignment refers to considerations and processes used to hire people and assign them specific responsibilities and job tasks. Improving the hiring process helps make sure the right (i.e., sufficiently experienced, skilled) worker is assigned to each task.
2. Information and practice is the act of providing all necessary information and development of skills needed to carry out work using safe practices. This includes communicating tasks, goals, and instructions clearly, as well as providing feedback, training, job aids, guided practice, and experience. This is the area that can often be improved through training and education. To be effective, training must supply many opportunities for skills practice.
3. Environment refers to all day-to-day influences on performance. These include (but are not limited to) equipment, floor plans, access to tools, and working relationships with co-workers and with safety and health professionals. Environmental problems include such situations as humidity problems in electrical work areas, slippery walkways, air pollution, etc., which are best solved by nontraining solutions.
4. Motivation/incentive solutions are any rewards by management for safe behavior, such as praise and commendations, bonuses, pay increases, and contest prizes. For example, a supervisor may offer a bonus to the employee with the best overall safety record, or a hundred days without an injury may be marked with a free lunch for all. Education may be used as a motivator, as well as peer pressure and negative pressures such as the threat of a loss of bonus, etc. Research shows that negative pressure does not produce results as well as positive pressure, but is is still frequently used.

The supervisor, manager, or team leader must carefully analyze the needs of his or her group and decide what solutions will best meet the needs. If training and education supplies the best answer, then a safety professional should be consulted or should set up the program.

Structured Training

When training is selected as the solution to a specific need or problem, the training program should be planned, designed, developed, and delivered in a structured manner. Its potential for solving the problem and changing behavior should be realistically evaluated.

Performance-based training is implemented to solve a specific on-the-job problem or encourage a specific behavioral change. Management can evaluate performance-based training by analyzing individual workers' performances. This training is directly related to the job the worker is expected to perform and linked to the organization's safety and health goals. It is measured and evaluated by the worker's observable performance, and created using a proven systematic approach of five phases.

1. Assessment. The safety professional consults with others and conducts research to determine whether training is the right solution and what knowledge, skills, and attitudes people need to develop.
2. Design. The designated safety trainer identifies instructional goals, chooses training media, and figures out the sequence in which workers will learn new information and skills. This work is based on needs assessment results, audience analysis, and specific performance-based objectives.
3. Materials acquisition and/or development. The designated safety trainer buys and/or creates training materials.
4. Delivery. The designated safety trainer delivers the training.
5. Testing and evaluation. The safety professional gets feedback from team members, leaders, and others involved in the performance-based training. Training is improved, based upon this feedback.

There are four major types of performance-based training.

1. Instructor-led training. This training is presented in a classroom-like setting. Everyone follows the pace set by the instructor, trainer, or facilitator. Participant materials are usually included.
2. Self-paced training. Employees usually complete this training on their own by working through a textbook and/or workbook. Self-paced training may also be done through computer-assisted instruction.
3. Computer-assisted instruction. In this version of

self-paced training, the learner is guided through the course by a computer, which provides the course content, guides learning activities, and administers tests. Computer-based training can also give the safety professional a quick list of who has completed the training, and some programs can provide tests and scores.

4. Structured on-the-job training. Similar to instructor-led training, structured on-the-job training is conducted at the learner's place of work. The instructor is usually a trained supervisor who acts as a coach or guide. This type of training may or may not include participant materials.

DEVELOPING THE TRAINING PROGRAM

If the safety professional lacks skills and experience in training development or instructional design, an internal or external instructional designer can be brought in to develop each training program, or the entire program can be outsourced, as a complete package, if it meets the needs. The development process should include a needs assessment, written performance objectives, selection of materials to fit the objectives and the audience, outline of content, timing, testing, and evaluation. The training program should be developed in the five-phase process, and should meet the same criteria, whether it is developed internally or selected from outside sources.

Phase 1—Assessment

To create an effective training program, the safety and health professional needs to assess current worker performance, comparing it to desired worker performance. He or she should also conduct a learner analysis. A needs assessment is important to the organization because it helps to:

• Distinguish between training and nontraining needs.
• Understand the problem before designing a solution.
• Save time and money by ensuring that solutions effectively address the problems they are intended to solve.
• Identify what factors will affect the training before it is developed.

Needs assessment is the process of determining the who, what, when, where, and why of the training need, after it is determined that a safety or health problem can be solved through training. Identifying the specifics of the training needs is the first phase of developing an effective training program.

The safety and health professional tries to determine at a general level who may need training, what

they should be trained to do, when they should be trained, where they should be trained, and why they need the training.

• Who does the job? (Which skilled or unskilled worker?)
• What do they do in their jobs? (Tasks.)
• When do they do the job? (All during a shift, in a set sequence, before or after certain other tasks?)
• Where do they do the job? (Area conditions; proximity to materials, doors, warehouse, docks, etc.)
• Why do they do the job? (What is the importance of this work?)
• How do they do the job? (Speed, work practices, safety measures in place or avoided, and any other qualifying characteristics that affect the need for training.)

A needs assessment is conducted primarily to determine the specifics of the problem or need that the training must address. It may show that nontraining solutions are also needed to address the issue completely. When a regulatory agency is not involved, the safety and health professional is responsible for determining whether training is needed at all, before defining the (perceived) training need. In initial stages of assessment, the many solutions that might solve the problem must be considered.

In the needs assessment phase, the safety and health professional should identify a training goal(s) in a statement that describes the general purpose and how the training will satisfy the safety and health need or solve the problem. Training goals are used to provide direction for learner, job, and setting analyses and to develop learning objectives. Training goals are also valuable tools for communicating the purpose of training to others. The training goal should clearly

• Describe how the training will satisfy the safety or health training and
• Communicate the purpose of the training.

While the training goals are being identified, the safety and health professional should also identify the characteristics of the learners (workers) who will be involved. Learner characteristics may be observed by study of general demographics, and learners' preferences, attitudes, knowledge, skills, and previous experiences. The results of this learner analysis will affect the training design. Results include specific traits that will affect design and delivery of the training (also called an audience description) and recommendations for how the training design should accommodate learners' specific traits and general adult learning needs. The steps in completing a learner analysis are:

1. Identify general traits of audience members, such

as job titles/positions and age.

2. Identify the learners' preferences regarding training.
3. Identify the learners' attitudes or feelings regarding training.
4. Identify the learners' knowledge or skills, especially as they relate to the content of the training.
5. Identify ways to adapt training and delivery to these characteristics.

Job and setting analyses are also conducted in the needs assessment phase. Job analysis is the process used to determine the procedures, decisions, knowledge, and skills required for a worker to perform a job function. Setting analysis is identifying specific characteristics of the training environment (where training will occur) that will affect training. Ideally, it is best to obtain setting information before designing the training. But since this is not always possible, the training course should be adaptable to a variety of settings.

Adult Learning Needs. A major factor in the success of training is the extent to which adult learners' needs are considered. Safety and health issues can be technical and difficult to learn. In addition, people may see safety and health training as a disruption in their busy workdays.

To keep the participants motivated and involved, the safety and health professional must allow for their learning needs. There are many adult learning principles. For example, one principle is based on the belief that each person learns best through one or a combination of three senses: hearing (11% of people are significant auditory learners), sight (83% are significant visual learners), or touch or activity (69% are significant kinesthetic learners). Training following this model would use media tailored to the learners' best learning modalities.

More generally applicable are the four needs common to all adult learners. This information is adapted from Clay Carr, a widely respected trainer and author of *Smart Training: The Manager's Guide to Training for Improved Performance*.

First, adults need to know why they are learning a particular topic or skill, because they need to apply learning to immediate, real-life changes. Second, adults need to be able to apply their extensive experience to all new learning. Third, they need to be in control of their learning. Fourth, they want to learn things that will make them more effective and successful. These principles should help shape any safety and health training developed or contracted.

Need to Know Why. Adults need to know why they are learning a particular topic or skill because they need to apply learning to immediate, real-life challenges. They learn best when:

- They can see clear demonstrations of how the training applies directly to their jobs.

- They can apply the new information or skill to solve problems immediately, during the training.
- They have opportunities to think about how they can use the new information or skill.

An example of meeting this need is a supervisor demonstrating how to operate a new machine safely then providing feedback while watching the workers run the machine.

Need to Apply Experience. Adults bring experience with them to all new learning. They learn best when:

- They can share their experiences during training.
- They can use comparison and contrast, relationships, and association of ideas.
- They have opportunities to think about how the new information or skill relates to their past, present, or future experiences.
- Their experience and opinions are taken seriously by the trainer.

For example, to meet these needs during a training session on personal protective equipment (PPE), the trainer should give participants time to talk about their past or current PPE practices and those of their co-workers at their facility.

Need to Control. Adults need to be in control of their learning. They learn best when:

- They choose (or at least influence) the training they receive.
- They are in a flexible learning environment.
- They are actively involved in the learning process rather than passive receivers of information. (The trainer guides and facilitates.)
- They have a chance to voice concerns and see them addressed. They take active part in group discussions, role plays, simulations, and other activities.

Meeting this need is illustrated by workers at a fluorescent light factory who are part of a safety and health committee. Their responsibilities include assessing workers' needs for training, identifying good training solutions, and evaluating training after it is delivered.

Need to Succeed. Adults want to learn things that will make them more effective and successful. They learn best when:

- They know why they are taking the training and how it will benefit them.
- They can ask and answer the questions, "What will this do for me? How will it help me get ahead? What hazards will it help me control?"

An example of meeting this need is training in a safety management techniques course that includes action planning for future personal and professional development.

Appropriate Level of Materials. Another consideration that must be made in meeting adult learning needs is the literacy level of the worker. Many workers read, use math skills, comprehend information, and solve problems at a low- or mid-literacy level. Low-literacy generally refers to workers whose abilities are at or below fourth-grade level. Mid-literacy refers to workers whose abilities are at fifth- through eighth-grade levels.

Low- or mid-literacy-level workers may have trouble understanding much of the information that is presented in complex textbooks, manuals, and lectures. The design and delivery of safety and health training programs must take this into account. For instance, training can be written at a fourth- or fifth-grade reading level or simply rely on less printed material. Use of symbols and illustrations in training and promotion should minimize the risk of the message not being understood.

Objectives. Writing performance objectives and evaluating learning progress are two of the most important steps in developing effective training programs. Performance objectives provide a structure for developing training. They are also important guides in the selection and development of course content, the selection and development of learning activities, and measurement of the learners' performance.

Objectives are important to the workers being training because they provide a target for performance (or behavior), help learners identify their focus, and tell learners how they will be evaluated. Objectives are especially critical in the safety and health arena because there is too much risk (of potential injury, illness, or loss of life) to workers to allow a hit-or-miss approach. Not determining the training objectives in advance may also waste a vast amount of time and money.

Unlike goals, objectives are measurable and observable. They differ from learning activities in that they describe the results, not how to achieve them. The four parts of an objective (sometimes referred to as the ABCD method of objective writing) are as follows:

- Audience: Always identify the audience (learners).
- Behavior: Identify what learners must do to demonstrate mastery.
- Conditions: Identify what learners will or will not be given in order to carry out the behavior.
- Degree: Specify how well the audience members must perform the behavior.

Generally, most people will assume that 100 percent is the standard performance if it is not specified otherwise in the objective. Performance can be measured in terms of quantity, quality, or both.

By detailing training objectives, the safety and health professional can develop programs that meet an organization's established safety and health goals. The objectives should provide specific information about what learners will be able to do as a result of the training.

Phase 2—Design

The training developer organizes and times the training program through outlining the content. To outline, the developer selects an objective, identifies what actions the worker must take to achieve it, and lists related topics that the worker must know to accomplish the actions. The developer then puts the content in order (perhaps with help from the safety and health professional or management) and estimates the amount of time needed for each point. The time available may be predetermined. If it is not enough, the trainer can either ask for more time or narrow the scope of the objectives.

Phase 3—Materials Acquisition and/or Development

The materials of an effective safety and health training program should provide (1) an introduction, (2) presentation of the information, and/or demonstration of actions (3) practice and feedback, and (4) a summary, evaluation, review, or transition. The introduction can be an overview, a rationale, a warm-up activity, questions, or a story or analogy. The information should be presented in a manner that helps workers organize and remember the important facts. Materials that can help achieve this goal include text, graphics, examples, job aids, checklists, charts and graphs, tables, data, reports, relevant articles such as tools and materials, a glossary, and a table of contents.

Practice and feedback should give workers the opportunity to try out their new knowledge and skills, and to receive a response and feedback about their performance. Materials that help guide the process of practice and feedback include activity directions, activity time allowance, activity worksheets, question-and-answer worksheets, business scenarios, videotapes of performance, performance checklists, and sample solutions.

The summary, evaluation, review, or transition phase should stress important training points, answer workers' questions, and relate the training experience to the next training topic or the job. Supporting materials for this phase include text, graphics, question-and-answer sheets, listing of next steps, and goal sheets for implementing new knowledge and skills on the job.

Phase 4—Delivery

Many companies regard orientation as an excellent

opportunity to begin training workers in safety policies and safe work practices. In fact, regulatory agencies require employers to provide such training. Safety and health training can be delivered through an orientation program and through written policies and procedures manuals. (See the section on New Employee Training and Orientation later in this chapter.)

Phase 5—Testing and Evaluation

Performance testing is one way the safety and health professional can measure whether learners have met the learning objective. Performance tests cannot be created until objectives have ben developed. There are three main types of performance tests: pretests, review tests (taken while training is in progress), and post-tests.

In safety and health training, tests are highly recommended when:

- The training involves a certification or qualification process.
- The organizational culture supports its use.
- The risks of not mastering the objectives include injury, death, or severe financial loss.
- The effectiveness of training may be questioned.
- Qualitative and quantitative data are needed pertaining to training and/or safety and health issues.

Evaluation is another way of determining whether a training has been successful. An evaluation might involve surveying participants immediately following the completion of a training and again at ongoing intervals (after six weeks, three months, etc.) about how well the training achieved its goals, types of learning methods and media, and perceived increase in productivity.

NEW-EMPLOYEE TRAINING AND ORIENTATION

Safety and health training begins when a new employee is hired. The person is usually open to ideas and information about how the company operates. From the first day, the new employee formulates opinions about management, supervisors, co-workers, and the organization. Human resources managers sometimes say they have never hired a worker with a bad attitude. While this may be an overstatement, many employees' negative attitudes are developed on the job.

Timeliness of instruction is key to the success of the orientation program. For example, it is generally agreed that new employees are more prone to work-related accidents. This fact is attributed to inexperience, lack of familiarity with procedures and facilities,

and zealousness to work. A significant number of workers are also injured after they've been on the job for four to six years. This is generally attributed either to a change in work duties when the employee is transferred or to worker complacency. Covering these issues in an ongoing training program can help to reduce the number of accidents.

The following subjects are suggested as part of the orientation program in the aspects that relate to safety and health:

- Company orientation: history and goals
- Organized labor agreements
- Safety and health policy statement
- Acceptable dress code
- Introductions to people
- Housekeeping standards
- Communications about hazards
- Personal protective equipment (PPE)
- Emergency response procedures (fire, spills, etc.)
- Unsafe acts/conditions reporting procedures
- Accident reporting procedures
- Near-miss accident reporting
- Accident investigation (supervisors)
- Lockout/tagout procedures
- Machine guarding
- Electrical safety awareness
- Ladder use and storage
- Confined space entry
- Medical facility support
- First aid/CPR
- Hand tool safety
- Ergonomic principles
- Eyewash and shower locations
- Fire prevention and protection
- Access to exposure and medical records
- Personal responsibility for safety

These subjects and many others are far too important to overlook or leave to casual learning. A formal program should be developed to give workers this information, and also to forge a strong link between all employees and the organization's safety and health policy.

Many of these subjects are left to the manager, supervisor, or team leader to cover when new employees arrive at the worksite. Unfortunately, it's hard to find time to cover them, and the importance of doing so is often not recognized until an accident occurs. Supervisors should continually update their information to avoid contradicting the training manual. They should also become involved in the development of training programs to make sure the program information is up-to-date and practical in relation to current work demands. If they disregard or contradict the training manual, the entire program and the company's image lose credibility with the workers.

The manager, supervisor, or team leader should reinforce the training program content by demonstrating how it will apply to workers' specific job assignments. For example, the new employee may have been trained on how to read the warning labels on chemical containers. The supervisor can enhance this information by touring the department to point out the hazardous materials and identify the protection provided by PPE and control measures.

In addition, all leaders can provide examples by their own conduct. "Lead by example" is a critical concept to the success of the safety and health program.

Regulatory Obligations

All regulatory agencies publish information about the employer's responsibility to provide training to employees. If the supervisor encounters difficulties with the final interpretation, the best solution is to communicate with the agency involved, and with the employer's legal counsel. Table 4–A lists the Occupational Safety and Health Administration (OSHA) references that reflect training requirements. Table 4–B shows the OSHA and Mine Safety and Health Administration (MSHA) training requirements. These regulatory obligations represent only the minimum requirements and are rarely comprehensive enough to provide a truly safe and healthful work environment.

Policies and Procedures Manuals

The use of written policies and procedures manuals has long been an efficient way to provide information to new employees. The company should make sure the manual is easy to understand, and comprehensive, and that all rules are enforced. The manual should cover such areas as first aid, PPE, electrical safety, and housekeeping, to name just a few. Participants can use the manual as a key reference during training. Manuals should be periodically reviewed and updated, as necessary, to ensure compliance with changes in regulations and other requirements.

TRAINING METHODS

There are various training methods from which trainers can choose when preparing a program. Each method has strengths and weaknesses and serves different needs. The methods selected will depend on the objectives to be met, the type of student participation, time allocated, facility being used, and equipment available. Different people learn at different speeds and feel most comfortable with different methods. Trainers must have the teaching skills to address these elements of human behavior. The following are the most common training techniques used in industry:

1. On-the-job training (OJT)
   a. job instruction training (JIT)
   b. coaching
2. Group methods
   a. conference
   b. brainstorming
   c. case studies
   d. incident process
   e. facilitated discussion
   f. role playing
   g. lecture
   h. question and answer
   I. simulation
3. Individual methods
   a. drill
   b. demonstration
   c. testing
   d. video-based training
   e. interactive computer-assisted training
   f. reading
   g. independent study
   h. seminars and short courses

On-the-Job Training

On-the-job-training (OJT) is widely used because the worker can produce during the training period. However, three considerations must be addressed. (1) The trainer must have good training skills. (2) A training program should make sure all workers are trained in the same way to perform their tasks in the safest and most productive manner. (3) The trainer and trainee must have enough time to cover the subject thoroughly.

A variation of OJT is job instruction training (JIT), also referred to as the four-point method. Instruction is broken down into four simple steps: preparation, presentation, performance, and follow-up. The trainer must know the job thoroughly, be a safe worker, and have the patience, skill, and desire to train.

In this four-point method, the trainer-trainee relationship works in the following ways:

1. Preparation. The trainer puts the workers at ease. He or she explains the job and determines what the worker currently knows about the subject. This stage also includes the preparation of the proper learning/working environment.
2. Presentation. The trainer demonstrates the work process. The student watches the performance and asks questions. The trainer should present the job procedures in sequence and stress all key points.
3. Performance. The worker performs the task under close supervision. The trainer should identify any discrepancies in the work performance and note good performance. The worker should explain the

(Text continues on page 53)

Table 4–A. Index to OSHA Training Requirements

Hazard	Part	Subpart	Section
Blasting or Explosives	1910.109	H	(d)(3)(i)(iii)
	1926.901	U	(c)
	1926.902	U	(i)
	1915.10	B	(a) thru (b)
	1916.10	B	(a) thru (b)
	1917.10	B	(a) thru (b)
Carcinogens			
4-Nitrobiphenyl	1910.1003	Z	(e)(5)(i) thru (ii)
alpha-Naphthylamine	1910.1004	Z	(e)(5)(i) thru (ii)
4, 4'-Methylene bis (2-chloroaniline)	1910.1005	Z	(e)(5)(i) thru (ii)
Methyl chloromethyl ether	1910.1006	Z	(e)(5)(i) thru (ii)
3, 3'-Dichlorobenzidine (and its salts)	1910.1007	Z	(e)(5)(i) thru (ii)
bis-Chloromethyl ether	1910.1008	Z	(e)(5)(i) thru (ii)
beta-Naphthylamine	1910.1009	Z	(e)(5)(i) thru (ii)
Benzidine	1910.1010	Z	(e)(5)(i) thru (ii)
4-Aminodiphenyl	1910.1011	Z	(e)(5)(i) thru (ii)
Ethyleneimine	1910.1012	Z	(e)(5)(i) thru (ii)
beta-Propiolactone	1910.1013	Z	(e)(5)(i) thru (ii)
2-Acetylaminofluorene	1910.1014	Z	(e)(5)(i) thru (ii)
4-Dimethylamino-azobenzene	1910.1015	Z	(e)(5)(i) thru (ii)
N-Nitrosodimethylamine	1910.1016	Z	(e)(5)(i) thru (ii)
Vinyl chloride	1910.1017	Z	(j)(1)(i) thru (ix)
Cranes and Derricks	1910.179	N	(m)(3)(ix)
	1910.180	N	(h)(3)(xii)
Decompression or Compression	1926.803	S	(a)(2)
	1926.803	S	(b)(10)(xii)
	1926.803	S	(e)(1)
Employee Responsibility	1910.109	H	(g)(3)(iii)(a)
	1926.609	U	(a)
Equipment Operations	1910.217	O	(f)(2)
	1926.20	C	(b)(4)
	1926.53	D	(b)
	1926.54	D	(a)
	1910.252	Q	(c)(6)
Fire Protection	1916.32	D	(e)
	1917.32	D	(b)
	1926.150	F	(a)(5)
	1926.155	F	(e)
	1926.351	J	(d)(1) thru (5)
	1926.901	U	(c)
Forging	1910.218	O	(a)(2)(i) thru (iv)
Gases, Fuel, Toxic Material, Explosives	1910.109	H	(d)(3)(i) and (iii)
	1910.111	H	(b)(13)(ii)
	1910.266	R	(c)(5)(i) thru (xi)
	1910.106	H	(b)(5)(vi)(v)(3)
	1916.35	D	(d)(1) thru (6)
	1926.21	C	(a) and (b)(2) thru (6)
	1926.350	J	(d)(1) thru (6)
General	1926.21	C	(a)
Hazardous Material	1915.57	F	(d)
	1916.57	F	(d)
	1917.57	F	(d)

Hazard	Part	Subpart	Section
Medical and First Aid	1910.94	G	(d)(9)(i) and (vi)
	1910.151	K	(a) and (b)
	1915.58	K	(a)
	1917.58	F	(a)
	1926.50	D	(c)
Personal Protective Equipment	1910.94	G	(d)(11)(v)
	1910.134	I	(a)(3)
	1910.134	I	(b)(1), (2) and (3)
	1910.134	I	(e)(2), (3) and (5)
	1910.134	I	(e)(5)(i)
	1910.161	K	(a)(2)
	1915.82	I	(a)(4)
	1915.82	I	(b)(4)
	1916.57	F	(f)
	1916.58	F	(a)
	1916.82	I	(a)(4)
	1916.82	I	(b)(4)
	1917.57	F	(f)
	1918.102	J	(a)(4)
	1926.21	C	(b)(2) thru (6)
	1926.103	E	(c)(1)
	1926.800	S	(e)(xii)
Pulpwood Logging	1910.266	R	(c)(5)(i) thru (xi)
	1910.266	R	(c)(6)(i) thru (xxi)
	1910.266	R	(c)(7)
	1910.266	R	(e)(2)(i) and (ii)
	1910.266	R	(e)(9)
	1910.266	R	(e)(1)(iii) thru (vii)
Powder-Actuated Tools	1915.75	H	(b)(1) thru (6)
	1916.75	H	(b)(1) thru (6)
Power Press	1910.217	O	(e)(3)
Power Trucks, Motor Vehicles, or Agricultural Tractors	1910.109	H	(d)(3)(iii)
	1910.109	H	(g)(3)(iii)(a)
	1910.178	N	(1)
	1910.266	R	(e)(9)
	1910.266	R	(e)(6)(viii)
	1928.51	C	(d)
Radioactive Material	1916.37	D	(b)
Signs—Danger, Warning, Instruction	1910.96	G	(f)(3)(viii)
	1910.145	J	(c)(1)(ii)
	1910.145	J	(c)(2)(ii)
	1910.145	J	(c)(3)
	1910.264	R	(d)(1)(v)
Tunnels and Shafts	1926.800	S	(e)(xiii)
Welding	1910.252	Q	(b)(1)(iii)
	1910.252	Q	(c)(1)(iii)
	1915.35	D	(d)(1) thru (6)
	1915.36	D	(d)(1) thru (4)
	1916.35	D	(d)(1) thru (6)
	1917.35	D	(d)(1) thru (6)

Table 4–B. OSHA and MSHA Training Requirements

The continued importance of training is evidenced by the requirements of both the Occupational Safety and Health Administration (OSHA) and the Mine Safety and Health Administration (MSHA).

OSHA REQUIREMENTS

Listed next are the major parts of the OSHA regulations (Title 29—Labor, *Code of Federal Regulations*) covering training requirements. (Table 4–A gives a convenient index of the types of hazards and the parts of the regulations requiring training to protect against the hazards.)

- Part 1910, Safety and Health Training Requirements for General Industry
- Part 1915–18, Safety and Health Training Requirements for Maritime Employment
- Part 1926, Safety and Health Training Requirements for Construction
- Part 1928, Occupational Safety and Health Requirements for Agriculture

MSHA REGULATIONS

The following is a summary of the training requirements under the MSHA Regulations, 30 *CFR*, PART 48—TRAINING AND RETRAINING OF MINERS (See §48.2 and §48.22 for definitions of terms used in Part 48)

Subpart A—Training and Retraining of Underground Miners

§48.1 Scope

The provisions of this subpart A set forth the mandatory requirements for submitting and obtaining approval of programs for training and retraining miners working in underground mines. Requirements regarding compensation for training and retraining are also included.

§48.3 Training Plans; Time of Submission; Where Filed; Information Required; Time for Approval; Method for Disapproval; Commencement of Training; Approval of Instructors

a. Each operator of an underground mine shall have an MSHA approved plan containing programs for training new miners, training newly-employed experienced miners, training miners for new tasks, annual refresher training, and hazard training for miners.

b. The training plan shall be filed with the District Manager for the area in which the mine is located.

d. The operator shall furnish to the representative of the miners a copy of the training plan two weeks prior to its submission to the District Manager. Where a miners' representative is not designated, a copy of the plan shall be posted on the mine bulletin board 2 weeks prior to its submission to the District Manager. Written comments received by the operator from miners or their representatives shall be submitted to the District Manager. Miners or their representatives may submit written comments directly to the District Manager.

e. All training required by the training plan submitted to and approved by the District Manager as required by this subpart A shall be subject to evaluation by the District Manager to determine the effectiveness of the training programs. If it is deemed necessary, the District Manager may require changes in, or additions to, programs. Upon request from the District Manager the operator shall make available for evaluation the instructional materials, handouts, visual aids and other teaching accessories used or to be used in the training programs. Upon request from the District Manager the operator shall provide information concerning the schedules of upcoming training.

f. The operator shall make a copy of the MSHA approved training plan available at the mine site for MSHA inspection and for examination by the miners and their representatives.

g. Except as provided in §48.7 (New task training of miners) and §48.11 (Hazard training) of this subpart A, all courses shall be conducted by MSHA approved instructors.

h. Instructors shall be approved by the District Manager.

i. Instructors may have their approval revoked by MSHA for good cause which may include not teaching a course at least once every 24 months. A decision by the District Manager to revoke an instructor's approval may be appealed by the instructor. . . . Such an appeal shall be submitted to the Administrator within 5 days of notification of the District Manager's decision. Upon revocation of an instructor's approval, the District Manager shall immediately notify operators who use the instructor for training.

j. The District Manager for the area in which the mine is located shall notify the operator and the miners' representative, in writing, within 60 days from the date on which the training plan is filed, of the approval or status of the approval of the training programs.

k. Except as provided under §48.8(c) (Annual refresher training of miners) of this subpart A, the operator shall commence training of miners within 60 days after approval of the training plan, or approved programs of the training plan.

l. The operator shall notify the District Manager of the area in which the mine is located, and the miners' representative of any changes or modifications the operator proposes to make in the approved training plan. The operator shall obtain the approval of the District Manager for such changes or modifications.

m. In the event the District Manager disapproves a training plan or a proposed modification of a training plan or requires changes in a training plan or modification, the District Manager shall notify the operator and the miners' representative in writing.

n. The operator shall post on the mine bulletin board, and provide to the miners' representative, a copy of all MSHA revisions and decisions which concern the training plan at the mine and which are issued by the District Manager.

§48.4 Cooperative Training Program

a. An operator of a mine may conduct his own training programs, or may participate in training programs conducted by MSHA, or may participate in MSHA approved training programs

Table 4–B. (Continued)

conducted by State or other Federal agencies, or associations of mine operators, miners' representatives, other mine operators, private associations, or educational institutions.

b. Each program and course of instruction shall be given by instructors who have been approved by MSHA to instruct in the courses which are given, and such courses and the training programs shall be adapted to the mining operations and practices existing at the mine and shall be approved by the District Manager for the area in which the mine is located.

§48.5 Training of New Miners; Minimum Courses of Instruction; Hours of Instruction

a. Each new miner shall receive no less than 40 hours of training as prescribed in this section before such miner is assigned to work duties. Such training shall be conducted in conditions which as closely as practicable duplicate actual underground conditions, and approximately 8 hours of training shall be given at the minesite.

b. The training program for new miners shall include the following courses:

1. Instruction in the statutory rights of miners and their representatives under the Act; authority and responsibility of supervisors.
2. Self-rescue and respiratory devices.
3. Entering and leaving the mine; transportation; communications.
4. Introduction to the work environment.
5. Mine map; escapeways; emergency evacuation; barricading.
6. Roof or ground control and ventilation plans.
7. Health.
8. Cleanup; rock dusting.
9. Hazard recognition.
10. Electrical hazards.
11. First aid.
12. Mine gases.
13. Health and safety aspects of the tasks to which the new miner will be assigned.
14. Such other courses as may be required by the District Manager based on circumstances and conditions at the mine.

c. Methods, including oral, written, or practical demonstration, to determine successful completion of the training shall be included in the training plan.

d. Upon proof by an operator that a newly employed miner has received the courses and hours of instruction set forth in paragraphs (a) and (b) of this section within 12 months preceding initial employment at a mine, such miner need not repeat the training, but the operator shall give and the miner shall receive and complete the instruction and program of training set forth in paragraph (b) of §48.6 (Training of newly employed experienced miners), and §48.7 (New task training of miners), if applicable, before commencing work.

§48.6 Training of Newly Employed Experienced Miners; Minimum Courses of Instruction

a. A newly employed experienced miner shall receive and complete training in the program of instruction prescribed in this section before such miner is assigned to work duties.

b. The training program for newly employed experienced miners shall include the following:

1. Introduction to work environment.
2. Mandatory health and safety standards.
3. Authority and responsibility of supervisors and miners' representatives.
4. Entering and leaving the mine; transportation; communications.
5. Mine map; escapeways; emergency evacuation; barricading.
6. Roof or ground control and ventilation plans.
7. Hazard recognition.
8. Self-rescue and respiratory devices.
9. Such other courses as may be required by the District Manager based on circumstances and conditions at the mine.

§48.7 Training of Miners Assigned to a Task in Which They Have Had No Previous Experience; Minimum Courses of Instruction

a. Miners assigned to new work tasks as mobile equipment operators, drilling machine operators, haulage and conveyor systems operators, roof and ground control machine operators, and those in blasting operations shall not perform new work tasks in these categories until training prescribed in this paragraph and paragraph (b) of this section has been completed. This training shall not be required for miners who have been trained and who have demonstrated safe operating procedures for such new work tasks within 12 months preceding assignment. This training shall also not be required for miners who have performed the new work tasks and who have demonstrated safe operating procedures for such new work tasks within 12 months preceding assignment. The training program shall include the following:

1. Health and safety aspects and safe operating procedures for work tasks, equipment, and machinery.
2. (i) Supervised practice during nonproduction.
   (ii) Supervised operation during production.
3. New or modified machines and equipment.
4. Such other courses as may be required by the District Manager based on circumstances and conditions at the mine.

b. Miners under paragraph (a) of this section shall not operate the equipment or machine or engage in blasting operations without direction and immediate supervision until such miners have demonstrated safe operating procedures for the equipment or machine or blasting operation to the operator or the operator's agent.

c. Miners assigned a new task not covered in paragraph (a) of this section shall be instructed in the safety and health aspects and safe work procedures of the task, prior to performing such task.

d. Any person who controls or directs haulage operations at a mine shall receive and complete training courses in safe haulage procedures related to the haulage system,

Table 4–B. (Continued)

ventilation system, firefighting procedures, and emergency evacuation procedures in effect at the mine before assignment to such duties.

e. All training and supervised practice and operation required by this section shall be given by a qualified trainer, or a supervisor experienced in the assigned tasks, or other person experienced in the assigned tasks.

§48.8 Annual Refresher Training of Miners; Minimum Courses of Instruction; Hours of Instruction

a. Each miner shall receive a minimum of 8 hours of annual refresher training as prescribed in this section.

b. The annual refresher training program for all miners shall include the following courses of instruction:

1. Mandatory health and safety standards.
2. Transportation controls and communication systems.
3. Barricading.
4. Roof or ground control and ventilation plans.
5. First aid.
6. Electrical hazards.
7. Prevention of accidents.
8. Self-rescue and respiratory devices.
9. Explosives.
10. Mine gases.
11. Health.
12. Such other courses as may be required by the District Manager based on circumstances and conditions at the mine.

d. Where annual refresher training is conducted periodically, such sessions shall not be less than 30 minutes of actual instruction time and the miners shall be notified that the session is part of annual refresher training.

§48.9 Records of Training

a. Upon a miner's completion of each MSHA approved training program, the operator shall record and certify on MSHA form 5000–23 that the miner has received the specified training. A copy of the training certificate shall be given to the miner at the completion of the training. The training certificates for each miner shall be available at the minesite for inspection by MSHA and for examination by the miners, the miner's representative, and State inspection agencies. When a miner leaves the operator's employ, the miner shall be entitled to a copy of his training certificates.

b. False certification that training was given shall be punishable under section 110 (a) and (f) of the Act.

c. Copies of training certificates for currently employed miners shall be kept at the minesite for 2 years, or for 60 days after termination of employment.

§48.10 Compensation for Training

a. Training shall be conducted during normal working hours; miners attending such training shall receive the rate of pay as provided in §48.2(d) (Definition of normal working hours) of this subpart A.

b. If such training shall be given at a location other than the normal place of work, miners shall be compensated for the additional cost, such as mileage, meals, and lodging, they may incur in attending such training sessions.

§48.11 Hazard Training

a. Operators shall provide to those miners, as defined in §48.2(a)(2) (Definition of miner) of this subpart A, a training program before such miners commence their work duties. This training program shall include the following instruction, which is applicable to the duties of such miners:

1. Hazard recognition and avoidance;
2. Emergency and evacuation procedures;
3. Health and safety standards, safety rules, and safe working procedures;
4. Use of self-rescue and respiratory devices
5. Such other instruction as may be required by the District Manager based on circumstances and conditions at the mine.

b. Miners shall receive the instruction required by this section at least once every 12 months.

c. The training program required by this section shall be submitted with the training plan required by §48.3(a) (Training plans: Submission and approval) of this subpart A and shall include a statement on the methods of instruction to be used.

d. The operator shall maintain and make available for inspection certificates that miners have received the hazard training required by this section.

e. Miners subject to hazard training shall be accompanied at all times while underground by an experienced miner, as defined in §48.2(b) (Definition of miner) of this subpart A.

§48.12 Appeals procedures

The operator, miner, and miners' representative shall have the right of appeal from a decision of the District Manager.

Subpart B—Training and Retraining of Miners Working at Surface Mines and Surface Areas of Underground Mines

The provisions of this subpart B set forth the mandatory requirements for submitting and obtaining approval of programs for training and retraining miners working at surface mines and surface areas of underground mines. Requirements regarding compensation for training and retraining are also included.

§48.23 Training Plans; Time of Submission; Where Filed; Information Required; Time for Approval; Method for Disapproval; Commencement of Training; Approval of Instructors

a. Each operator of a mine shall have an MSHA approved plan containing programs for training new miners, training newly-employed experienced miners, training miners for new tasks, annual refresher training, and hazard training for miners.

2. Within 60 days after the operator submits the plan for approval, unless extended by MSHA, the operator shall have an approved plan for the mine.

3. In the case of a new mine which is to be opened or a mine which is to be reopened or reactivated after the effective date of this subpart B, the operator shall have an approved plan prior to opening the new mine, or

Table 4-B. (Continued)

reopening or reactivating the mine unless the mine is reopened or reactivated periodically using portable equipment and mobile teams of miners as a normal method of operation by the operator. The operator to be so excepted shall maintain an approved plan for training covering all mine locations which are operated with portable equipment and mobile teams of miners.

b. The training plan shall be filed with the District Manager for the area in which the mine is located.

d. The operator shall furnish to the representative of the miners a copy of the training plan 2 weeks prior to its submission to the District Manager. Where a miners' representative is not designated, a copy of the plan shall be posted on the mine bulletin board 2 weeks prior to its submission to the District Manager. Written comments received by the operator from miners or their representatives shall be submitted to the District Manager. Miners or their representatives may submit written comments directly to the District Manager.

e. All training required by the training plan submitted to and approved by the District Manager as required by this subpart B shall be subject to evaluation by the District Manager to determine the effectiveness of the training programs. If it is deemed necessary, the District Manager may require changes in, or additions to, programs. Upon request from the District Manager the operator shall make available for evaluation the instructional materials, handouts, visual aids, and other teaching accessories used or to be used in the training programs. Upon request from the District Manager the operator shall provide information concerning schedules of upcoming training.

f. The operator shall make a copy of the MSHA approved training plan available at the minesite for MSHA inspection and examination by the miners and their representatives.

g. Except as provided in §48.27 (New task training of miners) and §48.31 (Hazard training) of this subpart B, all courses shall be conducted by MSHA approved instructors.

h. Instructors shall be approved by the District Manager

i. Instructors may have their approval revoked by MSHA for good cause which may include not teaching a course at least once every 24 months. A decision by the District Manager to revoke an instructor's approval may be appealed by the instructor. Such an appeal shall be submitted to the Administrator within 5 days of notification of the District Manager's decision. Upon revocation of an instructor's approval, the District Manager shall immediately notify operators who use the instructor for training.

j. The District Manager for the area in which the mine is located shall notify the operator and the miners' representative, in writing, within 60 days from the date on which the training plan is filed, of the approval or status of the approval of the training programs.

k. Except as provided under §48.28(c) (Annual refresher training of miners) of this subpart B, the operator shall commence training of miners within 60 days after approval of the training plan, or approved programs of the training plan.

l. The operator shall notify the District Manager of the area in which the mine is located and the miners' representative of any changes or modifications which the operator proposes to make in the approval training plan. The operator shall obtain the approval of the District Manager for such changes or modifications.

m. In the event the District Manager disapproves a training plan or a proposed modification of a training plan or requires changes in a training plan or modification, the District Manager shall notify the operator and the miners' representative in writing.

n. The operator shall post on the mine bulletin board, and provide to the miners' representative, a copy of all MSHA revisions and decisions which concern the training plan at the mine and which are issued by the District Manager.

§48.24 Cooperative Training Program

a. An operator of a mine may conduct his own training programs, or may participate in training programs conducted by MSHA, or may participate in MSHA approved training programs conducted by State or other Federal agencies, or associations of mine operators, miners' representatives, other mine operators, private associations, or educational institutions.

b. Each program and course of instruction shall be given by instructors who have been approved by MSHA to instruct in the courses which are given, and such courses and the training programs shall be adapted to the mining operations and practices existing at the mine and shall be approved by the District Manager for the area in which the mine is located.

§48.25 Training of New Miners; Minimum Courses of Instruction; Hours of Instruction

a. Each new miner shall receive no less than 24 hours of training as prescribed in this section. Except as otherwise provided in this paragraph, new miners shall receive this training before they are assigned to work duties. At the discretion of the District Manager, new miners may receive a portion of this training after assignment to work duties, provided that no less than 8 hours of training shall in all cases be given to new miners before they are assigned to work duties. The following courses shall be included in the 8 hours of training: Introduction to work environment, hazard recognition, and health and safety aspects of the tasks to which the new miners will be assigned. Following the completion of this preassignment training, new miners shall then receive the remainder of the required 24 hours of training, or up to 16 hours, within 60 days. Operators shall indicate in the training plans submitted for approval whether they want to train new miners after assignment to duties and for how many hours. In determining whether new miners may be given this training after they are assigned duties, the District Manager shall consider such factors as the mine safety record, rate of employee turnover and mine size. Miners who have not received the full 24 hours of new miner training shall be required to work under the close supervision of an experienced miner.

b. The training program for new miners shall include the following courses:

Table 4–B. (Continued)

1. Instruction in the statutory rights of miners and their representatives under the Act; authority and responsibility of supervisors.
2. Self-rescue and respiratory devices.
3. Transportation controls and communication systems.
4. Introduction to work environment.
5. Escape and emergency evacuation plans; firewarning and firefighting.
6. Ground control; working in areas of highwalls, water hazards, pits and spoil banks; illumination and night work.
7. Health.
8. Hazard recognition.
9. Electrical hazards.
10. First aid.
11. Explosives.
12. Health and safety aspects of the tasks to which the new miner will be assigned.
13. Such other courses as may be required by the District Manager based on circumstances and conditions at the mine.

c. Methods, including oral, written or practical demonstration, to determine successful completion of the training shall be included in the training plan.

d. Upon proof by an operator that a newly employed miner has received the courses and hours of instruction set forth in paragraphs (a) and (b) of this section within 12 months preceding initial employment at a mine, such miner need not repeat the training, but the operator shall give and the miner shall receive and complete the instruction and program of training set forth in paragraph (b) of §48.26 (Training of newly employed experienced miners) and §48.27 (New task training of miners), if applicable, before commencing work.

§48.26 Training of Newly Employed Experienced Miners; Minimum Courses of Instruction

a. A newly employed experienced miner shall receive and complete training in the program of instruction prescribed in this section before such miner is assigned to work duties.

b. The training program for newly employed experienced miners shall include the following:

1. Introduction to work environment.
2. Mandatory health and safety standards.
3. Authority and responsibility of supervisors and miners' representatives.
4. Transportation controls and communication systems.
5. Escape and emergency evacuation plans; firewarning and firefighting.
6. Ground controls; working in areas of highwalls, water hazards, pits, and spoil banks; illumination and night work.
7. Hazard recognition.
8. Such other courses as may be required by the District Manager based on circumstances and conditions at the mine.

§48.27 Training of Miners Assigned to a Task in Which They Have Had No Previous Experience; Minimum Courses of Instruction

a. Miners assigned to new work tasks as mobile equipment operators, drilling machine operators, haulage and conveyor systems operators, ground control machine operators, and those in blasting operations shall not perform new work tasks in these categories until training prescribed in this paragraph and paragraph (b) of this section has been completed. This training shall not be required for miners who have been trained and who have demonstrated safe operating procedures for such new work tasks within 12 months preceding assignment. This training shall also not be required for miners who have performed the new work tasks and who have demonstrated safe operating procedures for such new work tasks within 12 months preceding assignment. The training program shall include the following:

1. Health and safety aspects and safe operating procedures for work tasks, equipment, or machinery.
2. (i) Supervised practice during nonproduction.
   (ii) Supervised operation during production.
3. New or modified machines and equipment.
4. Such other courses as may be required by the District Manager based on circumstances and conditions at the mine.

b. Miners under paragraph (a) of this section shall not operate the equipment or machine or engage in blasting operations without direction and immediate supervision until such miners have demonstrated safe operating procedures for the equipment or machine or blasting operation to the operator or the operator's agent.

c. Miners assigned a new task not covered in paragraph (a) of this section shall be instructed in the safety and health aspects and safe work procedures of the task, prior to performing such task.

d. All training and supervised practice and operation required by this section shall be given by a qualified trainer, or a supervisor experienced in the assigned tasks, or other person experienced in the assigned tasks.

§48.28 Annual Refresher Training of Miners; Minimum Courses of Instruction; Hours of Instruction

a. Each miner shall receive a minimum of 8 hours of annual refresher training as prescribed in this section.

b. The annual refresher training program for all miners shall include the following courses of instruction:

1. Mandatory health and safety standards. The course shall include mandatory health and safety standard requirements which are related to the miner's tasks.
2. Transportation controls and communication systems.
3. Escape and emergency evacuation plans; firewarning and firefighting.
4. Ground control; working in areas of highwalls, water hazards, pits, and spoil banks; illumination and night work.
5. First aid.
6. Electrical hazards.
7. Prevention of accidents.
8. Health.
10. Self-rescue and respiratory devices.

Table 4–B. (Continued)

11. Such other courses as may be required by the District Manager based on circumstances and conditions at the mine.

d. Where annual refresher training is conducted periodically, such sessions shall not be less than 30 minutes of actual instruction time and the miners shall be notified that the session is part of annual refresher training.

§48.29 Records of Training

a. Upon a miner's completion of each MSHA approved training program, the operator shall record and certify on MSHA form 5000–23 that the miner has received the specified training. A copy of the training certificate shall be given to the miner at the completion of the training. The training certificates for each miner shall be available at the minesite for inspection by MSHA and for examination by the miners, the miners' representative and State inspection agencies. When a miner leaves the operator's employ, the miner shall be entitled to a copy of his training certificates.

b. False certification that training was given shall be punishable under section 110 (a) and (f) of the Act.

c. Copies of training certificates for currently employed miners shall be kept at the mine site for 2 years, or for 60 days after termination of employment.

§48.30 Compensation for Training

a. Training shall be conducted during normal working hours; miners attending such training shall receive the rate of pay as provided in §48.22(d)

b. If such training shall be given at a location other than the normal place of work, miners shall be compensated for the additional costs, such a mileage, meals, and lodging, they

may incur in attending such training sessions.

§48.31 Hazard Training

a. Operators shall provide to those miners, as defined in §48.22(a) (2) (Definition of miner) of this subpart B, a training program before such miners commence their work duties. This training program shall include the following instruction, which is applicable to the duties of such miners:

1. Hazard recognition and avoidance;
2. Emergency and evacuation procedures;
3. Health and safety standards, safety rules and safe working procedures;
4. Self-rescue and respiratory devices; and,
5. Such other instruction as may be required by the District Manager based on circumstances and conditions at the mine.

b. Miners shall receive the instruction required by this section at least once every 12 months.

c. The training program required by this section shall be submitted with the training plan required by §48.23(a) (Training plans: Submission and approval) of this subpart B and shall include a statement on the methods of instruction to be used.

d. In accordance with §48.29 (Records of training) of this subpart B, the operator shall maintain and make available for inspection, certificates that miners have received the instruction required by this section.

§48.32 Appeals Procedures

The operator, miner, and miners' representative shall have the right of appeal from a decision of the District Manager.

(Text continued from page 46.)

steps being performed. This ensures that the worker not only can perform the task but understands how and why the task is done. This stage continues until the trainer is satisfied with the worker's competence at the job.

4. Follow-up. The trainer and/or the supervisor must monitor the worker's performance to be sure the job is being performed as instructed and to answer any questions the worker might have.

Advantages of the JIT training method include the following:

1. The personal guidance makes it easier to motivate the worker.
2. The trainer can spot and correct problems as they occur.
3. Results of the training are evaluated immediately because the worker is doing the actual job on actual equipment. The work performed can be judged against reasonable standards.
4. Training is practical, realistic, and demonstrated under actual conditions. Workers can ask questions easily.

The most common other method of OJT is coaching by a co-worker, or the buddy system. It can be quite effective in certain situations, but it has the following drawbacks.

1. Trainers may be selected for their availability rather than for their ability in training. A co-worker who lacks basic skills can completely undermine a new employee's orientation. An ability to do the task well does not guarantee an ability to teach it.
2. Each trainer may have his or her own individualized methods of doing the tasks being taught. This lack of consistency in training can make it difficult to control hazards in the workplace, and can lead to many accidents.
3. Key elements of orientation may be overlooked in the training program, and the omissions may not be realized until an accident occurs.
4. Poor techniques or bad habits can spread from the coach to the trainee. Shortcuts or safety violations may be demonstrated as "the way we do it."
5. The training may not emphasize safety performance.

Job performance should never be separated from safety standards.

Group Methods

Group techniques encourage participation from a selected audience. These methods allow trainees to share ideas, evaluate information, and become actively involved in planning and implementing company policy. All types of group training require skilled facilitators to be successful.

Conference or Meeting Method. The conference or meeting method of training is widely used in business and industry for education-sharing purposes and because of the knowledge each participant brings to the group. The trainer/facilitator controls the flow of the session as the participants share their knowledge and experience. The skill of the facilitator can mean the success or failure of these sessions. Facilitators must use various techniques to draw information and opinions from members. The number of people involved should be limited to allow open discussion from all participants. The opinions of each member should be recorded, and a summary of the group conclusions provided to those who were involved, as well as to those who should be kept abreast of the information.

The conference technique is also a valuable method of problem solving. It is important that at the beginning of the conference, members identify their goals and expectations for the session; for example, to make recommendations only, not to establish policy or procedures. When a group is asked to make recommendations, they should be kept informed of the results of those recommendations.

Proper control and guidance of a conference can ensure its success and make it a gratifying experience for the participants.

Brainstorming. Brainstorming is a technique of group interaction that encourages each participant to present ideas on a specific issue. The method is normally used to find innovative approaches to issues. There are four ground rules:

1. Ideas presented are not criticized.
2. Free-wheeling creative thinking and building on ideas are positively reinforced.
3. As many ideas as possible should be presented quickly.
4. Combining several ideas or improving suggestions is encouraged.

A recorder should be selected to write down all the ideas presented. The moderator must control the flow of suggestions, cut off negative comments, and solicit ideas from each member.

Brainstorming allows ideas to be developed quickly, encourages creative thinking, and involves everyone in the process. The group can go beyond old stereotypes or the "Way it's always done."

Case Study. Case studies are written descriptions of actual or fictitious business decisions or problems that learners will use as a basis for demonstrating predetermined skills and knowledge. The goal of the activity is to develop group members' insight and problem-solving skills. The study normally is presented by defining what happened in a particular incident and the events leading up to the incident. The group is then given the task of determining the actual causes of problems, the significance of each element, and/or the acceptable solutions.

There are two distinct advantages to building case studies: first, case studies provide an opportunity for the learners to use skills and knowledge acquired during the course. Second, case studies can serve as an evaluation tool for trainers to measure the degree of proficiency attained during a course or module. Here are other key benefits:

- Students, during a case study, begin to internalize the critical principles being taught and retain the information for longer periods of time.
- Case studies emphasize practical or critical thinking skills.
- The student's perspective is broadened through an interaction with others.
- Case studies encourage reflection, application, and analysis.
- Case studies reinforce the value of discussion and interaction with others.

Planning, thinking, and adhering to the instructional objectives are paramount in designing an effective case. The key is to start at the end and work backwards toward the beginning. Ask the following questions:

- What questions must be answered?
- What skills or knowledge should be exercised?
- What specific performance objectives should be measured?
- What learning performance objectives are to be addressed by the case study?

Incident Process. This is a type of case study in which the group works with a written account of any incident. The group is allowed to ask questions about facts, clues, and details. The trainer provides the answers to the questions, and the group must assemble the information, determine what has happened, and arrive at a decision. The trainer must guide the group to prevent arguments and to prevent one or two members from dominating the discussion. This is a useful training method that encourages employee participation in the accident prevention program. Situations can be real cases in the company or developed from potential hazards that exist.

The moderator must be capable of controlling group process and progress and of preventing the group from missing the true or root causes. For example, an employee was struck in the eye by a foreign body and an investigation revealed the employee was not wearing safety glasses while operating a bench grinder. The group must seek the root cause of the accident and not settle for the common conclusion "the worker failed to wear eye protection." They must specify why the supervisor or management failed to enforce the proper procedures (assuming there was an established procedure). Why did management allow this lax supervision?

Facilitated Discussion. Facilitated discussion or dialogue is the management of discussion about the course content so that the learning objectives are met, the discussion flows logically from topic to topic, and the applications to the learners' jobs are made clear. Facilitated discussion requires that the trainer/facilitator have the skills to accept all ideas and contributions as valid, show how they relate to the course objectives, and manage the time element and the flow of information to meet the course objectives.

The benefits of facilitated discussion include the following:

* Ensures that the learners are involved and challenged
* Builds a bond between the trainer/facilitator that encourages the free exchange of ideas and information

Role Playing. This training method is effective for evaluating human relations issues. Members attempt to identify the ways people behave under various conditions. Although this technique is not an effective method of problem solving, it can uncover issues not previously considered. This method is particularly helpful in identifying and changing personnel issues such as poor morale or negative attitudes.

Lecture. With this method, a single person can impart information to a large group in a relatively short time. This method is normally used to communicate facts, give motivational speeches, or summarize events for workers. There is little time or opportunity for interaction by the attendees. Follow-up for these sessions must be well planned in advance to be successful.

Question and Answer (Q and A) Sessions. Normally Q and A sessions follow training periods after the trainer has summarized the material presented. Workers can use this method to clarify individual concerns or facts. However, workers will often need time to prepare and organize their thoughts before they can ask questions. In situations where they must absorb a large amount of material, allow them time to reflect on or apply the knowledge and to formulate questions. The trainer can plan to have a follow-up session or allow workers to present their questions personally. Question-and-answer sessions are helpful in clarifying issues of policy or changes in schedules or events.

Simulation. When actual materials or machines cannot be used, trainers can use a simulation device. This method is used effectively in aircraft pilot training, railroad engineer training, and other applications. Various methods are employed in management training programs as well, such as the "in-basket technique," "war games," and others. One simulation demonstrates the loss of eyesight to workers to encourage them to wear safety glasses on the job. Safety procedures must be carefully followed to prevent any actual accident.

Simulation is most effective when the workers can participate. Careful planning and attention to detail are required. The initial costs of these sessions can be high because of the equipment required and time involved in conducting training.

Individual Methods

With individual methods, employees can work at their own pace. For some learners, this is the most effective approach. The learner does, however, miss out on the give-and-take of ideas. A combination of group and individual methods provides the broadest variety of learning opportunities.

Drill. Using the elements of practice and repetition, this method of instruction is valuable for developing worker skill in fundamental tasks and for performing under pressure.

Workers required to perform in crisis situations should be trained under conditions that resemble the crisis as closely as possible. For example, when instructing workers in cardiopulmonary resuscitation (CPR), the trainer must try to instill the tension that workers will experience when they attempt to resuscitate a real victim.

Demonstration. As discussed in the section on JIT, the method of demonstration allows the trainer to perform the actual task and then have the worker repeat the performance. Trainers must be sure the job is preformed exactly as required to prevent workers from developing poor habits and performance standards (and supervisors must see that employees follow the designated procedure). If the conditions used in demonstrations are not similar to the actual workstation or equipment on the job, this method will yield few useful results.

Testing. This technique is normally used to determine if workers understand the necessary information and can apply the knowledge when required. Developing good tests is a skill that requires constant review to ensure that training objectives are being met. Poor tests can reduce workers' morale and undermine training objectives.

Video-Based Training. The use of videotapes does not eliminate the need for professional instruction, but can enhance a classroom presentation. Videos are available from the National Safety Council and numerous private companies. Trainers should screen the tapes to make sure they fulfill the needs of the training program.

Some distinct advantages to using a video in a training class include:

- Video offers the learner an opportunity to see examples of tasks and processes being performed correctly.
- The cost of producing training videos has gone up while the difficulty in producing them has gone down. Newer formats, such as HI-8 and S-VHS, offer lighter cameras and greater ease of use.

Interactive Computer-Assisted Training. Interactive computer programs allow the workers to receive information by reading or watching a video presentation and then respond to situations and questions. If the correct response is entered, the computer will advance the program to repeat the information and retest the workers. The system is valuable for several reasons:

1. Workers can work at their own pace.
2. Records can be automatically kept of all training. The amount and type of records maintained can be modified to meet the company requirements.
3. Correct answers are required before a worker can proceed to the next lesson, or remediation methods are built into the program.
4. Workers receive training as time is available in their schedule, rather than having to meet training schedules.
5. Instructions can guide workers step be step through a curriculum designed to meet the goals of the individual, the company and/or any regulatory obligations.
6. This type of training works extremely well for organizations with small workforces or those that cannot remove large groups of workers from their jobs at any one time.

Reading. Companies should provide employees with written safety materials such as monthly newsletters and supervision and safety magazines. In addition, organizations may establish a library where employees can research information on subjects such as work procedures, safety, leadership, health care, family or home safety, and other subjects of concern. It is important, however, that management does not assume everyone has the ability to read and comprehend all of the written material provided. Companies cannot replace instruction or training programs simply by handing an employee a training manual. Written material is meant to supplement or to serve as a reference for training.

Independent Study. Home study courses or correspondence courses can help workers to advance within the organization or to improve their knowledge of their jobs and industries. A major advantage of this method is that the worker does not lose any time from work and can complete the course at his or her own pace. Another advantage is the low cost of home-study programs. Normally they are centered around a textbook assignment, followed by self-tests using multiple choice, true/false, fill-in, or essay questions. Several courses also provide laboratory or performance materials such as television, radio, or computer repair programs that work on actual equipment. Some home-study programs come with videotaped presentations for workers to view.

The National Safety Council offers home-study programs for supervisor training—"Supervising for Safety" and "Protecting Workers' Lives."

Seminars and Short Courses. Seminars, short courses, and workshops for safety and leadership information and skills are offered by many colleges and universities as well as by insurance companies and private organizations. The Safety Training Institute of the National Safety Council offers on-site instruction for workers, supervisors, and management. Seminars, short courses, and workshops range from one-hour sessions to several days.

SUMMARY OF KEY POINTS

Key points covered in this chapter include:

- Training focuses mainly on behavior change, showing workers how to perform properly and how to apply their knowledge on the job. In some cases, however, nontraining solutions are more appropriate.
- Training and nontraining problems and solutions are categorized as selection and assignment, information and practice, environment, and motivation/incentive.
- Adult learners have special needs that trainers must recognize in order for the program to be effective. Performance-based instruction generally works well for adult learners.
- To design an effective training program, the safety professional must assess learners' (workers') needs, analyze their characteristics, develop specific objectives, create materials and schedules, and design appropriate testing and evaluation methods.
- Training begins with the orientation of each new employee. Written policies and procedures manuals help to meet employee training needs and conform to regulatory standards.
- Training methods include on-the-job training (OJT), job instruction training (JIT), group methods, and individual methods.

5

PROMOTING
SAFETY AND HEALTH

After reading this chapter, you will be able to:

- Identify ways a safety and health committee can get employees involved in the issues

- Conduct a safety and health meeting

- Describe how involvement in the job safety analysis process and in perception surveys, employee recognition, safety contests and posters, and suggestion systems can promote employee investment in safety and good health

- List the elements of safety training and first aid

- Perform a job safety analysis

Employee involvement is an essential building block of any workplace safety and health program. All employees need to understand their responsibility for their own and their co-workers' safety. They have an obvious interest in working with co-workers and employers to improve conditions, since their very lives and livelihoods are at stake. Ways to involve employees in promoting safety include safety and health committees, safety and health meetings, involvement in the job safety analysis process and in perception surveys, employee recognition, safety contests and posters, suggestion systems, and safety and health training.

SAFETY AND HEALTH COMMITTEES

Employee involvement can be required, but a cooperative spirit cannot. One of the best ways to create not only awareness, but acceptance, of safety and health goals is to get employees involved in running a safety and health committee. Involving employees in the decision-making process can earn their respect and lead to increasing their commitment to the company's safety and health goals.

Two styles of committees are used: employee and joint (both employee and management). Either type gives workers a chance to contribute to the overall safety and health of the organization, but the joint committee is usually more effective in a team-based organization. The existence, composition, and purposes of safety and health committees in many places is mandated by law or by labor union contracts. In addition, many committees include environmental issues and goals as part of their safety and health objectives.

Best-Practice Approach for a Joint Committee

The best-practice approach to developing a safety and health committee involves gathering equal representation by leaders and team members. The leaders or supervisors may be appointed by senior management, but cooperation is maximized when worker and management committee members volunteer or are elected by their peers.

The committee should have balanced representation from the various company departments or operating units. The size of the committee should be workable, with enough members to make the workload comfortable, but not so many that they become disorganized. Time commitment to the committee should be defined. Members can have staggered one-to two-year terms, with a quarter of the members being replaced every six months. Officers should also be rotated and should be nonmanagement members at least half the time. The committee should use the company's safety and health professional as its advisor, but all members should have extra training on safety and health issues as needed.

Committee guidelines should be established and followed. Annual goals should be identified and progress measured. The committee should develop a written policy and define member responsibilities and duties. Meetings should be held at least monthly, and should be short and structured. They should be coordinated with other business priorities and schedules. The safety and health professional should evaluate the committee's effectiveness periodically. The committee must be given time to do its work and recognized for its successes. Its success can be greatly enhanced with visible support and constant commitment from top management. The committee structure should facilitate communication with all of the employee population.

Functions of Safety and Health Committees

The safety and health committee usually has some combination of the following functions and roles.

1. Policy, practice, and process.
   - Makes safety and health policy
   - Communicates safety and health issues and solutions
   - Performs job safety analysis
   - Conducts reviews of data from audits and other sources to decide company safety and health approach
   - Investigates and acts on unsafe conditions or acts and occupational injuries or illnesses
   - Participates in or oversees motivational programs
   - Oversees emergency response preparedness and drills
2. Inspection and investigation.
   - Conducts facility inspections and audits
   - Recommends changes to production units
   - Investigates and acts on injuries or hazards
   - Recommends changes to policy
3. Promotion, education, and motivation.
   - Promotes interest in the safety effort
   - Ensures compliance with safety procedures
   - Develops good safety attitudes in employees
   - Develops and evaluates safety and health training programs
4. Problem analysis and solving.
   - Uses problem solving to focus on solutions to safety and health problems
   - Recommends solutions to appropriate persons

SAFETY AND HEALTH MEETINGS

No one structure is ideal for employee involvement in safety and health. To be effective, the system must fit comfortably into the organization it serves. In some companies, formal safety and health committees are not essential. Regular safety and health meetings with all employees are enough to create team spirit, and keep the issue of safety foremost in workers' minds. Every meeting should have the goal of motivating workers to take an interest in and responsibility for safety and health.

The first question the organizer must address is "What is the purpose?" The safety director or department should be able to provide a multitude of ideas, based either on company experience or on National Safety Council suggestions. (Many models are provided in the materials listed in the Bibliography at the end of this book.)

If meetings are to be held monthly, for example, themes can be assigned for a whole year of meetings in advance. In this way they can tie into each other and build on knowledge shared at the previous meeting. The organizer could also choose to alternate on-the-job safety with off-the-job safety in assigning program topics. For example, the April meeting might discuss ways to protect against violence in the workplace, and the May meeting might address domestic violence and provide information on stress management and local assistance resource groups.

Running Effective Meetings

Some practical steps can be taken to make meetings go more smoothly.

First, preparation should include at least the following items. A written agenda ensures that essential points are covered and keeps speakers from straying from the subject. Speakers should rehearse and time their remarks. The meeting organizer should preview visual aids and check equipment and exhibits to make sure everything is operating properly.

A comfortable location will help attendees get the most out of the meeting. People become irritated and unresponsive if they must stand too long, are too warm or cold, or feel ill at ease. Background noises or glaring lights can also distract people from the message. The organizer should appoint a committee member to ensure the location is in good order.

Second, the meeting itself should feature no more than three main ideas. If a meeting organizer tries to cover too much for the time available and the nature of the topics, participants simply won't be able to absorb everything. If there are too many topics to address, then coverage can be spread out over several meetings.

Keeping the session short and simple is key to keeping attention. An employee safety and health meeting should not run longer than 30 minutes without a compelling reason, such as an absorbing video.

Tips for Successful Meetings

1. Feature no more than three ideas.

2. Meet in a comfortable place.

3. Plan the agenda and visual aids.

4. Keep it short and simple.

Figure 5–1. Display these tips at the beginning of a meeting.

Only the most exceptionally appealing material justifies a longer session. (These guidelines apply only to meetings, not to instructional courses.)

Plenty of time should be set aside at the end of the meeting for feedback and questions. This lets the meetings fill two bills at once: (1) disseminating useful information, and (2) soliciting input from employees. Circulating the meeting agenda in advance gives attendees time to consider what points they want to raise.

Visual Aids

Safety and health meetings may incorporate all kinds of visual aids: films, videos, charts, models, exhibits, and displays of equipment. Nonprojected visual aids are easy to make. All they require are an easel, a large pad of paper, and some felt-tip markers. For example, the information in Figure 5–1 can be written on a flipchart to display at the beginning of a meeting.

Videos and films can also be produced in-house if the company has the resources, or they can be rented or bought from professional production companies. The National Safety Council offers sets of 35-millimeter slides and videos covering a wide variety of industrial safety subjects. Each set has a reading script, and many have audiocassettes.

It is important to remember that prepared visual aids are, as the name suggests, merely aids. Although they can be effective training and motivating devices, they should not be expected to communicate the entire message. Their job is to reinforce the discussion. Visual aids should be selected because they illustrate the meeting's main ideas, not simply because they are available.

Variations on Meetings

Often, each department coordinates its own meetings for its own members. When company policies require discussion or the topic is general causes of incidents or accidents, large meetings of several departments are also valuable. Some meetings may be purely motivational; some are held to present safety awards.

Production huddles, or tailgate conferences, are

instructional sessions about a specific job. When these sessions incorporate the safety message into the job routine, they can be among the most effective safety meetings of all. For example, public utility crews often start a job with a tailgate conference. The crew members gather around their truck and discuss the work, laying out the tools and materials required and choosing which part of the job each person should handle. Such sessions can be indispensable, especially in highly hazardous situations. Placing the safety discussion at the start of such meetings accomplishes two goals: (1) it places safety as a primary concern from the start, and (2) it ensures that safety is not overlooked due to time constraints.

JSA PROCESS AND SAFETY CONTACTS

Everyone has something to contribute to the improvement of safety and health in the workplace. Management must constantly provide employees with opportunities to participate in the protection of their own well-being.

Job safety analyses (JSAs) formally allow employees to participate in identifying hazards and unsafe practices in the workplace and modify work methods to prevent injury, illness, and incidents and improve productivity and product quality. The JSA process encourages safe work practices on a personal level by a process that relies on employee's expertise and knowledge about each job and its related tasks. When used as a tool for continuous improvement, the JSA facilitates better communication between line employees, supervisors, or team lenders, and senior managers.

Safety contacts are commonly initiated after job safety observations or as part of a JSA process. This type of contact is highly individual in nature, involving a one-on-one safety-related meeting or discussion between supervisor and employee. After observing the employee at work, the supervisor takes the opportunity to both recognize and reinforce safe job performance, and to offer constructive criticism and further instruction as necessary. The employee, on the other hand, is given the opportunity to voice opinions regarding his or her own performance and company safety and health rules, procedures, and programs in a forum where he or she is assured of being heard.

PERCEPTION SURVEYS

Perception surveys are a means of giving all employees a voice in the direction of the safety and health program. The corporate safety and health program cannot be truly effective unless it is perceived as having

value, and acted on accordingly, by employees. These surveys quantify the attitudes that influence acceptance and safe job performance. A company-wide or facility-wide perception survey can pinpoint program inadequacies and credibility gaps in management's commitment to safety. Moreover, because the survey is anonymous and administered to all levels, all employees have an equal voice—even those who might be unable or unwilling to voice dissatisfaction in other circumstances. Results of the survey must be made known to all participants in a timely manner.

EMPLOYEE RECOGNITION

Half of the process of continuous improvement is a matter of correcting weaknesses or deficiencies. The other half is a matter of building on strengths. Employees who perform their work safely and promote the company safety and health program certainly exemplify strengths. The value of such performance is reinforced when it is formally recognized by management and co-workers.

In some companies, individual employees' performance appraisals assess their contributions to safety and health. Although safe job performance should be an accepted part of their jobs, and an ongoing goal, there will be times when individual and group efforts merit public recognition. Each of the following contributions deserves some special attention:

- Implemented suggestions or recommendations
- Reduction of safety- and health-related losses
- Achievement of significant loss-free production hours
- Demonstrated high level of safety awareness and activity

Types of Recognition

Recognition may be as simple as posting names and pictures of employees and their outstanding safety suggestions on a bulletin board or in the company newsletter. Many companies also use more systematic means of recognition—for example, issuing certificates or plaques at annual safety and health awards ceremonies. Many also participate in structured award programs sponsored annually by national organizations like the National Safety Council and OSHA.

Every recognition action has value in reinforcing safe job performance and safety and health awareness. Recognition must be appropriate to the significance of the contribution. True milestone achievements deserve more than plaques and publicity. Financial rewards, engraved gifts, or gift certificates are a few examples of good motivators.

Management's actions can also be proof of its appreciation. For example, an Iowa firm's new-employee-involvement program took it from being a poor safety performer to "best in class" in five years. When one of its facilities completed a year of record production with only minuscule losses due to safety factors, the company's top executives flew hundreds of miles to personally and formally recognize the achievement. Another company adds a day's vacation for each employee when the company meets or exceeds its safety goals.

Safety Recognition Organizations

Several organizations have been established in the United States and Canada to recognize people who have avoided or minimized serious injury by using personal protective equipment (PPE). Such awards provide an excellent opportunity to publicize safety performance.

The Wise Owl Program honors industrial employees and students who have saved their eyesight by wearing protective eye equipment. Write to Coordinator, Wise Owl Program, National Society to Prevent Blindness, 500 Remington Road, Schaumburg, IL 60173, or call 847/843–2020.

The Golden Shoe Club rewards employees who avoided serious injury because they were wearing safety shoes. Address inquiries to Golden Shoe Club, c/o Hy-Test, Inc., 130 S. Canal Street, Chicago, IL 60606, 312/559–7424.

The Turtle Club honors people who escaped serious head injury because they were wearing a hard hat at the time of their accident. Address inquiries to the Turtle Club, 1898 Safety Way, Cynthiana, KY 41031–9303, Attn: Beverly Pierce.

The Golden Belt Club's award is presented to an employee or employee's family members who reduced their injuries in a car accident by wearing seatbelts. Address inquiries to the National Safety Council, Sales Dept., 1121 Spring Lake Drive, Itasca, IL 60143-3201, 630/285–1121.

SAFETY CONTESTS

Safety contests are popular with employees and employers alike. Each year, about 5 million workers in the United States participate in the industrial safety contests sponsored by National Safety Council members. These contests represent more than 9 billion workhours and award recognition to hundreds of people.

Safety contests can be one of the most effective ways to motivate employees to prevent accidents.

Contests are no substitute, however, for a good safety program. Their entire purpose, in fact, is to create interest, awareness, and participation in organized safety programs.

Company or Industry Contests

Competitions can be held between departments, plants, divisions, or a number of companies within an industry. In addition to nationwide contests sponsored by the National Safety Council, trade associations and local safety councils often have their own contests.

Some contests are based on incident experience over a 6- to 12-month period. Winners are chosen based on relative standards, performance improvement, or other predetermined factors. Many effective local contests are launched with major publicity campaigns and run for a specified time.

Contests can also be based on many different measurements of a safety program, including the following:

- Quantity and quality of safety meetings held
- Number of near-miss incident reports made
- Employee safety suggestions offered
- Job safety analyses completed and/or revised
- Incident investigation reports turned in complete and on time
- Safety training courses, seminars, and programs completed
- Employee safety recommendations made
- Safety inspections held with quality results
- Safety committee involvement maintained

Interdepartmental contests are the most challenging for supervisors and team leaders. (National intercompany contests are usually handled by safety directors.) Interdepartmental competition is so successful because workers are likely to have a personal stake in the standing of their own departments. When contests are conducted among dissimilar departments in one plant, differences in size, type of operation, and exposure to hazards must be considered. When two or more departments have perfect safety records, the one with the most workhours since the last recordable incident wins.

When departments have wide variations in the number of employees and the kind of work performed, teams of 20 to 50 people can participate in intergroup competition. Each team should be made up of a proportionate number of workers from high-, medium-, and low-hazard occupations. Colored buttons designate team members. Team names can be chosen by the group or drawn out of a hat.

Every employee on the winning team should receive some sort of recognition. The best prizes are those that everybody wins—certificates, trophies, plaques—even if the prizes have little intrinsic value. When only one person on the winning team receives the prize through a lottery, everyone else who contributed tends to feel disappointed.

For contest purposes, calculating occupational injury and illness incidence rates is based upon the system given in the U.S. Department of Labor publication, *A Brief Guide to Record-Keeping Requirements for Occupational Injuries and Illnesses.*

Departmental Contests

Less formal contests among departments can also be quite successful in drawing attention to safety and health programs. Contests can be held for good housekeeping, greatest improvement in housekeeping, or wearing of PPE. Housekeeping contest winners are determined by periodic, unbiased inspections.

Elaborate point systems are sometimes devised, so that, in addition to helping to win a department trophy, employees can accumulate points individually for their department's performance. Then they can select merchandise from a catalog, or other prizes, on the basis of the points their group accumulated.

SAFETY AND HEALTH POSTERS

Safety posters alert people to safe and unsafe practices, and remind them of the information whenever they pass by. Figure 5–2 shows examples of effective eye-catching posters.

Posters should be displayed in a prominent, high-visibility location that will not interfere with traffic. They should be centered at an average eye level, about 63 inches above the floor. They should be placed in well-lighted areas or mounted with their own light (though a flashing light should never be used in a production area).

A good size for the poster board is 22 inches wide by 30 inches long. Smaller boards can be used to hold just one poster. National Safety Council posters are available in two standard sizes: size A is 8 by 11 inches, and size B is 17 by 23 inches.

Poster boards or frames should be attractively painted and covered with glass. One board is usually desired in a particular workspace. Several panels may be appropriate in a washroom, locker room, or lunchroom. The materials posted should be displayed separately and kept free of clutter or other notices that could compete for attention.

Changing posters and display materials frequently helps to ensure that employees will notice them. A poster that stays in one place too long runs the risk of becoming invisible background. Occasionally a message must be permanently displayed, such as a poster describing correct use of fire equipment or respirators

or a warning of radioactive material. These posters should be mounted on heavy board under glass and reviewed regularly to make sure they are not out of date.

The National Safety Council's POP (point of problem) posters and safety stickers highlight special hazards. POP posters are 4 by 5 inches; stickers are slightly smaller. Both are self-sticking.

SUGGESTION SYSTEMS

Every employee has some contribution to make to improve safety and health. Employee suggestion systems should encourage all workers to contribute their ideas or concerns on an ongoing basis. The system should incorporate a process for identifying unsafe conditions/acts (anonymously if necessary) with suggestions for solutions. This gives all employees a chance to buy into the company's procedures by inventing or modifying them. It also provides the company with an important source of knowledge and creativity.

A well-organized suggestion system encourages employees to contribute their ideas or concerns to help the company improve work methods or reduce work hazards. The system can also stimulate participants' thinking about problem solving. Employees may be rewarded with a percentage of the savings resulting from increase in production or decrease in accidents, injuries, and work-related illnesses. Solutions to safety problems can be rewarded in proportion to the usefulness of the idea.

The company should provide special forms for submitting suggestions and set out suggestion boxes in convenient locations. Suggestions should be gathered frequently and acknowledged promptly. Management must accept or reject each suggestion as soon as possible. When a suggestion is not accepted, the employee should be told why.

To be successful, a suggestion system must have a clearly explained operating plan. Employees must know that the plan is fair, impartial, and potentially profitable. The results of useful suggestions should be publicized, along with the names of the employees who made them. However, employees should be able to make anonymous suggestions if they wish. Anyone who suggests that attention be paid to dangerous or unethical activities should receive protection from retaliation.

EMPLOYEE SAFETY AND HEALTH TRAINING

At minimum, all employees should receive training in emergency procedures, including first aid and disaster

Figure 5–2. Using a variety of National Safety Council posters and changing them periodically will assure that the safety message is noticed.

plans. Companies with good emergency response training plans have found that the benefits are twofold. This training not only helps to save lives and reduce the severity of injuries, but also makes employees more safety-conscious and less likely to have accidents in the first place. The value of emergency training is most obvious in operations that are far away from professional medical help. Workers find that the skills they learn can be indispensable in off-the-job emergencies as well.

Beyond emergency preparedness training, employees may need specific job-related hazard training such as chemical safety, or lockout/tagout training. General safety awareness education and training sessions may be necessary to raise the level of understanding and to identify common objectives and goals. Additional training may be in order for safety committee members and also for those who organize or coordinate safety meetings.

First Aid Courses and CPR

According to the American Medical Association, when a person is injured or suddenly becomes ill, the first things to do are the following six basic steps:

1. Maintain breathing.
2. Maintain circulation.
3. Prevent loss of blood.
4. Prevent further injury.
5. Prevent shock.
6. Summon professional medical help.

The more workers present on each shift who are trained in giving first aid, the better it is for everyone. First-aid training should be given in standard National Safety Council, Red Cross, or Bureau of Mines courses, and only people with such training should attend the emergency patients. Emergency first aid is limited to what is necessary to save a life or prevent further injury. First aid training must include bloodborne pathogen education to meet OSHA's requirement 29 *CFR* 1910.1030.

When breathing stops, any delay in restoring the flow of oxygen to the patient's brain and other organs can result in permanent damage or even death. Basic cardiopulmonary resuscitation (CPR) involves mouth-to-mouth respiration and external heart massage. It does not require equipment and can be accomplished by only one rescuer, if need be (though more helpers increase the chance of success).

While the trained rescuers work, a bystander can call a doctor, emergency medical services, or the police or fire department. All these numbers should be posted prominently in the workplace. Each work area should have its own first-aid kit, whose contents should be checked frequently and replaced after every use.

All employees should be taught not to assume that a disoriented or incoherent person is drunk or using illegal drugs. These symptoms may be due to a head injury, allergies, reactions to medication, diabetes, stroke or neurological problems, and other causes. Unconsciousness can also be caused by a number of factors, among them concussion, carbon monoxide poisoning, epilepsy, diabetic coma, heart trouble, or various medications. Only qualified medical personnel can diagnose and treat such problems accurately. The responsibility of first aid- and CPR-trained personnel is solely to do what they can to save the stricken person's life and prevent further injury.

Disaster Plans

Employees should be involved in making contingency plans for disasters ranging from severe weather to bomb threats. Their input is invaluable in assessing the risks to begin with, and they are likely to have creative suggestions for dealing with emergencies.

The company should have a plan of action for each possible crisis. For example, suppose a heavy rainstorm strands hundreds of customers in a store right before closing. Are supervisors and clerks prepared to handle the situation? How do they control the crowd? What if hail or wind starts breaking the store windows or the power lines go down? Even if the emergency lighting system comes on without a hitch, all employees must know the emergency plans and be prepared to act responsibly.

Advance planning is the key to handling emergencies. Often plans of action will need to be coordinated with government agencies. All plans should be rehearsed under realistic conditions. Figure 5–3 shows a sample individual instruction notice for general posting.

Most companies develop their own emergency manuals. Getting employees involved in this process helps promote safety and helps identify all the areas of concern. Manuals should include the following information, plus other items, depending on the risk assessment and the available resources:

1. Company policy, purposes, authority, principal control measures, and emergency organization chart showing positions and functions
2. Some description of the possible disasters, with a risk statement
3. A map of the facility, office, or store showing equipment, medical and first aid, fire control apparatus, shelters, command center, evacuation routes, and assembly areas
4. A list of cooperating agencies and how to reach them (This list should also be posted.)
5. A facility warning system and list of warnings that are sounded, if more than one type is used
6. Central communications system, including employees'

home phone numbers
7. Shutdown procedure, including security guards
8. Procedure for handling visitors and customers
9. A list of locally related and necessary items
10. A list of needed equipment and resources and where to find them

Disaster Drills. People should be chosen for crisis management positions based on their ability to handle extreme stress, rather than on their job titles. The smaller the chain of command, the more efficient it will be in an emergency. The emergency coordinator should disseminate information quickly and

---

EMERGENCY EXIT INSTRUCTIONS
MACHINE SHOP—DAY SHIFT

Read Carefully

The following persons will be in command in any emergency, and their instructions must be followed:

    CHIEF OF EXIT DRILL—H. C. Gordon, General Sup't.
    MACHINE SHOP EXIT DRILL CAPTAIN—R. L. Jones, Foreman
    MACHINE SHOP MONITORS—Dave Thomas and A. L. Smith

In event of FIRE in machine shop

✔ **NOTIFY THE GENERAL SUPERINTENDENT'S OFFICE**

✔ **PUT OUT THE FIRE, IF POSSIBLE**—If the fire cannot quickly be controlled, follow instructions given by Exit Drill Captain R. L. Jones or by the shop monitors. Leave by the exit door at the south end of the shop; if it is blocked by fire, use the door through the toolroom to the outside stairway.

In event of FIRE or EMERGENCY in other sections of building

The general alarm gong will ring for two 10-second periods as an "alert" signal. Continue work, but be on the alert for the "evacuation" signal, which will be a series of three short rings. At the evacuation signal:

   ✔ SHUT OFF ALL POWER TO MACHINES AND FANS

   ✔ TURN OFF GAS UNDER HEAT TREATING OVENS

   ✔ CLOSE WINDOWS AND CLEAR THE AISLES

   ✔ FORM A DOUBLE LINE IN THE CENTER AISLE AND FOLLOW MONITORS AND EXIT DRILL CAPTAIN TO EXIT—Walk rapidly, but do not run or crowd; do not talk, push, or cause confusion!

After leaving the building, do not interfere with the work of the plant fire brigade or the city fire department. Await instructions from the General Superintendent or your foreman.

Returning to the building

Return-to-work instructions will be given over the loudspeaker system or by telephone from the Superintendent's office.

Figure 5–3. This sample emergency instruction notice should be prominently posted. Some facilities will add after-hours phone numbers and alternate communications protocols.

accurately, and request any needed support as quickly as possible. In small companies without regular security staff or firefighters, operating people should be trained to cover these duties.

Involving employees in planning such responses ensures their wholehearted cooperation, which is crucial to the success of an emergency plan. They need to see that the plan is designed to protect them, not just company property. Management must show that it takes the plan seriously by supporting it with training, simulated disaster drills (paper drills), and full-scale dress rehearsals involving all personnel. Such drills provide essential feedback for improving the disaster response. They also give people the assurance they need to avoid panicking in the face of a real crisis.

Hazardous Materials. Response plans for spills of hazardous materials are mandated by OSHA. Its regulations cover requirements for monitoring instrumentation, site safety plans, respiratory and personal protective equipment (PPE), medical surveillance, engineering controls, work practices, training, and other operational functions.

OSHA spells out the specific classroom training and field experience that hazardous waste operations and emergency response (HAZWOPER) teams must have. First responders who will identify hazardous materials, perform basic diking and containment operations, and initiate evacuation must have at least eight hours of training in PPE, decontamination, chemistry, and toxicology.

Knowledgeable Response. Any emergency, from a toxic spill to an earthquake, requires workers to respond instantly and knowledgeably. Employees who have helped to design the emergency plan will be better organized and focused, because they are more familiar with it and more committed to carrying it out.

JOB SAFETY ANALYSIS

Job safety analysis (JSA) is a procedure used to review job methods and uncover hazards that (1) may have been overlooked in the layout of the facility or building and in the design of the machinery, equipment, tools, workstations, and processes; (2) may have developed after production started; or (3) resulted from changes in work procedures or personnel.

A JSA can be written as shown in Figure 5–4. In the left column, the basic steps of the job are listed in the order in which they occur. The middle column describes all hazards, both those produced by the environment and those connected to the job procedure. The right column gives the safe procedures that should be followed to guard against the hazards and to prevent potential accidents. (See Figure 5–5 for a list of instructions printed on the

back of JSA forms published by the National Safety Council.)

For convenience, both the JSA procedure and the written description are commonly referred to as JSA. Health hazards are also considered when making a JSA.

Benefits of JSA

The principal benefits of a JSA include:

- Giving individual training in safe, efficient procedures
- Making employee safety contacts
- Instructing the new person on the job
- Preparing for planned safety observations
- Giving pre-job instruction on irregular jobs
- Reviewing job procedures after accidents occur
- Studying jobs for possible improvement in job methods

A JSA can be done in three basic steps. However, before initiating this analysis, management must first carefully select the job to be analyzed.

Selecting the Job

A job is a sequence of separate steps or activities that together accomplish a work goal. Some jobs can be broadly defined by what is accomplished, for example, making paper, building a facility, and mining iron ore. On the other hand, a job can be narrowly defined in terms of a single action, such as turning a switch, tightening a screw, and pushing a button. Such broadly or narrowly defined jobs are unsuitable for JSA.

Jobs suitable for JSA are those assignments that a line supervisor may make. Operating a machine, tapping a furnace, and piling lumber are good subjects for job safety analyses because they are neither too broad nor too narrow.

Selection of jobs to be analyzed and establishment of the order of analysis should be guided by the following factors:

- *Frequency of Accidents.* The greater the number of accidents associated with the job, the greater its priority claim for JSA.
- *Rate of Disabling Injuries.* Every job that has had disabling injuries should be given a JSA, particularly if the injuries prove that prior preventive action was not successful.
- *Severity Potential.* Some jobs may not have a history of accidents but may have the potential for producing severe injury.
- *New Jobs.* Changes in equipment or in processes obviously have no history of accidents, but their accident potential may not be fully appreciated. A JSA of every new job should be made as soon as

JOB SAFETY ANALYSIS	JOB TITLE (and number if applicable): Banding Pallets			DATE:	☒ NEW
INSTRUCTIONS ON REVERSE SIDE		PAGE 1 OF 2 JSA NO. 105		00/00/00	☐ REVISED
	TITLE OF PERSON WHO DOES JOB: Bander	SUPERVISOR: James Smith		ANALYSIS BY: James Smith	
COMPANY/ORGANIZATION: XYZ Company	PLANT/LOCATION: Chicago	DEPARTMENT: Packaging		REVIEWED BY: Sharon Martin	
REQUIRED AND/OR RECOMMENDED PERSONAL PROTECTIVE EQUIPMENT:	Gloves - Eye Protection - Long Sleeves - Safety Shoes			APPROVED BY: Joe Bottom	

SEQUENCE OF BASIC JOB STEPS	POTENTIAL HAZARDS	RECOMMENDED ACTION OR PROCEDURE
1. Position portable banding cart and place strapping guard on top of boxes.	1. Cart positioned too close to pallet (strike body & legs against cart or pallet, drop strapping gun on foot.)	1. Leave ample space between cart and pallet to feed strapping - have firm grip on strapping gun.
2. Withdraw strapping and bend end back about 3".	2. Sharp edges of strapping (cut hands, fingers & arms). Sharp corners on pallet (strike feet against corners).	2. Wear gloves, eye protection & long sleeves - keep firm grip on strapping - hold end between thumb & forefinger - watch where stepping.
3. Walk around load while holding strapping with one hand.	3. Projecting sharp corners on pallet (strike feet on corners).	3. Assure a clear path between pallet and cart - pull smoothly - avoid jerking strapping.
4. Pull and feed strap under pallet.	4. Splinters on pallet (punctures to hands and fingers) Sharp strap edges (cuts to hands, fingers, and arms).	4. Wear gloves - eye protection - long sleeves. Point strap in direction of bend - pull strap smoothly to avoid jerks.
5. Walk around load. Stoop down. Bend over, grab strap, pull up to machine, straighten out strap end.	5. Protruding corners of pallet, splinters (punctures to feet and ankles).	5. Assure a clear path - watch where walking - face direction in which walking.
6. Insert, position and tighten strap in gun.	6. Springy and sharp strapping (strike against with hands and fingers).	6. Keep firm grasp on strap and on gun - make sure clip is positioned properly.

Figure 5-4. This sample of a completed JSA shows how hazards and safe procedures are identified to help reduce the occurrence of accidents.

the job has been created. Analysis should not be delayed until accidents or near-misses occur.

After the job has been selected, the three basic steps in making a JSA are:

1. Break the job down into successive steps or activities and observe how these actions are performed.
2. Identify the hazards and potential accidents. This is the critical step because only an identified problem can be eliminated.
3. Develop safe job procedures to eliminate the hazards and prevent the potential accidents.

Breaking the Job Down into Steps

Before the search for hazards begins, a job should be broken down into a sequence of steps, each describing what is being done. Avoid the two common errors: (1) making the breakdown so detailed that an unnecessarily large number of steps results, or (2) making the job breakdown so general that basic steps are not recorded.

To do a job breakdown, select the right worker to observe—an experienced, capable, and cooperative person who is willing to share ideas. If the employee has never helped out on a job safety analysis, explain the purpose—to make a job safe by identifying and eliminating or controlling hazards—and show him or her a completed JSA. Reassure the employee that he or she was selected because of experience and capability.

Observe the employee perform the job and write down the basic steps. Consider videotaping the job as it is performed for later study. To determine the basic job steps, ask "What step starts the job?" Then, "What is the next basic step?" and so on.

Completely describe each step. Any possible deviation from the regular procedure should be recorded because it may be this irregular activity that leads to an accident.

To record the breakdown, number the job steps consecutively as illustrated in the first column of the JSA training guide, illustrated in Figure 5–6. Each step tells what is done, not how.

The wording for each step should begin with an action word like "remove," "open," or "weld." The action is completed by naming the item to which the action applies, for example, "remove extinguisher," "carry to fire."

Check the breakdown with the person observed and obtain agreement of what is done and the order of the steps. Thank the employee for cooperating.

Identifying Hazards and Potential Accident Causes

Before filling in the next two columns of the JSA—Potential Hazards and Recommended Action or Procedure—begin the search for hazards. The purpose is to identify all hazards—both those produced by the environment and those connected with the job procedure. Each step, and thus the entire job, must be made safer and more efficient. To do this, ask these questions about each step:

- Is there a danger of striking against, being struck by, or otherwise making harmful contact with an object?
- Can the employee be caught in, by, or between objects?

INSTRUCTIONS FOR COMPLETING JOB SAFETY ANALYSIS FORM

Job Safety Analysis (JSA) is an important accident prevention tool that works by finding hazards and eliminating or minimizing them *before* the job is performed, and *before* they have a chance to become accidents. Use your JSA for job clarification and hazard awareness, as a guide in new employee training, for periodic contacts and for retraining of senior employees, as a refresher on jobs which run infrequently, as an accident investigation tool, and for informing employees of specific job hazards and protective measures.

Set priorities for doing JSA's: jobs that have a history of many accidents, jobs that have produced disabling injuries, jobs with high potential for disabling injury or death, and new jobs with no accident history.

Here's how to do each of the three parts of a Job Safety Analysis:

SEQUENCE OF BASIC JOB STEPS

Break the job down into steps. Each of the steps of a job should accomplish some major task. The task will consist of a *set* of movements. Look at the first *set* of movements used to perform a task, and then determine the next logical *set* of movements. For example, the job might be to move a box from a conveyor in the receiving area to a shelf in the storage area. How does that break down into job steps? Picking up the box from the conveyor and putting it on a handtruck is one logical set of movments, so it is one job step. Everything related to that one logical set of movements is part of that job step.

The next logical *set* of movements might be pushing the loaded handtruck to the storeroom. Removing the boxes from the truck and placing them on the shelf is another logical set of movements. And finally, returning the handtruck to the receiving area might be the final step in this type of job.

Be sure to list *all* the steps in a job. Some steps might not be done each time—checking the casters on a handtruck, for example. However, that task is a part of the job as a whole, and should be listed and analyzed.

POTENTIAL HAZARDS

Identify the hazards associated with each step. Examine each step to find and identify hazards—actions, conditions and possibilities that could lead to an accident.

It's not enough to look at the obvious hazards. It's also important to look at the entire environment and discover every conceivable hazard that might exist.

Be sure to list health hazards as well, even though the harmful effect may not be immediate. A good example is the harmful effect of inhaling a solvent or chemical dust over a long period of time.

It's important to list *all* hazards. Hazards contribute to accidents, injuries and occupational illnesses.

In order to do part three of a JSA effectively, you must identify potential and existing *hazards*. That's why it's important to distinguish between a hazard, an accident and an injury. Each of these terms has a specific meaning:

HAZARD—A potential danger. Oil on the floor is a *hazard*.
ACCIDENT—An unintended happening that may result in injury, loss or damage. Slipping on the oil is an *accident*.
INJURY—The *result* of an accident. A sprained wrist from the fall would be an injury.

Some people find it easier to identify possible accidents and illnesses and work back from them to the hazards. If you do that, you can list the accident and illness types in parentheses following the hazard. But be sure you focus on the *hazard* for developing recommended actions and safe work procedures.

RECOMMENDED ACTION OR PROCEDURE

Using the first two columns as a guide, decide what actions are necessary to eliminate or minimize the hazards that could lead to an accident, injury, or occupational illness.

Among the actions that can be taken are: 1) engineering the hazard out; 2) providing personal protective equipment; 3) job instruction training; 4) good housekeeping; and 5) good ergonomics (positioning the person in relation to the machine or other elements in the environment in such a way as to eliminate stresses and strains).

List recommended safe operating procedures on the form, and also list required or recommended personal protective equipment for each step of the job.

Be specific. Say *exactly* what needs to be done to correct the hazard, such as, "lift, using your leg muscles." Avoid general statements like, "be careful."

Give a recommended action or procedure for *every* hazard.

If the hazard is a serious one, it should be corrected immediately. The JSA should then be changed to reflect the new conditions.

Figure 5–5. Use these instructions for preparing a JSA.

- Is there a potential for a slip or trip? Can the employee fall on the same level or to another?
- Can strain be caused by pushing, pulling, lifting, bending, or twisting?
- Is the environment hazardous to safety or health? For example, are there concentrations of toxic gas, vapor, mist, fume, dust, heat, or radiation? (See discussion in the National Safety Council's *Fundamentals of Industrial Hygiene*, 4th edition.)

Close observation and knowledge of the particular job are required if the JSA is to be effective. The job observation should be repeated as often as necessary until all hazards and potential causes for accident or injury have been identified.

When inspecting a particular machine or operation, ask the question, "Can an accident occur here?" More specific questions include:

- Is it possible for a person to come in contact with any moving piece of machine equipment?
- Are rotating equipment, set screws, projecting keys, bolt heads, burrs, or other projections exposed where they can strike at or snag a worker's clothing?
- Is it possible to be drawn into the inrunning nip point between two moving parts, such as a belt and sheave, chain and sprocket, pressure rolls, rack and gear, or gear train?
- Do machines or equipment have reciprocating movement or any motion where workers can be caught on or between a moving part and a fixed object?
- Is it possible for a worker's hands or arms to make contact with moving parts at the point of operation where milling, shaping, punching, shearing, bending, grinding, or other work is being done?
- Is it possible for material (including chips or dust) to be kicked back or ejected from the point of operation, injuring someone nearby?
- Are machine controls safeguarded to prevent unintended or inadvertent operation?
- Are machine controls located to provide immediate access in the event of an emergency?
- Do machines vibrate, move, or walk while in operation?
- Is it possible for parts to become loose during operation, injuring operators or others?
- Are guards positioned or adjusted to correspond with the permissible openings?
- Is it possible for workers to bypass the guard, thereby making it ineffective?
- Do machines, equipment, and appurtenances receive regular maintenance?
- Are machines placed so operators have sufficient room to safely work with no exposure to aisle traffic?
- Is there sufficient room for maintenance and repair?

- Is there sufficient room to accommodate incoming and finished work as well as scrap that may be generated?
- Are the materials handling methods adequate for the work in process and the tooling associated with it?
- If tools, jigs, and other work fixtures are required, are they stored conveniently, where they will not interfere with the work?
- Is the work area well illuminated with specific point-of-operation lighting where necessary?
- Is ventilation adequate, particularly for those operations that create dusts, mists, vapors, and gases?
- Is the operator using personal protective equipment?
- Is housekeeping satisfactory with no debris or tripping hazards or spills on the floor?
- Are there places where employees have access to machines (for example, the backside)?
- Are energy sources heat controlled for protection?
- Are energy sources controlled for maintenance?

All of these questions will be of most value if they are incorporated into an inspection form that can be filled out at regular intervals. Even though a question may not at first seem to apply to a specific operation, on closer scrutiny it may be found to apply. Using a checklist is a good way to make sure nothing is overlooked.

To complete column 2 of the JSA, the analyzer should list all potential causes for accidents or injury and all hazards yielded by a survey of the machine or operation. Record the type of potential cause of accidents and the agents involved. For example, to note that the employee might injure a foot by dropping a fire extinguisher, write down "struck by extinguisher."

Again, check with the observed employee after the hazards and potential causes of accidents have been recorded. The experienced employee will probably offer additional suggestions. You should also check with others experienced with the job. Through observation and discussion, the analyzer can develop a reliable list of hazards and potential accidents.

Developing Solutions

The final step in a JSA is to develop a recommended safe job procedure to prevent the occurrence of accidents. The principal solutions are as follows:

1. Find a new way to do the job.
2. Change the physical conditions that create the hazards.
3. Change the work procedure.
4. Reduce the frequency (particularly helpful in maintenance and materials handling).

To find an entirely new way to do a job, determine the work goal of the job, and then analyze the various ways of reaching this goal to see which way is safest.

JOB SAFETY ANALYSIS WORK SHEET
JOB: Using a Pressurized Water Fire Extinguisher

WHAT TO DO (Steps in sequence)	HOW TO DO IT (Instructions) (Reverse hands for left-handed operator.)	KEY POINTS (Items to be emphasized. Safety is always a key point.)
1. Remove extinguisher from wall bracket.	1. Left hand on bottom lip, fingers curled around lip, palm up. Right hand on carrying handle palm down, fingers around carrying handle only.	1. Check air pressure to make certain extinguisher is charged. Stand close to extinguisher, pull straight out. *Have firm grip, to prevent dropping on feet.* Lower, and as you do remove left hand from lip.
2. Carry to fire.	2. Carry in right hand, upright position.	2. Extinguisher should hang down alongside leg. (This makes it easy to carry and reduces possibility of strain.)
3. Remove pin.	3. Set extinguisher down in upright position. Place left hand on top of extinguisher, pull out pin with right hand.	3. Hold extinguisher steady with left hand. Do not exert pressure on discharge lever as you remove pin.
4. Squeeze discharge lever.	4. Place right hand over carrying handle with fingers curled around operating lever handle while grasping discharge hose near nozzle with left hand.	4. Have firm grip on handle to steady extinguisher.
5. Apply water stream to fire.	5. Direct water stream at base of fire.	5. Work from side to side or around fire. After extinguishing flames, play water on smouldering or glowing surfaces.
6. Return Extinguisher. Report Use.		

Figure 5–6. This JSA worksheet shows how to break down a job and analyze hazards and safe procedures.

Consider work-saving tools and equipment.

If a new way cannot be found, then ask this question about each hazard and potential accident cause listed: "What change in physical condition (such as change in tools, materials, equipment, layout, or location) will eliminate the hazard or prevent the accident?"

When a change is found, study it carefully to find other benefits (such as greater production or time saving) that will accrue. These benefits are good selling points and should be pointed out when proposing the change to higher management.

To investigate changes in the job procedure, ask the following questions about each hazard and potential accident cause listed: "What should the employee do—or not do—to eliminate this particular hazard or prevent this potential accident?" "How should it be done?" Because of his or her experience, in most cases the supervisor can answer these questions.

Answers must be specific and concrete if new procedures are to be any good. General precautions—"be alert," "use caution," or "be careful"—are useless. Answers should precisely state what to do and how to do it. This recommendation—"Make certain the wrench does not slip or cause loss of balance"—is incomplete. It does not tell how to prevent the wrench from slipping. Here, in contrast, is an example of a recommended safe procedure that tells both what and how: "Set wrench properly and securely. Test its grip by exerting a light pressure on it. Brace yourself against something immovable, or take a solid stance with feet wide apart, before exerting full pressure. This prevents loss of balance if the wrench slips."

Some repair or service jobs have to be repeated often because a condition needs repeated correction. To reduce the need for such repetition, ask "What can be done to eliminate the cause of the condition that makes excessive repairs or service necessary?" If the cause cannot be eliminated, then ask "Can anything be done to minimize the effects of the condition?"

Machine parts, for example, may wear out quickly and require frequent replacement. Study of the problem may reveal excessive vibration is the culprit. After reducing or eliminating the vibration, the machine parts last longer and require less maintenance.

However, reducing the frequency of a job contributes to safety only in that it limits the exposure. Every effort should still be made to eliminate hazards and to prevent accidents through changing physical conditions or revising job procedures or both.

A job that has been redesigned may affect other jobs and even the entire work process. Therefore, the redesign should be discussed not only with the worker involved, but also co-workers, the supervisor, the facility engineer, and others who are concerned. In all cases, however, check or test the proposed changes with those who do the job. Their ideas about the hazards and proposed solutions can be of considerable value. They can judge the practicality of proposed changes and perhaps suggest improvements. Actually these discussions are more than just a way to check a JSA. They are safety contacts that promote awareness of job hazards and safe procedures.

SUMMARY OF KEY POINTS

Key points covered in this chapter include:

- Employee involvement is an essential building block of any workplace safety and health program. Ways to involve employees include safety and health committees, meetings, employee recognition, safety contests and posters, and suggestion systems.
- In a joint safety and health committee, about half the members should be workers and half management. The workers should be either elected by their peers or volunteers. Members should have staggered one- and two-year terms.
- Safety and health committees are usually responsible for (1) policy, practice, and process, (2) inspection and investigation, (3) promotion, education, and motivation, and (4) problem analysis and solving.
- Safety and health meetings should motivate workers to take an interest in and responsibility for safety and health. Program topics should be designed to be useful and appealing.
- Each meeting should feature no more than three ideas and should be held at a comfortable location. There should be a written agenda, circulated to attendees beforehand. The meeting should be short and simple. Visual aids can add interest and help to fix information in people's memories. Production huddles should incorporate the safety message into the job routine.
- Recognizing employees' contributions to safety

and health shows them they are valued and encourages them to contribute more. Recognition may be as simple as publishing names, pictures, and safety suggestions of outstanding workers in the company newsletter. Or it may involve plaques issued at annual ceremonies or participation in NSC or OSHA award programs. Recognition should be in proportion to the significance of the contribution.
- Safety contests exist to create safety program participation. Contests can be held between departments, plants, or divisions, or among a number of companies in an industry. They can reward everything, from the quantity and quality of safety meetings held, to the number of job safety analyses completed, to the greatest improvement in housekeeping.
- Safety and health posters alert people to safe and unsafe practices and remind them of this information. They should be displayed in high-visibility locations that will not interfere with traffic. Changing posters can help to ensure that employees will notice them.
- Suggestion systems should encourage all employees to contribute their ideas or concerns on an ongoing basis. They might be rewarded with a percentage of the savings resulting from an increase in production or decrease in accidents, injuries, and work-related illnesses. Suggestions should be gathered frequently and acknowledged promptly.
- At least several workers in each department on each shift should have first-aid training. They should know how to perform CPR, maintain circulation, and prevent loss of blood, shock, and further injury. Each work area should have a well-stocked first-aid kit. Emergency numbers should be posted.
- Employees should be involved in making contingency plans for all possible types of disasters. Companies should consider developing an emergency manual. Plans should be supported with training, simulation drills, and full-scale dress rehearsals involving all personnel. Response plans for hazardous materials spills are mandated by OSHA.

6

SAFETY, HEALTH, AND
ENVIRONMENTAL AUDITING

After reading this chapter, you will be able to:

- Describe the objective and purpose of formal audits

- List the steps of a safety, health, and environmental audit

- Assess the auditor's report and follow up on it

- Describe how to prepare for an inspection, how often and what to inspect

- Write, and do the follow-up on, the inspection report

Knowing where and how specific safety policies and programs are succeeding or failing is crucial to continuous improvement. To measure how an organization is doing, formal, nonroutine audits or informal, routine inspections are conducted. An audit is a methodical examination of a facility's procedures and practices to verify whether they comply with legal requirements, internal policies, and good practices. Safety, health, and environmental audits usually cover the full range of pollution control, occupational safety, process safety, industrial hygiene, occupational health and medicine, and waste management.

Audits are usually performed by objective parties from outside the department or outside the company. External auditors may be hired by a company or they may be government regulators. An inspection is usually conducted by the department/facility supervisor or team leader.

FORMAL AUDITS

It is a well-known saying in business that, "What gets measured gets done." In the case of safety, health, and environmental audits, it might be more accurate to say, "What gets measured gets targeted for improvement," because a good audit serves as an educational tool.

Some companies place their audit program within a core corporate group, most often the corporate safety, health, and environmental staff. Others locate the program in the internal audit regulatory affairs, production or operations, or legal department. Some hire outside auditors to guarantee independent, objective findings. Most in-house safety, health, and environmental audit teams include individuals with technical expertise, knowledge of regulations, facility experience, and expertise in auditing procedures and techniques.

Objectives and Purposes

Reasons for developing a safety, health, and environmental auditing program range from (1) measuring compliance with specific regulations to (2) benchmarking with established goals, to (3) identifying potentially hazardous conditions for which standards do not exist. An auditing program may be established for the following reasons:

- Determine and document compliance status
- Improve overall safety, health, and environmental performance at operating facilities
- Assist facility management
- Increase the overall level of safety, health, and environmental awareness
- Provide a basis for awards
- Accelerate the development of safety, health, and environmental management sement control systems

- Improve the safety, health, and environmental risk management system
- Protect the company from potential liabilities
- Develop a basis for optimizing safety, health, and environmental resources
- Assess facility management's ability to achieve safety, health, and environmental goals (i.e., measure the outcome versus the goal)

Though these objectives all address compliance, they can produce differences in the audit program scope and focus, including different organizational, geographic, locational, functional, and compliance boundaries. Organizational boundaries indicate which of the company's operations are included in the audit program. Geographic boundaries address how broadly the program applies (state, national, international). Locational boundaries designate what territory is included in a specific audit (usually a particular facility).

A number of functional areas can be included. Most audit programs cover employee safety and industrial hygiene, air and water pollution control, and solid and hazardous waste management. Many now include occupational medicine and wellness, fire and loss prevention, process safety, and product safety. Compliance boundaries define the standards against which the facility is measured. They may include federal, regional, state, and local laws and regulations; corporate or division policies, procedures, standards, and guidelines; local facility operating procedures; or standards established by an outside group such as an industry or trade association.

The Audit Process

Certain basic activities are common to most audit programs. Some activities are undertaken before the on-site audit (planning), some during the audit fieldwork (understanding systems, assessing controls, gathering and evaluating evidence), and others after the field audit has been completed (reporting the results and following up). See Figure 6–1 for the key steps in the audit process. Virtually all audits involve gathering information, analyzing facts, making judgments about the status of the facility, and reporting the results to management. A team approach is generally used.

Preplanning. Activities that precede the audit include scheduling the visit, selecting the audit team, and gathering and reviewing background information. Some companies audit all facilities on a repeat cycle (for example, once a year). They often select facilities for audit on the basis of risk. Facility managers may be given one to six months' notice before an audit, although a few companies also conduct surprise audits.

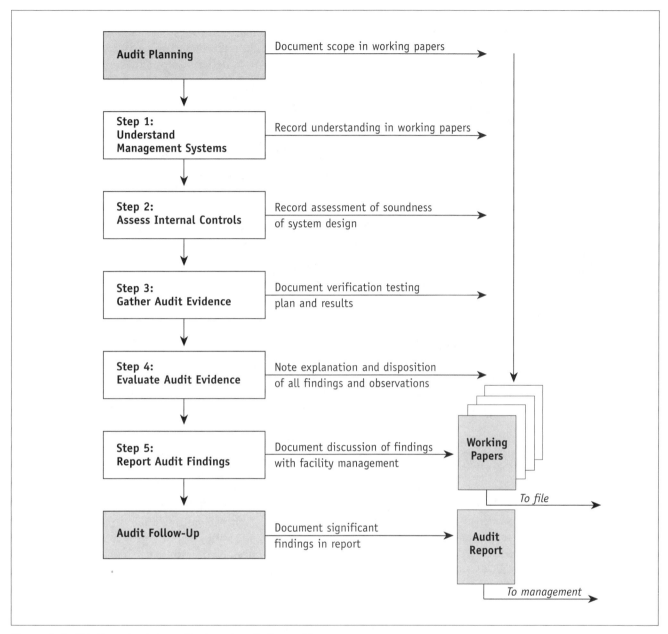

Figure 6–1. This flow chart shows the key steps in the audit process.

Background information is gathered on regulatory requirements, corporate policies, and facility characteristics (organization, processes, layout). Some companies' audit teams visit the facility in advance to learn about facility processes, and safety, health, and environmental management systems, and to brief the facility staff on the audit's objectives.

Fieldwork. Advance information is collected and reviewed, and an audit plan is prepared that outlines the required fieldwork steps, how they are to be accomplished, who will do them, and in what sequence.

Step 1. Understand management systems. Sources of information include background information

staff discussions, facility tours, questionnaires, and in-depth interviews.

Step 2. Assess the strengths and weaknesses of internal controls. Indicators include clearly defined responsibilities, an adequate authorization system, capable personnel, documentation, and internal verification.

Step 3. Gather audit evidence through inquiry, observation, and verification testing. Since audits usually serve as a check on the safety, health, and environmental management systems, and are not a substitute for it, auditors normally take a sample to verify items.

Step 4. Evaluate audit evidence. Auditors usually

evaluate their observations in the course of the audit, but most audit teams take a few hours after the audit is over to jointly discuss their findings. Their goals are to determine whether the audit objectives have been met, and to discover their significance.

Step 5. Report audit findings. The formal reporting process begins with an exit meeting between the audit team and facility personnel. The written audit report usually describe the audit's purpose and scope, outlines the approach, and identifies key participants. It describes the facility's overall compliance with regulations and with company policies and procedures. Some reports describe the facility's history, or predict its ability to handle a crisis. Some recommend ways to address the deficiencies identified.

Follow-Up. Most companies have formal procedures for responding to the audit report. The action plan assigns responsibility for finding solutions to correct deficiencies, establishing timetables, and taking corrective action. Responses to the audit report are usually prepared by the facility manager, and then sent to line management and the audit program manager for review and approval.

Working with Outside Auditors

Usually, an audit is announced in advance, so all team members can be prepared to answer the auditors' questions and offer assistance. The auditor should contact the supervisor or team leader first, in case any background information is needed. This is especially important when conditions have changed temporarily due to construction, maintenance, equipment downtime, or employee absence.

The supervisor should offer to accompany the auditor, although many times auditors prefer to work alone. The auditor will consult the supervisor after the audit to discuss the safety recommendations he or she proposes. The supervisor should correct any problems quickly. (The auditor's report may reflect such corrections.)

Supervisors and team leaders should avoid viewing the auditor's report as a personal criticism. The purpose of audits is fact-finding, not fault-finding. The process should always be conducted with a high degree of professionalism, and not be allowed to degenerate into issues of personality. Good auditors are friendly but firm, and most importantly fair.

Employees should understand that the audit is a way to improve safety, not a personal investigation of their work habits. They should be told that the auditor may need to observe them closely while they work, in order to gain a clearer understanding of how tasks are performed. Auditors should always ask an employee's permission to watch him or her work.

The main benefits of an audit are identify areas in which to improve safety, health, and the environment; save money; provide important measurements; and tell the organization what it is doing well. Another benefit can occur when a supervisor believes certain operations are unsafe: the supervisor can convince the auditor that such changes will help make the workplace safer and the auditor will include them in his or her working papers and perhaps in the final report.

Voluntary Protection Programs

OSHA's Voluntary Protection Programs (VPP) are designed to foster a cooperative relationship between management, labor, and OSHA that promotes effective safety and health management. Approved VPP facilities have either designed and implemented safety and health programs that meet all OSHA criteria for effective safety and health management (the Star Program); demonstrated potential and willingness to achieve Star status (the Merit Program); or demonstrated alternative approaches to safety and health management that meet the purposes of Star criteria (the Demonstration Program).

Interested organizations are asked to complete a concise self-assessment checklist as the initial step toward VPP participation. Once a formal application has been submitted and approved, nonenforcement OSHA personnel visit the site to conduct a document review, a facility walk-through, and formal interviews. Those candidates that are finally approved for VPP are publicly designated as participants in the appropriate program category and removed from programmed inspection lists.

This summary of the VPP application process is greatly simplified. The criteria for participation are intentially detailed and rigorous. The application itself involves paperwork and documentation that require a considerable commitment of staff time and effort. The OSHA assessment is thorough, and although citations are not issued, the site inspection can result in time-limited corrective "recommendations" for noted deficiencies. Successful applicants are exempted from programmed audits, but compliance remains mandatory, and VPP participants are periodically reassessed to confirm that they are maintaining VPP's high standards.

Yet, the sustained benefits of participation– improved morale and motivation, community recognition, program improvement, and continued cooperative assistance from OSHA—can far outweigh the temporary disadvantages. Involvement in VPP, or similar voluntary programs, should be duly considered by management. At the very least, the internal and external assessments involved in the application process can contribute significantly to program evaluation. Successful application can signal the achievement of a new milestone in the continuous improvement of safety and health management.

INSPECTIONS

The basic purposes of safety, health, and environmental inspections are to ensure compliance with standards and to help evaluate supervisors' safety, health, and environmental performance activities. Inspections are a regular element of standard operating procedures. In the process of conducting a safety, health, and environmental inspection, the supervisor may detect potential hazards that require immediate correction or precautions to prevent accidents. Just as inspections of the manufacturing process are important to quality control, safety, health, and environmental inspections are vital to loss control.

Prompt correction of substandard or hazardous conditions detected in an inspection shows everyone that the company is serious about accident prevention. A supervisor or team leader who finds that workers are not following safety, health, and environmental procedures should verify that the procedures are current and retrain the workers in safety, health, and environmental policies and guidelines.

A safety, health, and environmental inspection program should answer the following questions:

- What items need to be inspected?
- What aspects of each item need to be examined?
- What conditions need to be inspected?
- How often must items be inspected?
- Who will conduct the inspection?

Frequency of Inspection

When should a safety, health, and environmental inspection be conducted? Some supervisors might answer, "The third Friday of every month." A better answer would be, "Every time I go through my department." The goal is continuous inspection; every member of the department should constantly be on the lookout for hazards on the job. Safety, health, and environmental inspections should be part of every phase of production, service, etc. Required environmental inspections are defined in EPA standards, e.g., weekly inspections of hazardous waste storage areas.

In addition to continuous inspections, formal inspections should be made routinely at least once a month or as required by standards (Figure 6–2). The supervisor or team leader who conducts the inspection should have a checklist like the one in Figure 6–3.

Checklists or documentation should be completed in full, dated, and signed by the inspectors. Action items should be corrected and recorded as completed.

There are three types of scheduled inspections: (1) Periodic inspections are inspections of specific items conducted weekly, monthly, semiannually, or at other given intervals. (2) Intermittent (special) inspections or assessments are performed at irregular intervals. An accident in another department that involves machinery similar to one's own would lead to an intermittent inspection of the equipment. (3) General inspections are designed to include all areas that do not receive periodic inspections—for example, parking lots, sidewalks, and fences.

The frequency of inspections can be determined by the answers to these five questions:

1. *Are there required inspections?* Regular inspection of certain equipment may be mandated by regulation, by a manufacturer's recommendation, or by OSHA or other government agencies. Inspections of such equipment should be documented.
2. *What is the loss severity potential of the problem* (human, equipment, facilities, products, services)? Example: A frayed wire rope on an overhead crane block has the potential to cause a much greater loss than a defective wheel on a wheelbarrow. The rope needs to be inspected, therefore, more frequently than the wheel.
3. *What is the potential for injury to employees?* If the item or a critical part failed, how many employees would be endangered and how frequently? For example, a stairway that is used all the time needs to be inspected more frequently than one that is used rarely.
4. *How quickly can the item or part become hazardous?* Equipment and tools that get heavy use become damaged or defective, or wear out faster, than those that are not often used. In addition, an item in one location may be exposed to greater potential damage than an identical item in a different location.
5. *What is the past record of failures?* Maintenance and production records and accident investigation reports can provide valuable information both about how often items have failed, and what results have occurred, in terms of injuries, damage, delays, and shutdowns.

What to Inspect

Many different types of inspection checklists are available. They vary in length from a few items to

(*Text continues on page 82.*)

Figure 6–2. Supervisors should make formal inspections routinely at least once a month or as required by standards. Areas to check include housekeeping (*top*), dates of service or maintenance (*left*), and improper procedures, such as leaving a cover off an electrical box (*right*).

Supervisor's Facility and Administrative Inspection Checklist

Building/Department _____

This checklist is intended only as a guide in reviewing general facility and administrative items. Only unsatisfactory items and their location need be identified by a check. Those items identified as unsatisfactory should be targeted for corrective action.

FACILITY AND OPERATIONS	Check If Action Required	Location/Comments/Action Required
Machinery and Equipment		
General safeguarding provided and in place	_____	_____
Operators properly attired (no loose clothing, jewelry)	_____	_____
Point of operation safeguarding provided and functioning properly	_____	_____
Proper tools provided for cleanup and adjustments	_____	_____
Other:	_____	_____
Materials Handling and Storage		
Manual materials handling equipment in good condition	_____	_____
Powered materials handling equipment in good condition	_____	_____
Hazardous and toxic materials handled, stored, and transported in accordance with regulatory requirements	_____	_____
Storage areas properly illuminated	_____	_____
Cylinders transported and stored in upright position; properly secured	_____	_____
Shipping/receiving areas in good condition	_____	_____
Racking and other storage procedures followed	_____	_____
Wheel chocks and restraining devices available/ functioning properly	_____	_____
Other:	_____	_____
Hand and Portable Power Tools		
Correct tools provided	_____	_____
Hand tools and power equipment in good condition	_____	_____
Guards are in place, adjusted properly	_____	_____
Grinding wheel tool rest is within $\frac{1}{8}$ in. of wheel	_____	_____
Stored tools are locked and/or secured	_____	_____
Electrical tools GFCI protected	_____	_____
Electrical tools and receptacles grounded	_____	_____
Other:	_____	_____

Figure 6–3. Supervisor's Facility and Administrative Inspection Checklist.

Fire Protection

Portable fire extinguishes:
- Provided as required
- Inspected as marked
- Location identified

Locations are readily accessible

Alarm system tested (as required)

Fire doors in good operating condition

Exits marked and accessible

Fire detectors working

Other:

Electrical

Outlet boxes covered

Electric cords properly placed

Outlet circuits properly grounded

Portable electric tools:
- GFCI protected
- Double insulated
- Grounded as required

Switches in clean, closed boxes

Switches properly identified

Circuit fuses, circuit breakers identified

Motors are clean, free of oil, grease, and dust

Approved extension cords in good condition

Other:

Housekeeping/Maintenance

Work areas maintained in clean and orderly condition

Floors, aisles, work areas free of obstruction, slipping and tripping hazards

Washrooms and change facilities clean and well maintained

Tools, equipment, and materials properly stored when not in use

Waste materials stored in appropriate containers and disposed of in a safe manner

Scheduled maintenance

General ventilation systems

Local exhaust systems (paint booths, welding areas, etc.)

Machinery (lubrication, belts, servicing, etc.)

Other:

Figure 6–3. (Continued)

Personal Protective Equipment

Equipment (determined by exposure)
- Head protection _____ _____
- Eye protection _____ _____
- Ear protection _____ _____
- Foot protection _____ _____
- Clothes _____ _____
- Hand protection _____ _____
- Respiratory protection _____ _____

Personal protective equipment procedure in place _____ _____

Other: _____ _____

Administrative

Training records:
- Safety and health orientation _____ _____
- Hazard communication (right to know) _____ _____
- Safe operating procedures _____ _____
- Confined space entry procedures _____ _____
- Lockout/tagout _____ _____
- Evacuation emergency response _____ _____
- Equipment/vehicle operation _____ _____
- Fire protection equipment use _____ _____
- Other _____ _____

Plans:
- Disaster preparedness _____ _____
- Chemical emergencies/spills _____ _____
- Fire/evacuation _____ _____
- Emergency medical _____ _____
- Equipment maintenance _____ _____
- Other _____ _____

Records/reports
- Injury/illness _____ _____
- Accidents/incidents _____ _____
- MSDSs _____ _____
- Inspection summaries _____ _____
- Noise surveys _____ _____
- Equipment service logs _____ _____
- Other: _____ _____

Other:
- OSHA required postings _____ _____
- Emergency phone listings _____ _____
- Required labeling _____ _____
- Defective equipment procedure in place _____ _____

Completed by: _____

Date: _____

Route to:
○ Maintenance   ○ Engineering   ○ Other

Figure 6–3. (Concluded)

hundreds. Compare the checklist in Figure 6–3 to the one that follows:

1. Environmental factors (illumination, dust, gases sprays, vapors, fumes, noise)
2. Hazardous chemicals and materials (explosives, flammables, acids, caustics, toxic materials, or by-products), including labeling, signage, plackards, MSDSs
3. Production and related equipment (mills, shapers, presses, borders, lathes)
4. Power equipment (steam and gas engines, electric motors)
5. Electrical equipment (switches, fuses, breakers, outlets, cables, extension and fixture cords, grounds, connectors, connections)
6. Hand tools (wrenches, screwdrivers, hammers, power tools)
7. Personal protective equipment (PPE) (hard hats, safety glasses, safety shoes, respirators)
8. Personal service and first-aid facilities (drinking fountains, washbasins, soap dispensers, eyewash fountains, first-aid supplies, stretchers)
9. Fire protection and extinguishing equipment (alarms, water tanks, sprinklers, fire doors, stand-pipes, extinguishes, hydrants, hoses)
10. Walkways and roadways (ramps, docks, side-walks, aisles, vehicle ways)
11. Elevators, electric stairways, and lifts (controls, wire ropes, safety devices)
12. Working surfaces (floors, ladders, scaffolds, cat-walks, platforms, sling chairs)
13. Materials-handling equipment (cranes, dollies, conveyors, hoists, forklifts, chains, ropes, slings)
14. Transportation equipment (cars and trucks, rail-road cars, front-end loaders, helicopters, motorized carts and buggies)
15. Warning and signaling devices (sirens, crossing and blinker lights, Klaxons®, warning signals)
16. Containers (scrap bins, disposal receptacles, carboys, barrels, drums, gas cylinders, solvent cans)
17. Storage facilities and areas, both indoor and outdoor (bins, racks, lockers, cabinets, shelves, tanks, closets), including labels, signs, plackards, MSDSs
18. Structural openings (windows, doors, stairways, sumps, shafts, pits, floor openings)
19. Buildings and structures (floors, roofs, walls, fencing)
20. Grounds (parking lots, roadways, sidewalks)
21. Loading and shipping platforms
22. Outside structures (small, isolated buildings)
23. Miscellaneous (any items that do not fit in the preceding categories)
24. Emergency equipment (fire extinguishes, deluge showers, spill kits, telephones, etc.)
25. Environmental controls (pollution control devices, etc.)

Generally, longer checklists are keyed to OSHA standards. They are useful in determining which standards or regulations apply to individual situations. Once the relevant standards are identified, a checklist can be custom-tailored and computerized for easy follow-up. Checklists serve as reminders of what to look for, and document what has been covered in past inspections. They provide direction, and permit easy, on-the-spot recording of all findings and comments. If an inspection is interrupted, checklists provide a record of what has already been covered. An inspector who does not have a printed checklist should at least carry a notebook to jot down items during continuous inspections.

It is important to remember that checklists are merely an aid to the inspection process, not an end in themselves. A safety, health, and environmental inspection means much more than simply checking items off the list. Any hazard observed during an inspection (even if it is not on the list) must be recorded, and also corrected. Any hazard posing imminent danger to employees must be corrected on the spot or the area must be closed down until the hazard is corrected. Delay could lead to an incident.

High-Risk Areas. Some items—such as floors, stairways, housekeeping procedures, fire hazards, electrical installations, and chains/ropes/slings—are especially high-risk areas in a department or company. They should be inspected often.

Floors should be inspected carefully, no matter what material they are made of, especially those that are slippery or in areas with heavy traffic. Critical factors include:

- Is the surface damaged or wearing out too rapidly?
- Is the floor material shrinking?
- Are there slippery areas?
- Are there holes or unguarded areas?
- Are there indications of cracks, sagging, or warping?
- Are replacements necessary due to deterioration?
- Is shoring up, strengthening, resurfacing, or reattaching necessary?
- Is structural investigation indicated?

Stairs must be clear, sturdy, slip-resistant, and must have adequate safety rails, treads, and enclosures. They should never be used for storage. Stairways should be checked for the following:

- Are treads and risers in good condition and of uniform height and width?
- Are handrails secure and in good condition?
- Is lighting on stairs bright enough?
- Are access areas at the top and bottom open and free of clutter?
- Is the overall structure sturdy and strong enough for its purpose?

General housekeeping throughout the facility should be checked regularly. Aisles should be marked off with painted lines and kept clear; burned-out lights should be replaced promptly; locks, handles, door frames, appliances, and furnishings should be kept in good repair; tools stored in organized storage; and general cleanliness and good order maintained.

Fire is one of the greatest hazards, especially to industrial facilities. All fire protection equipment (including sprinkler systems, alarms, extinguishes, standpipes, and hoses) should be inspected often and kept in working order. All exits from the building and emergency lighting systems should be routinely inspected and tested. Fire protection equipment must be in the right place and not blocked or obscured.

Electrical installations should be in compliance with the *National Electrical Code,* ANSI/NFPA 70-1996, published by the National Fire Protection Association. In this book, refer also to Chapter 16, Electrical Safety.

A qualified expert should inspect chains, wire and fiber ropes, slings, and other equipment that is subject to severe strain in handling heavy equipment and materials. Careful records of each inspection should be maintained.

Work Practices. Another important purpose of a safety, health, and environmental inspection is to observe work practices. Are employees following specific safety, health, and environmental procedures and training instructions as they do their jobs? The best guide for determining how a job should be done is a job safety analysis.

The following questions highlight some work practices problems that should be inspected.

1. Are machines or tools being used without authorization?
2. Is equipment being operated at unsafe speeds?
3. Are guards or other safety devices being removed or rendered ineffective?
4. Are defective tools or equipment being used, or are tools or equipment being used unsafely?
5. Are employees using their hands or bodies instead of tools or push sticks to move items in hazardous situations?
6. Is overloading or crowding occurring? Are workers failing to store materials properly?
7. Are materials being handled in unsafe ways? For example, are employees lifting loads improperly?
8. Are employees repairing or adjusting equipment while it is in motion, under pressure, or electrically charged?
9. Are employees not using (or not using properly) PPE or other safety devices?
10. Are unsafe, unsanitary, or unhealthful conditions being created by workers' poor personal hygiene—

for example, smoking in unauthorized areas or using compressed air to clean their clothes?
11. Are employees standing or working under suspended loads, scaffolds, shafts, or open hatches?

Preparation

Inspections should be scheduled when there is a maximum opportunity to view operations and work practices with a minimum of interruptions. Although the areas and routes should be planned in advance, the day and time on which formal inspections are done should be varied, in order to check the widest possible variety of conditions.

It is a good idea to review all accidents that have occurred in the area in the past. The inspector should review copies of previous inspection reports. This review will reveal whether earlier recommendations to remove or correct hazards were followed.

Results and Reports

The supervisor or team leader should discuss the results of inspections with the employees. If poor work habits have been observed, the inspector should point them out to employees immediately and explain the correct ways to do the work.

Communicating good news as well as the bad is essential. Many supervisors forget to mention the positive actions and practices that take place in their departments. People who follow good work practices should be encouraged, and their success applauded, rather than ignored. Comments like, "I'm glad to see that you always check the condition of your tools before using them," can give workers positive reinforcement. An inspector who finds it necessary to correct someone's actions should be sure to acknowledge any resulting improvement in the person's work patterns.

A clearly written report must follow each formal inspection. The report should specify the department or areas inspected (identifying the boundaries or location, as needed) and the date and time of the inspection.

One way to begin the report is to copy hazards identified on the last report that have not been completely corrected, numbering each item consecutively. The inspector should specify the recommended corrective action, the date by which the hazard should be corrected, and the person responsible for fixing it.

The report should describe in detail each hazard uncovered during the current inspection. Machines and specific operations should be identified, and locations should be precisely described. Instead of merely noting "poor housekeeping," for example, the report should give clear details: "Empty pallets left in aisles, slippery spots on floor from oil leaks, a ladder lying across empty boxes, scrap piled on floor around machines." Instead of noting "guard missing," the report should read "guard missing on shear blade of No. 3 machine, SW corner of Bldg. D."

Follow-Up

The supervisor or team leader should follow up inspection reports to correct any problems that were discovered. If a problem can be resolved internally, workers should be assigned to fix it promptly. For example, if materials are not properly stacked, employees should be asked to restack them safely. If the condition cannot be corrected by team members, the leader should write a maintenance work-order request.

Resolving underlying causes of hazards is as important as their immediate correction. For example, wiping up an oil spill removes the immediate problem, but it does not eliminate the cause. If the oil leak came from a forklift truck, the truck must be identified and repaired, to prevent further oil spills.

If permanent correction of the problem will take time, the supervisor should consider temporary measures that will help to prevent an accident. Roping off the area, tagging or locking out equipment, and posting warning signs are examples of intermediate actions. They are not ideal, but they are steps in the right direction. They can prevent injuries or damage, while team members seek permanent solutions to the problem.

If the inspector discovers other dangerous conditions during follow-up work, they should be reported immediately to the appropriate person or management. The report should include recommendations for removing or correcting the conditions. If, for any reason, it is impossible to recommend the optimal, best-practice, long-term solutions, then intermediate or short-term solutions must be recommended, and identified as such. If identified, the best-practice solution should also be listed, and its advantage described, along with recommendation for its being applied as soon as possible, when conditions permit.

General categories into which recommendations might fall include the following:

- Devise a better process
- Relocate a process
- Redesign a tool or piece of equipment
- Provide personal protective equipment (PPE)
- Improve training
- Improve maintenance procedures

Key inspection results should be shared so that others can learn and avoid similar problems.

SUMMARY OF KEY POINTS

Key points covered in this chapter include:

- Safety, health, and environmental audits are key to continuous improvement. They usually cover everything from occupational safety and industrial hygiene to process and product safety, occupational health and medicine, and pollution control.
- An audit program can fulfill many different objectives, though most of them involve legal and regulatory compliance. Programs may differ in organizational, geographic, locational, functional, and compliance boundaries.
- The audit process involves planning, fieldwork (understanding systems, assessing controls, gathering and evaluating evidence), the audit report, and follow-up. Virtually all audits gather information, analyze facts, make judgments about the status of the facility, and report the results to management.
- When outside auditors are used, supervisors and employees should avoid viewing their reports as personal criticism or scrutiny of their individual work habits. They should understand that the purpose of the audit is to improve safety and health and to protect them, not punish them.
- The safety, health, and environmental inspection program should address the issues of how often inspections should occur, what items should be inspected, what aspects of each item need to be examined, what conditions should be inspected, and who will conduct the inspection. Supervisors or team leaders should conduct both continuous and periodic safety, health, and environmental inspections, as well as intermittent and general inspections. The timing of inspections should be staggered to check the widest possible variety of conditions.
- High-risk items should be inspected often. They include floors, stairways, exits and entrances, housekeeping procedures, fire hazards, electrical installations, and chains, ropes, and slings.
- The supervisor should discuss the results of inspections with employees, making sure to give positive reinforcement for good work practices. The written report should describe in detail each hazard uncovered during the inspection. Response to reports should include immediate follow-up to correct problems and resolve their underlying causes.

ACCIDENT
INVESTIGATION

7

After reading this chapter, you will be able to:

- Describe the purpose and objectives of accident reporting

- List and use the proper accident investigation steps

- Preserve evidence

- Select and interview accident witnesses

- Fill out an accident report

Accident investigation is an essential step towards making changes that can prevent recurrence of accidents from similar causes. If anything positive results from an accident, it is the opportunity to determine the causes and how to eliminate them. Thorough accident investigation can identify hazardous activities and conditions within an organization. By definition, this knowledge is gained after, rather than before, the occurrence of an accident.

Accident investigation involves more than merely filling out forms. Well-managed organizations insist on quality accident investigations, just as they insist on efficient, quality production. Part of the performance evaluation for a supervisor or team leader may be based on how well accident investigation is handled.

ACCIDENT REPORTING

A good accident investigation program is based on an effective, thorough accident notification plan. Many organizations make the mistake of requiring notification and investigation of only "serious" accidents. All incidents must be reported as soon as possible—whether they did or could have resulted in personal injuries, illnesses, or property damage. For example, a tool falling off a platform that is missing toeboards might not hurt anyone. However, it just as easily could have fallen on someone's head, inflicting a serious injury. Regardless of the outcome, this kind of incident should be reported, the causes investigated, and corrective action taken to prevent recurrence. When the causes are removed, accidents can be prevented. To help establish an effective accident reporting system, make sure the subject is covered in the new-employee orientation program.

Figure 7–1 shows that accidents are not just events that cause injuries or property damage but can be "near misses" that result in neither injuries nor damage. A supervisor or team leader has a personal interest in every incident that occurs in his or her department. In the interest of prevention, team leaders and supervisors must learn the contributing causes for every incident so they can be eliminated. (See a sample report form for near misses in Chapter 1, Safety Management, Figure 1–2.)

FINDING CAUSES

The purpose of accident investigation is to determine causes and recommend corrective actions to eliminate or control these hazards. Accident investigation should be aimed at fact-finding rather than fault-finding; otherwise, the investigation can do more harm than good. The emphasis should be on identifying all

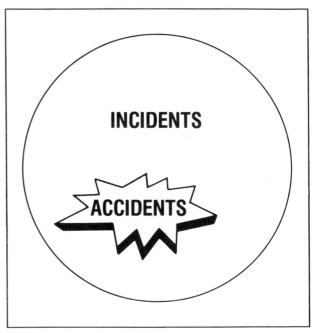

Figure 7–1. Accidents are a part of a group of events that adversely affect the completion of a task. All of the events in this group are incidents; those that do not result in injury and/or property damage are sometimes called "near misses," but they represent significant warnings of potential accidents.

possible causes, not on who could be blamed for the accident. An attempt to place blame damages the investigator's credibility and generally reduces the amount and accuracy of information received from workers. This does not mean oversights or mistakes should be ignored, nor that personal responsibility should not be determined, when appropriate. It means that the investigation should be concerned only with the facts. To do a quality job of investigating accidents, the investigator must be objective and analytical.

There are many factors that combine to cause accidents. The theory of multiple causation states that it is the random combination of various factors that results in accidents.

A report must do more than just identify unsafe acts or hazardous conditions surrounding an incident. These may be only contributing causes but not the root causes of a specific accident. For example, an accident report might state that the incident was caused by an oily patch on the floor. Instead of merely noting there was oil on the floor, the supervisor must determine how and why the oil got there. If there was an equipment lubrication leak, had the condition been previously reported? If poor maintenance was the cause, had that been reported? If so, when? If not, why not? These are the kinds of questions that must be asked and answered (i.e., who, what, where, when, how, and why).

Generally, the procedures followed during an investigation are designed to elicit clues to other problems. Inadequate maintenance, poorly designed equipment, untrained employees, and lack of policy enforcement or standard procedures (management control) are all causative factors (Figure 7–2).

Taking Immediate Action

The safety and health of employees and visitors must be the primary concern when an accident occurs. If an injury or illness results, make sure the affected persons get immediate medical attention and take steps to provide for emergency rescue or evacuation, as necessary. In addition, take any actions that will prevent or minimize the probability of further accidents happening as a result of the initial accident.

Securing the Accident Site

After rescue and damage control are completed, the accident site must be secured for the duration of the investigation. It is essential to barricade or isolate the scene with ropes, barrier tapes, cones, and/or flashing lights to warn people or otherwise restrict access to the area. In extreme cases, guards may be posted to make sure no unauthorized persons enter the scene.

Nothing should be removed from the site without approval of the person in charge of the investigation. The site should be maintained, as much as possible, just as it was at the time of the accident to help investigators identify and examine evidence.

Preserving Evidence

Time is critical when investigating accidents. The faster the investigator gets to the accident scene, the less chance there is that details will be lost. A prompt and careful investigation will help to answer the questions of who, what, where, when, how, and why. For example, witnesses will remember more facts, chemical spills will not have had time to evaporate, and dust and debris will remain as it was.

Preserving evidence at the accident scene ultimately makes the investigative process much less frustrating (Figure 7–3). Observing and recording evidence—such as instrument readings, control panel settings, and details of weather and other environmental conditions—can greatly improve investigation results. Evidence can be preserved on film, recorded on tape, diagramed, or sketched. Detailed notes must accompany any pictures or drawings.

Recording Visual Evidence

Visual evidence should be recorded by using cameras, videocameras, drawings, and notes. The camera is one of the most valuable tools for studying accidents. A self-developing camera is often preferred by those with no photographic experience. Photographs can be studied in detail and at length to look for things initially overlooked. General and specific scenes should be photographed in order to make comprehensive visual records. No one can predict in advance which data will be most useful, so take photographs from many different angles. An old saying for accident photographers is "overshoot and underprint." This means take every possible photo that might be needed, and make necessary enlargements after the proofs have been studied. In addition, accurate, complete, and annotated sketches and diagrams of the accident scene should be made.

Photographs of objects involved in the accident should be identified and measured to show the proper scale and perspective. A ruler or coin can be placed next to objects photographed close up to demonstrate the object's size. Accurate measurements of the area, equipment, and materials involved can be vital to accident investigations, with or without photographs.

EFFECTIVE USE OF WITNESSES

If witnesses are found and interviewed promptly, they can serve as the best information source about an accident. The following sections offer some useful tips for finding and interviewing witnesses.

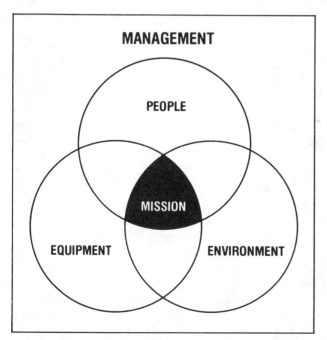

Figure 7–2. This diagram shows a basic system in which people, equipment, and the environment are managed to accomplish a mission. Sometimes they work together in unplanned ways to produce accidents.

Figure 7–3. The supervisor should record all details of the accident scene, including sketches of the scene, as soon after the accident as possible.

Identifying Witnesses

Avoid restricting the search for witnesses to those who saw the accident happen. Anyone who heard or knows something about the event may offer useful information. Ask witnesses to identify and document the names of others who were in the area, so that everyone can be contacted.

Interviewing Process

Witnesses should be interviewed one at a time in privacy, as soon after the accident as possible. The accuracy of people's recall is highest immediately following an event. A prompt interview minimizes the possibility of a witness subconsciously adjusting his or her story. If too much time elapses, many things can cloud a person's memory. For example, hearing other opinions or reading stories about the accident can influence a witness. A person with a vivid imagination can "remember" situations that did not actually occur. Make sure to set aside time at the beginning of the investigation to interview as many witnesses as you can.

Whenever possible, interviews should be conducted at the accident site. This gives witnesses opportunity to describe and point out what happened. Also, being at the scene can spark a person's memory. Tactful, skilled investigation will usually elicit greater cooperation from employees. Remind witnesses that you are interested in the facts of the accident, not in placing blame. Such reassurance can dispel people's fear that they may incriminate themselves or others.

How to Interview. When conducting a witness interview, establish a relaxed atmosphere. Be a good listener. Witnesses should be allowed to tell their stories without interruption or prompting. More detail can be sought after the full story has been told. Interruptions can derail a person's train of thought, influence answers, and inhibit responses. To elicit the most information, open-ended questions without leading words are best. For example, asking "Where was he standing?" is preferable to "Was he standing there?" Remember that the purpose is to determine the facts of the accident and to gain as much information from witnesses as possible. Questions should cover the specifics of who, what, where, when, how, and why.

Figure 7–4. Interview witnesses alone and in a private place. If a worker shows you how the accident happened, be sure power is off and all safety procedures are followed so that the accident cannot happen again.

Take notes and record employee statements, as unobtrusively as possible, for later review. In some cases, it may be best to wait until the employee has finished explaining what happened before making notes or recording details. If the employee's statement is recorded, get his or her permission and have it typed for the employee's review, corrections, and signature. After the witness is through, always repeat the information as it is understood. This feedback technique reduces misunderstandings and often leads to further clarifications.

Employees who have been directly involved with the accident should be contacted first, followed by eyewitnesses, and those who were nearby. Always use discretion in situations where employees have been injured. Fellow workers might be in shock, unable to speak coherently, or have no memory of the accident. Sometimes it is better to postpone interviews until the employee or a witness who has been involved is in a more stable physical or emotional condition. A part of the emergency should include

care for traumatized co-workers.

It is helpful to solicit ideas from the interviewed employees regarding ways to prevent a recurrence of the accident. Often they will have the best suggestions. This will also help them to feel involved in the investigation and provide an opportunity to recognize their willingness to participate. Be sure to record these as suggestions made before causes have been determined.

The interviews should end on a positive note. Thank each individual for his or her time, information supplied, and ideas offered.

Reenacting the Accident. The investigator may want to ask employees to show "what they mean or how it happened." In some cases, reenacting an accident can provide valuable information about how and why it occurred (Figure 7–4). Expert investigators have learned to use this technique with caution, however, so the reenactment doesn't cause another accident!

Before asking people to reenact the scene, follow these steps:

1. Ask employees to explain what happened first. This preliminary explanation should give additional insight about the accident.
2. Make sure that the witnesses thoroughly understand that they are to go through only the general motions of the event under close supervision. They should not repeat the exact actions that produced the original event.

Selecting a Location. When it is not possible to interview people at the accident site, choose another location free from distractions and away from other witnesses. Privacy is essential. When people are discussing what they saw take place, they can influence other witnesses. You may want to avoid having people revise their stories. Do not pressure or influence witnesses. If possible, do not use your office to conduct interviews; employees may find the supervisor's office intimidating, inhibiting, or distracting.

ACCIDENT INVESTIGATION REPORTS

The purpose of accident reporting is to alert and inform management and other concerned people about the circumstances surrounding an accident. The report should record in clear, concise language all appropriate facts of the accident. Write down all causal factors that might have led to the event. These factors will be assigned to one or more of the following causal categories: equipment, environment, personnel, and management.

Here are some typical questions to help you identify the accident causes: Was there a procedure in place to

ACCIDENT INVESTIGATION REPORT

Case Number

Company .... **Address** ....

Department .... **Location (if different from mailing address)** ....

---

1. Name of injured

2. Social Security Number

3. Sex ☐ M ☐ F

4. Age

5. Date of accident

---

6. Home address

....

7. Employee's usual occupation

8. Occupation at time of accident

9. Length of employment

☐ Less than 1 mo.   ☐ 1-5 mo.

☐ 6 mo. - 5 yr   ☐ More than 5 yr.

10. Time in occup. at time of accident

☐ Less than 1 mo.   ☐ 1-5 mo.

☐ 6 mo. - 5 yr   ☐ More than 5 yr.

11. Employment category

☐ Regular, full-time   ☐ Regular, part-time

☐ Temporary   ☐ Seasonal   ☐ Non-Employee

12. Case numbers and names of others injured in same accident

....

13. Nature of injury and part of body

---

14. Name and address of physician

....

15. Name and address of hospital

....

16. Time of injury

A. _____ a.m.
        p.m.

B. Time within shift

C. Type of shift

17. Severity of injury

☐ Fatality

☐ Lost workdays—days away from work

☐ Lost workdays—days of restricted activity

☐ Medical treatment

☐ First aid

☐ Other, specify_____

---

18. Specific location of accident

....

On employer's premises?   ☐ Yes   ☐ No

19. Phase of employee's workday at time of injury

☐ During rest period   ☐ Entering or leaving plant

☐ During meal period   ☐ Performing work duties

☐ Working overtime   ☐ Other _____

---

20. Describe how the accident occurred

---

21. Accident sequence. Describe in reverse order of occurrence events preceding the injury and accident.
Starting with the injury and moving backward in time, reconstruct the sequence of events that led to the injury.

A. Injury event

B. Accident event

C. Preceding event #1

D. Preceding event #2, 3, etc.

Figure 7–5. Complete an Accident Investigation Report after you have gathered all possible evidence. See the text for information on how to complete this form.

22. **Task and activity at time of accident**	23. **Posture of employee**

General type of task

Specific activity

Employee was working:

☐ Alone ☐ With crew or fellow worker ☐ Other. specify

24. **Supervision at time of accident**

☐ Directly supervised ☐ Indirectly supervised
☐ Not supervised ☐ Supervision not feasible

25. **Causal factors.** Events and conditions that contributed to the accident.
Include those identified by use of the Guide for Identifying Causal Factors and Corrective Actions.

26. **Corrective actions.** Those that have been, or will be, taken to prevent recurrence.
Include those identified by use of the Guide for Identifying Causal Factors and Corrective Actions.

Prepared by	**Approved**	
Title	**Title**	**Date**
Department Date	**Approved**	
Developed by the National Safety Council ©**1995 National Safety Council**	**Title**	**Date**

Figure 7–5. (Continued)

detect the hazardous condition? Was the correct equipment, material, or tool readily available and in safe condition? If so, was it used according to established procedure?

Responsibility for carrying out recommended corrective actions and a timetable for their completion should be established. Completion reports should be required as a means for assuring the recommended actions have been carried out.

Filling Out a Report

The following instructions apply to using the Accident Investigation Report shown in Figure 7–5.

This report is designed primarily for investigation of accidents involving injuries. However, it also can be used to investigate occupational illnesses arising from a single exposure (for example, dermatitis caused by splashed solvent or a respiratory condition caused by the release of a toxic gas). In property damage accidents, simply write "D.N.A." (does not apply) across Items 1 through 17 and fill out the remainder of the form.

All questions on this form should be answered. If no answer is available, or the question does not apply, the investigator should indicate this on the form. Answers should be complete and specific. Supplementary sheets can be used for other information, such as drawings and sketches, and should be attached to the report. A separate form should be completed for each employee who is injured in a multiple-injury accident.

The report form meets the record keeping requirements specified in OSHA Form 101. The individual entries are explained below.

Department. Enter the department or other local identification of the work area to which the injured is assigned (for example, maintenance shop or shipping room). In some cases, this may not be the area in which the accident occurred.

Location. Enter the location where the accident occurred if different from the employer's mailing address.

1. Name of injured. Record the last name, first name, and middle initial.
2. Social Security number.
3. Sex.
4. Age. Record current age.
5. Date of accident. Or date of initial diagnosis of illness.
6. Home address.
7. Employee's usual occupation. Give the occupation to which the employee is normally assigned (for example, assembler, lathe operator, clerk, etc.).
8. Occupation at time of accident. Indicate occupation in which the injured is working at the time of

accident, if other than his or her usual occupation.
9. Length of employment. Check appropriate box to indicate how long the employee has worked for the organization.
10. Time in occupation at time of accident. Record the total time employee has worked in occupation indicated in Item 8.
11. Employment category. Indicate injured's employment category at the time of accident (for example, regular, part time, temporary, or seasonal).
12. Case numbers and names of others injured in same accident. For reference purposes, names and case numbers of all others injured in the same accident should be recorded here.
13. Nature of injury and part of body. Describe exactly the kind of injury, or injuries, resulting from the accident, and the part, or parts, of the body affected. For an occupational illness, give the diagnosis and the body part, or parts, affected.
14. Name and address of physician.
15. Name and address of hospital.
16. Time of injury. In part B, indicate as closely as possible the hour and minute that the injury-producing accident occurred.
17. Severity of injury. Check the highest degree of severity of injury. The options are listed in decreasing order of severity.
18. Specific location of accident. Indicate whether accident or exposure occurred on employer's premises. Record the exact location of the accident (for example, at the feed end of No. 2 assembly line, at support column No. 32 in the warehouse, or in the employees' lunch room). Attach a diagram or map if it would help to identify the exact location.
19. Phase of employee's workday at time of injury. Indicate what phase of the workday the employee was in when the accident occurred, in which hour of the shift the injury occurred (for example, first hour). In part C, record the type of shift (for example, rotating or straight day). If "other," be specific.
20. Describe what happened. What physical things and procedures were involved? Provide a complete, specific description of what happened. Tell what the injured and others involved were doing prior to the accident; what relevant events preceded the accident; what objects or substances, operations, equipment, tools, etc., were involved; how injury occurred and the specific object or substance that inflicted the injury; and what, if anything, happened after the accident. Include only facts obtained in the investigation. Do not record opinions or place blame.
21. Accident sequence. Provide a breakdown of sequence of events leading to the injury. This

breakdown enables the investigator to identify additional areas where corrective action may be taken.

In most accidents, the accident event and the injury event are different. For example, suppose a bursting steam line causes burns to an employee's hands or a chip of metal strikes an employee's face during a grinding operation. In these cases the accident event—the steam line bursting or the metal chip flying up—is separate from the injury event—the steam burning the employee's hands or the chip cutting the employee's face. The question is designed to draw out this distinction and to record other events that led to the accident event.

There also may be events preceding the accident event that, although not accident events themselves, contributed to the accident. These preceding events can take one of two forms. They can be something that happened that should not have happened, or something that did not happen that should have happened. The steam line, for example, may have burst because of excess pressure in the line (preceding event #1). The pressure relief valve may have been corroded shut, preventing the safe release of excess pressure (preceding event #2). The corrosion may not have been discovered and corrected because a regular valve inspection/test was not carried out (preceding event #3).

To determine whether a preceding event should be included in the accident investigation sequence, the investigator should ask whether its occurrence (if it should not have happened) or nonoccurrence (if it should have happened) affected the event sequence that produced the accident and brought about injury.

Take enough time to carefully think through the sequence of events leading to the injury and record them separately in the report. The information found in the Finding Causes section, earlier in this chapter, can be used to help identify management system defects that may have contributed to events in the accident sequence. By identifying such defects, management may help prevent other types of accidents in addition to the one under investigation. (See Item 25, below.)

For example, identifying failure to detect a faulty pressure relief valve would lead to a management review of all equipment inspection procedures. This review could prevent other accidents that might have resulted from failure to detect faulty equipment in the inspection process.

Additional sheets may be needed to list all of the events involved in the accident sequence.

22. Task and activity at time of accident. In parts A and B, first record the general type of task the employee was performing when the accident occurred (for example, pipe fitting, lathe maintenance, or operating a punch press). Then record the specific activity in which the employee was engaged when the accident occurred (for example, oiling shaft, bolting pipe flanges, or removing material from the press). In part C, check the appropriate box to indicate whether the injured employee was working alone, with a fellow worker, or with a crew.

23. Posture of employee. Record the injured's posture in relation to the surroundings at the time of the accident (for example, standing on a ladder, squatting under a conveyor, or standing at a machine).

24. Supervision at time of accident. Indicate whether, at the time of accident, the injured employee was directly supervised, indirectly supervised, or not supervised. If appropriate, indicate whether supervision was not feasible at the time.

25. Causal factors. Record causal factors (events and conditions that contributed to the accident) that were identified by use of the Finding Causes section earlier in this chapter, discussed under Item 21, above.

26. Corrective actions. Describe recommended corrective actions to be taken after the accident to prevent recurrence, including immediate temporary or interim actions (for example, removed oil from floor) and permanent actions (for example, repaired leaking oil line).

NOTE: Users may add other data to the form to fulfill local or corporate requirements. Types of data that might be added include information typical of a particular industry or organization. For example, an establishment regulated by the Mine Safety and Health Administration (MSHA) may wish to add a MSHA identification number, required training, and training received.

SUMMARY OF KEY POINTS

Key points covered in this chapter include:

- The purpose of accident investigation is to determine causes and recommend corrective actions to eliminate or control hazards and prevent recurrence of accidents. All incidents should be investigated, not only those that cause serious injury or property damage. The investigation should emphasize fact finding, not finding fault.

- An organization should set up a plan for responding to accidents. The plan should specify what should be done in case of an accident, list names and phone numbers of people to call, and assign specific responsibilities to all persons involved in the Emergency Response Plan.

- When an accident occurs, supervisors should first see that injured persons get immediate medical attention, then secure the accident site for the investigation duration. Evidence can be preserved on film, recorded on tape, diagrammed, or sketched.
- Witnesses are often the best source of information about an accident. Supervisors should speak with anyone who was in the accident area, not only with those who actually witnessed the event.
- Witnesses should be interviewed one at a time as soon after the accident as possible to ensure accurate recall. The supervisor may ask witnesses to reenact, with appropriate safeguards, how an accident happened. Interviews should be conducted in a sensitive manner and in a comfortable location to avoid intimidating witnesses or influencing their stories.
- Accident reports are designed to inform management and other concerned people about the circumstances surrounding an accident. Causal factors leading to an accident generally fall into one or more of the following categories: equipment, environment, personnel, and management. These reports must be filled out as completely and thoroughly as possible.

INDUSTRIAL HYGIENE

After reading this chapter, you will be able to:

- Describe various chemical, physical, ergonomic, or biological health hazards commonly encountered on the job

- Recognize environmental health hazards in one or more of these four categories in your work area

- List basic methods of controlling various harmful environmental hazards and stresses

- Describe the concept and purpose of exposure limits

- Assure that standard operating procedures are established to help ensure a safe, healthy work environment

In addition to safety responsibilities, supervisors—together with management and safety personnel—must make sure the work area is free from conditions that could be detrimental to health. Understanding the basics of industrial hygiene can assist the supervisor in providing a safe and healthy work environment. This chapter provides information to help you recognize environmental health hazards in your work area. You can then request assistance from industrial hygienists, who work with medical, safety, and engineering personnel to eliminate or safeguard against such hazards.

Industrial hygienists define their work as "the anticipation, recognition, evaluation, and control of environmental conditions that may have adverse effects on health, that may be uncomfortable or irritating, or that may have some undesired effect upon the ability of individuals to perform their normal work." It is possible to group these environmental conditions or hazards into four general categories—chemical, physical, ergonomic, and biological. Their effects are often systemic and latent. Each category is discussed in some detail in this chapter.

CHEMICAL HAZARDS

Chemical compounds in the form of solids, liquids, and gases may cause health problems by inhalation (breathing), skin absorption or contact, ingestion (eating or drinking), or injection.

Inhalation

The major route of entry for employee exposure to chemical compounds is through the inhalation of airborne contaminants. Contaminants inhaled into the lungs can be in the form of gases, vapors, and solid or particulate matter. Particulate matter includes dust, mists, fumes, fibers, smoke, or aerosols.

Absorption

Absorption through the skin can occur quite rapidly if the skin is punctured, cut, or abraded. Unfortunately, many compounds that exist either in liquid or gaseous form, or both, can be absorbed through intact skin. Some are absorbed through the hair follicles while others penetrate by dissolving into the fats and oils of the skin.

Examples of chemical compounds that can be hazardous by skin absorption are alkaloids; phenols; lead acetate; salts of lead, arsenic, and mercury; nitrobenzene; nitrotoluene; aniline; and nitroglycerine. Other harmful compounds include triorthocresylphosphate, tetraethyl lead, and parathion and related organic phosphates. Compounds such as toluene and xylene that are good solvents for fats may also be absorbed through the skin.

Ingestion

Ordinarily, people do not knowingly eat or drink harmful materials. However, toxic compounds capable of being absorbed from the gastrointestinal tract into the blood—for example, lead oxide—can create serious exposure problems if people working with these substances are allowed to eat, drink, apply cosmetics, or smoke in their work areas. Careful and thorough wash-ups should be required before eating or smoking, and at the end of every shift. When necessary, workers should change their clothes before leaving work to avoid contaminating their home environment or other clean work areas.

Physical Classification of Airborne Materials

Because inhalation of airborne compounds or materials is a common problem, the supervisor should know the physical classifications of these substances.

Dusts. These are solid particles generated by handling, crushing, grinding, rapid impact, detonation, and decrepitation (breaking apart by heating) of organic or inorganic materials, such as rock, ore, metal, coal, wood, and grain. Dust is a term used in industry to describe airborne solid particles that range in size from 0.1 to 25 μm (micrometers or microns) in diameter (μm = 1/10,000 cm = 1/25,000 in.; μm is the abbreviation for micrometer).

A person with normal eyesight can detect individual dust particles as small as 50 μm in diameter. Dust particles below 10 μm in diameter cannot be seen without a microscope. High concentrations of suspended small particles may look like haze or smoke.

Dusts settle to the ground under the influence of gravity. The larger the particle, the more quickly it settles. Particles larger than 10 μm in diameter settle quickly while those under 10 μm remain suspended in air for much longer. Particles less than 5 μm, called "respirable dusts," can penetrate into the inner recesses of the lungs. Nearly all larger particles are trapped in the nose, throat, trachea, or bronchi from which they are either expectorated or swallowed.

Some larger-sized particles can also cause difficulty, however. Ragweed pollen, which ranges from 18–25 μm in diameter, can trigger an allergic reaction known as hay fever when particles enter the upper respiratory tract. Other allergenic dusts, as well as some bacterial and irritant dusts, can also cause respiratory problems in workers exposed to them.

Dust may enter the air from various sources. It may be dispersed when a dusty material is handled—for example, when lead oxide is dumped into a mixer, when talc is dusted on a product, or where asbestos-containing acoustical and/or fireproofing materials are being removed (Figure 8–1). When solid materials are reduced to small sizes in such processes as grinding, crushing, blasting, shaking, and drilling, the mechanical action of the grinding or shaking device can disperse the dust formed. When dusty materials are handled or transported, dust may be dispersed to other facility areas, unless good controls are used.

Fumes. A fume is formed when volatilized solids, such as metals, condense in cool air. The solid particles that make up a fume are extremely fine, usually less that 1.0 μm in diameter. In most cases, hot material reacts with air to form an oxide. Examples are lead oxide fume from smelting and iron oxide fume from arc welding.

A fume also can be formed when a material such as magnesium metal is burned or when welding or gas cutting is done on galvanized metal. Gases and vapors are not fumes, though they are often incorrectly called that.

Smoke. Smoke is made up partly of carbon or soot particles less than 0.1 μm in size that result from incomplete combustion of such carbonaceous materials as coal or oil. Smoke generally contains liquid droplets as well as dry particles. Tobacco, for instance, produces a wet smoke composed of minute tarry droplets. The particle size in tobacco smoke is about 0.25 μm in diameter.

Aerosols. Liquid droplets or solid particles fine enough to be dispersed and to remain airborne for a prolonged period of time are called aerosols. If inhaled, some of these may irritate or injure workers' mucus membranes, eyes, noses, throats, and lungs.

Mists. Mists are suspended liquid droplets generated by chemicals condensing from the gaseous

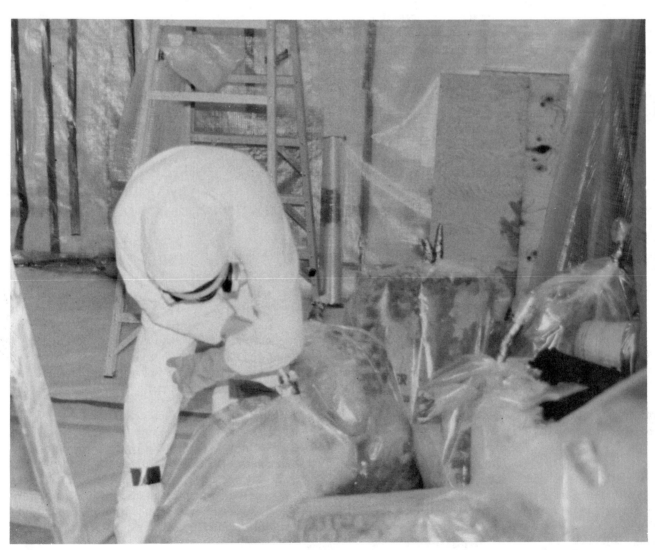

Figure 8–1. To prevent dispersion of asbestos-containing dusts, a worker in an asbestos removal enclosure double-bags all materials, taping each bag securely before disposal. Note that the worker is wearing protective clothing and a filter respirator.

to the liquid state, or by a liquid breaking into a dispersed state by splashing, foaming, or atomizing. Mist is formed when a finely divided liquid is suspended in the atmosphere. Examples are the oil mist produced during cutting and grinding operations, acid mists from electroplating, acid or alkali mists from pickling operations, spray-paint mist from spraying operations, and condensation of water vapor into fog or rain.

Gases. Usually gases are formless fluids that occupy the space or enclosure in which they are confined. Gases can be changed to the liquid or solid state only by the combined effect of increased pressure and decreased temperature. Gases spread out, or diffuse, into the surrounding atmosphere easily and readily. Examples are welding gases, internal combustion engine exhaust gases, and air.

Vapors. Gaseous forms of substances that are normally in solid or liquid state (at room temperature and pressure) are called "vapors." Vapors can be changed back to the solid or liquid state either by increasing the pressure or by decreasing the temperature. Evaporation is the process by which a liquid is changed into the vapor state and mixed with the surrounding atmosphere. Solvents with low boiling points—for example, acetone—will volatilize (evaporate) readily at room temperature.

Hazards Involved. The hazard associated with breathing a gas, vapor, or mist usually depends upon the solubility of the substance. For example, if the compound is very soluble—such as ammonia, formaldehyde, sulfuric acid, or hydrochloric acid—it is rapidly absorbed in the upper respiratory tract and does not penetrate deeply into the lungs. Consequently, the nose and throat become so irritated that a person is driven out of the exposure area before he or she is in much danger of toxicity from the chemical. (Toxicity is the ability of a substance to produce disease or physical harm.) Nevertheless, exposures even for brief periods to high concentrations of these compounds can produce serious health effects.

Compounds that are not soluble in body fluids can penetrate deeply into the lungs before the person can sense exposure. Thus, a serious hazard can be present but not be immediately recognized. Examples of such gases are nitrogen dioxide and phosgene. The immediate danger from these compounds in high concentrations is acute edema (abnormal accumulation of serous fluid in body tissues or serous cavities) or, possibly, pneumonia or circulatory impairment.

However, numerous chemical compounds do not follow the general solubility rule. They are not especially soluble in water and yet are irritating to the eyes and respiratory tract. They can cause lung damage and even death under the right conditions. One example is acrolein, which belongs to the chemical family called aldehydes.

Particulates

To evaluate particulate exposures properly, you must know the chemical composition, particle size, dust concentration in air, method of dispersion, and many other factors described in this section. With the exception of certain fibrous materials, dust particles must usually be smaller than 5 μm to enter the alveoli or inner recesses of the lungs. Although a few particles up to 10 μm in size may enter the lungs, nearly all larger particles are trapped in the nasal passages, throat, larynx, trachea, and bronchi, from which they are expectorated, or swallowed, entering the digestive tract.

Dusts. Most industrial dusts consist of particles that vary widely in size, with small particles greatly outnumbering large ones. Consequently, with few exceptions, when dust can be seen in the air around an operation, there are probably more invisible than visible particles present. The main hazard to personnel occurs when dust becomes airborne. Also, airborne dusts can be flammable and potentially explosive.

A number of occupational diseases result from exposures to such nonmetallic dusts as asbestos. Although specialized knowledge and instruments are needed to determine the severity of a hazard, you should be trained to recognize a hazard and to ask for expert help in controlling it.

A process that produces dust fine enough to remain suspended in the air should be regarded as hazardous until proven safe. An air-monitoring survey of airborne chemicals present will determine employee exposure levels and the overall relative safety. Processes that can generate excessive exposure include:

- Abrasive blasting or machining
- Bagging and handling of dry materials, i.e., grains, cement
- Ceramic coating
- Dry mixing
- Metal forming
- Grinding

Consider, for example, free silica, which can cause a damaging lung disease, silicosis. Silica dust is produced in hard-rock mines, and by quarrying and dressing granite. Grinding castings that contain mold sand also represent a silica hazard.

Methods of drilling rock with power machinery produce more dust than old-fashioned hand methods. This dust, however, is controlled by applying water to the drill bit so that the dust forms a slurry instead of being suspended in air.

Processes in which materials are crushed, ground, or transported are also potential sources of dust. They

Figure 8–2. This exhaust hood over the die head of a plastics extruder effectively reduces harmful fumes and gases close to the source.

should either be controlled by use of wet methods or should be enclosed and ventilated by local exhaust. Points where conveyors are loaded or discharged, transfer points along the conveying system, and heads or boots of elevators should be enclosed and, usually, exhaust-ventilated.

Built-in dust controls must be checked periodically to see that no one cancels their effectiveness by tampering with or improperly using them. Required respiratory equipment must be worn by trained workers who need supplementary protection.

Fumes. Welding and other hot operations produce fumes, which may be harmful under certain conditions. For example, arc welding volatizes metal that then condenses—as the metal or its oxide—in the air around the arc. In addition, the rod coating is in part volatized. Because they are extremely fine, these fumes are readily inhaled.

Highly toxic fumes, such as those formed when welding galvanized metal or structures painted with red lead, may rapidly produce severe symptoms of toxicity. Fumes should be controlled by using good local exhaust ventilation or by protecting the welder with approved respiratory protection equipment (Figure 8–2).

When brass is poured, zinc volatizes from the molten mass and oxidizes in air to produce zinc fume, which (in high concentrations) may produce the disease known as "metal fume fever." If lead is present, it will also become airborne. As a result, brass foundry workers may incur either chronic lead poisoning if lead is present in minute amounts, or acute poisoning if it is present in substantial amounts.

Most soldering operations, fortunately, do not require temperatures high enough to volatize an appreciable amount of lead. However, some lead in the molten solder is oxidized by contact with air at the surface. If this oxide, often called dross, is mechanically dispersed into the air, it may produce a severe lead-poisoning hazard.

In operations where this condition may happen—for example, soldering or lead battery making—preventing occupational poisoning depends upon scrupulous housekeeping to prevent the lead oxide from dispersing into the air. You can enclose melting pots, dross boxes, and similar operations, and provide exhaust ventilation to capture airborne contaminants.

Gases. Gases which are used or generated in many industrial processes may be toxic. For example, welding in the presence of chlorinated solvent vapors can produce phosgene, a very toxic gas that causes respiratory distress and damage. Propane-fueled forklift trucks or any process or equipment that burns fuel or other organic materials has the potential to generate carbon monoxide. Carbon monoxide prevents the body from absorbing oxygen, causing headache, nausea, confusion, dizziness, and in severe cases, coma and death. In addition, carbon monoxide and other toxic gases like hydrogen sulfide and methane may be found in confined spaces where organic substances have deteriorated.

Assessing Respiratory Hazards. When a respiratory hazard exists or is suspected, the actual airborne concentration of the contaminant(s) must be measured by an industrial hygienist. While conditions are sometimes similar in different areas within one facility or in different facilities within an industry, the degree of respiratory hazard must be assessed in each case by scientifically valid methods, such as air sampling. Some typical industrial processes and the types of air contamination generated are listed in Table 8–A.

You must also be aware of the hazards of oxygen deficiency. Oxygen deficiency results when work area air contains less than the normal amount of oxygen found in the atmosphere, about 21%. OSHA regulations define immediately dangerous to life or health (IDLH) as starting at 19.5% for confined space.

You should sample the air for toxic chemicals and test for oxygen and flammable gases before allowing workers to enter any confined space. Many gases are heavier than air and thus may remain present even in confined spaces that are open to the atmosphere. Many gases are odorless and colorless, which makes detection unlikely unless appropriate air-sampling equipment is used.

Air sampling may be necessary around workers and process equipment to make certain that hazard controls are working. For example, you may need to sample for ozone and other hazardous gases to document a welder's exposure. Industrial hygiene, safety, or engineering personnel should provide you with information on gases that might be used or generated in your work areas.

The degree of hazard involved is an important factor when analyzing respiratory conditions. Some hazards, such as gases and vapors, can produce an immediate threat to life and health when present in high concentrations; and oxygen deficiency, by its very nature, is automatically dangerous to life and health.

Respiratory Protection

It is important to understand the basic concepts of respiratory protection to adequately protect your workers. A competent industrial hygienist or safety professional should decide what kind of respiratory protection is needed. However, the more you know about respiratory hazards, and the types and selection of respirators, the better prepared you will be to make sure that employees are protected. Selection of the proper respirator is discussed in Chapter 9, Personal Protective Equipment. See the National Safety Council's Data Sheet 734, *Respiratory Protective Equipment,* available from the NSC library.

Liquid Chemicals: Solvents

Liquid chemicals typically being used include fuel or fuel additives, pesticides, lubricants, detergents and cleaning agents, or degreasing or processing solvents. Solvents are perhaps the most widespread class of chemicals in manufacturing. Many of these solvents evaporate readily into the air; therefore, their use can pose immediate exposure problems.

Solvents can be further categorized as aqueous or organic. Aqueous solvents are those that readily dissolve in water. Many acids, alkalis, or detergents, when mixed with water, form aqueous solvent systems.

The term *solvent,* however, is commonly used to mean organic solvents. Many of these chemicals do not mix easily with water but dissolve other organic materials such as greases, oils, and fats. Important types and specific examples of organic solvents include aliphatic, cyclic, aromatic, halogenated, esters, ketones, alcohols, ethers, glycols, aldehydes, freons; and hexane, gasoline, turpentine, benzene, trichloroethylene, ethyl acetate, acetone, formaldehyde, methanol, ethyl ether, and ethylene glycol.

Organic solvents generally have some effect on the central nervous system. With short-term (acute) exposures, they may cause nervous system depression, in

Table 8–A. Potentially Hazardous Operations and Air Contaminants

Process Types	Contaminant Type	Contaminant Examples
Hot Operations		
Welding	Gases	Chromates
Chemical reactions	Particulates(p)	Zinc and compounds(p)
Soldering	(dusts, fumes, mists)	Manganese and compounds(p)
Melting		Melting oxides(p)
Molding		Carbon monoxide(g)
Burning		Ozone(g)
		Cadmium oxide(p)
		Fluorides (p)
		Lead (p)
		Vinyl chloride (g)
Liquid Operations		
Painting	Vapors(v)	Benzene (v)
Degreasing	Gases (g)	Trichloroethylene (v)
Dipping	Mists (m)	Methylene chloride (v)
Spraying		Hydrochloric acid (m)
Coating		Sulfuric acid (m)
Etching		Hydrogen chloride (g
Cleaning		Cyanide salts (m)
Dry cleaning		Chromic acid (m)
Pickling		Hydrogen cyanide (g)
Plating		TDI, MDI (v)
Mixing		Hydrogen sulfide (g)
Galvanizing		Sulfur dioxide (g)
Chemical reactions		Carbon tetrachloride (v)
Solid Operations		
Pouring	Dusts(d)	Cement
Mixing		Fibrous glass
Separations		
Extraction		
Crushing		
Conveying		
Loading		
Bagging		
Pressurized Spraying		
Cleaning parts	Vapors(v)	Organic solvents(v)
Applying pesticides	Dusts(d)	Chlordane(m)
Degreasing	Mists(m)	Parathion(m)
Sand blasting		1,1,1-trichloroethylene (v)
Painting		Methylene chloride(v)
		Quartz (free silica)
Shaping Operations		
Cutting	Dusts(d)	Asbestos
Grinding		Beryllium
Filing		Uranium
Milling		Zinc
Molding		Lead
Sawing		
Drilling		

(Reprinted with permission from *Occupational Exposure Sampling Strategy Manual,* NIOSH Pub. No. 77–172.)
Abbreviations: d = dust, g = gases, m = mists, p = particulates, v = vapors.

which the victim experiences dizziness, feelings of intoxication and nausea, and a decrease in muscular coordination. Higher levels or more prolonged exposure may cause loss of consciousness, coma, and, in some cases, death.

Other effects of solvents vary. Some can cause long-term damage to the liver or other organs, or affect the worker's reproductive ability. A few have been found to cause cancer. Some research suggests that long-term (chronic) exposure to low levels of certain solvents may cause long-lasting effects, including chronic headaches, inability to concentrate, memory loss, and reduced nervous system and muscular functioning.

As the supervisor, you can assist the industrial hygienist and safety professional in controlling exposure to such chemicals. Substantial exposures, fortunately, can be controlled. Spray-painting booths can be ventilated and degreasing tanks can be exhausted. These will not ordinarily present serious hazards as long as the equipment works. It is the job of the supervisor, engineering, human resources, and safety personnel to make sure that controls and personal protective equipment (PPE) are properly maintained and used at all times. Even small exposures on jobs that come up infrequently or that involve small amounts of solvents not covered by standard operating procedures can produce significant exposure hazards.

What is most important is not how much solvent is used at the job site, but the actual degree of exposure by inhalation or skin absorption. A close check must be kept on all minor, as well as major, uses of solvents. These chemicals should be issued only after determining that they can be used properly and safely. Industrial hygienists or safety professionals can provide information about which solvents to use and for what purposes. The supervisor or team leader is responsible for seeing that this information is used and that employees follow the recommendations.

Selection and Handling. Getting the job done without hazard to employees or property is dependent upon the proper selection, application, handling, and control of solvents and an understanding of their properties.

One problem encountered with chemical selection is the similarity of names for the various kinds of solvents in the chlorinated hydrocarbon family. Perchloroethylene, trichloroethylene, trichloromethane, and dichloromethane are a few examples of such solvents that have similar-sounding names, yet possess characteristic hazards.

Hazard Communication. Many state regulations and the federal Hazard Communication Standard (29 *CFR* 1910.1200) require that management provide information about chemical hazards to the workforce.

The role of a supervisor includes making sure this information is provided to employees. Many of these regulations require:

- An inventory and assessment of chemical hazards in the workplace
- Development and use of labels that describe chemical hazards and protective measures to use
- Material Safety Data Sheets (MSDSs) that detail chemical hazard and precaution information
- Training on identifying hazards, including specific chemicals or groups of chemicals with which employees work, reading labels, and understanding MSDSs (Figure 8–3).
- A written program that describes how the company intends to accomplish these tasks and provides documentation that workers have been trained

Many companies have developed programs using in-house or outside experts to inform employees about general classes of chemicals and chemical hazards. Supervisors and other departmental personnel are often expected to provide specific information on the chemical hazards found in the department, and measures that should be used to control them. Such measures could include engineering controls (ventilation or isolation of a hazardous substance), work-practice controls such as housekeeping programs, or PPE programs. The company should train supervisors to recognize the hazards and to tell their workers how they can best protect themselves.

If the supervisor finds a new chemical for which training has not been provided, or which lacks labels, the supervisor will need to discuss it with the industrial hygienist, safety professional, or chemist, as well as with the administrator of the hazard communication program. They may also need to contact purchasing to request notifying the supplier that an appropriate label and MSDS are needed.

Degree of Hazard Severity. The severity of hazard in the use of organic solvents depends on the following facts:

- How the solvent is used
- How much solvent is used
- Type of job operation (which determines how the workers are exposed)
- Work pattern
- Duration of exposure
- Operating temperature
- Exposed liquid surface
- Ventilation efficiency
- Evaporation rate of solvent
- Pattern of air flow
- Concentration of vapor in workroom air
- Housekeeping

The solvent hazard, therefore, is determined not only by the toxicity of the solvent itself, but by the conditions of its use—who, what, when, where, and how long. Precautionary labeling and/or the MSDS should indicate the major hazards and safeguards (Figure 8–3).

For convenience, job operations employing solvents may be divided into three categories:

1. *Direct contact.* This occurs when skin, usually the hands, contacts the solvent directly. Emergency repair of equipment, spraying or packaging volatile materials without ventilation, cleanup of spills, and manual cleaning using cloths or brushes wetted with solvent are examples of jobs in which employees may have direct contact with the solvent.

   Remember that many solvents can penetrate the skin and be absorbed into the body. Solvents are a leading cause of industrial dermatitis or skin disease. It also is important to note that many solvents can penetrate various glove materials. Thus, properly selected chemical resistant gloves and clothing must be used to prevent solvents from soaking through and contacting the skin.

2. *Intermittent or infrequent contact.* This occurs when the solvent is contained in a semi-closed system where exposure can be controlled. Examples are: paint spraying in an exhaust-ventilated spray booth, vapor degreasing in a tank with local lateral slot exhaust ventilation, and charging reactors or kettles in a batch-type operation in which the worker is exposed only at infrequent intervals.

3. *Minimal contact.* This is characterized by remote operation of equipment totally isolated from the work area. This type of operation includes directing chemical facility operations from a control room, mechanical handling of bulk-packaged materials, and other operations where the solvent is contained in a closed system and is not discharged to the atmosphere in the work area.

It is unfortunate that the term "safety" solvent has been applied to some proprietary cold degreasers, because the term is not precise and is subject to various interpretations. For example, a "safety" solvent may be considered by some users as nondamaging to surfaces being cleaned. Other users may consider it to be free from fire or toxicity hazards. Depending on conditions of use, neither of these criteria can be met by a so-called "safety" solvent.

These solvents are prepared as mixtures of halogenated hydrocarbons and petroleum hydrocarbons to be used for cold degreasing. The halogenated hydrocarbons are effective grease and oil solvents and generally are not flammable. They also are relatively expensive and may be toxic under adverse conditions. The petroleum hydrocarbons are effective solvents

Figure 8–3. An MSDS must list the potential hazards, required safeguards, and emergency actions.

METHANOL

┤ POTENTIAL HAZARDS ├

HEALTH HAZARDS
Poisonous; may be fatal if inhaled, swallowed or absorbed through skin.
Contact may cause burns to skin and eyes.
Runoff from fire control or dilution water may cause pollution.

FIRE OR EXPLOSION
Flammable/combustible material; may be ignited by heat, sparks or flames.
Vapors may travel to a source of ignition and flash back.
Container may explode in heat of fire.
Vapor explosion and poison hazard indoors, outdoors or in sewers.
Runoff to sewer may create fire or explosion hazard.

┤ EMERGENCY ACTION ├

Keep unnecessary people away; isolate hazard area and deny entry.
Stay upwind; keep out of low areas.
Self-contained breathing apparatus and chemical protective clothing which is specifically recommended by the shipper or producer may be worn but they do not provide thermal protection unless it is stated by the clothing manufacturer. Structural firefighter's protective clothing is not effective with these materials.
Isolate for 1/2 mile in all directions if tank car or truck is involved in fire.
CALL CHEMTREC AT 1-800-424-9300 FOR EMERGENCY ASSISTANCE. If water pollution occurs, notify the appropriate authorities.

FIRE
Small Fires: Dry chemical, CO_2, Halon, water spray or standard foam.
Large Fires: Water spray, fog or standard foam is recommended.
Move container from fire area if you can do it without risk.
Dike fire control water for later disposal; do not scatter the material.
Cool containers that are exposed to flames with water from the side until well after fire is out. Stay away from ends of tanks.

Withdraw immediately in case of rising sound from venting safety device or any discoloration of tank due to fire.

SPILL OR LEAK
Shut off ignition sources; no flares, smoking or flames in hazard area.
Do not touch spilled material; stop leak if you can do it without risk.
Water spray to reduce vapors; but it may not prevent ignition in closed spaces.
Small Spills: Take up with sand or other noncombustible absorbent material and place into containers for later disposal.
Large Spills: Dike far ahead of spill for later disposal.

FIRST AID
Move victim to fresh air and call emergency medical care; if not breathing, give artificial respiration; if breathing is difficult, give oxygen.
Remove and isolate contaminated clothing and shoes at the site.
In case of contact with material, immediately flush skin or eyes with running water for at least 15 minutes.
Keep victim quiet and maintain normal body temperature.
Effects may be delayed; keep victim under observation.

W.H. BRADY CO., SIGNMARK® DIV. CATALOG NO. 93565

and are inexpensive, but they are flammable (flash points below 100 F) and can also be toxic.

To combine the best qualities of each solvent, manufacturers mix them in an attempt to produce a cold degreaser that has a flash point higher than that of the flammable petroleum hydrocarbon alone. Such a mixture, however, can present both fire and toxicity hazards, depending on the evaporation rates of the solvents used. If the flammable liquid is more volatile than the nonflammable solvent, the vapors from the mixture can be highly flammable. Conversely, if the nonflammable solvent is more volatile, it can evaporate to leave a flammable liquid.

Supervisors must keep these considerations in mind when using such solvents. Never forget that, under the right conditions, the vapor from a "safety" solvent can be just as deadly and flammable as the vapor from an extremely toxic solvent. If there is a chance that workers might be exposed to these hazardous chemicals in the work area, have the air monitored for contaminant levels.

PHYSICAL HAZARDS

The second category of environmental factors or hazards involves physical agents. This category includes such hazards as noise, ionizing and nonionizing radiation (including visible radiation and lasers), temperature extremes, and pressure extremes. You should be alert to these hazards because they can have immediate or cumulative effects on employee health.

Noise

Noise (unwanted sound) is a form of vibration that can be conducted through solids, liquids, or gases. The effects of noise on people include:

- Psychological effects. Noise can startle, annoy, and disrupt concentration, sleep, or relaxation.
- Interference with verbal communication, and as a consequence, interference with job performance and safety.
- Physiological effects. Noise-induced hearing loss, aural pain, or even nausea (when exposure is severe). Some research links long-term overexposure to noise to circulatory problems and heart attack.

Noise Measurement. A source that emits sound waves produces minute changes in air pressure. We usually measure these pressure changes with sound-level meters and noise dosimeters. The human ear can hear sound over a wide range of pressures, with the ratio of highest to lowest pressures about 10,000,000 to 1.

To deal with the problem of this huge pressure range, scientists have developed the decibel (dB) scale, which is logarithmic. Decibels are not linear units like miles or pounds. Rather, they are points on a sharply rising curve. Thus, 10 decibels is 10 times greater than one decibel; 20 decibels is 100 times greater (10 x 10); 30 decibels is 1,000 times greater (10 x 10 x 10); and so on. The rustle of leaves is rated at 20 dB. Typical office noise has a level of about 50 dB. A vacuum cleaner runs at about 70 dB, while a typical milling machine from 4 ft (1.2 m) away is rated at 85 dB. The sound of a newspaper press is about 95 dB, a textile loom is 105 dB, a rock band is about 110 dB, a large chipping hammer is 120 dB, and a jet engine registers about 160 dB.

The pitch or frequency of sound is also important in noise measurement because the ear hears some frequencies better than others. Our ears tend not to hear very low or very high frequencies as well as those in the middle range. Thus, noise-measuring instruments are designed to mimic the response of the human ear on what is called the "A" weighting scale. As a result, standards for noise are written for the dBA scale. The OSHA Noise Standard requires that noise monitoring instruments be capable of measuring noise between 80 and 130 dBA. Noise above 90 dBA is to be avoided to prevent injury and unprotected exposure should not be permitted. Workers exposed to noise above 90 dBA should use hearing protection or have their exposure time reduced, as discussed below. Typically, if you have to shout to be heard you are in a noise zone.

Measurement of workplace noise can be quite complex and usually requires the services of a trained safety professional or industrial hygienist. They can make decisions regarding engineering controls or other methods to deal with the noise hazards. They will also furnish data to the company's hearing conversation program administrator, who can decide what types of hearing protection are needed. Details and definitions of noise, and its measurement and control, are given in the National Safety Council's *Fundamentals of Industrial Hygiene* and *Noise Control: A Guide for Employees and Employers*.

Factors in Hearing Loss. If the ear is subjected to high levels of noise for sufficient time, hearing loss may occur. A number of factors can influence noise exposure effects. Among these are:

- Variation in individual susceptibility
- Total energy (loudness) of the sound
- Frequency distribution of the sound
- Other characteristics of noise exposure, such as whether it is continuous, intermittent, or made up of a series of impacts
- Total daily duration of exposure
- Length of exposure in the noisy environment

Table 8–B. Permissible Exposures

Duration per Day Hours	Sound Level dB(A)*
8	90
6	92
4	95
3	97
2	100
1 1/2	102
1	105
3/4	107
1/2	110
1/4	115

*Sound level in decibels as measured on a standard level meter operating on the A-weighting network with slow meter response.

Criteria have been developed to protect workers against hearing loss. OSHA 29 *CFR* 1910.95(a) and (b), Occupational Noise Exposure, which sets allowable noise levels based on the number of hours of exposure. This standard limits unprotected exposure to eight hours at 90 dBA, four hours at 95 dBA, and so on (Table 8–B). Because noise may affect some workers' hearing more than others, it is good practice to require hearing protection when workers are exposed to levels above 85 dBA, regardless of length of exposure.

The OSHA standard requires employers to reduce noise exposures with administrative and engineering controls, where feasible. The standard also requires audiometric testing, training, initial monitoring, regular remonitoring, and remonitoring whenever changes in production, processes, or controls increase noise exposure.

Some individuals may experience hearing loss below 90 dBA. To further protect worker hearing, OSHA enforces its Hearing Conservation Amendment to the noise standard, which requires employers to:

- Provide annual audiometric tests to all employees exposed to noise over 85 dBA (for 8 hours)
- Offer optional hearing protection to workers exposed above 85 dBA and to make protection mandatory where noise exposures exceed 90 dBA (for 8 hours)
- Ensure that workers with existing hearing loss wear protection when exposed to noise levels above 85 dBA (for 8 hours)
- Provide additional training in:

  - Effects of noise on hearing
  - Proper selection, fitting, use, and care of hearing

protection, and advantages and disadvantages of each type
- Explanation of the purpose and methods of the hearing test

The regulation is very specific concerning the ways a hearing test is conducted. To be successful, the test requires the close cooperation of the supervisor. If possible, you should schedule employees so their hearing can be tested before they are exposed to noise on the job or at least 14 hours after exposure. If this is not possible and employees must be called off the job, make certain that workers selected for hearing tests wear ear plugs or muffs properly before their test. This will help to prevent confusing a temporary detrimental effect on hearing with a more serious long-term effect. Tests must be conducted in a noise-free environment, such as a testing booth (Figure 8–4).

Audiograms. The first audiogram is called the baseline audiogram, because it is the basis of comparison for subsequent yearly audiograms required to assess hearing. Employees must be notified if a significant shift in hearing is discovered.

As a supervisor, you may be asked to ensure that employees wear their hearing protection, both before audiometric testing and on a regular basis. One of the best ways to achieve employee cooperation is to set an example by wearing protection yourself wherever it is required. This is true also for managers, visitors, and other employees who visit noisy areas.

Hearing Protection. The most commonly used hearing protection equipment includes ear plugs, canal caps, and ear muffs. Ear plugs are available in preformed and disposable types. Many styles in both plastic and foam or other material are available. Ear plugs can also be custom-made to fit an individual employee's ear.

Canal caps are typically cone-shaped caps attached to a headband. They are most often used where the employee goes into noisy areas for brief periods, as they allow for easy insertion and removal.

Ear muffs fit over the entire outer ear. How effective they are depends on the tightness of the seal between the ear and muff. The earpiece of eyeglasses, for example, can create a significant noise "leak" if it passes between the muff and the wearer's skin.

Each type and brand of protection has been given a "Noise Reduction Rating" (NRR), which describes how much an employee's exposure can be reduced by wearing the device. Not all types offer the same protection. In general, custom-molded plugs offer the greatest NRR. Ear plugs offer the next best NRR, with muffs and caps offering somewhat less. It is important to note that not all brands of one type of protector offer the same noise reduction.

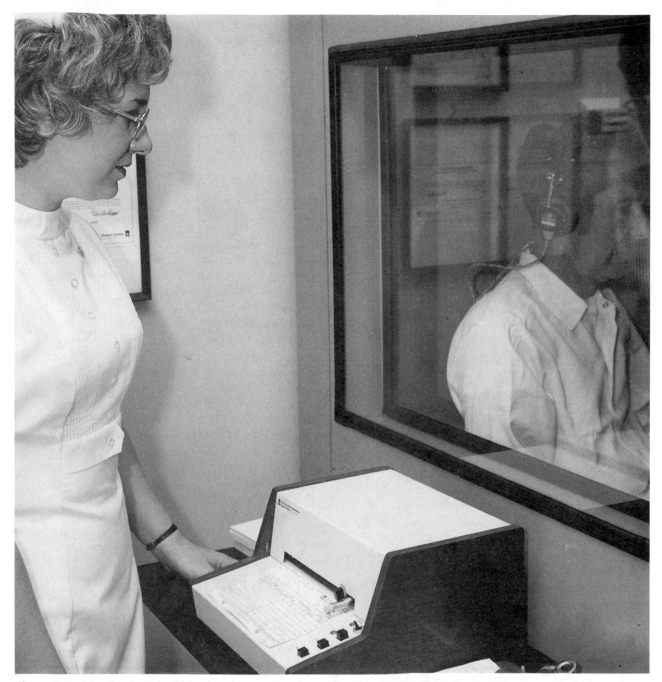

Figure 8–4. A hearing test should be done in an enclosed, soundproof booth by a trained technician.

Noise reduction ratings were developed in a laboratory setting. In the actual work situation, ear plugs, muffs, or caps may not provide the expected degree of protection. Therefore, many health and safety personnel select hearing protection devices that provide an extra margin of safety. They try to provide hearing protection that will reduce an employee's exposure below 85 dBA. (However, hearing protection often is not the sole answer to noise exposure. At sound levels over 90–100 dBA, engineering controls must be considered to reduce exposure.)

In selecting protection, the company and employee should also consider the advantages and disadvantages of the different types of hearing protection for their workplace. For example, it might not be appropriate for a mechanic or someone whose hands are routinely soiled by work to wear foam or other insert plugs. The workers will transfer grit and grime to the plug which can contribute to ear canal irritation and possible infection.

Hearing protectors must be comfortable when worn properly, or employees will likely find ways to avoid wearing them. Even ear plugs, which seem like simple devices, need to be seated properly each time they are worn. They must be inserted fully into the ear canal, not just "propped" at the opening. The administrator of the hearing conservation program should be responsible for ensuring that employees receive proper training and are comfortable with the hearing protection selected. It is up to the supervisor to ensure that employees wear hearing protection when needed. The supervisor must check for use of hearing protection when doing routine inspections, and also be a good role model by wearing the protection regularly. Another way to get employees to use their hearing protection is to get them involved in the selection process and get their "buy-in" on the protection chosen for their jobs.

Ionizing Radiation

Although ionizing radiation is a complex subject, a brief discussion is included here because it is one of the more dangerous physical agents. Radiation is a form of energy. Familiar forms of radiant energy include light, which enables us to see; and infrared, which we feel as heat. The relationship among the different forms of electromagnetic radiation is shown in Figure 8–5.

Ionizing radiation refers only to those types that involve ionization (a change from the normal electrical balance in an atom). Only the forms of radiation discussed below possess sufficient energy to change this electrical balance, which can produce damage in living cells. Other forms of radiation that are not ionizing may also cause damage, but usually to a lesser extent and by a different mechanism.

Certain chemicals, when exposed to radiation, may form hazardous compounds or more hazardous forms of an already-toxic substance. For example, when oxygen is electrified, ozone is formed. Ozone is highly irritating to nose, throat, and lung tissues. When working with chemicals and ionizing radiation, it is important to know what reactions between the two can take place.

Gamma Radiation. This type of radiation, produced from radioactive materials and X-radiation, is highly penetrating and can damage body tissues. The amount of damage depends, in part, on the energy of the radiation and on the relative sensitivity of the tissue.

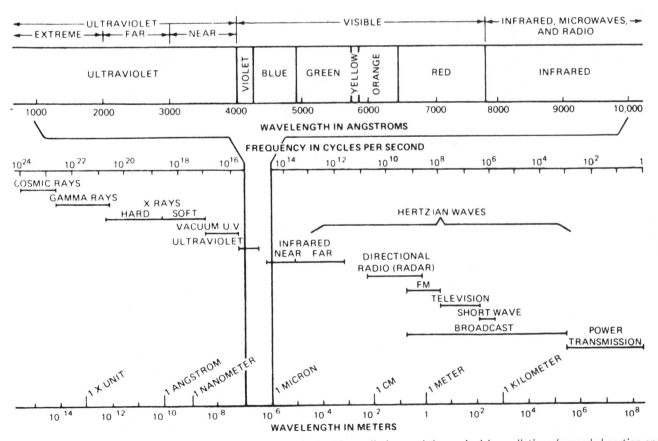

Figure 8–5. The electromagnetic spectrum, encompassing the ionizing radiations and the nonionizing radiations (expanded portion and right). Top portion expands spectrum between 10^{-7} and 10^{-8} m. Note: cycles per second (cps) = hertz (Hz).

Alpha and Beta Radiation. These types of ionizing radiation take the form of particles. Alpha particles are heavy and do not travel far in air. Although alpha particles are usually stopped by the skin, they can do considerable internal damage if they are inhaled or ingested. Beta particles are lighter and thus travel farther. If a beta source is left on the skin, it can damage the tissue. Proper shielding material is necessary to protect employees from beta radiation.

High-Energy Protons and Neutrons. In handling of some types of radioactive materials and in operation of high-powered particle accelerators, high-speed protons and high-speed and thermal neutrons can be created. A harmful dose is dependent on both the number of these particles and their energy distribution. Exposure potential must be evaluated by a health physicist (radiation specialist).

X-Radiation. This hazard is produced by high-potential electrical discharge in a vacuum. It should be anticipated in evacuated electrical apparatus operating at a potential of 10,000 volts or more. Although X-radiation is certain to be produced in such apparatus, it is dangerous only if it penetrates the apparatus jacket and enters the inhabited part of the workroom.

If a worker is overexposed to radiation of a comparatively low-voltage X-ray tube, dermatitis of the hand is generally the first result. It is characterized by rough, dry skin; a wartlike growth; and dry, brittle nails. Continued exposure and more penetrating X-rays can cause bone destruction.

Beta sources and X- and gamma radiation are found in some gages to measure the thickness of various materials, or in beta-source thickness gages. Sealed gamma sources and X-ray generators are used as various types of level gages, for example, to measure the liquid level inside a tank. They are also used to X-ray welds at joints. Gamma radiation is also used to sterilize some health care products.

These types of radiation cause injury by ionizing tissue in which they are absorbed. Such injuries differ widely in location and extent, but the determining factors are the ability of the particular radiation to penetrate through the tissues, and the amount of ionization produced by a given amount of radiation. Alpha radiation, for instance, is classed as 20 times more biologically active than X-rays because alpha particles are heavy, and can greatly disrupt and ionize tissue molecules with which they come in contact.

Radiation Protection. There are three general methods for preventing injuries from any penetrating radiation. First, workers can be separated from the hazard by distance sufficient to reduce radiation received to below the maximum permissible dose. Second, the time of exposure can be limited so workers will not receive a harmful dose. Third, specific shielding materials can be utilized to protect workers from exposure.

Rules for working with ionizing radiation are set by industrial hygienists or health physicists. They must be strictly followed at all times to avoid the chance of radiation injuries and illnesses.

Nonionizing Radiation

Electromagnetic radiation has varying effects on the body, depending largely on the particular wavelength of the radiation involved. Following, in approximate order of decreasing wavelength and increasing frequency, are some hazards associated with different regions of the nonionizing electromagnetic radiation spectrum.

Low Frequency. The longer wavelengths—including power frequencies, broadcast, radio, and short-wave radio—can produce general heating of the body. The health hazards of low-frequency radiation were previously though to be minimal. Now data suggest that exposure to radiation at electric power frequencies may be associated with adverse health effects, including cancer. Scientists have not yet discovered how or why these effects occur. Protection from electric field exposure can be provided by shielding and by limiting access to affected areas.

Microwaves. These are wavelengths of 3 m to 3 μm (100 to 100,000 megahertz, MHz). They are found in radar, communications, and diathermy applications. Microwave intensities may be sufficient to cause significant heating of tissues.

The effect is related to wavelength, power intensity, and time of exposure. Generally, longer wavelengths will produce a greater temperature increase in deeper tissues than will shorter wavelengths. How- ever, for a given power intensity, workers are less aware of the heat from longer wavelengths because it is absorbed beneath the skin.

An intolerable rise in body temperature, as well as localized damage, can result from exposure to wavelengths of sufficient intensity and time. In addition, flammable gases and vapors may ignite when they are inside metallic objects bombarded by a microwave beam.

Power intensities from microwaves are given in units of watts per square centimeter (w/cm^2). Areas with a power intensity greater than 0.01 w/cm^2 should be avoided. In such areas, dummy loads should be used to absorb energy output while equipment is being operated or tested. If a dummy load cannot be used, adjacent populated areas should be protected by adequate shielding.

Infrared Radiation (IR). This type of radiation does not penetrate much below the superficial layer of the skin, so its major effect is to heat the skin tissues immediately below this layer. Longer wave infrared is completely absorbed in the surface layer of skin. Mid-range infrared can cause skin burns and increased skin pigmentation. Short-wavelength IR radiation from furnaces can cause eye problems known as "glass blower's cataract" or "heat cataract." Exposed workers need to wear appropriate eye protection gear with darkened filtered lenses (see Chapter 9, Personal Protective Equipment).

Ultraviolet (UV) and Visible Radiation. These two types of radiation do not penetrate appreciably below the skin. Their effects are essentially to heat the surface. However, high-intensity visible radiation can severely damage the retina of the eye, even causing blindness in a relatively short exposure time. This is the reason behind insistent warnings to the public not to stare directly at the sun during periods of partial solar eclipses. The automatic defense mechanism of closing the eyes or looking away usually prevents damage. It should be emphasized that this defense mechanism is not effective against visible light emitted by lasers.

The most common exposure to ultraviolet radiation is from direct sunlight. Ultraviolet radiation is responsible for "suntan" and "sunburn." People may develop tumors on exposed skin areas. These tumors occasionally become malignant.

The effects of ultraviolet light are generally more of a problem because ultraviolet can produce a severe burn with no warning, and excessive exposure can result in significant damage to the lens of the eye. Because ultraviolet radiation in industry is found around electrical arcs, they should be shielded by materials opaque to UV light.

Electrical welding arcs and germicidal lamps are the most common strong producers of ultraviolet radiation in industry. The ordinary fluorescent lamp generates a good deal of ultraviolet light inside the bulb, but the UV is essentially absorbed by the glass bulb and its fluorescent coating.

Some industrial materials also influence effects of ultraviolet radiation from the sun on human skin. After exposure to compounds such as cresols, for example, the skin is exceptionally sensitive to the sun. Even short exposure in late afternoon when the sun is low is likely to produce a severe sunburn. Other compounds minimize the effects of UV rays. Some are used in certain protective creams, for example, sunblock found in various suntan lotions.

Visible Radiation or Lighting. Visible radiation, which falls about midway in the electromagnetic spectrum, should concern the supervisor because it can affect both the quality and accuracy of employees' work. Good lighting invariably results in increased product quality with less spoilage and increased production. Good lighting contributes to sanitary, clean, and tidy operations (Figure 8–6).

Proper Lighting. Good lighting is the result of several factors: the amount and color of the light, direction and diffusion, and the nature of illuminated surfaces. A dark-gray, dirty surface may reflect only 10% or 12% of a light source while a light-colored, clean surface may reflect more than 90%.

Lighting should be bright enough for easy viewing of work tasks and directed so that it does not create glare. The level should be high enough to permit efficient sight. The *Accident Prevention Manual for Business & Industry* and *Fundamentals of Industrial Hygiene* list the illumination levels and brightness ratios recommended for various occupational tasks by the Illuminating Engineering Society. The IES's *Practice for Industrial Lighting* is designated as American National Standard ANSI/IES RP7.

A major problem associated with lighting is glare—brightness within the field of vision that causes discomfort or interferes with proper vision. The brightness can be caused either by direct or reflected light. To prevent glare, keep the light source well above the line of vision, or shield it with opaque or translucent material. You can dull reflections by having such surfaces matte-finished (dull) rather than polished or smooth.

The video display terminals of computers are especially likely to be affected by glare, which can be reduced by dimming overhead lights and using a task light instead, by placing the terminal parallel to windows, or by installing an antiglare screen.

Another visual problem occurs when an object is too bright for its surrounding field. A highly reflecting white paper in the center of a dark, nonreflecting surface, or a brightly illuminated control handle or dial on a dark or dirty machine are two examples. The contrast can irritate the eyes. To prevent this condition, keep surfaces light or dark with little differences in surface reflectivity. Instead, use color contrasts to set objects apart from their backgrounds.

Although it is generally best to provide even, shadow-free light, some jobs require contrast lighting. In these cases, keep the general or background light well diffused and free of glare, and add a supplementary source of light directed to cast shadows as desired. Too much contrast in brightness between the work area and its surroundings can be as distracting and harmful as direct or reflected glare.

Lasers. The word laser is an acronym for *Light Amplification by Stimulated Emission of Radiation*. Lasers emit beams of coherent light of a single color (or wavelength). Contrast this with conventional light sources that produce random, disordered light-wave mixtures of various wavelengths. A laser is made up of light waves that are nearly parallel to each other (collimated), and that are all traveling in the same direction. Electrons of atoms capable of emitting excess energy in the form of visible light are "pumped" full of enough energy to force them into higher-energy-level orbits.

Figure 8–6. Good lighting promotes increased safe production and good housekeeping.

These electrons quickly return to their normal-energy-level orbits, but, in so doing, they emit a photon of visible light. The light waves given off are directed to produce the coherent laser beam.

Lasers are extremely versatile, and new applications for this invention are continually being developed. They are used for welding microscopic parts, for welding heat-resistant exotic metals, for communications instruments, for many kinds of guidance systems, for various surgical procedures, in chemistry laboratories, in bar-code readers, and in numerous other applications.

Biological Hazards. The eye is the organ most vulnerable to injury induced by laser energy. This is because the cornea and lens focus the parallel laser beam onto a small spot on the retina. Make sure workers are instructed never to observe a laser beam or its reflection directly, or to look at it with an optical aid, such as a binocular or a microscope.

The work area should contain no reflective surfaces (such as mirrors or highly polished furniture) as even a reflected laser beam can be hazardous. Suitable shielding to contain the laser beam should be provided. Other safety measures may also be necessary, such as safety interlocks and lockout procedures, storing combustible solvents and other materials in proper containers, shielding them from laser beams or induced electrical sparks, and safeguarding all potential electrical hazards. The fact that infrared radiation of certain lasers may not be visible to the naked eye contributes to the potential hazard. Eyes must be protected. Lasers generating in the ultraviolet range of the electromagnetic spectrum produce corneal burns rather than retinal damage because of the way the eye handles ultraviolet light.

Other factors that influence the degree of eye injury induced by laser light include the following:

- Pupil size—the smaller the pupil diameter, the smaller the amount of laser energy permitted to the retina
- Power of the cornea and lens to focus the incident light on the retina
- Distance from the energy source to the retina
- Energy and wavelength of the laser
- Pigmentation of the subject
- Focal point of light on the retina
- Divergence of the laser light
- Presence of scattering media in the light path

Control of exposure to laser radiation should include use of protective devices built into the laser equipment, training employees in eye hazards, and use of protective eyewear as necessary. To develop a program of controls for the more hazardous lasers may require the help of a laser specialist.

Temperature Extremes

General experience shows that extremes of temperature affect the amount of work people can do and the manner in which they do it. In industry, people are more often exposed to hazards associated with high temperatures than with low temperatures.

The body is continuously producing heat through its metabolic processes. Since these processes are designed to operate efficiently within only a narrow temperature range, the body must dissipate excess heat as rapidly as it is produced. Body temperature is regulated by a complex set of thermostatic controls that react quickly to significant changes in internal and external temperatures.

Sweating. Sweating is the most important of the temperature-regulating and heat-dissipating processes. Almost all parts of the skin are provided with sweat glands that excrete a liquid (mostly water and a little salt) to the body surface. This process goes on continuously, even under conditions of rest. In an individual who is resting and not under stress, the sweating rate is approximately 1 liter per day, which is evaporated in the air as rapidly as it is excreted. Under the stress of heavy work or high temperature, the sweating rate may increase to as much as 4 liters (approximately one gallon) in 4 hours. Up to 10 to 12 gm (150 to 190 grains) of salt per day will be lost with the water. Both water and salt loss must be replaced promptly to ensure good health. Such stress is especially dangerous for those with heart problems because loss of salt and water can affect the heart's ability to function.

If sweat evaporates as rapidly as it forms, and if heat stress does not exceed the maximum sweating rate of which the body is capable, an individual can maintain the necessary constant temperature. Rate of evaporation depends on the moisture content and temperature of the surrounding air (heat and humidity), and the rate of air movement. The industrial hygienist must consider all these factors when determining the effect of the thermal environment on humans.

Radiant Heat. Radiant heat is electromagnetic energy that does not heat the air it passes through. It affects the body's ability to remain in equilibrium with its surroundings. For instance, an object, such as a glass window, may be far below the body temperature of a nearby worker. As a result, a large amount of heat will be radiated from the person near the window, who may feel chilled even if the air in the immediate environment is fairly warm. Conversely, if an object in the surroundings, such as a furnace wall, is above body temperature, people can receive considerable heat by radiation. They will find it difficult to keep cool even if the room is air conditioned.

Thus, when radiant heat strikes an object (such as a person), it is absorbed and produces the sensation of heat. Merely blowing air around offers no relief from the source. The only protection is to set up a barrier. Placing radiant-reflective shielding between the heat source and the worker is usually the most cost-effective way to reduce employee exposure. Use of heat-reflective personal protective clothing may be considered when exposures are short and intermittent.

Heat Conduction and Convection. If heat can be conducted through clothing and dissipated into the air, this will cool the body to some degree. Conduction usually is not an important means of cooling, however, because conductivity of clothing and heat capacity of air are usually low.

Conduction and convection become an important means of heat loss when the body is in contact with a good cooling agent, such as water. For this reason, when people are exposed to cold water, they become chilled much more rapidly than when exposed to air at the same temperature.

Air movement cools the body by conduction and convection. More importantly, moving air removes the saturated air layer around the body (which is formed rapidly by evaporation of sweat) and replaces it with a fresh layer, capable of accepting more moisture.

Effects of High Temperature. Effects of high temperature are counteracted by the body's attempt to keep the internal temperature down through increasing the heart rate. The capillaries in the skin dilate to bring more blood to the surface so that the cooling rate increases and, gradually, the body temperature stabilizes.

If the thermal environment is tolerable, these measures will soon restore the body's equilibrium to the point where the heart rate and the body temperature remain constant. If this equilibrium is not reached by the time the body temperature is about 102 F (38.9 C), corresponding to the sweating rate of about 2 liters per hour, there is imminent danger of heatstroke. Intermittent rest periods for people exposed to extreme heat reduces this danger.

Heatstroke or Sunstroke. This is not necessarily the result of exposure to the sun. It is a condition caused by exposure to an environment in which the body is unable to cool itself sufficiently. Sweating stops, and the body can no longer rid itself of excess heat. As a result, the body soon reaches a point where the heat-regulating mechanism breaks down completely and internal temperature rises rapidly.

The symptoms are hot, dry skin (which may be red, mottled, or bluish); severe headache; visual disturbances; rapid temperature rise; and confusion, delirium, and loss of consciousness. The condition is recognizable by a flushed face and high temperature. The victim should be removed from the heat immediately and cooled as rapidly as possible, usually by being wrapped in cool, wet sheets. This is a medical emergency, and the supervisor must get assistance. Studies show that for people admitted to hospital emergency rooms for heatstroke, the higher the body temperature upon admission, the higher the mortality rate. Since the condition can be fatal, medical help should be obtained immediately.

Heatstroke is a much more serious condition than heat cramps or heat exhaustion, which are discussed next. An important factor is excessive physical exertion. The only method of control is to reduce the surrounding temperature or to increase the ability of the body to cool itself so that body temperature does not rise. Heat shields, discussed under the next section, Preventing Heat Stress, are of value here.

Heat Cramps. These may result from exposure to high temperature for a relatively long time, particularly if accompanied by heavy exertion and excessive loss of salt and moisture from the body. Even if the moisture is replaced by drinking plenty of water, an excessive loss of salt may provoke heat cramps or heat exhaustion.

Heat cramps are characterized by the cramping of the muscles of either the skeletal system or the intestines. In either case, the condition may be relieved in a few hours under proper treatment, although soreness may persist for several days.

Heat Exhaustion. This may result from sustained physical exertion in a hot environment, especially if a worker has not been acclimatized to heat and has not replaced water lost in sweating. Symptoms include relatively low temperature, pallor, weak pulse, dizziness, profuse sweating, and cool, moist skin.

Preventing Heat Stress. Most heat-related health problems can be prevented, or, at least, the risk can be reduced. The following basic precautions help to reduce heat stress.

- **Acclimatization.** Acclimatize workers to heat by giving them short exposures, followed by gradually longer periods of work in the hot environment. New employees and workers returning from an absence of two weeks or more should have a five-day period of acclimatization. This period might begin with 50% of the normal workload and time exposures the first day and gradually build up to 100% on the fifth day.
- **Mechanical Cooling.** A variety of engineering controls, including general ventilation and spot cooling by local exhaust ventilation at points of high heat production, may be helpful. Evaporative cooling, mechanical refrigeration, and cooling fans are other ways to reduce heat, along with eliminating steam leaks. Making equipment modifications, using power tools to reduce manual labor, and using personal cooling devices or protective clothing (Figure 8–7) are other ways to reduce heat exposure for workers.
- **Procedures.** Work practices can also be tailored to prevent heat disorders. Companies can provide a period of acclimatization for new workers and those returning from vacation. They can make plenty of cool drinking water available, as much as a quart per worker per hour (Figure 8–8). Companies should train first aid workers to recognize and treat heat stress and should post the names of trained staff in a location known to all workers. Always consider individual workers' physical conditions when determining their fitness for working in hot environments. Older workers, obese workers, and workers on certain types of medication are at greater risk for heat-stress illnesses.

  If you alternate work and rest periods, with longer rest periods in a cool area, you can help workers avoid heat strain. If possible, schedule heavy work during cooler parts of the day and provide appropriate protective clothing. Allow workers to stop for rest if they become extremely uncomfortable in the heat.
- **Rehydration.** Employee education is vital so that workers are aware of the need to replace fluids and salt lost by sweating, and so they can recognize dehydration, exhaustion, fainting, heat cramps, heat exhaustion, and heatstroke as heat disorders. Workers should be advised to drink water beyond the point of thirst. Adding salt to the diet is not usually recommended or needed for a worker who is acclimatized to heat. Workers should also be informed of the importance of weighing themselves daily before and after work to avoid dehydration. If the worker has health conditions aggravated by this stressor, the worker's physician must be consulted.

  Companies may want to supply a cooled carbohydrate electrolyte beverage to encourage rehydration. Such a beverage should have six to eight percent carbohydrate to provide entry and stimulate fluid absorption as well as some sodium to help maintain the desire to drink and offset loss of sodium from sweating. A number of carbohydrate electrolyte beverages are on the market.

  Workers should be educated about the negative effects of high-carbohydrate (above 10%) diet and caffeinated beverages. The former will tend to decrease fluid absorption, and the latter will increase fluid loss through urination. Alcoholic beverages

Figure 8–7. Jobs with high radiant heat loads often require reflective clothing. This first-response firefighting team is properly equipped with heat-protective clothing.

should also be avoided. (See also Extreme Climactic Conditions in Chapter 3, Human Performance Management.)

Dermatitis

Skin problems in industry account for the single largest cost of all hygiene problems. Individual skin disorders usually can be traced to exposure to some kind of chemical compound, or to some form of physical abrasion or irritation of the skin. Although rarely a direct cause of death, skin disorders cause much discomfort and are often hard to cure (Figure 8–9). Even normally harmless substances can cause irritations of varying severity in sensitive skin.

Sources of Hazards. Causes of occupational skin disorders are classified in these ways:

1. Mechanical agents—friction, pressure, trauma
2. Physical agents—heat, cold, radiation
3. Chemical agents—organic and inorganic subdivided according to action on the skin as primary irritants

Figure 8–8. To help prevent heat disorders, cooled drinking water should be easily accessible and workers should be trained to recognize symptoms of heat stress.

or sensitizers
4. Biological agents—bacteria, fungi, parasites
5. Plant poisons—several hundred plants and woods, such as poison ivy

There are two general types of skin reactions: primary irritation dermatitis and sensitization dermatitis. Practically all persons suffer primary dermatitis from mechanical, physical, or chemical agents. Brief contact with a high concentration of a primary irritant or prolonged exposure to a low concentration will result in inflammation. Allergic reaction is not a factor in these conditions. Sensitization dermatitis, on the other hand, is the result of an allergic reaction to a given substance. Once sensitization develops, exposure even to small amounts of the material may cause symptoms. Some substances can produce both types of dermatitis. Examples are organic solvents, formaldehyde, and chromic acid.

Occupational Acne. Organic-based cutting fluids are frequently responsible for occupational acne. When the skin becomes infected, oil pimples or boils form that resemble adolescent acne. Dirty oil, poor materials handling, and addition of germicides in the fluid increase the possibility of occupational acne.

To reduce the possibility of this condition, keep machines and areas around them clean. Substitute an aqueous-based cutting fluid for the organic cutting fluid. Change cutting fluids at regular intervals. Have employees use barrier creams or gloves, aprons, or face shields. Encourage employees who work in these areas to shower or thoroughly wash at the end of each shift. Work clothes should be changed before going home. Abrasive cleaners or brushes should be avoided. Instead, workers should use warm water, mild soap, and a soft brush. They should be discouraged from wearing oil-soaked clothing. Some sensitive skins may require special cleaners.

Certain factors help account for some skin disorders. Any investigation of occupational skin disease should take into consideration:

1. Degree of perspiration
2. Personal cleanliness
3. Preexistence of skin disorders
4. Allergic conditions
5. Diet

Atmospheric Pressures

From the beginning of caisson work, dating back to about 1850, it has been recognized that those who work under greater-than-normal pressures are subject to various disorders. The main effect, decompression sickness (commonly known as the bends), results from the release of nitrogen bubbles into the bloodstream and surrounding tissues during decompression. The bubbles lodge at joints and under muscles, causing severe cramps. To prevent this condition, decompression is carried out slowly and by stages so the nitrogen can be eliminated without forming bubbles.

Deep-sea divers are supplied with a mixture of helium and oxygen for breathing. Since helium is an inert diluent and is less soluble in blood and tissue than is nitrogen, it presents a less formidable decompression problem.

Common troubles encountered by workers operating under compressed air are pain and congestion in the ears caused by inability to ventilate the middle ear properly during compression and decompression. As a result, many workers under compressed air suffer from temporary hearing loss. The cause is considered to be obstruction of the eustachian tubes that prevents proper equalization of pressure on the middle ear. Occasionally, permanent hearing loss may result.

The effects of low pressure are much the same as the effects of decompression from high pressure. If pressure is reduced too rapidly, decompression sickness and ear disturbances similar to, if not identical with, the divers' conditions may result.

Persons working at reduced pressure are also subject to oxygen deprivation, which can have serious and insidious effects upon the senses and judgment. There is considerable evidence showing that exposure for 3 to 4 hours to an altitude of 9,000 ft (2.7 km) above sea level, without breathing an atmosphere enriched with oxygen, can result in severely impaired judgment. Even if pure oxygen is provided, the altitude should be limited to that giving the same partial pressure of oxygen as air at 8,000 ft (2.4 km).

Reduced pressure is not the only condition under which oxygen deprivation may occur. Deficiency of oxygen in the atmosphere of confined spaces is commonly experienced in industry. (Normal air contains about 21% oxygen by volume.) The oxygen content of any tank or other confined space should be monitored before workers are allowed to enter. Instruments, such as the oxygen analyzer, are commercially available for this purpose (Figure 8–10).

The first physiological signs of a deficiency of oxygen (anoxia) are increased rate and depth of breathing. Oxygen concentrations of less than 19% by volume cause dizziness, rapid heartbeat, and headache. One should never enter or remain in areas where tests have indicated such low concentrations unless wearing self-contained, supplied-air respiratory equipment. Oxygen concentrations less than 16% can cause loss of consciousness and death.

In an oxygen-deficient atmosphere, workers may find it hard to move and have a hazy lack of concern about the imminence of death. In cases of sudden entry into areas containing little or no oxygen, the individual usually has no warning symptoms, but immediately loses consciousness and has no recollection of the incident, if rescued and revived.

Atmospheric contaminants are discussed in this chapter in the sections, Threshold Limit Values®, and Standard Operating Procedures. Entry into confined spaces also requires compliance with lockout/tagout procedures that should be used to reduce hazards from chemical, mechanical, electrical, and other physical hazards. These are discussed in Chapter 11, Machine Safeguarding.

ERGONOMIC STRESSES

The third category of environmental factors is ergonomic. Involved are human reactions to posture, force, frequency duration, vibration, mechanical stresses, and temperature. This subject is discussed in greater detail in Chapter 10, Ergonomics.

The term *ergonomics* literally means the customs, habits, and laws of work. According to the International Labor Office, ergonomics is:

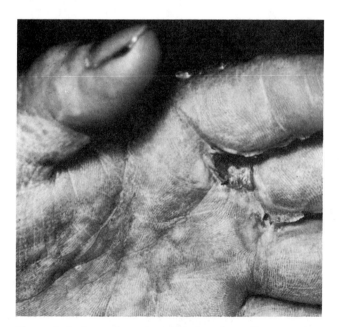

Figure 8–9. Eczematous dermatitis is a form of contact dermatitis caused by working with organic solvents. It is one of the most prevalent types of dermatitis.

Figure 8–10. This portable, hand-held oxygen indicator with digital display and remote galvanic sensor operates by diffusion and measures oxygen on a 0–100 percent-by-volume range. (Courtesy MSA.)

. . .the application of human biological science in conjunction with the engineering sciences to achieve the optimum mutual adjustment of man and his work, their benefits being measured in terms of human efficiency and well-being.

The ergonomics approach goes beyond productivity, health and safety. It includes consideration of the total physiological and psychological demands of the job upon the workers.

The human body can endure considerable discomfort and stress and can perform many awkward and unnatural movements, but only for a limited time. When unnatural conditions or motions continue for prolonged periods, they may exceed workers' physiological limitations. To ensure a continuously high level of performance, work systems must be tailored to human capacities and limitations.

Biomechanics is that phase of engineering devoted to the improvement of the worker-machine-task relationship in an effort to reduce operator discomfort and fatigue. Biotechnology, a broader term, encompasses biomechanics, human factors engineering, and

engineering psychology. To arrive at biotechnological solutions to work-stress problems, the sciences of anatomy, physiology, psychology, anthropometry, and kinesiology need to be brought into play. Areas of concern are:

- Strictly biomechanical aspects—the consideration of stress on muscles, bone, nerves, and joints
- Sensory aspects—the consideration of eye fatigue, odor, audio signals, tactile surfaces, and the like
- External environment aspects—the consideration of lighting, glare, temperature, humidity, noise, atmospheric contaminants, and vibration
- The psychological and social aspects of the working environment

Information obtained from studying these factors can be translated into tangible changes in work environments. Reducing fatigue and stress with redesigned hand tools, adjustable chairs and workbenches, better lighting, control of heat and humidity, and noise reduction are definitely rewarding improvements.

In the broad sense, the benefits that can be expected from designing work systems to minimize physical stress on workers are:

- More efficient operation
- Fewer accidents
- Lower cost of operation
- Reduced training time
- More efficient use of personnel

See Chapter 10, Ergonomics, for more detailed information on ergonomic stresses.

BIOLOGICAL HAZARDS

Biological hazards include any viruses, bacteria, fungi, or any other living organisms capable of causing a disease in humans.

Tuberculosis

Tuberculosis and other bacterial infections are classified as occupational diseases when they are contracted by nurses, doctors, or healthcare workers caring for tuberculosis patients, or during autopsy or laboratory work where the bacilli may be present.

Fungal Infections

A number of occupational infections are common to workers in agriculture and in closely related industrial jobs. Grain handlers who inhale grain dust are likely to come in contact with some of the fungi that may contaminate grain. Some of these fungi can and do flourish in the human lung, causing a condition called "farmer's lung." Fine, easily spread fungi, such as rust and smut, can be contacted once the covering of infected grain is broken and may produce sensitization in the exposed individuals.

One example, coccidioidomycosis or Valley Fever, is a disease caused by a fungus present in the soils of southwestern states and the San Joaquin Valley in California. It causes a lung disease in humans called Valley Fever, which may be either mild or severe. Because the spores are found in the soil, Valley Fever outbreaks sometimes occur during excavation and construction work.

Byssinosis

Byssinosis occurs in individuals who have experienced prolonged exposure to heavy air concentrations of cotton dust. Flax dust also has been incriminated.

The exact mode of action of the cotton dust is unknown, but one or more of these factors may be important: (1) toxic action of microorganisms adhering to the inhaled fibers, (2) mechanical irritation from the fibers, and (3) allergic stimulation by the inhaled cotton fibers or adherent materials. It takes several years of exposure before manifestations are noticed.

Anthrax

Anthrax is a highly virulent bacterial infection. In spite of considerable effort in quarantining infected animals and in sterilizing imported animal products, this disease remains a problem. With prompt detection and modern methods of treatment, it is less likely to be fatal than it was a few years ago. Vaccination of high-risk workers and use of personal protective equipment can prevent infection.

Q Fever

There have been reports of infection with the rickettsial organism (Q Fever) among meat and livestock handlers. It is similar to, but apparently not identical with, tick fever, which has been known for many years. Q Fever may come from contacting freshly killed carcasses or droppings of infected cattle. Probably in the latter case, transmission is by inhalation of the infectious dust. Prevention undoubtedly depends upon recognition and elimination of the disease in the animal host. Workers should use personal and respiratory protection when working in dusty and contaminated environments, and should receive immunization.

Brucellosis

Brucellosis (undulant fever) has long been known as an infection produced by drinking unpasteurized milk from cows suffering from Bang's disease (also called infectious abortion). Since it can also be contracted by handling the animals or their flesh and is transmitted by swine and goats, it is an occupational ailment of slaughterhouse workers, as well as of farmers, although more common among the latter. Here also, preventive measures are primarily proper animal testing and control to eradicate the disease. Personal protection when handling infected materials is important.

Leptospirosis

Leptospirosis can infect livestock and wild animals. It causes a flu-like illness and, in severe cases, liver damage. Hunters and people working in rice paddies may come in contact with water containing liquid animal waste. Control includes immunization of animals and appropriate personal protection for workers.

Psittacosis

Psittacosis is a risk for poultry workers and poultry-processing workers. This disease causes a mild flu-like illness and conjunctivitis. Sanitation and personal hygiene are important in prevention.

Lyme Disease

Lyme disease is caused by bacteria that live in the gut of deer ticks. It causes flu-like symptoms and may cause a bulls-eye-shaped rash. If left untreated, it can lead to rheumatoid arthritis. Employees who work outdoors, especially in woods or brush, or who have potential contact with wildlife, or who handle stock or domestic animals are most vulnerable. Prevention requires awareness of the signs and symptoms, often difficult to identify at the time of the bite. Precautions should include covering exposed skin with clothing, tucking pant legs into boots, and surveying the body for ticks on leaving the area. Any ticks found should be removed carefully, using protective gloves. The tick should not be squeezed or crushed, as this could release the bacteria. First aid should be administered to the wound and contact made to the worker's physician for follow-up treatment.

HIV

Acquired immune deficiency syndrome (AIDS), caused by the Human Immunodeficiency Virus or HIV, is usually contracted during sexual contact with an infected person, during intravenous drug use involving shared contaminated needles, before or during birth from an infected mother, or by contact with infected body fluids. It cannot be transmitted by casual contact with an infected person in the workplace. The virus is present in the blood and other body fluids of infected persons, and can be a hazard for workers who might have contact with these fluids: medical and emergency rescue personnel, hospital and laboratory workers who may have contact through cuts or breaks in the skin.

The Centers for Disease Control have developed guidelines for workers in health-care and hospital settings and OSHA has a regulation for exposure to bloodborne pathogens covering all potentially affected employees. The CDC guidelines specify that all patients should be assumed to be infectious for HIV and other bloodborne illnesses (such as hepatitis B). Precautions to avoid the spread of bloodborne disease include:

- Proper handling and disposal of all needles in impervious containers labeled as biohazardous
- Prohibiting recapping of needles and other measure to avoid needlestick injuries
- Using protective personal resuscitation clothing, including impervious gloves, aprons, boots, face shields, resuscitation masks, and goggles in some situations
- Using resuscitation devices instead of mouth-to-mouth techniques
- Adopting appropriate sterilization, disinfection, and housekeeping methods where potential for exposure to blood or body fluids is involved. In addition, the OSHA standard has requirements for a written exposure control plan, employee training, and medical surveillance.

Upper Respiratory Tract Infections

Workers exposed to dusts from such vegetable fibers as cotton, bagasse (sugar cane residues), hemp, flax, or grain may develop upper respiratory tract infections ranging from chronic irritation of the nose and throat to bronchitis, complicated by asthma, emphysema, or pneumonia, or a combination of these. Dust from any of these fibers may also be a source of allergens, histamine, and toxic metabolic products of microorganisms.

THRESHOLD LIMIT VALUES

Guidelines called Threshold Limit Values® (abbreviated as TLV®s) have been established for airborne concentrations of many chemical compounds. Supervisors should understand the meaning of TLV®s and the terminology in which concentrations are expressed.

Concept and Purpose

The basic idea of the TLV® is fairly simple. TLV®s represent exposure levels under which it is believed that nearly all workers may be repeatedly exposed

without adverse effect. However, because individual susceptibility varies widely, exposure of some individuals at (or even below) the threshold limit may not prevent discomfort, aggravation of a preexisting condition, or occupational illness. In addition to the TLV®s set for chemical compounds, there are limits for physical agents, such as noise, microwave radiation, and heat stress.

The TLV® may be a time-weighted average value that would be acceptable for an 8-hour exposure. For some substances, such as an extremely irritating one, an eight-hour time-weighted average concentration would not be acceptable, so a ceiling value is established. The ceiling limit means that at no time during the 8-hour work period should the airborne concentration exceed that limit. It is important to understand that TLV®s are not intended to be fine lines between safe and unsafe conditions. Considerable judgment needs to be used when deciding on an appropriate exposure level in a particular work situation. Many companies choose to set an in-house "action level" of one-half or less of the TLV®. When the concentration of the chemical exceeds the "action level," further air monitoring is often conducted, and measures are taken to reduce employee exposure.

As more toxicological information has become available, the wisdom of this practice is clear. Many of the TLV®s have been reduced as additional epidemiological and laboratory studies become available that show health effects at lower exposure levels. The National Institute for Occupational Safety and Health (NIOSH) also has set Recommended Exposure Levels (RELs) for a number of substances; many of these RELs are below the Threshold Limit Values®.

Establishing Values

Threshold Limit Values have been established for more than 600 substances, as well as for heat and cold stress, vibration, radiation, and biological exposure indexes. The TLV®s are developed and reviewed by the American Conference of Governmental Industrial Hygienists, a voluntary organization composed of industrial hygienists, toxicologists and medical personnel who work in government, educational institutions, and private industry.

The data for establishing a TLV® come from animal studies, human studies, and industrial experience. The limit may be selected for one or several reasons. A substance may be very irritating to the majority of people exposed to concentrations above a given level. Or the substance may be an asphyxiant. Other reasons for establishing a limit might be that the chemical compound is an anesthetic, or fibrogenic, or can cause allergic reactions or malignancy. Some TLV®s are established because, above a certain airborne concentration, a substance poses a nuisance to workers.

The concentrations of airborne materials capable of causing problems are quite small. Consequently, industrial hygienists use special terminology to define these concentrations. They often talk in terms of parts (of contaminant) per million (parts of air) (ppm) when describing the airborne concentration of a gas or vapor. If measuring airborne particulate matter, such as dust or fume, the term milligrams (of dust) per cubic meter (of air), or mg/m3 is used to define concentrations. For materials composed of fibers, such as asbestos or fibrous glass, industrial hygienists may use the number of fibers present in one cubic centimeter of air being tested. In this case, the descriptive term becomes the number of fibers per cubic centimeter or fibers/cc.

As an example of the small concentrations involved, the industrial hygienist commonly samples and measures substances in the air of the working environment in concentrations ranging from 1 to 100 ppm. Some idea of the magnitude of these concentrations can be appreciated when one realized that 1 in. in 16 miles, one cent in $10,000, one oz of salt in 62,500 lb of sugar, one oz of oil in 7,812.5 gal of water—all represent one ppm.

STANDARD OPERATING PROCEDURES (SOP)

One of your primary functions as supervisor is to assist in developing standard operating procedures for jobs that require industrial hygiene controls, and to enforce these procedures once they have been established. Common jobs involving maintenance and repair of systems for storing and transporting fluids or entering confined spaces for cleaning and repairs are controlled almost entirely by the immediate supervisor.

Industrial Hygiene Controls

Finally, the supervisor should be aware of the various kinds of controls used to maintain a healthy work environment. The type and extent of these controls depend on the physical, chemical, and toxic properties of the air contaminant, the evaluation made of the exposure, and the operation that disperses the contaminant. The extensive controls needed for lead oxide dust, for example, would not be needed for limestone dust, since limestone is less toxic.

General methods of controlling workplace hazards include the following:

1. Substitution of a less harmful material for one that is dangerous to health. Make sure a professional carries out this task so that a more hazardous material is not mistakenly substituted.
2. Change or alteration in a process to minimize worker contact.
3. Isolation or enclosure of a process or work operation to reduce the number of persons exposed, or isolation of the workers in remote locations.
4. Wetting down dust during working operations and cleanup to reduce generation of dust.
5. Local exhaust ventilation at the point of contaminant generation and dispersion.
6. General or dilution ventilation with clean air to provide a less contaminated atmosphere. (Do not use dilution ventilation improperly for toxic chemicals; local exhaust ventilation is the way to control these.)
7. Personal protective equipment, such as special clothing, eye, and respiratory protection. (These are discussed in Chapter 9, Personal Protective Equipment.)
8. Good housekeeping, including cleanliness of the workplace, proper waste disposal, adequate washing and eating facilities, potable drinking water, and control of insects and rodents.
9. Special control methods for specific hazards, such as reduction of exposure time, film badges and similar monitoring devices, and continuous sampling with preset alarms to detect intake of toxic materials.
10. Training and education to supplement engineering controls. (Training in work methods and in emergency response procedures—medical, gas release, etc.—to teach who should and should not respond in specific emergencies.)
11. Medical surveillance programs to monitor employee health before, during, and after exposure.

The supervisor plays an important role in making sure that controls are effective. You must observe and study process changes, shielding or enclosures, use of wet methods, local exhaust ventilation, and general ventilation carefully and systematically to make sure that they are functioning properly, or to determine if maintenance or repair work is needed. Even more important is the part you must play in enforcing the use of personal protective equipment, such as hearing protection and respirators, in areas where they should be worn. If this is not done, your company can be cited and penalized by regulating agencies.

Good housekeeping, special monitoring devices, and safety and hygiene training and education are dependent on your full cooperation. In fact, as a supervisor, you are a key person where safety and industrial hygiene controls are concerned. Without your help and guidance, these controls are of questionable value in ensuring a safe, healthy workplace.

SUMMARY OF KEY POINTS

Key points covered in this chapter include:

- The more supervisors know about industrial hygiene, the better they can fulfill the objectives of their job: to ensure safe production. Supervisors should know how to identify health hazards and take steps to correct them in four key areas of industrial hygiene: chemical, physical, ergonomic, and biological hazards.

- Chemical hazards exist in the form of dusts, fumes, smoke, aerosols, mists, gases, vapors, and liquids. They can cause health problems when workers inhale, absorb (through the hair or skin), contact, inject, or ingest one or more of these substances.

- The major exposure hazard to chemical compounds is through inhalation of airborne contaminants in gases, vapors, and particulate matter. Supervisors should know the physical classifications of these substances.

- The hazard associated with breathing an airborne contaminant depends on the nature of the chemical. Symptoms can range from mild irritation of the nose and throat to serious injuries deep within the lungs. Concentration of airborne contaminants must be measured by an industrial hygienist. Supervisors should also be alert to the hazards of an oxygen-deficient environment. Workers can guard against respiratory hazards though use of respiratory protective equipment and reducing their exposure to the hazards.

- Absorption of chemicals can occur through breaks in the skin or through direct contact with intact skin. Solvents, classified as aqueous or organic, are among the most widely used liquid chemicals in industry and present the greatest hazard to workers. Solvent hazard is determined not only by toxicity of the chemical but by the conditions of its use.

- Adverse effects of solvents vary from mild skin irritation, to more serious disorders of the central nervous system and major organs, to the formation of cancerous tumors. Even low-level exposure over time may cause long-lasting effects on workers' health. Preventive and protective measures include the proper selection, application, handling, and control of solvents; a working knowledge of each chemical's risks to the workers; a sound training program to educate workers in chemical hazards; and an effective program of engineering controls, work-practice controls, and personal protective equipment.

- Toxic chemicals can be absorbed from the gastrointestinal tract into the blood. Employees working with toxic materials should not eat, drink, smoke, or apply cosmetics in the work area, should wash their hands after every job, and maintain a clean environment.

- Physical hazards include noise, ionizing and nonionizing radiation, temperature, and pressure extremes.

- Noise, unwanted sound, can have harmful psychological and physiological effects on workers as well as interfere with communication on the job. Effects vary, based on individual differences among workers and conditions on the job. Supervisors should try to reduce or eliminate noise hazards as well as take protective and preventive measures to ensure workers' health.

- According to OSHA Noise Standard, noise above 90 dBA is harmful to workers. When exposed to these (or higher) noise levels, employees must either be given hearing protection equipment, have exposure time reduced, or both, to ensure proper working conditions. Companies are required to administer annual audiometric tests to all workers exposed to noise over 85 dBA. Hearing protection equipment includes ear plugs, canal caps, and ear muffs. Supervisors should make sure these devices fit workers properly and are correctly used.

- Ionizing radiation involves a change in the normal electrical balance in an atom. The most common forms which may be encountered on the job include gamma radiation, alpha and beta radiation, high-energy protons and neutrons, and X-radiation. Chemicals also can interact with ionizing radiation to form highly toxic substances. This type of radiation causes injury by ionizing the tissues in which they are absorbed. Injuries can range form mild skin irritations to radiation poisoning, to the formation of various types of cancers. Workers can be protected from ionizing radiation by separating or shielding from the hazard or by reducing time of exposure.

- Nonionizing radiation includes low-frequency, microwaves, infrared, and ultraviolet and visible radiation. These types of radiation heat the tissues in which they are absorbed. Injuries range from mild irritation to severe burns and cancerous tumors as well as possible damage to the eyes. Protective measures include shielding workers from the radiation source and providing adequate protective gear.

- Visible radiation or light can affect both quantity and accuracy of employees' work. Proper lighting in the work area is the result of the amount and color of light, direction and diffusion, and nature of illuminated surfaces. Lighting should be bright enough to produce easy viewing and directed so that it does not create glare. Supervisors should work to eliminate or reduce glare off all reflective surfaces and VDT terminals. Visible light tools such as lasers, directly, can injure the retina of the eye. Protective measures include shielding laser sources, training workers in eye hazards, and use of protective eyewear.

- Temperature extremes present health hazards when the body cannot heat or cool itself adequately to overcome the extreme. Most work hazards are associated with high temperatures. The body adapts to high temperatures by sweating, radiating heat, and through heat conduction or convection. If the body cannot overcome the extreme, a worker may suffer heatstroke, heat cramps, or heat exhaustion. Supervisors and workers must know the appropriate emergency medical steps to counteract these potentially life-threatening conditions.

- Workers can be protected against high temperatures by allowing them to acclimatize to the hot environment, providing plenty of fluids, allowing intermittent rest periods in a cooler environment, and providing proper work clothing and protective gear.

- Occupational dermatitis, or skin disorders, are generally caused by mechanical, physical, chemical, or biological agents, and plant poisons. Skin reactions are classified as primary irritation dermatitis and sensitization dermatitis. Mechanical, physical, and chemical agents produce primary irritation dermatitis. Sensitization dermatitis, on the other hand, is the result of an allergic reaction to a given substance. Occupational acne is caused by certain cutting fluids. Protective and preventive measures include scrupulous housekeeping, showering or washing at the end of each shift, use protective clothing, and changing work clothes before leaving the job.

- Atmospheric pressures related to high altitude or underwater work can cause decompression sickness, hearing problems, and oxygen deprivation. An oxygen-deficient environment can impair workers' judgment or cause sudden loss of consciousness and even death in a short time. Workers should be allowed sufficient time to decompress from underwater or high-altitude work and be equipped with proper independent air supplies when entering oxygen-deficient environments.

- Ergonomics involved creating the best "fit" between worker and job, including such factors as work environment (seating, desks, lighting, noise, and so on); body postures and movements; and psychological and social aspects. The disciplines of biomechanics and biotechnology are used to solve work-stress problems on the job. Minimizing ergonomic stress on workers can raise efficiency and reduce costs, number of accidents and training time.

- Biological hazards include any virus, bacterium, fungus, or other living organism capable of causing disease in humans. Biological hazards in workplaces include tuberculosis, fungus infections, byssinosis, anthrax, Q fever, brucellosis, leptospirosis, psittacosis, Lyme disease, HIV and upper respiratory tract infections. Supervisors must know how to prevent organisms from entering the workplace, what steps to take if an infection occurs, and how to protect workers.

- Threshold Limit Values (TLV®s) have been established for airborne concentrations of many chemical compounds. TLV®s represent conditions under which it is believed that nearly all workers may be repeatedly exposed without adverse effect. These values have been established for more than 600 substances and may be used to calculate exposure times for employees who work in areas where airborne contamination exists.

- Supervisors are key personnel in creating and supporting effective industrial hygiene controls. They must make use of substitution, engineering controls, good housekeeping measures, special monitoring devices, and safety/hygiene training and education to help ensure a safe, healthy workplace.

PERSONAL
PROTECTIVE EQUIPMENT

9

After reading this chapter, you will be able to:

- Develop an effective personal protective equipment (PPE) program

- Overcome employees' objections to using personal protective equipment and educate them about its importance

- Know the types of protective equipment available for the head, face, eyes, ears, respiratory tract, torso, and extremities

- Select the best protective equipment to fit the hazard involved

When a hazard is identified in the workplace, every effort should be made to eliminate it so that employees are not harmed. Elimination thus becomes the first thrust of control. For instance, if there is a milling operation that produces large quantities of steel chips that fly all over an aisleway, the first approach is to determine whether the hazard can be engineered out of the operation. The engineering control to be taken, or of eliminating the operation, is either to enclose the operation or provide a deflector to keep the chips from flying in the aisle. Another way to reduce or control the hazard is to isolate the process. Can that dip tank be relocated? Can the spray booth be enclosed? Can that noisy machine operation by isolated, separated from the rest of the facility?

A second approach is to control the hazard by administrative control. For example, a milling operation might be done only at certain times of the day, after the shift ends. This reduces the flying chip hazard to employees by reducing the incidental traffic in the aisle. Now, in spite of these two approaches and because the milling operation must be done and employees will be present, it will be necessary to use personal protective equipment (PPE). For example, safety glasses may be required in the work area.

Too often PPE usage is considered the last thing to do, in the scheme of hazard control. Personal protective equipment can provide that added protection to the employee even when the hazard is being controlled by other means. In the milling operation, a deflector virtually eliminates flying chips, but because there may be a slight chance of flying chips, affected employees should wear eye protection.

In some situations, the only available protection will be the use of PPE. Often, in emergencies, PPE will be required. PPE should be considered one aspect of the overall safety plan.

CONTROLLING HAZARDS

People who must work in hazardous areas should use PPE, which can protect a person from head to toe. To develop an effective PPE program, supervisors should:

- Be familiar with requirements of government standards.
- Conduct a workplace assessment to identify hazards, and document the need for PPE.
- Be familiar with the safety equipment on the market to protect against specific hazard(s).
- Know the company procedures for paying for and maintaining the equipment.
- Develop an effective education and training program for employees required to wear the proper protective equipment.

- Review all material safety data sheets (MSDSs) that require personal protective equipment for protection against hazardous chemicals and materials.
- Establish an industrial hygiene evaluation procedure to determine whether PPE meets MSDS recommendations.

As the supervisor, once you know what and when personal protective equipment is needed, you need to make sure employees understand the need for PPE and when to use it. If workers fail to see the reason for using protective equipment, you must impress on them its value and help them to recognize the need for it (Figure 9–1). This is a function of a good education and training program and is required by OSHA (29 *CFR* 1910.132–139). Getting some workers to use protective equipment may be one of the toughest jobs you face. The more you involve employees in the PPE assessment and selection process, the greater the likelihood they will cooperate and comply with the requirements.

To fulfill this responsibility, supervisors should be familiar with government regulations, be able to recognize hazards, and know the equipment needs the and vendor so that they are thoroughly familiar with all types of protective equipment. Get help from the safety and hygiene personnel to keep up to date. Even more, allowing employees to participate in equipment selection will enhance their acceptance of it.

Local safety meetings are a good source of PPE information. Equipment is usually on display and manufacturers' or distributors' representatives are handy. Hundreds of vendors exhibit at the National Safety Congress and Exposition held each fall. Company safety or human resources departments can also be helpful. A wealth of information on safety equipment is published—manufacturers' catalogs, trade literature, and the equipment issue of *Safety & Health*, published each March.

The rule to follow when specifying or buying safety equipment is to insist on the best and to deal only with reputable vendors. Do not take a chance on inferior items just because they may be less expensive. Protective equipment should conform to standards where they apply. Standards for protective equipment are discussed later in this chapter.

Finally, workers will want to know how easily protective equipment can be cleaned and maintained. This is particularly important in the case of hearing and respiratory protection.

Check with safety or industrial hygiene personnel who can specify the appropriate equipment. They should make sure that it fits correctly and that there is a good cleaning and maintenance program so that equipment is always sanitary. By using common sense, empathizing with the users, and understanding

Figure 9–1. Supervisors need to motivate employees to routinely wear properly fitted personal protective equipment. Here the super visor is talking to a new employee about obtaining a better fitting pair of safety glasses.

the basic principles about protective equipment, you can overcome any objections.

Selling the Need for PPE

The OSHA PPE standard requires selling workers on the need for personal protective equipment. If your employees can be made to see the need for such protection, your job is much easier. Involve the affected employees in every step of the PPE process to generate a sense of ownership in the program.

The workplace assessment process will define the need for PPE. When considering personal protective equipment, picture the human body. Starting with the head, analyze all hazards or types of accidents that could possibly occur. For example, the head is vulnerable to injuries such as bumps and abrasions, so think of what types of protection are needed: hard hats, bump caps? Move on to the eyes, ears, throat and lungs (respiratory protection), arms and legs, torso, hands and toes. In this way, you will be reminded of all types of hazards that can occur and the protection that is required.

Another important consideration is availability and convenience of equipment. Make it simple, available, popular, and clean.

Cost of Equipment

Companies differ in their personal protective equipment payment policies. Usually companies that have "clean room" environments will furnish laundered coveralls, aprons, smocks, and other garments.

Personal items that can be used off the job, such as safety shoes and prescription safety glasses, are often purchased on a shared-cost basis with employees paying part of the cost. Some firms maintain their own stores of PPE and others have routine scheduled vendor visits, such as safety shoe mobiles, to make equipment available to employees. Items considered necessary to the job are often supplied free to the employee.

Companies should be concerned with controlling costs, but not at the expense of safety. In the long run it can be less expensive to purchase necessary equipment than to pay for workers' injuries caused by the lack of proper PPE.

Overcoming Obstacles

One of the biggest obstacles supervisors face is overcoming the objections of some workers who have to wear protective equipment. Many of the objections you will encounter are quite similar, whether employees are talking about eye, ear, or head protection; respiratory protective equipment; or protective clothing such as vests, aprons, gloves, and even safety shoes.

To understand why employees do not wear the prescribed protective equipment, try to be objective and to see the entire picture.

First, proper fit and comfort are important. Make sure that safety glasses, ear protectors, respirators, aprons, gloves, and shoes are properly fitted. Proper fit is facilitated by having more than one size and more than one brand available for employees' selection of the best fit. No one wants to wear equipment that does not feel comfortable, that is not clean, or that constantly needs adjusting (Figure 9–2).

Second, appearance should be considered. If the piece of equipment does not look attractive—for example, big clumsy safety shoes as opposed to snug steel-toed moccasins or wing-tip dress shoes—workers may not wear it. The best situation gets workers involved in the selection process.

HEAD PROTECTION

Safety hats or helmets are needed on jobs where a person's head is menaced by falling or flying objects or by bumps (Figure 9–3). American National Standard

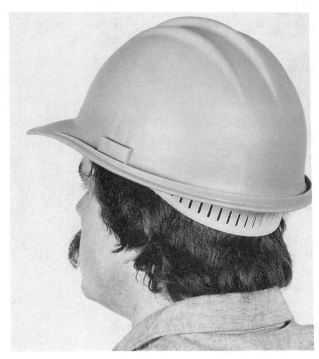

Figure 9–3. Safety helmets displace the impact of blows to the head from falling or flying objects or bumps. This safety helmet also has a rain trough. (Courtesy E.D. Bullard Company.)

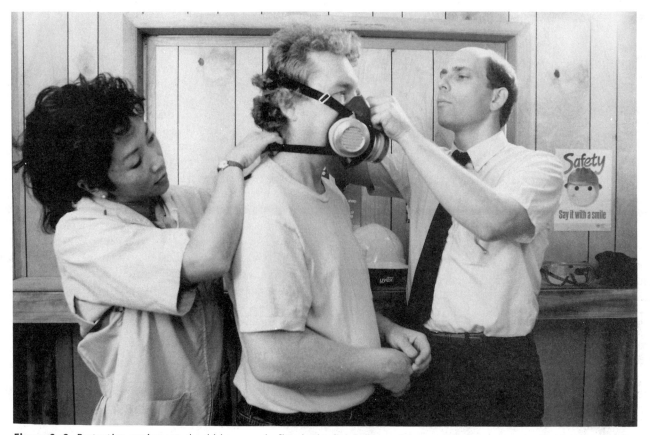

Figure 9–2. Protective equipment should be properly fitted. The fit of this canister respirator is being checked by a technician.

Z89.1–1986, *Protective Headwear for Industrial Workers,* gives specifications for protective hats.

Impact resistance is essential. Where contact with energized circuits is possible, only helmets that meet the requirements of Class B, ANSI Z89.1, should be worn. These helmets should have no conductive fittings passing through the shell. Class B hard helmets are tested at 20,000 volts.

A brim around the helmet provides the most complete protection for the head, face, and back of the neck. In a situation where the brim would get in the way, a helmet with a cap visor may have to be worn. Helmets should not be worn backwards unless approved by the manufacturer, because the bill is designed to deflect.

Another type of head protection is known as "bump hats" or "bump caps." These are used only in situations where the hazard is limited to bumping the head on some obstruction. Bump hats should never be used on construction sites, shipyards, or other locations where more dangerous hazards are present. This type of head gear does not meet the requirements of ANSI Z89.1.

Figure 9–4. This rescue worker needs a safety helmet with a 3–point chin strap to ensure the helmet does not fall off as he climbs. (Courtesy E.D. Bullard Company.)

Types and Materials

Plastic, molded under high pressure, is most frequently used for safety helmets. It resists impact, water, oil, and electricity. Fiberglass impregnated with resin is preferred because of its high strength-to-weight ratio, high dielectric strength, and resistance to moisture. Helmets with reflective trim are available on special order. They give added protection against accidents at night or in darkened areas.

The hard outer shell of the helmet is supported by a suspension. A cradle is attached to a headband that keeps the shell away from the head and provides some degree of protection against falling objects. This suspension must be adjusted to ensure a good fit. Keep extra cradles and headbands on hand, and replace a cradle or headband whenever it shows signs of deterioration or soiling.

Bands, cradles, and shells can be washed in warm, soapy water and rinsed or steam cleaned. Thirty days is the recommended maximum interval between cleanings. Make sure helmets and suspensions are thoroughly cleaned and inspected before reissuing them. The suspension should be replaced at least once a year.

Never attempt to repair the shell of a helmet once it has been broken or punctured. A few extra shells should be kept on hand as replacements. Do not let people drill holes in their safety helmets to improve ventilation or let them cut notches in the brims. Such practices destroy the ability of the helmet to protect the wearer.

Auxiliary Features

Liners for safety head gear are available for cold weather use. Do not let workers remove the safety helmet suspension in order to wear the helmet over parka hoods. This practice completely destroys the protection given by the helmet and has led to tragic results.

A chin strap is useful when the wearer may be exposed to strong winds, such as those on oil derricks (Figure 9–4).

An eye shield of transparent plastic can be attached to some helmets. The shield is secured under the brim and lies flat against it when not in use. Full face shields should be used where needed.

Brackets to support welding masks or miners' cap lamps are available on some safety helmets. Other mounting accessories for hearing, eye, and face protection are available (Figure 9–5).

Standard colors of safety helmets are white, yellow, red, green, blue, brown, and black. Other colors are available on special order. Colors that go all the way through the laminated shell are permanent. Painting is not recommended because paint may contain solvents that can make the shell brittle.

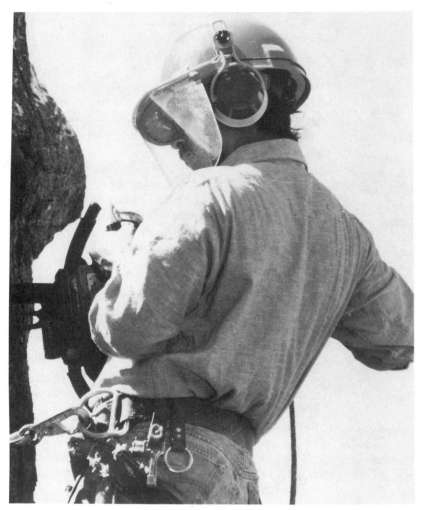

Figure 9–5. This tree trimmer uses a safety helmet with a 2-point chin strap and attachments for a safety shield and muff-type hearing protectors. (Courtesy E.D. Bullard Company.)

Often distinctive colors or designs are used to designate the wearer's department or trade, especially in companies where certain areas are restricted to selected employees. Avoid applying stickers, which could hide cracks or defects.

Obtaining User Acceptance

In some cases, workers' personal preferences in appearance and dress may clash with safety regulations. For example, hair styles may be incompatible with wearing safety helmets. As the group's leader, you will have to convince people that it is more important (indeed, it may be required) to wear helmets than to have stylish hairdos.

In certain industries, such as food-processing, workers are required not only to wear bump caps but to cover all facial hair—beards and mustaches—with hair nets. This is a requirement of the FDA and is done for sanitary reasons.

In overcoming objections and convincing people to use PPE, you must be prepared to counter all types of complaints. In the case of safety helmets, for instance, workers may complain that in winter the helmets are too cold. A liner or skull cap large enough to come down over the ears can take care of this objection. The suspension must not be removed, however, some people may complain that helmets give them headaches. This is a possible effect until the employee becomes accustomed to wearing the helmet. In general, however, there is no physiological reason for a properly fitted helmet to cause a headache.

Although safety helmets are slightly heavier than ordinary hats, the difference is unnoticed after a while, particularly if the headband and suspension are properly adjusted. Workers should not be bothered by helmets unless there are unique physiological or psychological reasons for their complaints.

FACE PROTECTION

Many types of personal protective equipment shield the face (and sometimes the head and neck) against

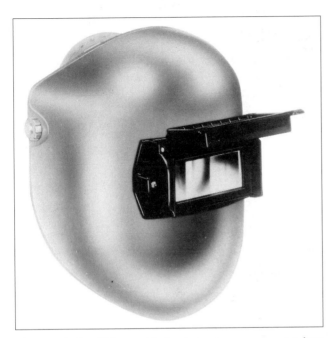

Figure 9–6. This welder's helmet protects against splashes of molten metal and radiation. (Courtesy Sellstrom Manufacturing Co.)

impact, chemical or hot metal splashes, heat radiation, and other hazards.

Face shields of clear plastic protect the face of a person who is sawing or buffing metal, doing sanding or light grinding, or handling chemicals. The shield should be slow burning, and must be replaced if warped or scratched. Eye protection should be worn under the face shield. A regular replacement schedule must be set up because plastic tends to become brittle with age.

A face shield that protects against light grinding may not protect against certain chemicals. The headgear and shield should be easy to clean and be adjustable to size and contour of the head. Many types permit the wearer to raise the shield without removing the headgear.

Plastic shields with dichroic coating or metal screen face shields deflect heat and still permit good visibility. They are used around blast furnaces, soaking pits, heating furnaces, and other sources of radiant heat.

Helmets

Welding helmets protect eyes and face against splashes of molten metal and radiation produced by arc welding (Figure 9–6). Helmets should have proper filter glass to keep ultraviolet and visible rays from harming the eyes. A list of shade numbers recommended for various operations is give in Table 9–A. Workers' eyes vary due to age, general health, and the care given them. Two people may require different shaded lenses when doing the same

Table 9–A. Selection of Shade Number for Welding Filters

Welding Operation	Suggested Shade Number
Shielded Metal-Arc Welding, up to 5/32 in. (4 mm) electrodes	10
Shielded Metal-Arc Welding, 3/16 to 1/4 in. (4.8 to 6.4 mm) electrodes	12
Shielded Metal-Arc Welding, over 1/4 in. (6.4 mm) electrodes	14
Gas Metal-Arc Welding (Nonferrous)	11
Gas Metal-Arc Welding (Ferrous)	12
Gas Tungsten-Arc Welding	12
Atomic Hydrogen Welding	12
Carbon Arc Welding	14
Torch Soldering	2
Torch Brazing	3 or 4
Light Cutting, up to 1 in. (25 mm)	3 or 4
Medium Cutting, over 1 to 6 in. (26 to 150 mm)	4 or 5
Heavy Cutting, over 6 in. (150 mm)	5 or 6
Gas Welding (Light), up to 1/8 in. (3.2 mm)	4 or 5
Gas Welding (Medium), 1/8 to 1/2 in. (3.2 to 12.7 mm)	5 or 6
Gas Welding (Heavy) over 1/2 in. (12.7 mm)	6 or 8

*The choice of filter shade may be made on the basis of visual acuity and may therefore vary widely from one individual to another, particularly under different current densities, materials, and welding processes. However, the degree of protection from radiant energy afforded by the filter plate or lens when chosen to allow visual acuity will still remain in excess of the needs of eye filter protection. Filter plate shades as low as shade 8 have proven suitably radiation-absorbent for protection from the arc-welding processes.

In gas welding or oxygen cutting where the torch produces a high yellow light, it is desirable to use a filter lens that absorbs the yellow or sodium line in the visible light of the operation (spectrum).

Notes:

1. All filter lenses and plates shall meet the test for transmission of radiant energy prescribed in ANSI Z 87.1–1989, *Practice for Occupational and Educational Eye and Face Protection*.

2. All glass for lenses shall be tempered, substantially free from striae, air bubbles, waves, and other flaws. Except when a lens is ground to provide proper optical correction for defective vision, the front and rear surfaces of lenses and windows shall be smooth and parallel.

3. Lenses shall bear some permanent distinctive marking by which the source and shade may be readily identified.

Source: *U.S. Code of Federal Regulations*, Title 29–Labor, Chapter XVII, Part 1910, Occupational Safety and Health Standards, §1910.252(e)(2)(i), and Part 1926, Safety and Health Regulations for Construction, §1926.102(a)(5).

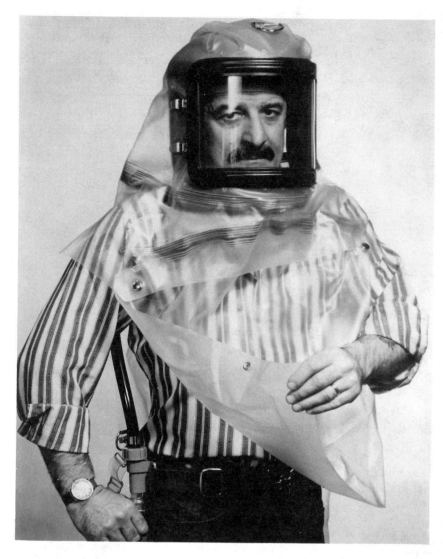

Figure 9–7. This air-supplied hood with plastic window shields against chemical splashes. (Courtesy AO Safety Products.)

job. Cracked or chipped filter lenses must be replaced. Otherwise, they may permit harmful rays to reach the welder's eyes.

The shell of the helmet must resist sparks, molten metal, and flying particles. It should be a poor heat conductor and a nonconductor of electricity. Helmets that develop pinholes or cracks must be discarded. Headgear should permit workers to use both hands and raise the helmet to position their work.

Most types of helmets have a replaceable, heat-treated glass or plastic covering to protect the filter lens against pitting and scratching. Some have a lift-front glass holder that permits the worker to inspect work without lifting or removing the helmet.

Proper eye protection should be worn under helmets to protect welders from flying particles when the helmets are raised. The spectacle type with side shields is recommended as minimum protection from flying materials, adjacent work, or from the popping scale of a fresh weld. Welders' helpers should also wear the required eye protection while assisting in a welding operation or while chipping flux away after a bead has been run over a joint.

Welding Shields and Goggles

A hand-held shield can be used where helmet protection is not needed, such as for inspection work, tack welding, and other operations requiring little or no welding. Frame and lens construction are similar to that of the helmet.

Welding goggles are available with filter glass shades up to No. 8. If darker shades are required, then complete face protection is needed because of the danger of skin burns. When shades darker than No. 3 must be used, side shields or cup goggles are recommended.

Hoods and Face Shields

Acid-proof hoods that cover head, face, and neck are

worn to protect against splashes from corrosive chemicals (Figure 9–7). This type of hood has a window of glass or plastic that is securely joined to the hood to prevent acid from seeping through. Chemical-resistant hoods made of vinyl, fiberglass, rubber, neoprene, plastic film, or impregnated fabrics are available for resistance to different chemicals. Consult manufacturers to find the protective properties of each material.

Hoods with an air supply should be worn when high levels of toxic fumes or dusts are present. These hoods provide a supply of clean, breathing-quality air, which excludes toxic materials and maintains worker comfort. To support the air hose, the worker should wear a harness or belt.

Transparent face shields, supported from a head harness or headband, are used when there is limited exposure to direct splashes of corrosive chemicals. Splash-resistant cup goggles should be worn under the shield for eye protection.

EYE PROTECTION

Industrial operations expose the eyes to a variety of hazards, such as flying objects, splashes of corrosive liquids or molten metals, dusts, and harmful radiation. Eye injuries not only disable a person but they often disfigure the face. Per-injury cost is high to both employee and employer.

Flying objects such as metal or stone chips, nails, or abrasive grits cause most injuries. Prevent

Figure 9–8. Fog ban safety goggles with high impact-resistant polycarbonate lenses. (Courtesy General Bandages, Inc.)

Blindness America lists the other chief causes of eye injury as:

- Abrasive wheels (small flying particles)
- Corrosive substances
- Damaging visible or thermal radiation
- Splashing metal
- Poisonous gases or vapors

Operations in which hardened metal tools are struck together, where equipment or material is struck by metal handtools, or where cutting action causes particles to fly, usually require the user and other workers nearby to wear eye protection. The hazard can be minimized by using nonferrous, "soft" striking tools, and by shielding the job with metal, wood, or canvas deflectors. Safety goggles or face shields should be worn when woodworking or cutting tools are used at head level or overhead with the chance of particles falling or flying into the eyes (Figure 9–8).

Occasionally, the need for eye protection is overlooked on potentially hazardous jobs. These include cutting wire and cable, striking wrenches, using hand drills, chipping concrete, removing nails from scrap lumber, shoveling material to head level, working on the leeward side of the job, using wrenches and hammers overhead, and other jobs where particles or debris may fall. Make sure workers on these jobs wear proper protective eye gear with sufficient side shields and brow guards.

Contact Lenses

Because contact lenses are popular among many workers, you should understand the related potential safety problems. Where there are appreciable amounts of dust, smoke, irritating vapors, or liquid irritants that could splash into the eyes, contact lenses usually are not recommended. A judgment must be made as to the degree of potential hazard. Workers should wear proper safety spectacles or goggles over their contacts where any potential eye hazard exists.

One rarely noticed regulation under OSHA prohibits the wearing of contact lenses in contaminated areas with respirators; see 29 *CFR* 1910.134(e)(5)(ii).

Prevent Blindness America, through its Advisory Committee on Industrial Eye Health and Safety, has issued the following position statement:

Except for work environments in which there are significant risks of ocular injury, employees may be allowed to wear contacts on the job. However, contact lens wearers must conform to management policy regarding contact lens use. When formulating policy on contact lens use,

OSHA and NIOSH guidelines should be used. Prevent Blindness America offers a list of recommendations on contact lens use in the workplace. To obtain these recommendations and for further information, contact Prevent Blindness America, 500 East Remington Road, Schaumburg, IL 60173. (847) 843–2020.

Goggles

Goggles and other kinds of eye protection are available in many styles, along with the protective medium of heat-treated or chemically treated glass, plastic, wire screen, or light-filtering glass. Supervisors should be familiar with the various forms of eye protection and should know which ones are the best for each job.

Types and Materials. Cover goggles are frequently worn over ordinary spectacles. Lenses that have not been heat- or chemically treated are easily broken. Cover goggles protect against pitting, as well as breaking.

These goggles include the cup type with heat-treated lenses and the wide-vision type with plastic lenses. Both are used for heavy grinding, machining, shipping, riveting, working with molten metals, and similar heavy operations. They offer the advantage of being wide enough to protect the eye socket and to distribute a blow over a wide area.

Spectacles without side shields may be worn only when it is unlikely that particles will fly toward the side of the face. However, spectacles with side shields are recommended for all industrial uses. Frames must be rigid enough to hold lenses directly in front of the eyes. These glasses should be fitted by a specialist.

Goggles with soft vinyl or rubber frames protect eyes against splashes of corrosive chemicals and exposure to fine dusts or mists. Lenses can be heat-treated glass or acid-resistant plastic. For exposures involving chemical splashes, these goggles are equipped with baffled ventilators on the sides. For vapor or gas exposures, they must be nonventilated. Some types are made to fit over spectacles.

Dust goggles should be worn by employees who work around noncorrosive dusts—for example, in cement and flour mills. The goggles have heat-treated or filter lenses; wire-screen ventilators around the eye cup provide air circulation.

Welders' goggles with filter lenses are available for such operations as oxyacetylene welding, cutting, lead burning, and brazing. Table 9–A is a guide for the selection of the proper shade numbers. These recommendations may be varied to suit individual needs.

All goggle frames should be of corrosion-resistant material that will neither irritate nor discolor the skin, that can withstand sterilization, and that are flame-resistant or nonflammable. Metal frames should not be worn around electrical equipment or near intense heat.

To permit the widest range of vision, goggles should be fitted as close to the eyes as possible without the eyelashes touching the lens. The lenses should have no appreciable distortion or prism effect. Some specifications are covered by ANSI Z87.1–1989, *Practice for Occupational and Educational Eye and Face Protection.*

Care of Eye Protection Equipment. Eye protection equipment must be sterilized before being reissued to different employees. The proper procedure is to disassemble the equipment and wash it with soap or detergent in warm water. Rinse thoroughly. Immerse all parts in a solution containing germicide, deodorant, and fungicide. Do not rinse, but hang to dry in air. When dry, place parts in a clean, dust-proof container—a plastic bag is acceptable. Many companies have stations that dispense cleaning liquid and tissues to encourage frequent cleaning after goggles get smudged or dirty. Carrying cases or storage cabinets also help promote care of eye protection equipment.

Replace Defective Parts. If a lens has more than just the most superficial scratch, nick, or pit, it should be replaced. Such damage can materially reduce protection afforded the wearer.

Some companies keep extra goggles in the main supply department, while others keep them in individual departments, along with a stock of parts.

Glasses and goggles should never be placed with the lenses facing down because they could become scratched, pitted, or dirty. They should not be kept in a bench drawer or tool box, unless they are in a sturdy case.

Cleaning stations or materials should be easily available. A lens "fog" problem can usually be eliminated by use of one of the many commercial antifog preparations. In hot weather, workers can wear a sweatband to keep perspiration off their goggles. Although it takes a little effort to keep the goggles clean, this should not be an excuse for workers to discard them and risk losing an eye.

Visitors to areas requiring eye protection should be provided with ANSI-approved eye protection, the same as that used by employees. Avoid the use of non-approved eye protection, which could mistakenly be used by employees.

EAR PROTECTION

Excessive noise should be reduced by engineering changes and administrative controls, whenever possible, as outlined in Chapter 8, Industrial Hygiene. Hearing protection should be used only as a last resort.

Figure 9–9. This selection of hearing protection devices includes ear muffs, ear plugs, and Swedish wool.

Under OSHA, where the sound levels exceed an 8-hour time-weighted average of 85 dB measured on the A scale, a continuing, effective hearing conservation program shall be administered. This level may be increased slightly as the duration of exposure decreases. (Review the discussion in Chapter 8, Industrial Hygiene.)

It takes specialized equipment and trained personnel to analyze a noise exposure and to recommend the types of hearing protection that will be most effective (Figure 9–9). The best hearing protection is the one that is accepted by the individual and is worn properly. Properly fitted protectors can be worn continuously by most people and will provide adequate protection against most industrial noise exposures.

Insert Ear Protectors

Insert protectors are, of course, inserted into ear canals and vary considerably in design and material. Materials used are pliable rubber and soft or medium plastic. Rubber and plastic are popular because they are inexpensive, easy to keep clean, and give good performance.

Another type of ear plug popular with workers is one that is molded to fit each ear. After being allowed to set, it will hold its form. Because each person's ear canal is shaped differently, these plugs become the property of the individual to whom they were fitted. The plugs, of course, must be molded and fitted by a trained, qualified professional.

Some pressure is required to fit a rubber or plastic earplug into the ear canal so that the noise does not leak around the edges (as much protection as 15 dB could be lost). As a result, points of pressure develop, and they may cause discomfort. Even though plugs can be fitted individually for each ear, a good seal cannot be obtained without some initial discomfort. However, if the earplugs are made of soft material and kept clean, they should not create any lasting problems. Avoid earplugs composed of hard, rigid materials as they could injure the ear canal.

Skin irritations, injured ear drums, or other harmful side effects of PPE are rare when hearing protection is properly designed, well fitted, and kept clean. They should cause no more difficulty than does a pair of well-fitted safety goggles. If fitting continues to be a problem, casts can be made of the ear canals, and a plastic plug can be custom-made for each.

Muff Devices

Cup or muff devices cover the external ear to provide an acoustic barrier. The effectiveness of these devices varies with the size, seal material, shell mass, and suspension of the muff as well as with the size and shape of workers' heads. Muffs are made in a universal type of in specific head, neck, or chin sizes. Hearing protection kits that can be used with hard helmets are also available.

In the past few years, some employees have listened to radios with headphones. You, as the supervisor, should make sure that these are not worn in the shop. In addition to the possibility of damage to the ears from excessive noise levels, they are not safe in a working environment. People may not be able to hear oncoming facility trucks, vehicular traffic, other workers, or any number of industrial noises. Most Air Force bases have banned their use for this reason.

As supervisor, you should be aware of noise hazards

Figure 9–10. These employees are wearing filter and cartridge respirators while taking an air sample.

in your department. You should also be able to demonstrate the correct use of hearing protection when other forms of control are not practical.

RESPIRATORY PROTECTION

Respirators can be regarded as emergency equipment, or equipment for occasional use. Of course, if contaminants are present, they should be removed at the source, or the process should be isolated. Since leaks and breakdowns do occur, however, and since some operations expose a person only briefly and infrequently, respiratory protection should be available. Workers must be medically fit, fit-tested, and instructed and trained in its proper use and its limitations.

Types of Equipment

Respiratory equipment includes air-purifying devices (mechanical filter respirators, chemical cartridge respirators, combination mechanical filter and chemical cartridge respirators, and masks with canisters, Figure 9–10) and air-supplied devices (airline respirators, Figure 9–11, and self-contained breathing apparatus, Figure 9–12).

Air-Purifying Devices. Air-purifying devices remove contaminants from air prior to its entering the respiratory system. They can be used only in environments containing sufficient oxygen to sustain life. Air-purifying devices are only effective in the

limited concentration ranges for which they are designed, and must never be used where contaminant levels exceed the respirator manufacturer's accepted protection factor. These respirators generally consist of a soft, resilient facepiece and some kind of replaceable filtering element. Several types of air-purifying respirators, however, are available as completely disposable units. Various chemical filters can be employed to remove specific gases and vapors, while mechanical filters remove particulate contaminants. Air-purifying devices are never used in environments that are immediately dangerous to life or health.

Mechanical Filter Respirators. Mechanical filter respirators protect against exposure to nuisance dusts, mists, and fumes. Examples of nuisance dusts are aluminum, cellulose, cement, flour, gypsum, and limestone. Another type of filter respirator is approved for toxic dusts, such as lead, asbestos, arsenic, cadmium, manganese, selenium, and their compounds.

Figure 9–11. This is a type C supplied-air respirator and a 5-minute compressed air self-contained escape breathing apparatus to use in the event of a primary air source failure. (Courtesy National Draeger.)

Figure 9–12. This compressed air breathing apparatus is designed to provide respiratory protection during entry into and escape from atmospheres immediately dangerous to life or health.

Protection against mists—for example, chromic acid, and exposure to such fumes as zinc and lead—is given by mechanical filter respirators specifically approved for such exposures. The filter, usually made of paper or felt, should be replaced frequently. If it becomes clogged, it restricts breathing or becomes inoperative. This condition can happen as frequently as several times a shift. A mechanical filter respirator is of no value as protection against chemical vapors, injurious gases, or oxygen deficiency. To use it under these conditions is a serious mistake.

Chemical Cartridge Respirators. Chemical cartridge respirators have either a half-mask facepiece or a full-mask facepiece connected to one or more small containers (cartridges) or sorbent, typically activated charcoal or soda lime (a mixture of calcium hydroxide with sodium or potassium hydroxide) for absorption of low concentrations of certain vapors and gases.

The life of the cartridges can be relatively short. For protection against mercury vapors, the nominal container life is 8 hours. After use, the cartridges must be discarded. These respirators must not be used in atmospheres immediately dangerous to life or health, such as those deficient in oxygen.

Gas Masks. Gas masks consist of a facepiece or mouthpiece connected by a flexible tube to a canister (a very large cartridge). Inhaled air, drawn through the canister, is cleaned chemically. Unfortunately, no one chemical has been found that removes all contaminants.

Gas masks have definite limitations on their effectiveness. Both gas concentration and length of time influence this. Gas masks, like chemical cartridge respirators, do not protect against oxygen deficiency. When the canister is used up, it should be removed and replaced by a fresh one. Even if they have not been used, canisters should be replaced periodically. Manufacturers will indicate the maximum effective life.

Gas masks must be quickly available for emergencies. For example, in an ice facility where an ammonia leak is likely, masks should be placed either just inside or just outside the exit doors so that they can be reached quickly. Masks should be stored away from moisture, heat, and direct sunlight. They should be inspected regularly.

Hose Masks. Hose masks, with or without a blower, should not be used in atmospheres immediately dangerous to life or health.

Careful selection of air–purifying cartridges and canisters must be conducted to match the contaminants to the correct filter or cartridge. Cartridges and canisters are color-coded according to the type of exposure.

Supplied-Air Devices. Supplied-air devices deliver breathing air through a hose connected to the wearer's facepiece. The air source used is monitored frequently to make sure it does not become contaminated—for example, with carbon monoxide.

The air-line respirator can be used in atmospheres not immediately dangerous to life or health, especially where working conditions demand continuous use of a respirator. Each person should be assigned his or her own respirator.

Air-line respirators are connected to a compressed air line. A trap and filter should be installed in the compressed air line ahead of the mask to separate oil, water, grit, scale, or other matter from the air stream. When line pressures are over 25 psi (170 kPa), a pressure regulator is required. A pressure release valve, set to operate if the regulator fails, should be installed.

To get clean air, keep the compressor intake away from any source of contamination, for example, internal combustion engine exhaust. The compressor should have a carbon monoxide alarm in order to guard against the carbon monoxide hazard either from overheated lubricating oil or from engine exhaust. The most desirable air supply is provided by a nonlubricated or externally lubricated medium-pressure blower—for example, a rotary compressor.

If workers need to move from place to place, they may find that the air hose is a nuisance. You should be aware that using the hose will reduce its efficiency. Care must also be exercised to prevent damage to the hose; for example, it should not be permitted to be in oil.

Abrasive Blasting Helmets. Abrasive blasting helmets are a variety of an air-line respirator designed to protect the head, neck, and eyes against the impact of the abrasive, and to give a supply of breathing air. The air quality requirements are the same as those described for air-line respirators.

The helmet should be covered both inside and out with a tough, resilient material. This increases comfort and still resists the abrasive. Some helmets have an outer hood of impregnated material and a zippered inner cape for quick removal. The helmet should contain a glass window, protected by a 30- to 60-mesh fine wire screen or plastic cover plate. Safety glass, used to prevent shattering under a heavy blow, should be free of color and glass defects.

A lighter weight helmet is approved for less hazardous work, such as spraying paint, buffing, bagging, and working in clean rooms (Figure 9–11).

Self-contained breathing apparatus (SCBA) offers protection for various periods of time by providing a portable, clean, tested air supply, which is worn by the user. Pressure-demand is the most common SCBA type.

The wearer of a self-contained breathing apparatus is independent of the surrounding atmosphere; therefore, this kind of respiratory protective equipment can be used in environments where air contaminants are immediately dangerous to life and health. This equipment is frequently used in mine rescue work and in firefighting (Figure 9–12). Because of the extreme hazard, no one wearing self-contained breathing apparatus should work in an oxygen-deficient atmosphere unless other people similarly equipped are standing by, ready to give help.

Selecting the Respirator

Air contaminants range from relatively harmless substances to toxic dusts, fumes, vapors, mists, and gases that may be extremely harmful. Respiratory protection is required in certain areas or during certain operations when engineering controls are not available to reduce airborne concentrations of contaminants to a safe level. Engineering controls might not be present because they are technically unfeasible, or the hazardous operation might be done infrequently, making controls impractical. Respiratory protection is also needed while engineering controls are being implemented.

Your first step in selecting a respirator is to determine the chemical or other offending substance in the environment and to evaluate the extent of the hazard. Conducting a hazard assessment should help you with this information and will help you choose the respirator that best protects against the particular hazard. Before respiratory equipment is ordered, discuss the type of exposure you have identified with the company safety or industrial hygiene department, and with manufacturers and dealers.

Protection factors are a measure of the overall effectiveness of a respirator. These factors, based on tests and on professional judgment, range from 5 to 10,000. The maximum use concentration for a respirator is determined by multiplying the Threshold Limit Values (TLV®s) of the substance to be protected against by the protection factor. For example, a respirator with a protection factor of 10 for acetic acid would protect a worker against concentrations up to 10 times its TLV. Since the TLV of acetic acid is 10 ppm, the worker would be protected in atmospheres containing acetic acid concentrations as high as 100 ppm. (TLVs were explained in Chapter 8, Industrial Hygiene.)

The second step is to make sure workers are medically fit to wear a respirator. Routine medical surveillance, including spirometry, is required to ensure the workers have adequate lung and heart functioning to wear a respirator that can place physical demands on the worker.

Then, you must make sure the respirator fits workers properly. Fitting respirators to employees involves trying on several types and sizes and conducting fit checks and fit tests to determine the right respirator for the worker. This process could take one to two hours. In addition, training and education are needed. You should explain why it is essential to wear the respiratory protection, how it works, and, in general, why workers should wear the proper respirator for the operation in question. Get the affected employees involved in the entire process to facilitate their compliance.

Only respirators approved by the National Institute for Occupational Safety and Health (NIOSH) and the Mine Safety and Health Administration (MSHA) are acceptable. A NIOSH-MSHA-approved respirator is assigned an approval number prefixed by the letters TC, indicating that it has been tested and certified by NIOSH and MSHA for the air contaminant at the concentration range stated.

This approval ensures that the design, durability, and workmanship of the equipment meet minimum standards. Technicians have tested the respirator for worker safety, freedom of movement, vision, fit, and comfort of facepiece and headpiece; the ease with which the filter and other parts can be replaced; tightness of the seal against dust; freedom from leakage; and resistance to air flow when the wearer is inhaling and exhaling.

Cleaning the Respirator

Respirator facepieces and harnesses should be cleaned and inspected regularly. If several persons must use the same respirator, it should be disinfected after each use in order to comply with OSHA regulations. Methods of disinfection and cleaning should be obtained from the equipment manufacturer.

The supervisor or a designated employee should inspect respirators at intervals to check for damage or improperly functioning parts, such as headbands or valve seats.

If the respirator is removed at intervals, dust settles on it. When the respirator is replaced, these dust particles are transferred to the skin and cause irritation. To prevent this, the wearer should keep the respirator on at all times when in a contaminated atmosphere.

BODY PROTECTION

The most common protection for the abdomen and trunk is the full apron. Aprons are made of various materials. Leather or fabric aprons, with padding or stays, offer protection against light impact and against sharp knives and cleavers, such as those used in packinghouses. Heat-resistant coats and aprons are often used by those who work around hot metal or other sources of intense conductive heat.

An apron worn near moving machinery should fit snugly around the waist. Neck and waist straps should be either light strings or instant-release fasteners in case the garment is caught. There should be a fastener at each end of the strap to prevent severe friction burns should the strap be caught or drawn across the back of the neck. Split aprons should be worn on jobs that require mobility on the part of the worker. Fasteners draw each section snugly around the legs.

Welders are often required to wear leather vests or capes and sleeves, especially when doing overhead welding, as protection against hot sparks and bits of molten metal. On jobs where employees must carry heavy, angular loads, pads of cushioned leather or padded duck are used to protect the shoulders and back from injury.

Harnesses

Harnesses with lifelines attached should be worn by those who work at high levels or in closed spaces and by those who work where they may be buried by loose material or be injured in confined spaces. As of January 1, 1998, safe belts will be allowed as a positioning device only. This discussion, however, does not pertain to vehicular seat belts or linemen's belts.

Normal use involves comparatively light stresses applied during regular work—stresses that rarely exceed the static weight of the user. Emergency use means stopping an individual when he or she falls. Every part of the harness may be subjected to an impact loading many times the weight of the wearer.

Harnesses distribute evenly the shock of an arrested fall (Figure 9–13). This shock is distributed over shoulders, back, and waist, instead of being concentrated at the waist. The harness permits a person to be lifted with a straight back rather than bent over a waist strap. This makes rescue easier if the victim is unconscious, buried, or must be lifted out through a manhole. Wherever a job requires use of a self-contained breathing apparatus or supplied-air device, a harness and lifeline should also be used to assist in escape.

When workers are at risk of falling, the safety harness is designed to distribute the impact force over the legs and chest as well as the waist. A shock absorber or decelerating device, which brings the falling person to a gradual stop, lessens the impact load on both the equipment and the person. To prevent a fall, the line should be tied off overhead and should be as short as movements of the worker will permit.

Materials. Harnesses are made of natural or synthetic materials by most manufacturers. Nylon lanyards and harnesses, although widely used, may not be satisfactory under some chemical conditions. Special types of harnesses are available for certain environments—wax-treated to resist paint solvents or mildew, neoprene-impregnated to resist acids and oil products. Be sure to consult with your safety department so you obtain the right equipment.

Harnesses are designed to give a little when stress loaded. If a person does fall and stresses a harness, the harness should be taken out of service and replaced. Several kinds of devices that attach the harness to a lifeline serve to reduce the stress of a fall; these minimize both the injury to the user and the damage to the harness.

Inspection. Harnesses and lines should be examined each time before using. At least once every three months, harnesses should be checked by a trained inspector.

Harnesses should be inspected for worn and torn fibers. When a number of the outer fibers are worn or

Figure 9–13. A Worker who has fallen is stopped by a fall arrest system. (Courtesy Rose Manufacturing Co.)

cut through, the harness should be discarded.

Hardware should be examined and worn parts replaced. If the harness is riveted, each rivet should be carefully inspected for wear around the rivet hole. Dirt and dust should be brushed carefully from harnesses to prevent fiber damage. Broken or damaged stitching may require replacement.

Harnesses may be washed in warm, soapy water, rinsed with clear water, and dried by moderate heat. If these harnesses are worn under unusual conditions, or if a dressing is to be used, consult the manufacturer.

The outer surface of rope lines should be examined for cuts and for worn or broken fibers. Rope should be examined for breaks, discoloration, and deterioration. If the rope shows any of these signs, it should be discarded.

A steel wire rope should be examined for broken strands, rust, and kinks that may weaken it. Ropes must be kept clean, dry, and rust free. They should be lubricated frequently, especially before use in acid atmospheres or before exposure to salt water. After such use, wire rope should be carefully cleaned and again coated with oil.

Lifelines

Wire rope should not be used for lifelines where falls are possible, unless a shock-absorbing device is also used. The rigidity of steel greatly increases the impact loading. Steel is, furthermore, hazardous around electricity.

Knots reduce the strength of all ropes. The degree of loss depends upon the type of knot and the amount of moisture in the rope. More information is given in Chapter 15, Materials Handling.

Protection Against Chemicals

Protecting workers in and around chemicals is of paramount importance. Many chemical substances to which employees are exposed can be irritating and result in serious burns on the body, hands, arms and legs. You, as the supervisor, have the responsibility of evaluating operations using chemicals and determining what protective equipment or proper clothing is needed. (See the discussion of chemical hazards in Chapter 8, Industrial Hygiene.)

A workplace assessment should include an evaluation of the need for protection against chemicals.

If you need help for this task, contact your safety department, insurance company safety engineer, industrial hygiene department, protective equipment supplier, or a chemist. The best way to protect people is by making sure they wear protective clothing designed specifically for the chemicals involved. Keep in mind, however, that the best course of action may be to substitute the chemical with a less toxic one. If the chemical cannot be eliminated, replaced, or isolated, then your only recourse is protective clothing.

Protection Against Ionizing Radiation

Ionizing radiation, dangerous because of its serious biological effects, need not be feared if it is respected and if suitable precautionary measures are taken. The National Safety Council's *Fundamentals of Industrial Hygiene* has much information on the topic and many references. (Also see Chapter 8, Industrial Hygiene, for a more detailed discussion of avoiding radiation hazards.)

Standards for protection against radiation, specified in OSHA regulations 1910.96 and .97, reference the Nuclear Regulator Commission regulations. These are spelled out in the *Code of Federal Regulations*, Title 10, "Energy," Part 20.

Rubber gloves for handling radioactive materials, disposable clothing, suits with supplied air, and approved respiratory devices are some of the specialized equipment needed. Under no circumstances should contaminated clothing be worn into clean areas. Thorough washing with soap and water is usually the best general method for decontamination of the hands and other parts of the body, regardless of the contaminant.

Protective Clothing

Ordinary work clothing, if clean, in good repair, and suited to the job, may be considered safe. "Protective clothing" refers to garments designed for specific, hazardous jobs where ordinary work clothes do not give enough protection against such injuries as abrasions, burns, and scratches.

Guidelines. Good fit is important. Trousers that are too long must be shortened to proper length, preferably without cuffs. If cuffs are made, they should be securely stitched down so the wearers cannot catch their heels in them. Cuffs should never be worn near operations that produce flying embers, sparks, or other harmful matter.

Long or loose sleeves, gloves, and loose-fitting garments (especially about the waist) create a hazard because they are easily caught in moving (or revolving) machine parts.

Clothing soaked in oil or flammable solvent is easily ignited and is a definite hazard. Remember that skin irritations are often caused by continued contact with clothing that has been soaked with solvents or oils (see Dermatitis in Chapter 8, Industrial Hygiene).

Materials. A number of protective materials are used in making clothes to protect workers against various hazards. Supervisors should be familiar with these materials.

Aluminized and reflective clothing has a coating that reflects radiant heat. Aluminized or glass fiber is used for heavy-duty suits, and aluminized fabric for fire-approach suits (Figure 8–7).

Flame-resistant cotton fabric is often worn by people who work near sparks and open flames. Although the fabric is durable, the flame-proofing treatment may have to be repeated after one to four launderings.

Flame-resistant duck, used for garments worn around sparks and open flames, is lightweight, strong, and long-lasting. However, it is not considered adequate protection against extreme radiant heat.

Glass fiber is used in multilayered construction to insulate clothing. The facing is made of glass cloth or of aluminized fabric.

Impervious materials (such as rubber, neoprene, vinyl, and fabrics coated with these materials) protect against dust, vapors, mists, moisture, and corrosives. Rubber is used often because it resists solvents, acids, alkalis, and other corrosives. Neoprene resists petroleum oils, solvents, acids, alkalis, and other corrosives.

Leather protects against light impact and against sparks, molten metal splashes, and infrared and ultraviolet radiation.

Synthetic fibers (such as acrylics and low-density polyethylene) resist acids, many solvents, mildew, abrasion and tearing, and repeated launderings. Because some fabrics generate static electricity, garments should not be worn in explosive or high oxygen atmospheres, unless they are treated with an antistatic agent.

Water-resistant duck is useful for exposures to water and noncorrosive liquids. When it is aluminum-coated, the material also protects against radiant heat.

Wool may be used for clothing that protects against splashes of molten metal and small quantities of acid and small flames.

Permanent-press fabrics are used in a wide variety of protective clothing. However, care must be taken to determine whether or not the fabrics are flammable.

PROTECTING EXTREMITIES

Workers' extremities are highly vulnerable to injury in most work environments. Protective clothing and gear can reduce the number and severity of injuries workers suffer each year. As supervisor, you should educate and train all employees in the importance of using such protection. Injuries to or losses of fingers, toes, even parts of an arm or leg can happen with frightening suddenness. A good hazard assessment will define the type of protection needed and help workers understand the choices and that their best protection is prevention. This section discusses protective equipment for arms, hands, and fingers and for feet and legs.

Arms, Hands, Fingers

Fingers and hands may be subject to cuts, scratches, bruises, chemical exposure, and burns. Although fingers are hard to protect (because they are needed for practically all work), they can be shielded from any common injuries with such proper protective equipment as the following:

1. Heat-resistant gloves protect against burns and discomfort when the hands are exposed to sustained conductive heat.
2. Metal mesh gloves, used by those who work constantly with knives, protect against cuts and blows from sharp or rough objects.
3. Rubber gloves are worn by electricians. They must be tested regularly for dielectric strength.
4. Rubber, neoprene, and vinyl gloves are used when handling chemicals, etc. Neoprene and vinyl are particularly useful when petroleum products are handled.
5. Leather gloves are able to resist sparks, moderate heat, chips, and rough objects. They provide some cushioning against blows. They are generally used for heavy-duty work. Chrome-tanned leather or horsehide gloves are used by welders (Figure 9–14).
6. Chrome-tanned cowhide leather gloves with steel-stapled leather patches or steel staples on palms and fingers are often used in foundries and steel mills.
7. Cotton fabric gloves are suitable for protection against dirt, slivers, chafing, or abrasion. They are not heavy enough to use in handling rough, sharp, or heavy materials.
8. Heated gloves are designed for use in cold environments, such as deep freezers, and can be part of a heated-clothing system. Other types designed for such work are insulated with foam, and most such gloves are waterproof.

Specially made electrically tested rubber gloves, worn under leather gloves to prevent punctures, are used by linemen and electricians who work with high-voltage equipment. A daily visual inspection and air test (by mouth) must be made. Make sure that rubber gloves extend well above the wrist so that there is no gap between the coat or shirt sleeve and glove. The shortest glove is 14 in. (42 cm). (See National Safety Council Occupational Safety and Health Data Sheet 598, *Flexible Insulating Protective Equipment for Electrical Workers*, available in the NSC library.)

People are willing to wear gloves when the dangers of not wearing them are present. With gloves, workers do not have to worry about cutting their hands, and they can grip materials better. As a result, production increases. Still, if employees become lax, you should remind them often to wear their gloves.

In some instances, however, gloves may be a hazard, particularly when worn around certain machining operations, such as a drill press. The material may become caught in machine parts. Use of gloves in these areas should be prohibited.

Hand leathers or hand pads are often more satisfactory than gloves for protecting against heat, abrasion, and splinters. Wristlets or arm protectors are available in the same materials as gloves.

Feet and Legs

Foot Protection. About a quarter of a million disabling occupational foot injuries take place each year. This points to the need for foot protection in most industries, and the need for supervisors to see that their workers wear this gear. All safety shoes and boots should have toes reinforced with a toe cap. The three classifications are shown in Table 9–B. Protective shoes or boots also have other protective features, as well as reinforced toes. Protective footwear also includes insulating and conducting shoes or boots made to help protect workers from temperature extremes and from electrical shock.

The responsibility for proper care of protective footwear ordinarily rests with the employees. Letting

Figure 9–14. Welding gloves with self-extinguishing fleece lining. (Courtesy Elliot Glove Company.)

Table 9–B. Minimum Requirements for Safety-Toe Shoes

Classification	Compression Pounds	Impact Foot Pounds	Clearance Inches
			Men
75	2,500	75	$^{16}/_{32}$
50	1,750	50	$^{16}/_{32}$
30	1,000	30	$^{16}/_{32}$
			Women
Note: 1 pound force = 4.45 newtons			$^{15}/_{32}$
1 foot pound force = 1.36 joules			$^{15}/_{32}$
1 inch = 2.54 centimeters			$^{15}/_{32}$

Source: American National Standard Z41–1983, *Safety-Toe Footwear*.

employees select their shoe style encourages cooperation.

Metal-free shoes, boots, and other footwear are available for use where there are specific electrical hazards or fire and explosion hazards.

"Congress" or *gaiter-type shoes* are used to protect people from splashes of molten metal or from welding sparks. This type can be removed quickly to avoid serious burns. These shoes have no laces or eyelets to catch molten metal.

Reinforced or inner soles of flexible metal are built into shoes worn in areas where there are hazards from protruding nails and when the likelihood of contact with energized electrical equipment is remote, as in the construction industry.

For wet work conditions, in dairies and breweries, rubber boots and shoes, leather shoes with wood soles, or wood-soled sandals are effective. Wood soles have been so commonly used by workers handling hot asphalt that they are sometimes called "pavers' sandals."

Foot protection with metatarsal guards should always be worn during operations where heavy materials, such as drums, pig iron, heavy castings, and timbers, are handled. They are recommended whenever there is a possibility of objects falling and striking the foot above the toe cap. Metal foot guards are long enough to protect the foot back to the ankle, and may be made of heavy gauge or corrugated sheet metal. They are usually built onto the shoe.

Supervisors must often overcome workers' objections to wearing foot protection. Protective footwear used to be hot and heavy, and people often complained that they were uncomfortable. Current designs now make protective footwear comfortable, practical, and attractive. The steel cap weighs about as much as a pair of rimless eyeglasses or a wristwatch. The toe box is insulated with felt to keep the feet from getting too hot or cold. Some shoes have inner soles of foam latex with tiny "breathing cells."

Some people object to wearing safety shoes

because they do not cover the smallest toes. However, studies show that 75% of all toe fractures happen to the first and second toes. In most accidents, the toe box takes the load of the impact for the entire front part of the foot.

Another objection commonly voiced is that if the toe box were crushed, the steel edge would cut off the toes. Accidents of this type are rare. In the majority of cases, safety shoes give sufficient protection. Freak accidents, against which there is no sure protection, may happen. But a blow that would crush the toe cap would certainly smash someone's toes if he or she were not wearing foot protection.

Leg Protection. Leggings that encircle the leg from ankle to knee, and have a flap at the bottom to protect the instep, protect the entire leg. The front may be reinforced to give impact protection. Such guards are worn by persons who work with molten metal. Leggings should allow fast removal in case of emergency. Hard fiber or metal guards are available to protect shins against impact.

Padding can protect parts of the legs from injury. Knee pads protect employees whose work, like cement finishing or tile setting, requires much kneeling. Ballistic nylon pads are often used to shield thighs and upper legs against injury from chainsaws.

Foot and leg protectors are available in many different materials. The type selected depends on the work being done. Where molten metals, sparks, and heat are the major hazards, heat- or flame-resistant materials or leather is best. Where acids, alkalis, and hot water are encountered, natural or synthetic rubber or plastic, resistant to the specific exposure, can be used.

SUMMARY OF KEY POINTS

Key points covered in this chapter include:

- Supervisors must be familiar with government regulations regarding personal protective equipment (PPE), be able to recognize and assess hazards requiring PPE use, know the best types of equipment for particular hazards workers face, and educate and train employees in proper equipment use and maintenance.

- Supervisors can overcome employee objections to PPE by making sure equipment fits properly, looks appealing, and is easy to clean and repair or replace. Employees should be involved and then they can be sold on the need for PPE and how PPE can protect the body.

- Protective equipment may be purchased by the company, jointly paid for by company and employees. The best rule is to buy the best equipment and deal only with reputable firms.

- Head protection is worn to prevent head injuries from falling or flying objects or from bumps against stationary objects. Helmets can be insulated against electric shock and contain a hard outer shell with inner suspension or cradle, and headband. Helmets should be kept clean and replaced if damaged. Auxiliary features include liners, chin straps, eye shields, brackets, and standard or special-order colors. Helmets should be adapted to fit comfortably.

- The most common causes of eye injuries are flying objects, particles from abrasive wheels, corrosive substances, visible or thermal radiation, splashing metal, and poisonous gases or fumes. Workers can be protected either by shielding from the hazard and/or by providing safety goggles and face shields. Employees wearing contact lenses should be particularly careful in contaminated areas. Goggles are the most common eye protection gear, and supervisors should know which types are the best for each job. All eye-wear gear should be sterilized after use and defective parts replaced. Workers should be trained in proper use, handling, and care of protective eye-wear.

- Face protection should shield the face (and sometimes head and neck) from impact, chemical or hot metal splashes, heat radiation, and other hazards. Equipment includes helmets, shields and goggles, and hoods (covering head, face, and neck). Some hoods are provided with independent air supplies for work around toxic fumes or dusts. Gear should be selected for the particular hazard, be easy to clean, and adjustable to fit in size and contour. Supervisors should set up regular replacement schedules, because plastic gear tends to become brittle with age.

- Hearing protection equipment must be provided to workers exposed to noise levels above 85 dBA on an 8-hour time-weighted average to reduce noise to an acceptable level. Equipment includes insert ear protectors and muff devices. These protectors must be designed and fitted properly and kept clean or they can cause ear infections and hearing loss. Cup or muff devices cover the external ear; their effectiveness depends on size, shape, seal material, shell mass, suspension of the muff, and how well they fit. Workers should be thoroughly trained in use and care of protective hearing equipment.

- Respiratory equipment can be regarded as emergency equipment or gear for occasional use on the job. This equipment includes air-purifying devices and supplied-air devices. To select proper respiratory equipment, supervisors must first determine the hazard in the environment, evaluate the extent of hazard, and choose the respirator that best protects against that hazard. Workers should be trained in proper use and maintenance and gear should be inspected and tested frequently.

- The most common protection for abdomen and trunk is full aprons, harnesses, lifelines, suits and equipment designed for radiation and chemical hazards, and protective clothing. Harnesses can contain the worker or help to distribute the shock of an arrested fall. Belts, harnesses, and lifelines must be cleaned and inspected frequently. Equipment must meet all standards and be worn when required.

- Protective equipment items for arms, hands, and fingers are usually gloves, wristlets, and arm protectors made of various materials designed to protect against cuts, scratches, bruises, and burns. Supervisors should know which type is best for a particular hazard: heat-resistant, metal mesh, rubber, leather, chrome-tanned cowhide, cotton, or thermal chemical-resistant gloves. Gloves should be inspected regularly and kept clean.

- Suits and equipment for radiation and chemical hazards include gloves, disposable clothing, suits with supplied air, and respiratory devices. Employees exposed to radiation hazards must be monitored continuously to ensure safety levels. This clothing is designed for specific, hazardous jobs where ordinary work clothes do not give enough protection.

- Protective shoes, leggings, and knee, thigh, and leg pads provide protection for the feet and legs. All protective shoes are reinforced with a toe cap. Supervisors should see that protective shoes are chosen for the particular hazard and that they fit properly. They include metal-free footwear; "congress" or gaiter-type shoes; reinforced shoes, boots, and other footwear for wet work; and shoes with metatarsal guards. Leggings must protect workers' legs and permit rapid removal in case of emergency. Hard fiber or metal guards protect shins, and nylon or cotton padding protects knees, thighs, and upper legs. Supervisors must educate employees in proper use and maintenance of protective foot and leg gear.

10

ERGONOMICS

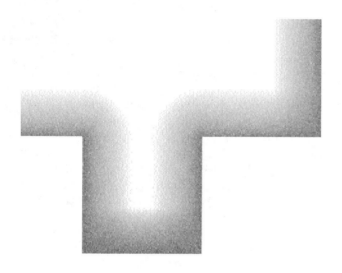

After reading this chapter, you will be able to:

- Describe various issues related to ergonomics and be alert for symptoms among your workers

- Explain the principles and goals of ergonomics and how physiology, anthropometrics, and biomechanics are used to solve work-stress problems

- Analyze how the workspace can be engineered to match workers' body characteristics

- Give examples of how to reduce ergonomic problems related to hand work, hand tools, whole-body vibration, video display terminals, and other work-related stresses

As a group leader or supervisor, you may be asked to understand and use some of the principles of ergonomics. Ergonomics is the study of how people interact with their work—in other words, how to create a good match between employees and the workstations and tools with which they work. The goal of an ergonomics program is to minimize accidents and illnesses due to chronic physical and psychological stresses, while maximizing productivity, quality, and efficiency. Many corporations have found that effective ergonomics programs have reduced operating costs, improved quality, and increased production and worker morale.

You should be familiar with ways to recognize problems that may be resolved using ergonomic principles, evaluate them, and suggest methods to control the risks. Many types of controls can be applied, from eliminating the problem, to changing work practices, to designing new tools, to training workers in how to use tools and equipment more effectively.

This chapter covers causes and symptoms of ergonomic problems, understanding the field of ergonomics, and applying basic principles to eliminate or reduce work-stress problems.

WHAT ARE POSSIBLE CAUSES?

Physical stress may arise when workstations, equipment, or tools do not fit the worker well. These stresses can cause long-term damage to muscles, nerves, and joints. Most illnesses due to ergonomic causes occur because of forceful or repetitive work activities, mechanical stress, temperature, lighting, or because workers are required to assume awkward postures over a period of time. Although it may take a physician to diagnose a worker's illness, you should be alert to any employee complaints of pain, tingling, numbness, swelling, or other discomforts. These symptoms may indicate a work-related problem.

Back, shoulder, and neck strains and sprains can be results of exposure to stresses. Hand, wrist, and other arm symptoms are also often referred to as cumulative trauma disorders (CTDs). The following are some of the more common types of CTDs. (You may also use the term "repetitive strain injuries.")

Tenosynovitis. In this disorder tendons in the wrist and fingers become sore and inflamed because of repetitive motion or awkward postures. Symptoms include pain, swelling, cracking sounds, tenderness, and loss of function. This disorder is caused by poor workstation design, problems with tool design, or changes in work habits.

Tendinitis. Tendinitis is an inflammation of the tendon that occurs when a muscle/tendon unit is repeatedly tensed. Tendinitis causes symptoms similar to those of tenosynovitis. This disorder typically occurs in the shoulder, wrist, hands, or elbow.

Thoracic Outlet Syndrome. A disorder of the shoulder that affects nerves in the upper arm, thoracic outlet syndrome is usually caused by doing tasks overhead for long periods of time. Loss of feeling on the little finger side of the hand and arm, with pain, weakness, and deep, dull aching, may occur.

Ganglion Cysts. These cysts are associated with cumulative trauma or repetitive motion. They appear on the wrist as bumps containing synovial fluid and can be surgically removed.

Carpal Tunnel Syndrome. This condition is caused by excessive overextension or twisting of the wrists, especially where force is used. It affects the median nerve, which runs through a bony channel in the wrist called the carpal tunnel. Symptoms include burning, itching, prickling or tingling feelings in the wrist or thumb and first three fingers. When the condition is severe, some of the thumb muscles may become weaker, and there may be overall weakness in the hand.

DeQuervain's Disease. This is an inflammation of the sheath of the tendons to the thumb. Scarring in the sheaths restricts the thumb's movement.

Trigger Finger Syndrome. This is another form of tendinitis caused by repetitive flexing of the fingers against vibrating resistance. Eventually the tendons in the fingers become inflamed, causing pain, swelling, and a loss of dexterity. Trigger finger may result from using a power tool with too large a handle.

Epicondylitis. Also known as tennis elbow, epicondylitis is the inflammation of tissues on the inner (thumb) side of the elbow. It may be caused by violent or highly repetitive action that requires downward rotation of the forearm and wrist.

UNDERSTANDING ERGONOMICS

Many different fields of study have contributed to industry's understanding of ergonomics. A few of these are:

- Anatomy and physiology of the human body
- Anthropometrics: study of differences in the size of bodies and body parts in different groups of humans
- Biomechanics: study of the way work activities produce forces on muscles, nerves, and bones
- Psychology: how people recognize and respond to signals in the environment
- Industrial design and engineering: design of workplaces, tools, and processes

In approaching an ergonomic problem, an organization may bring together specialists in these areas to analyze the ways people interact with machines, tools, work methods, and workspaces. Safety professionals, industrial hygienists, and occupational physicians and nurses may add their expertise to help create solutions. Management and employees also have important roles in providing information, helping to evaluate work situations, and giving feedback on changes.

One way you as the supervisor can help in the effort is by keeping track of possible indicators of workplace problems. These include:

- Apparent trends in injuries, illnesses, and accident record keeping
- Incidence of cumulative trauma disorders
- Excessive use of sick days, high turnover rate
- Employees changing their workstations, e.g., adding padding to hand tools or equipment, using makeshift platforms to stand on, making changes in the work flow
- Poor product quality

Some of the causative factors you should look for include:

- Reliance on incentive pay systems that motivate people to work faster than is safely possible
- High overtime and increased work rate
- Manual materials handling and other repetitive tasks
- Work requiring awkward postures
- High amount of hand force required for tasks
- Mechanical stress, e.g., hands or forearms resting on sharp table edges
- Raised elbows, bent wrists, hands
- Grasping or pinching objects
- Exposure to temperature extremes
- Use of vibrating tools

As supervisor, you should assist in the review of any new manufacturing processes, designs, and major equipment purchases. In this way you will be able to point out potential difficulties in how workers may actually use equipment in the workplace and work with the person doing the job.

Physiology

Work tasks need to be designed to match employees' physical capacities. A person's ability to do work at a given level is usually determined by the ability of her or his respiratory and cardiovascular systems to deliver oxygen to working muscles and to make use of energy derived from food.

One measure of capacity for manual work is maximum oxygen uptake. In industry, maximum effort is usually required only for brief periods, such as when an employee must lift heavy loads onto a hand truck. Over an 8–hour shift, however, the amount of energy used is usually well below the maximum capacity of the worker.

For instance, light work is generally associated with a heart rate of 90 or fewer beats/minute; and heavy work, about 120 beats/minute. Strenuous work produces a heart rate of 140–160 beats/minute or more, and cannot be sustained without rest or cessation of effort. At strenuous work loads, waste products of energy consumption build up in the blood and muscles, producing fatigue or a sensation of soreness.

A person's capacity for manual work is also limited by muscle strength; by flexibility, or lack of it, in the joints; and by the strength of the spinal column. The science of biomechanics provides measures of the amount of force put on these parts when employees work in different positions. It helps to determine which positions make best use of muscular strength.

The muscle's ability to perform is also affected by the way it is used. Activity can be either static or dynamic. A dynamic activity, like walking, is created by rhythmic contraction and relaxation of the muscles. Dynamic work allows muscles to rest during the relaxation phase, so that the blood can supply oxygen and remove waste products. In contrast, static work is caused by the worker holding one object or body part in one position for an extended period. (Standing still is a static posture, for example.) The muscle is in a fixed and locked position, blood vessels are compressed, and blood flow is reduced; the muscle tires rapidly. A static muscle consumes more energy. It can also place more stress on tendons and joints, which may result in back, shoulder, and neck pain or inflammation of the tendons.

It is important to remember that heart rate (and the worker's ability to do the job) may also be affected by temperature, humidity, and the worker's age and fitness level. While engineers and work physiologists usually are responsible for assessing the physiological demands of a job, you need to understand the basics of how work demands affect body functioning.

Anthropometrics

Anthropometric studies provide information that can be used to design workspaces to match body dimensions. The idea that there is an average-sized person or worker is obsolete. Even two people who are of "average" height, say 5 ft 8 in. (1.7 m), may have very different body dimensions. Also, a person who has an average arm length might have "long" legs, or one with average leg length might have "long" arms.

Studies of military and civilian personnel have been used as the basis for tables of information on body sizes. These have defined the extremes of body and body part sizes in a particular population. This information is described by percentiles. For example, regarding height, a 5th-percentile female is taller than 4% of the female population and shorter than 95% of the females in the group. A 95th-percentile male is taller than 94% of the male population and shorter than the tallest 5%.

These percentiles are useful to engineers who try to design a workspace to fit as many in the population as possible. Usually they try to make the workstation adjustable so that it can fit anyone from the 5th percentile female to the 95th percentile male.

Biomechanics

The science of biomechanics explains human body characteristics in mechanical terms. These models allow us to describe the amount of force produced when a muscle works to move an object of a specific weight. The models may describe the arm as a level, for example. They are often illustrated as simple line drawings which show the different forces acting on the muscle and on the object. While these seem over-simplified, they are useful in helping to identify stressful postures. Calculations are derived from the laws of physics and are complex, because often more than one muscle is involved.

Looking at the Workplace

Many workplace operations have potential for causing injury or illness, in different ways. The amount of physical stress a job produces is often determined by the following:

- Posture the worker must assume to do his or her job
- Force needed to perform the task
- Number of repetitions of a stressful task
- Length of time doing the task

The next sections will discuss these principles of ergonomics and ways to apply them in the workplace.

MATERIALS MOVEMENT

When materials are moved, the appropriate equipment should be used. This can include mechanical systems such as conveyors, hand trucks, and carts to reduce manual lifting and carrying. Wherever possible, people should be engineered out of the flow of materials handling. Chapter 15, Materials Handling and Storage, discusses this topic in more detail.

Manual Materials Handling

Manual materials handling refers to the human activity necessary to move an object. This often means lifting and carrying but can also describe pushing, pulling, and shoving. Look for the following situations that may cause injury:

- Lifting from the floor or lifting while twisting
- Lifting heavy weights or bulky objects
- Repetitive lifts
- Lifting above shoulder height or while seated
- Pushing or pulling loads
- Bending while moving objects

Guidelines for Safe Materials Handling

Making manual materials handling easier may mean changing one or more of the following factors:

- Where possible, modify the object to make movement easier. Handholds should be provided to allow use of a power grip that will make carrying more efficient. In a power grip, the object is clasped between the flexed fingers. This technique allows less muscle force to be used than does a pinch grip, in which the hands cannot be placed entirely around the object. Cardboard cartons, for example, should have openings that can accommodate large gloved hands if needed. Surfaces of openings or handles should be smooth to avoid mechanical irritation of the skin (Figure 10–1). The center of gravity should be below the handholds and close to the body.
- Avoid lifting or carrying large, unwieldy, or heavy objects. Mechanical lifting and transporting means should be used wherever possible. These can include equipment such as conveyors, automatic or gravity-feed devices, load-leveling pallets, hoists, and lift trucks.
- Where mechanical means cannot be used, seek assistance from another person in lifting heavy or unwieldy objects.
- Where feasible, split loads in containers into smaller loads. Containers can be made of lighter materials. Shapes can be changed to allow objects to be handled closer to the body.

The weight workers can lift is influenced by many factors. Various calculations have been developed to help determine acceptable and maximum loads for each worker. These calculations are usually based on *Applications Manual for the Revised NIOSH Lifting Equation*, published by the National Institute for Safety and Health in 1994. The guidelines are modeled on lifting an object with two hands, in front of the body, with good handholds and footing, and no twisting required. The NIOSH formula takes into account the distance of the load from the body, the location of the object before and at the end of the lift, and the frequency of lifting.

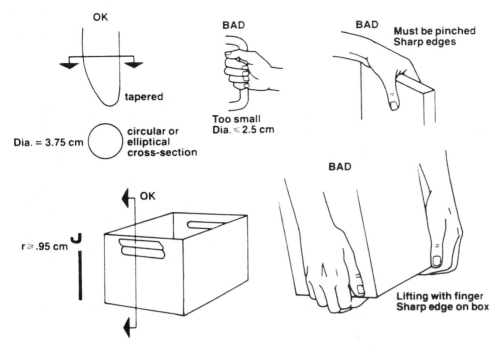

Figure 10–1. Handles should be smooth and shaped to permit a power grip.

Many industrial lifts do not meet these criteria. In general, the following guidelines apply to all lifts:

- Lifts should be kept between knuckle and shoulder height.
- The lift should begin as close to the body as possible. Sometimes objects are bulky or must be placed over or in containers or on shelving far back on counters. Job redesign should be considered in these cases, especially if a heavy object is involved. The longer the reach required to grasp the object, the less weight can be lifted. Lifting with arms outstretched puts far more stress on the back than lifting with the object close to the body.
- The height of the object above the floor at the start of the lift should be between shoulder and knuckle (mid-thigh). Placing objects on the floor or other location that requires stooping should be minimized. Rearrange the workplace so starting and ending points of the lift are close together. Avoid stacking items above shoulder height. Store loads so they will be easy to retrieve, and avoid deep shelves. Use spring-loaded bottoms for bins or gravity-feed bins to bring the load into easy reach.
- Avoid or minimize repetitive lifting tasks. Design tasks so that twisting while lifting or carrying is not required.
- The location in which the task is performed should be set up for optimal materials handling. Footing should be comfortable and should not be slippery or sticky. The space should have adequate room to work, but be small enough to limit the need to reach, stretch, or carry. Make sure workers have good lighting and clear sight lines.
- Eliminate the need for pushing or pulling by using conveyors, slides, or chutes. When objects must be pushed or pulled, push whenever possible.
- Carts should have large coasters to make the job easier. Limit ramp slope to 10 degrees or less.
- The worker must be considered as part of the system. Training in correct methods using ergonomics can sometimes help to reduce the risk of injury/illness. Training will probably not be the entire answer, however, especially where jobs with moderate or severe risk of injury due to lack of ergonomics are involved. Also consider a comprehensive back injury-prevention program that includes ergonomic principles of engineering, work methods, and administrative controls.

It is therefore generally better to design a job for safe lifting and to train people to work safely.

Some experts believe that lifting training is appropriate, although there is still controversy over what constitutes proper lifting techniques. The older rule was to lift with a straight back from a bent-knee position. However, some workers' leg muscles and knee joints may not be strong enough to lift this way repetitively.

Some experts advocate other styles of lifting, such as the "kinetic lift," the "one-handed lift," or the "two-hand squat lift," but it is not clear that these will work for all people. Another approach includes improving workers' physical fitness through stretches and warm-ups.

Therefore, while there are no comprehensive rules for safe lifting, rules described in the previous section may apply. Additional guideline are summarized below:

- Eliminate manual lifting and lowering from the task, where feasible.
- Be in good physical shape. Workers who are not used to lifting and vigorous exercise should not attempt difficult lifting tasks.
- Think before acting. Place material in a convenient place and within easy reach. Use the handling aids available.
- Get a good grip on the load. Test the weight before trying to lift it. If it is too heavy or bulky, get an assistant, mechanical lifting aid, or both.
- Get the load close to the body with feet close to the load. Stand in a stable position with feet pointing in the direction of movement. Lift mostly by straightening the legs.
- Don't lift with sudden or jerky movements.
- Don't twist the back or bend sideways.
- Don't lift or lower with arms extended.
- Don't heave a heavy object.
- Don't lift a load over an obstacle.

WORKSPACE AND BODY CHARACTERISTICS

Because each person has unique body characteristics (anthropometrics), the workspace must be flexible or adjustable to allow for these differences. Ideally, all work platforms, tables, and carts should be adjustable. If that is not possible, the workspace should allow for the tallest person, with adjustable chairs, foot rests, or adjustable platforms used to accommodate shorter workers.

Equipment controls should be placed between shoulder and waist level. The normal work surface should be just below the elbow; the surface should be higher for precise work, lower for heavy work.

As discussed previously, workstation surfaces and leg room need to be designed for the largest potential user, with adjustments to accommodate smaller individuals. However, reaching distances in the workstation need to be designed for the smaller worker. For example, at a seated assembly operation that involves reaching into bins for parts, the chairs and work-surface adjustments should be provided so that short individuals with short arms will not have to over-reach, putting stress on shoulders and arm muscles. Infrequent reaching behind or to the sides is acceptable but should not be required on a routine basis.

The workspace for the hands should be between hip and chest height in front of the body. Most work should be done just below elbow height. In heavy work, the hands should be lower, while in fine assembly work that involves precise visual inspection, the worker will need a higher surface, perhaps with rests for forearms. Work-surface height should allow the arms to be kept low and close to the body.

Work objects should be located close to the front edge of the work surface to prevent the employee from having to bend over and lean across the surface to grasp items. Surfaces with no sharp edges but rounded or padded edges are preferred. Sufficient leg room must be allowed for seated operators.

Any visual symbols or displays should be placed in front of the body and below the eye level, between 10 degrees and 40 degrees below the horizontal line of sight. The most important information for the task should be placed in the worker's viewing area.

Posture

The workstation should be designed to reduce static effort. Workers should be reminded to change positions frequently, and to keep the body in what is called the neutral posture:

- Body is relaxed, with arms hanging loosely at sides
- Wrists are neutral
- Shoulders are relaxed
- Elbows are close to the body

Working surfaces and seats should be designed to eliminate the need to work with a bent spine. In using ergonomics to identify problems, the supervisor should investigate workstations where workers must:

- Bend necks forward more than 15 degrees
- Lean forward or twist sideways
- Crouch over their work
- Work with arms above head or out, away from body
- Use awkward wrist positions, especially if work is repetitive or requires forceful movements

Standing Work

Many jobs can be performed with less effort when the worker is standing. However, prolonged standing can create stresses on the legs and lower back. Workers who stand for a long time on hard concrete surfaces should be supplied with padded antifatigue mats. The height of the work surface should usually be 2 to 6 in. (5–15 cm) below the level of the worker's elbow when the arm is hanging in a relaxed posture (Figure 10–2).

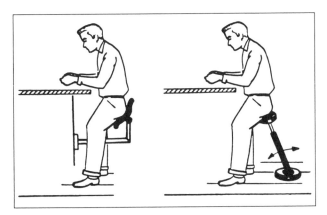

Figure 10–2. "Stand-seats" help relieve leg and lower back stress. (Courtesy *American Industrial Hygiene Association Journal.*)

Resting one foot on a foot rest while standing can help reduce low-back stress. Provide adjustable footrests where possible. When workers alternate between sitting and standing, a stool that leans with the body is sometimes useful.

Seating

Industrial seating is often poorly designed or uncomfortable. On the assembly line, providing a chair or stool is frequently an afterthought. As a result, workers tend to rig their own stools, or salvage discarded chairs, cushions, or car seats to use at workstations.

The same care should be given to providing comfortable seating on the shop floor as to providing correct chairs and equipment to office workers. Obviously, not all jobs will allow a worker to sit, but whenever possible, the following guidelines should be considered:

1. The weight-bearing seat area should be large enough to accommodate the person without putting pressure on the buttocks or thighs [at least 18.2 in. (46.2 cm) wide]. Seat surface should be comfortably but firmly upholstered to distribute pressure evenly. Upholstery should be a woven fabric to reduce sweating. The seat should allow the worker to change his or her seated posture. The front of the seat should be rounded. It should not produce pressure on the back of the thighs or calves, which can occur when the seat is too deep. To relieve subsequent discomfort, workers may sit too far forward, causing them to lean forward in a static posture. They may slouch, which will put stress on the back. Generally, seats should be between 15 and 17 in. (38–43 cm) deep.

2. The seat height should be adjustable within a range of approximately 15 to 20 in. (38–51 cm) if the work surface is at a standard desk height.

Adjustments of seat height will have to be made upward where the work surface is higher. For example, on an assembly line where the work surface is 32 in. (81 cm) high, the seat should be adjustable from 20 to 26 in. (51–66 cm) off the floor. For higher seats, foot rests must be provided to prevent legs from dangling, which can cause circulatory problems. Foot rests must also be used for shorter workers at any seat height where their legs may dangle.

3. A minimum 8–in. (20–cm) vertical clearance should be provided under the work surface, with 26 in. (66 cm) forward clearance for feet and legs. Foot rests should be between 1 and 9 in. (2.5–23 cm) high, and be slanted, with a depth of about 12 in. (30 cm).

4. The back should be supported in the lumbar region. Preferable, the backrest should be adjustable both horizontally and vertically. The support should maintain the back in it natural S-shaped position. The backrest should be at least 6 to 9 in. (15–23 cm) high and 12 to 14 in. (30–36 cm) wide. Too small a backrest will provide inadequate support and too large a backrest may prevent workers from moving as required by their work cycle (Figure 10–3).

Displays and Controls

A worker's interface with the job may be through a display console. On these jobs, workers are required to receive information through continual sensory signals, to process the information mentally, and to make a decision or take action. The display must be arranged so communication is clear, concise, and easily understood. The controls must also be arranged to minimize error.

Basic ergonomic and human factors principles of control and display design help ensure that this critical process is performed in the best manner. Controls should be within easy reach of the operator while he or she views displays or the field of operation. They should be easily located by touch, with adequate spacing between controls.

Good sensory input (primarily involving sight and hearing) is a key principle, and requires proper lighting, visual stimuli, sound levels, and sound patterns. Urgent information is best conveyed by auditory signals, such as bells or buzzers. Complicated information can be displayed visually. A combination of visual and auditory signals often improves the effectiveness of alerting the operator. For example, a buzzer could alert an operator to an emergency situation while a visual display would give details of the emergency.

Because a person's capacity for perceiving information is limited, it is important to group and simplify

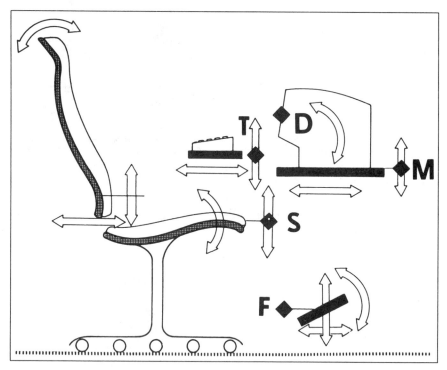

Figure 10–3. Adjustment features of a VDT office workstation. Key: S = seat height; T = table; F = footrest; D = monitor; M = support. (Reprinted with permission from Kroemer, 1985.)

displays and controls. The following design concepts can help an operator to interpret and make decisions:

- Displays (gauges, meters, etc.) should be grouped so that normal ranges are all in one area and pointers are in the same direction. The most frequently used displays and controls should be in the center of the visual field. A deviation from normal is then easy to detect because it stands out. Controls should be clearly labeled and color coded, if appropriate, to distinguish machines, operation, and department.
- Symbols should be bold and simple. Use broad, rotating pointers on dials. The pointer should touch the dial gradations but should not obscure the numbers. The simpler and fewer the dial gradations, the better.
- Controls should be grouped with the related display and should follow the worker's usual expectations. In the United States, for instance, switches are usually moved *up* to activate electrical equipment and *down* to turn them off. Use distinctive shapes to aid in identification by touch.
- Controls should have adequate space between them for efficient handling.
- Controls must be arranged so workers will be able to shift their bodily positions during the workday.
- Switches must be guarded with cowls to prevent accidental operation.

- Hand controls should be used where precision and speed are required; foot controls may be used where greater strength is needed.
- Workers should have time to make decisions about their work. For more complicated jobs, special training in decision-making may be needed. Give operators immediate feedback on their decisions and actions.

HAND WORK AND USE OF TOOLS

Most jobs require some manipulation of objects with the hand. Although the human hand is a highly evolved grasping tool, it cannot always withstand the stresses of rapid, difficult, and/or repetitive occupational tasks. Cumulative trauma problems are especially likely to occur where there are less than 30 seconds per cycle, or where more than 1,000 parts per shift are handled or processed.

Principles of Reducing Job Stress

The following principles should be considered when determining how much stress the job may produce:

- Activities should be performed in what ergonomists call the "natural range of movement." For the upper limbs, a work posture that allows the

Figure 10–4. This worker can keep her hands in a natural, "handshake" position while working.

arms and wrists to remain in a natural posture as in ordinary activities is ideal. This means avoiding awkward wrists and, when necessary, keeping elbows at right angles when using the arms to apply force to the proper application. It also means avoiding extreme inward or outward rotation of the arms and wrists; instead, keep the wrist and hands in the "handshake" position (Figure 10–4).

- Repetitive motion, duration, and the amount of force needed to perform work should also be minimized. Jobs that require extremes of these conditions may put the worker at greatly increased risk of cumulative trauma disorders.

Ways to reduce the effect of task repetitiveness and forcefulness include:

- Reducing the weight of the object (Figure 10–5), picking up fewer objects at a time, lifting the object with two hands instead of one, or changing the size and shape of the objects.
- Planning rotation to jobs where demands on muscles and nervous system are different.
- Alternating use of limbs.
- Allowing workers to pace themselves, taking short breaks as needed.
- Enlarging the job to include a variety of tasks.
- Using mechanical aids that perform in one motion the many tasks a worker might perform. A pneumatically operated lug wrench at a tire installation operation on an auto assembly line, for example, can reduce repetitive forceful wrist

motion for the operator.
- Using jigs and fixtures to reduce the amount of holding required by assembly line workers.
- Modifying the work surface to angle the object so that it is easier to work on.
- Using foot pedals that do not require the worker to continually depress them during operations.
- Avoiding impulse or impact forces on parts of the body. For example, the worker should use the right tool rather than hammer small parts together with his or her palms.

Good Tool Design and Use

Most workers use hand tools for part of their jobs. As the time spent using tools increases, good tool design becomes more important. Improper design can result in inefficiency and cumulative trauma disorders. Tools that require effort to hold or operate, have awkward handles that cause the wrist to be bent, or produce excessive pressure or vibration may injure the user.

Static loading of the muscles will occur where the tools or arms have to be held without support. This can lead to fatigue and sometimes cause a disorder in susceptible individuals. At times, bending a tool, such as a straight soldering iron, will reduce the need to keep the arm lifted. To reduce forces on the fingers, the tool should be activated by a bar or grip switch instead of a single finger trigger.

Minimize awkward postures, especially bent wrists.

Figure 10–5. This reaming tool is supported above the work to relieve stress on hands, arms, and shoulders.

Because grip strength is greatest when the wrist is straight, tools designed with bent handles may allow the worker to keep the wrist in a more natural posture. However, each job must be analyzed to ensure that the right tool is being used.

Tool handles should be long enough to extend past the palm of the user's hand. The surfaces should be broad enough to distribute pressure evenly, and should be padded or slip-resistant.

Tools such as hammers and screwdrivers should have handles about 1.5 in. (3.8 cm) in diameter. Those for fine work should be about half an inch in diameter. Handle spans, on two-handled tools, should be in the 2.5 to 3.5 in. (6.3–8.9 cm) range. They should be long enough to avoid pressure on the flesh of the palm or at the base of the thumb. Textured, but not highly grooved, surfaces are preferred for hand tools. This is all dependent on the individual using the tool.

Purchase tools that allow the worker to use a power rather than a pinch grip. Avoid using tools that expose workers to mechanical pinch points. Also, avoid tools with form-fitting handles that may pinch a larger worker's flesh and may reduce a smaller worker's efficiency. Hand-tool handles should be insulated against electricity, heat, and cold and fit the worker's hand. Avoid repetitive use of hand tools that put pressure on the back of the hand. For example, conventional scissors have been redesigned to remove the oval handle that creates pressure on the backs of the fingers.

Whenever possible, use power tools that are balanced with a neutral buoyancy tool balancer. These items may be better than a balancer that automatically reels the tool upward because they reduce the amount of reaching for and pulling the tool.

All surfaces with which arms, elbows, and other body parts are in contact should be rounded or padded to avoid mechanical contact stresses on the nerves.

Use of power tools that vibrate excessively can produce a disease known as Raynaud's syndrome or vibration white finger. This disease affects the ability of the blood to circulate in the hands, leading to numbness, stiffness, pain, and loss of strength. The affected fingers appear white. If the worker is not removed from exposure, severe tissue damage can occur.

Hand-arm vibration at frequencies below 1,000 hertz (cycles per second) should be avoided. Where workers use hand tools that vibrate, hazards may be reduced by decreasing exposure time or by dampening or isolating equipment vibration.

Driving a fastener into material with a power tool transfers torque to the worker's hand when the fastener bottoms out. The worker experiences this as a snapping action, which is a potential cause of stress. This stress can be reduced by using slip clutches, torque limiters, or torque-absorbing tools; by keeping the torque setting low, by mounting the tool on an articulating arm, or by providing an extra handle so the worker can use two hands to help counter the torque effort.

The grip of a tool can be dampened by making it out of flexible materials. Vibration-damping gloves may be used, but if they are too large or bulky, the worker will need to use extra force to hold onto the tool. Other tool-handling guidelines include:

- Reducing tool weight and handle size so the worker does not need to use excessive grip strength to hold onto the tool (or increasing handle size—depending on the worker)
- Counter-balancing tools so their weight will not twist the tools out of the workers' hands
- Using tools that can be held in one hand and guided by the other hand
- Balancing and replacing worn-out parts
- Using vibrating tools less frequently or rotating personnel to other jobs

WHOLE-BODY VIBRATION

Whole-body vibration can also create physical stress. Driving a truck, bus, or car, or operating large vibrating production machinery can produce discomfort and health stresses. When in contact with an object that vibrates at very low frequencies, the body, or body parts, will resonate (pick up the vibration). Workers exposed to very low frequency (1–20 Hz) vibrations may experience difficulty breathing, pains in the chest and abdomen, backache, headache, muscular tension, and other problems. Workers exposed to chronic excessive vibration may suffer spinal problems and intestinal complaints. Strong vibration may impair vision, mental processes, and ability to complete skilled movements (Grandjean, p. 299).

Heavy vibrating tools such as jack hammers can cause damage to bones, joints, and tendons, as well as Raynaud's syndrome. The vibration from large equipment can be reduced by:

- Mounting the tool on springs or compression pads
- Using structural materials that produce less vibration. These will also possibly reduce noise generated by the equipment.

The person can also be isolated from the vibrating source. For instance, for truck driving or other seated tasks, you can provide cushioning or springs to insulate the seat from the vibrating surface. For standing tasks at large industrial equipment, make sure workers have a vibration-absorbing (rubber or vinyl) floor mat.

VIDEO DISPLAY TERMINALS

Video display terminals (VDTs) are widely used in offices and are becoming more common in industrial situations. They provide an example of the need for good displays and controls.

The VDT screen conveys information for the worker. It is important that the screen be free from glare. Shielding windows with drapes or blinds can help to reduce glare. Reflected glare from shiny objects can be decreased by using dark and/or matte finishes. Glare from lighting can be reduced by avoiding overhead lighting, lowering the light level where possible, and adding task lighting on documents as needed.

- Screens should tilt in order for the worker to comfortably look down on the screen. The top of the screen should not be higher than eye level.
- The screen should be adjustable in placement on the desk. A comfortable viewing distance is usually between 6 and 20 in. (15–51 cm), but this varies with each individual.
- Keyboard height and placement should be adjustable, as should seat and desk height. The employee should not remain in the same position for long periods—several breaks in keyboard work should be encouraged.

LIGHTING, NOISE, AND HEAT

Lighting must be appropriate to the task. For example, low light may be acceptable in storage areas, while high lighting may be needed for fine visual tasks. The amount of light needed will depend on the job, the operator's age and eyesight, and other factors. A full discussion of the technical aspects of lighting levels is beyond the scope of this book. However, the same types of guidelines apply to the general workplace as to the VDT operations.

- Provide adequate but not excessive light. Excessive light can create eye fatigue.
- Provide adjustable blinds at windows.
- Avoid direct or reflected light sources in the worker's field of vision.
- Shade or diffuse all light sources. Light should be reflected down from the ceiling, not up from the floor.
- Use task lighting where extra illumination is required.
- Ensure sufficient contrast for written material (i.e., use a black ink rather than a pencil).
- Increase the size of small critical details.
- Put the task material perpendicular to the operator's line of sight (Kodak, vol. 1, p. 225).

Noise and heat stress are discussed in Chapter 8, Industrial Hygiene. When considering workplace ergonomics, remember that too little heat can be as detrimental as too much. In cold, workers' muscles stiffen, making them less able to make precise movements, and making muscles and nerves more vulnerable to injury.

SUMMARY OF KEY POINTS

Key points covered in this chapter include:

- Ergonomics is the study of how people interact with their work, or how to create a good match between the employee and the workstation and tools with which the person works. The goal of ergonomics is to minimize accidents and illnesses due to chronic physical and psychological stresses on the job and maximize productivity and efficiency. Supervisors must be familiar with the principles of ergonomics to help recognize problems involving ergonomics, evaluate them, and suggest methods to control the risks.

- Problems may arise when workstations, equipment, or tools do not fit the workers well. These stresses may cause immediate or long-term damage to muscles, nerves, and joints. Most injuries/illnesses are caused by forceful or repetitive motion activities or because workers are required to assume awkward postures over a period of time. These disorders are referred to as cumulative trauma disorders and include tenosynovitis and tendinitis (an inflammation of the tendons in shoulder, arm, elbow, wrist, or hands).

- Ergonomics draws on anatomy and physiology, anthropometrics, biomechanics, and psychology to solve work-related problems. Anthropometrics provides information used to design workspaces to match body dimensions. Workspaces are generally designed for the tallest or shortest worker and adjusted for individual differences. Biomechanics explains characteristics of the human body in mechanical terms. These models help to identify stressful postures and to develop countermeasures to relieve stress. In looking at the workspace, supervisors should be aware that the amount of job-produced physical stress is often determined by the posture a worker must assume, the force needed to perform the task, and the number of repetitions required by a stressful task.

- To help companies reduce or eliminate problems, supervisors can provide key information by keeping track of trends in injuries and accidents, incidence of cumulative trauma disorders, sick leave and turnover rates, employee adjustments to workstations and equipment, and product quality levels. Ergonomics can alert supervisors to possible causes of problems, including high overtime and increased work rate, manual materials handling, repetitive tasks, mechanical stress on the body, temperature extremes, and vibrating tools.

- According to ergonomic principles, work tasks should be designed to match the employee's physical capacity for the job. This capacity can be measured by maximum oxygen uptake or by heat produced by the body. Job demands that exceed an employee's maximum capacity; that put strain on tendons, muscles, or joints; or that disregard environmental factors (heat, humidity, etc.) can cause injuries.

- Ergonomic principles can be applied to manual materials handling to eliminate or reduce stress-related illnesses. Workers should be taught the correct ways to lift, push, or pull objects; to use mechanical means wherever feasible; to ask for assistance; and to modify workloads to make moving objects safer.

- Because each person has unique body characteristics, the workspace must be flexible to allow for these differences. Ergonomic principles applied to workplace conditions include designing furnishings and equipment to keep the body in a neutral posture as much as possible; providing relief when employees must stand to do their work; designing proper seating to prevent lower back and leg problems; and providing display consoles that allow easy viewing, maximize information obtained, and permit rapid, nonstressful operation.

- Ergonomic principles can also be applied to reduce job stress in hand work and when using hand tools. Activities should be performed as much as possible in the "natural range of motion," avoiding awkward postures and bent wrists, and keeping elbows at right angles. Repeated motions should be minimized, and the amount of force required to do the work reduced as much as possible. Using the proper tool for the job may minimize cumulative trauma disorders, and tools should always be kept in good condition.

- Whole-body vibration, such as riding a truck or operating large vibrating tools, can create physical stress. Ergonomic principles can be used to reduce this stress by dampening the vibration and/or isolating the person from the vibrating source.

- In addition, video display terminals must be free of glare and set at the proper height and angle to prevent neck and shoulder problems. Keyboard height and placement should be adjustable, as should seat and desk height.

- Lighting, noise, and heat must all be controlled and appropriate to the work environment or workers may suffer from work-related problems. Proper lighting is particularly important to prevent fatigue and chronic eye disorders.

HAZARD
COMMUNICATION

After reading this chapter, you will be able to:

- Describe the intent and definitions of the Hazard Communication Standard (HCS), OSHA 29 *CFR* 1910.1200

- Recognize various types of warning labels that must be used on all containers of hazardous materials

- Describe Material Safety Data Sheets (MSDSs) and know how to generate and maintain a chemical inventory

- List the employer's and employee's training and safety obligations under the HCS.

- Discuss the elements of a hazard communication program

The purpose of this chapter is to introduce the requirements of the Hazard Communication Standard (HCS) and methods of compliance. Best practices that exceed the standard are also discussed.

HAZARD COMMUNICATION STANDARD REQUIREMENTS

The federal HCS, issued by the U.S. Department of Labor, Occupational Safety and Health Administration (OSHA), regulates an employer's duty to communicate chemical hazard information to employees who might be exposed to these chemicals in the workplace. The HCS, known as OSHA 29 *CFR* 1910.1200, became effective November 25, 1988, and applies to all workplaces where hazardous chemicals are used or stored. The overall intent and purpose of this regulation is to reduce the number and frequency of workers' occupational illnesses and injuries caused by unprotected exposure to hazardous chemicals.

Definitions

Figure 11–1 gives the definitions that OSHA has found to be significant in the interpretation of the intent of the HCS and the information that is required as part of the written HazCom program.

The HCS is a "performance-oriented" standard. This approach allows the employer great flexibility when setting up workplace programs designed to meet the following mandates (intents) of the standard:

The written program must include the following:

- Hazard determination/inventory
- MSDS availability
- Labeling and warning signs
- Training and communication of information to employees
- A written hazard communication program for the specific workplace

Employer and Employee Responsibilities

OSHA is responsible for enforcing this standard in states that have not developed an HCS of their own (the state-developed HCSs are required to be as stringent, or more stringent, than the federal standard). Federal OSHA must approve all HCSs that are state developed and administered. In states with

their own HCSs, the state OSHA agency, or state Department of Health, Safety, and the Environment, is responsible for enforcement of the standard.

Best Practices

The HazCom program is usually a part of the company's safety, health, and environmental compliance department. The employees in charge of these areas are usually responsible for developing the HazCom program. However, as with any other safety, health, and environmental compliance programs, the final responsibility for ensuring that the programs are appropriately followed rests with all management and nonmanagement employees.

ELEMENTS OF A HAZARD COMMUNICATION PROGRAM

A company hazard communication program includes written information on hazardous chemicals, proper labeling, usage of MSDSs, employee training, and a written hazard communication plan. Each aspect of the hazard communication program must meet or exceed federal HCS requirements. Best practices suggestions offer guidelines on developing and implementing an effective program.

Written Hazard Communication Program

The company's written program provides employers and employees with specific policies and procedures to follow when handling hazardous substances in the workplace.

The HCS requires that the employer must develop, implement, and maintain a written hazard communication program for the workplace. This written document must be kept at the workplace, and must, at the very least, describe how the employer will meet the responsibilities of establishing the following elements:

- Employee training program
- Labeling and warning sign procedures or systems
- MSDS collection, storage, and review
- List of the hazardous chemicals known to be present in the workplace, using an identity referenced on the appropriate MSDS
- Methods the employer will use to inform employees of the hazards of nonroutine tasks
- Methods the employer will use to inform employees of the hazards associated with chemicals contained in unlabeled pipes in work areas

ACUTE EXPOSURE: Exposure is usually of short duration and high concentration. Symptoms will usually appear within 24 hours.

ASSISTANT SECRETARY: The Assistant Secretary of Labor for Occupational Safety and Health, U.S. Department of Labor, or designee.

ARTICLE: A manufactured item formed to a specific shape or design that has an end-use function that depends wholly or partly on its shape or use during end use; and that does not release, or otherwise result in exposure to, a hazardous chemical, under normal conditions of use.

CHEMICAL: Any element, chemical compound, or mixture of elements and/or compounds.

CHEMICAL MANUFACTURER: An employer with a workplace where chemical(s) are produced for use or distribution.

CHEMICAL NAME: The scientific designation of a chemical according to the naming system developed by the International Union of Pure and Applied Chemistry (IUPAC) or the Chemical Abstracts Service (CAS), or a name which will clearly identify the chemical for the purpose of conducting a hazard evaluation.

CHRONIC EXPOSURE: Symptoms are usually delayed, cumulative, and result from repeated exposure to low-level concentrations of hazardous chemicals.

COMBUSTIBLE LIQUID: Any liquid having a flashpoint at or above 100 F (37.8 C), but below 200 F (93.3 C). The exception is a mixture having components with flashpoints of 200 F, or higher, the total volume of which makes up 99% or more of the total volume of the mixture.

COMMON NAME: Any designation or identification such as a code, trade, brand, or generic name, or a code number used to identify a chemical other than by its chemical name.

COMPRESSED GAS: A gas or mixture of gases having, in a container, an absolute pressure exceeding 40 psi at 70 F (21.1 C); or a gas or mixture of gases having, in a container, an absolute pressure exceeding 104 psi at 130 F (54.4 C), regardless of the pressure exceeding 40 psi at 100 F (37.8 C) as determined by ASTM D–323–72.

CONTAINER: Any bag, barrel, drum, bottle, box, can, cylinder, reaction vessel, storage tank, or the like, that contains a hazardous chemical. For the purposes of the HCS, pipes, piping systems, engines, fuel tanks, and other operating systems in a vehicle are not considered to be containers.

DESIGNATED REPRESENTATIVE: Any individual or organization to whom an employee gives written authorization to exercise said employee's rights under this standard. A recognized or certified collective bargaining agent (such as a union representative) shall be treated automatically as a designated representative without regard to written employee authorization.

DIRECTOR: The Director, National Institute for Occupational Safety and Health, U.S. Department of Health and Human Services, or designee.

DISTRIBUTOR: A business, other than a chemical manufacturer or importer, that supplies hazardous chemicals to other distributors or employees.

EMPLOYEE: A worker who may be exposed to hazardous chemicals under normal operating conditions, or in foreseeable emergencies. Workers such as office workers or bank tellers who encounter hazardous chemicals only in non-routine, isolated instances, are not included in this definition of "employee."

EMPLOYER: A person engaged in a business where chemicals are either used, distributed, or are produced for use or distribution, including a contractor or subcontractor.

EXPLOSIVE: A chemical that causes a sudden, almost instantaneous release of gas, pressure, or heat, when subjected to sudden shock, pressure, or high temperature.

EXPOSURE OR EXPOSED: The instance of being exposed, or an employee who is exposed to, or has the potential to be exposed to, hazardous chemicals in the course of employment, through any route of entry (absorption, inhalation, ingestion, skin contact, injection) to the body.

FLAMMABLE: A chemical that falls into one of the following categories: aerosol flammable, gas flammable (includes forming a flammable mixture with air at a 13% concentration of the chemical, or less, AND forming a flammable range, when mixed with air, that is wider than 12% by volume, regardless of the lower limit); liquid flammable; and solid flammable.

FLAMMABLE LIQUID: Any liquid having a flashpoint below 100 F (38 C).

FLASHPOINT: The minimum temperature at which a liquid gives off vapor in sufficient concentration to ignite when tested by the Tagliabue Closed Tester, Pensky-Martens Closed Tester, or Setaflash Closed Tester methods. Organic peroxides, which undergo autoacceleration thermal decomposition, are excluded from any of the flashpoint determination methods mentioned here.

FORESEEABLE EMERGENCY: Any potential occurrence such as, but not limited to, equipment failure, rupture of containers, or failure of control equipment, which could result in an uncontrolled release of a hazardous chemical into the workplace.

HAZARDOUS CHEMICAL: any chemical that is a physical or health hazard.

HAZARD WARNING. Any words, pictures, symbols, or combination thereof, appearing on a label or other appropriate form of warning, which conveys the hazard(s) of the chemical(s) in the container(s).

HEALTH HAZARD: A chemical for which there is

Figure 11–1. Key Terms of the Hazard Communication Standard

statistically significant evidence, based on at least one study conducted in accordance with established scientific principles, that acute or chronic health effects may occur in exposed employees. The term "health hazard" includes chemicals that are carcinogenic, toxic, or highly toxic, reproductive toxins, irritants, corrosives, sensitizers, hepatotoxins, nephrotoxins, neurotoxins, hematopoietic agents, and agents that can damage the lungs, skin, mucous membranes, or eyes.

IDENTITY: Any chemical or common name for the chemical that is indicated on the MSDS. The identity used shall permit cross-references to be made among the required list of hazardous chemicals, the label, and the MSDs.

IMMEDIATE USE: The hazardous chemical will be used, or under the control of, only the person who transfers it from a labeled container, and only within the work shift in which it was transferred.

IMMEDIATELY DANGEROUS TO LIFE AND HEALTH (IDLH): The maximum level of a substance from which, in the event of respirator failures, a person can escape within 30 minutes without any impairing symptoms or irreversible health effects.

IMPORTER: The first business, with employees and within the Customs Territory of the United States that receives hazardous chemicals produced in other countries for the purpose of supplying them to distributors or employees within the United States.

LABEL: Any written, printed, or graphic material, displayed on or affixed to, containers of hazardous chemicals.

MATERIAL SAFETY DATA SHEETS (MSDSs): Written or printed material concerning a hazardous chemical that is prepared in accordance with Paragraph (g) of the HCS.

MIXTURE: A combination of two or more chemicals, in which the combination is not wholly or partly the result of a chemical reaction.

ORGANIC PEROXIDE: An organic, highly reactive compound that contains the bivalent -O-O- structure, and which may be considered to be a structural derivative of hydrogen peroxide, where one or both of the hydrogen atoms has been replaced by an organic radical.

OXIDIZER: A chemical, other than a blasting agent or explosive, that initiates or promotes combustion in other materials, thereby causing a fire either of itself, or through the release of oxygen or other gases.

PERMISSIBLE EXPOSURE LIMIT (PEL): Allowable air concentration of a substance in the workplace 8 hours a day, 40 hours a week, as established by NIOSH and enforced by OSHA.

PHYSICAL HAZARD: A chemical for which there is scientifically valid evidence that it is a combustible liquid, a compressed gas, explosive, flammable, an organic peroxide, an oxidizer, pyrophoric, unstable, or water reactive.

PRODUCE: To manufacture, process, formulate, or repackage.

PYROPHORIC: A chemical that will ignite spontaneously in air at a temperature of 130 F (54.4 C) or below.

RESPONSIBLE PARTY: A person who can, if necessary, provide additional information on the hazardous chemical and appropriate emergency procedures.

SPECIFIC CHEMICAL IDENTITY: The chemical name, Chemical Abstracts Service (CAS) Registry Number, or any other information that reveals the precise chemical designation of the substance.

THRESHOLD LIMIT VALUE (TLV®): Refers to airborne concentrations of substances, and represents conditions under which is believed that nearly all workers may be repeatedly exposed day after day without adverse effect. TLVs are recommended guidelines established by the American Conference of Governmental Industrial Hygienists (ACGIH).

THRESHOLD LIMIT VALUE CEILING (TLV-C): The ceiling level of the exposure that should never be exceeded. This value has been established as the maximum level to be used in computing the TWA and STEL limits. The ceiling value is under the IDLH limit for a given substance.

THRESHOLD LIMIT VALUE SHORT-TERM EXPOSURE LIMIT (TLV-STEL): An exposure level safe to work in for short periods of time (15 minutes) 4 times a day maximum, with at least 60 minutes between exposures. No irritation or other adverse effects should be experienced. This level is higher than the TLV-TWA for a given substance.

THRESHOLD LIMIT VALUE TIME-WEIGHTED AVERAGE (TLV-TWA): The time-weighted average concentration of a substance for a normal 8–hour workday and a 40–hour workweek, to which nearly all workers may be repeatedly exposed, day after day, without adverse effect.

TRADE SECRET: Any confidential formula, pattern, process, device, information, or compilation of information that is used in an employer's business, and that gives the employer an opportunity to obtain an advantage over competitors who do not know or use it.

UNSTABLE (REACTIVE): A chemical that, in the pure state or as produced or transported, will vigorously polymerize, decompose, condense, or become self-reactive under conditions of shocks, pressure, or temperature.

USE: To package, handle, react, or transfer

WATER-REACTIVE: A chemical that reacts with water to release a gas that is either flammable or presents a health hazard.

WORK AREA: A room or defined space in a workplace where hazardous chemicals are produced or used, and where employees are present.

WORKPLACE: An establishment, job site, or project at one geographical location, containing one or more work areas.

Figure 11–1. (Continued)

- Methods the employer will use to provide outside employees (who work for another employer) with a copy of MSDSs for the work area
- Methods the employer will use to inform outside employees about any precautions that must be taken in the workplace
- Methods the employer will use to inform outside employees of the labeling and warning sign system used in the workplace

A copy of the written HazCom program must be kept at the workplace and be available upon request to:

- An OSHA compliance officer or other OSHA representative, or
- An investigator from the National Institute for Occupational Safety and Health (NIOSH), or employees and their designated representatives, and
- All employees.

When developing the HazCom written document, employers must ensure that the policies and procedures cover the exact chemicals and processes that exist at their particular workplace. It is not acceptable to take a "generic" HCS outline, put the company's name on it, and call it "the company's written HazCom program." However, the employees can use a generic outline to ensure that all mandatory sections are addressed in the company-specific HazCom program. (A good reference source would be the OSHA booklet *Hazard Communication Guidelines for Compliance*, available through the government printing offices.)

The written HazCom program must be reviewed and updated whenever a chemical process or change occurs in the workplace. Likewise, any time the written program is altered significantly, the company must hold information/training sessions that explain the changes in the written program to all affected employees.

Best Practices

When designing the training program that must accompany the company HazCom program, the employer may have to develop different levels of training for different areas of the workplace. For example, although office employees may not come into direct contact with hazardous chemicals during their routine job assignments, they may need to know what types of hazardous chemicals are used in the workplace, and what to do in case of a spill,

fire, or other emergency involving the hazardous chemicals. An employee who is directly involved with handling or using hazardous chemicals must receive training covering the chemical and physical hazards associated with the chemical; how to safely use the chemical, MSDSs, and personal protective equipment (PPE); labeling requirements; and other practical aspects of a complete HazCom program. Whenever employee training sessions are held, the meetings, subjects covered, and attendance should be documented through meeting minutes and attendance sheets.

Implementing the HazCom program usually requires several key steps:

- First, the company should take an inventory (Figure 11–2) of all chemicals in the workplace (except those exempted from the standard, as discussed below. In addition to listing the chemical names, best practice would include documenting the information required for environmental reporting, such as specific chemical name, storage location, annual quantity used, and maximum quantity stored. The inventory process should include procedures for updating the inventory and notification of affected employees and management.)
- Second, management should define the employee training program. The written HazCom training plan should include how employees will be informed of the requirements of the HCS, the title of the person(s) doing the training, the availability and location of the written HazCom program, a training schedule, the method of communicating the training schedule to the employees, and documentation of the content of, and attendees at, the various training sessions.
- Third, a specific employee, such as facility supervisor, should be given responsibility for distributing the hazardous chemical information to outside contractors, and for ensuring that the contractor signs a document stating that the necessary information has been received. The facility supervisor should also request, and approve from the contractor, a list of chemicals (and copies of MSDSs) that the contractor will be using at the facility. Any hazard information should be shared with affected employees.
- Fourth, supervisors should document any new job assignments where an employee might be exposed to hazardous chemicals in pipes, or while performing a nonroutine job task. Appropriate HCS training should then be given.

CHEMICAL INVENTORY

CHEMICAL NAME	Toxic/Hazardous Components	Area Where Used/Produced	Dates Use/Production Began Ceased	Quantity in Area	Quantity in Storage and Location	Substance Category

INVENTORY
An inventory of hazardous chemicals should be prepared on a master chart such as the one shown here.

Figure 11–2. Sample Chemical Inventory Form

ACETONE

DANGER!
EXTREMELY FLAMMABLE. HARMFUL IF SWALLOWED OR INHALED. CAUSES IRRITATION.

Keep away from heat, sparks, and flame. Avoid contact with eyes, skin, and clothing. Avoid breathing vapor. Keep in tighlty closed container. Use with adequate ventilation. Wash thoroughly after handling.

PRECAUTIONARY STATEMENTS: Contact with skin has a defatting effect, causing drying and irration. Overexposure to vapors may cause irritation of mucous membranes, dryness of mouth and throat, headache, nausea, and dizziness.

FIRST AID PROCEDURES: If inhaled, remove to fresh air. If not breathing, give artificial respiration. If breathing is difficult, give oxygen. In case of contact, immediately flush eyes or skin with plenty of water for at least 15 minutes. Flush skin with water. If swallowed, and victim is conscious, immediately induce vomiting.   CAS NO. [67-64-1]

Hazard Rating		Shipping Classification

 Consult MSDS for further health and safety information.

Figure 11–3. Sample hazardous waste label. (Label courtesy of Lab Safety Supply Inc., Janesville, WI.)

Periodically management should review the effectiveness of the written HazCom program. Regularly, all changes in chemicals, processes, emergency response procedures, equipment, and MSDSs should be entered into the written program, and the new information communicated to all employees. This new information may be communicated through training sessions and written summaries posted on employee bulletin boards.

The company can test program effectiveness through the use of drills, management walk-throughs, and question sessions. The results of all drills, walk-throughs, and question sessions can be critiqued by supervisory staff and employee representatives. Summaries of these critiques can be posted on employee bulletin boards. Management should immediately include any changes to the HazCom plan adopted during the critiques in the written plan. This information should be passed along to all employees.

Labeling

All containers and drums of hazardous chemicals must be labeled accurately, and labels must be prominently displayed. Workers must be trained to understand the labels and know how to interpret first aid or emergency instructions. Supervisors must know the company's labeling policies and procedures.

HCS Requirements. The HCS establishes specific labeling requirements to ensure the proper labeling of hazardous chemicals (Figure 11–3). The requirement places responsibility for creating an appropriate label at the source of the hazardous chemical, namely, the chemical manufacturer, importer, or distributor (Figures 11–2 and 11–4).

The employer, however, is responsible for maintaining the appropriate warnings on every container of hazardous chemicals used or stored in the workplace (Figure 11–5).

Chemical manufacturers, importers, and distributors of hazardous chemicals are required to ensure that, unless specifically exempt, each container of hazardous chemical that they ship is labeled, tagged, or marked with:

- Identity of the hazardous chemical(s)
- Appropriate hazard warnings
- Name and address of chemical manufacturer, importer, or other responsible party.

The following types of materials are exempt, or

LABEL REQUEST LETTER

(Date)

(Name)
(Title)
(Company Name)
(Address)
(City, State, ZIP Code)

Dear Sir:

The OSHA Hazard Communication Standard (29 CFR 1910.1200) requires companies that sell hazardous materials to label containers with the identity of the chemical, the name and address of the manufacturer, importer, or other responsible party, as well as appropriate hazard warning(s). Please provide a quantity of labels for _____ which we purchase
(Name of Product)
from your company.

Your prompt attention to this request will be appreciated.

Sincerely,

Any supplier improperly labeling hazardous chemicals should be notified in writing and all correspondence dated and retained in an appropriate file.

Figure 11–4. Sample Label Request Letter

RESPONSIBILITY FOR HAZARD DETERMINATION

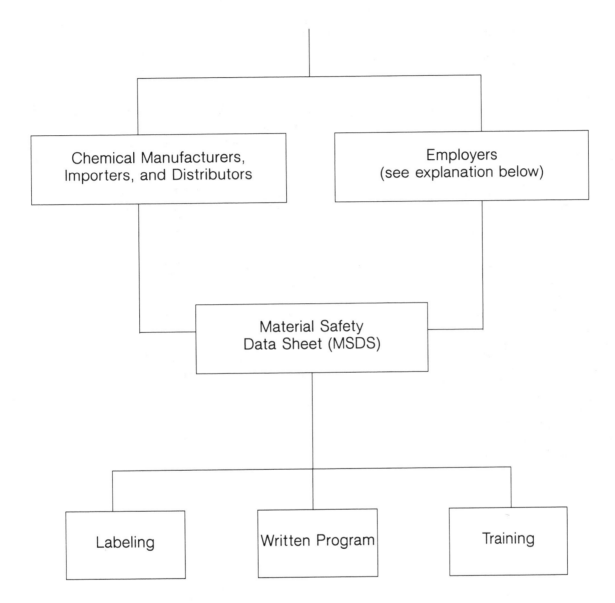

IMPORTANT NOTE!
Employers may rely on the hazard determination of their suppliers or, at their option, make their own hazard determination. The employer, however, is responsible for the accuracy of Material Safety Data Sheets, labeling, the written hazard communication program, and employee training.

Chemical manufacturers or importers (or employers who choose to evaluate chemicals) are required to describe the procedures they use to determine the hazards of the chemicals they evaluate. These procedures must be in writing and are to be made available, on request, to employees, their designated representatives, and designated OSHA officials. The written description may be incorporated into your required written hazard communication program.

Figure 11–5. Responsibility for Hazard Determination

partially exempt, from the labeling requirements. (It should be noted that even if a hazardous material is exempt from the HC labeling requirements, it still may need to meet other requirements specified by the HCS:

Exempt from All Requirements.

- Any hazardous waste subject to the EPA Resource Conservation and Recovery Act (RCRA)
- Tobacco or tobacco products
- Wood or wood products (but not wood dust)
- Food, drugs, cosmetics
- Food, drugs, or cosmetics intended for personal consumption by employees while in the workplace
- Any consumer product or hazardous substance, (as defined in the Consumer Product Safety Act and Federal Hazardous Substances Act, respectively) that an employer can demonstrate is used in the workplace in the same manner as in normal consumer use, so that workers are not exposed to the substance any longer or more often than any other consumers would be
- Any drug (as defined in the Federal Food, Drug, and Cosmetic Act) that is in solid, final form for direct administration to the patient
- A manufactured article, formed to a precise shape or design during manufacturing, whose end use function(s) is completely or partially dependent on its precise shape or design. Under normal conditions of use, the item must not be able to release or otherwise expose workers to hazardous chemicals.

Exempt from HCS Labeling—Not from Other HCS Requirements.

- Pesticides (as defined by the Federal Insecticide, Fungicide, and Rodenticide Act—FIFRA), when the product is subject to FIFRA and EPA labeling requirements. Containers used for mixing at a loading site do not require labels under the HCS labeling requirements, as long as the mixture is to be used within 12 hours, and the FIFRA labeling and warning sign system is in use in the mixing/loading area
- Any food, food additive, color additive, drug, cosmetic, or medical or veterinary device (including articles that are components of these devices), as defined in the Food and Drug Act, that are subject to the labeling requirements under the Food and Drug Act (labeling requirements apply only when the drug is packaged, and not before)
- Any distilled spirits, wine, or malt (as defined in the Federal Alcohol Administration Act and regulations issued under the act) when subject to labeling requirements issued under the act by the Bureau of Alcohol, Tobacco, and Firearms
- Any consumer product or hazardous substance (as defined in the Consumer Product Safety Act and

the Federal Toxic Substances Act, respectively) when subject to a consumer product safety standard or labeling requirement of those acts by the Consumer Product Safety Commission

Three specific circumstances exist in which the HCS exempts the employer from labeling a hazardous chemical container in the workplace:

1. Portable containers do not have to be labeled when they contain hazardous chemicals that were transferred from labeled containers, and the chemicals in the unlabeled containers are intended to be used immediately (during the same work shift) by the person who made the transfer. Best practice would include labeling all containers to avoid confusion and mistakes, especially in the event of an emergency.
2. Employers may affix signs, placards, process sheets, batch tickets, operating procedures, or other such written materials, to individual process containers instead of affixing labels, provided:

- The container to be labeled by this alternate method is clearly identified.
- The alternative labeling method contains the same chemical identity and hazard information found on the appropriate MSDS and on the list of in-house hazardous chemicals.

3. If existing labels already convey the required information, chemical manufacturers, distributors, importers, and employers do not have to relabel the chemical to comply with the HCS.

Employers are required to ensure that, unless specifically exempted as described above, all containers, including those used to transfer hazardous chemicals, as well as original containers, are labeled, tagged, or marked with:

- The identity of the chemical
- The appropriate hazard warning(s) including target organ effects

All labels must be legible and in English (labels in another language may be added to, but not substituted for the English language label).

Employers do not have to relabel an original container, unless the container label is obscured, hard to read, defaced, or missing. Employers are not to remove or deface existing labels on incoming containers of hazardous materials.

A description of the methods chosen to meet the labeling requirements (employers can order printed labels or produce their own), should become a part of the written HC program. If the employer uses a numerical coding system for the in-house identification of

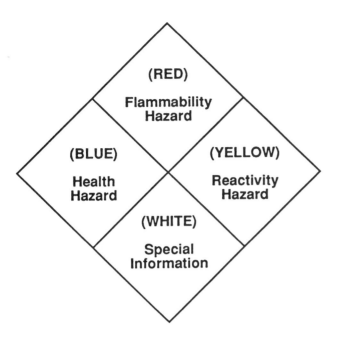

Figure 11-6. NFPA 704 M Hazard Identification System

hazardous materials (often used when the labeling system is integrated with the chemical inventory system), then he or she must make sure employees are trained to understand and use the numerical system, and to obtain more information from the MSDSs.

If the hazardous chemical is regulated by OSHA in a chemical-specific health standard, the chemical manufacturer, importer, distributor, or employer must ensure that the labels or other forms of warning used are in accordance with the requirements of that standard. Competent individuals, familiar with the hazards of the chemical and with the language and interpretation of the law, should prepare the information to appear on the labels. Sources of information for creating in-house labels include MSDSs, ANSI Guidelines, and software programs.

Best Practices Suggestions. Employers can use any in-house labeling and hazard warning sign method(s) they want, as long as they comply with the HCS requirements. Two systems in common use are the National Fire Protection Association (NFPA) program and the Hazardous Material Identification System (HMIS). These two labeling and warning sign systems are easy to use. They usually refer to the information on the MSDS, and require no special printing or manufacturing procedures.

The HMIS is considered an appropriate method for complying with HCS labeling provisions. The NFPA system is used primarily as a guide by emergency response teams when choosing fire fighting equipment, techniques, and evacuation procedures at fires involving hazardous chemicals.

In both NFPA and HMIS systems, the category "health hazards" refers to the ability of the material to cause personal injury from absorption, ingestion, inhalation, or injection into the body. The NFPA health hazards were developed mainly for reference to acute exposures, i.e., fire fighting, not everyday chronic exposure. The category "flammability hazards" indicates how easily the material burns. The category "reactivity" (instability) is concerned with the ability of the material to release energy when it reacts with itself or other materials.

NFPA Labeling System. This system uses three main categories to identify hazards of a chemical: health, flammability, and reactivity (stability). The degree of severity is indicated numerically by five divisions ranging from 0 to 4. A "4" rating indicates a severe hazard, while a "0" means that no special hazard exists. The NFPA label is diamond-shaped, and includes four smaller different-colored diamonds within the large diamond. Flammability information appears in the red diamond at the top of the large diamond label. Health appears in the blue diamond on the left side (as it faces the reader). Reactivity (stability) appears in the yellow diamond on the right. The white diamond at the bottom is used to indicate whether the chemical is water-reactive, or whether other special conditions exist (Figure 11-6).

HMIS Labeling System. This is a comprehensive and uniform labeling system designed to alert employees to chemical hazards to which they are, or may be, exposed in the workplace. The hazard communication portion of the HMIS system communicates information regarding chemical identity, health, flammability, and reactivity hazards; and proper personal protective equipment.

The chemical identity is conveyed by chemical or common name, a code number, or other descriptive term. The identity process in use should serve as a direct link to the appropriate MSDS. According to the HCS, English must be the primary language appearing on the label, although *additional* languages also can be used. As in all labeling systems, the same name or identifier used the chemical on the HMIS label must also appear on the corresponding MSDS label.

Numerical ratings communicate the acute health, flammability, and reactivity hazards of the chemical. An alphabetic or picture code designates the type of personal protective equipment that should be used. An (*), or other symbol, is used to communicate health hazards on the HMIS label.

The same numerical identification system as described in the NFPA system is used in the HMIS (a rating of "0" indicates fairly low- to nonexistent hazard; a rating of "4" indicates severe or extreme hazard). As a rule, any category with a "2" or higher rating is considered very dangerous, or as having the potential to create a very dangerous situation. The basic HMIS label is a colored rectangle that provides:

- Identification of the chemical
- Acute health, flammability, and reactivity ratings
- Chronic health hazard information
- Personal protective equipment designation

The rectangle is broken into five (5) rectangles, with each rectangle identified by its specific color and written category name. The top white rectangle is blank and is for writing in the chemical name/identification. Next is the blue for health. The next is red for flammability. Next is yellow for reactivity/stability. The last rectangle is white for personal protective equipment.

More Best Practices Suggestions. The labeling and inventory systems can be integrated, provided that the HazCom program contains a written section describing exactly how the two systems are to be blended. This written policy should be a component of the HazCom program and should be made available in the workplace at all times.

A common way to integrate labeling and inventory systems is to create a log book or computer database in which every chemical is assigned its own page or section, usually in alphabetical order. The chemical's identification appears at the top of the page along with a corresponding inventory control number. A completed NFPA, HMIS, or other label for the chemical is secured (or linked) to each inventory page. An envelope of completed labels for the chemical is attached to the back of each page.

Thus, every time the chemical enters the facility, the employee in charge of placing the chemical in inventory can turn to the appropriate chemical page, note the amount and destination of the chemical, and immediately affix the appropriate chemical label. This method provides a record of the destination of the chemical and ensures that employees will label the chemical appropriately. Even if the chemical does not need to be relabeled, an employee can still log it in, and affix the inventory control number to its existing label.

This type of inventory and labeling system also can be used to keep track of incoming MSDSs. The employee can place a check mark next to the chemical name indicating whether an MSDS has been received. If not, the employee can send a letter to the distributor, requesting the appropriate MSDS.

Like any other method, a labeling system is only good if employees use it consistently. Supervisors should make visual checks or all process, storage, transfer, and use areas each month. Employees must be encouraged to express their concerns and preferences about the labeling system so that any changes

can incorporate employees' observations.

A labeling program is useless if employees do not know the meaning of the hazard information on the labels. Training for use of the labeling system is critical and should consist of at least the following elements:

- Audiovisual aids to explain and illustrate the labeling system being used in the workplace
- Wallet cards that provide basic information needed to understand the hazard ratings used on the workplace hazard labels. Employees should be encouraged to carry the cards at all times.
- Wall posters to explain the numerical identification system being used. The wall posters can also display personal protective equipment codes and be hung in appropriate work areas when in use for training sessions.
- Written material that explains the workplace labeling system in easily understood terms and format. This information should be supplied in a second language also, if needed, as long as the primary language is English.
- Raw materials information for employees that explains the procedure and criteria for assigning chemical hazard ratings
- Implementation procedures describing the necessary steps for understanding and implementing the hazard label system
- Training sessions on how to read and interpret the labels, to give specific information regarding product hazards, and information on personal protective equipment.

Material Safety Data Sheets

MSDSs are one source of information on hazardous chemicals. Employers should have relevant MSDSs on hand, and all employees should be trained in their use and know where the sheets are kept in the workplace. MSDSs should be accessible to all employees at all times. Employers are responsible for requesting updates from manufacturers and importers.

HCS Requirements. Chemical manufacturers and importers must obtain or develop an MSDS for each hazardous chemical they produce or import. Employers also must have an MSDS for each hazardous chemical the company uses. Each MSDS must be in English and contain at least the following information:

- The chemical identity/name used on the label—with a few exceptions, including trade secrets
- Physical and chemical characteristics of the hazardous chemical

- Physical hazards of the chemical
- Health hazards of the chemical
- The primary routes of entry into the body
- The OSHA permissible exposure limit (PEL), American Conference of Governmental Industrial Hygienists (ACGIH) Threshold Limit Value, and any other exposure limit used or recommended
- Whether the chemical is listed in the National Toxicology Program (NTP) Annual Report on Carcinogens (latest edition) or has been found to be a potential carcinogen in the International Agency for Research on Cancer (IARC) Monographs (latest edition), or by OSHA, or ACGIH.
- Any generally applicable precautions for safe handling and use, known to the chemical manufacturer, importer, or employer preparing the MSDS
- Any generally applicable control measures known to the chemical manufacturer, importer, or employer preparing the MSDS
- Emergency and first aid procedures
- The date of preparation of the MSDS or the last change to it
- The name, address, and telephone number of the chemical manufacturer, importer, employer, or other responsible part preparing or distributing the MSDS, who can provide additional information on the hazardous chemical and appropriate emergency procedures.

If no relevant information is found for any given category on the MSDS, then the preparer should mark the category to indicate that no information was found. When complex mixtures have similar hazards and contents, one MSDS may apply to all these similar mixtures. The chemical manufacturer, importer, or employer preparing the MSDS must ensure that the recorded information accurately reflects the scientific evidence used in making the hazard determination.

Chemical manufacturers or importers must provide distributors and employers with an appropriate MSDS with their initial shipment and with the first shipment after an MSDS is updated. In turn, distributors must provide the MSDSs and update information to other distributors and employers. The employer should maintain copies of the required MSDSs for each hazardous chemical in the workplace, and make sure they are readily accessible to employees during each work shift when workers are in their job areas.

MSDSs may be kept in any form, including operating procedures, and may be designed to cover groups of hazardous materials in a work area where it may be more appropriate to address the hazards of a process rather than individual hazardous chemicals. MSDSs also must be readily available, upon request, to employees' designated agency representatives, to the Assistant Secretary of OSHA, and to the NIOSH director.

Best Practices Suggestions. Several attempts have been made to standardize the forms used for MSDSs, but none has been adopted successfully on a national or international scale. Therefore, the MSDSs that a company receives or produces are usually adopted by the company producing or distributing the chemical.

Employers who use hazardous chemicals supplied by a manufacturer, importer, or distributor do not need to produce an MSDS for the hazardous chemical. However, the employer may choose to produce a company MSDS for the supplied chemical or may have to produce an MSDS of a mixture or solution used in the workplace. If either situation occurs, the employer must ensure that a competent person fills out the MSDS and that the employer's name, address, and telephone number appear on the sheet.

MSDSs must be current, updated as necessary, and easily accessible to employees at all times. It is a common practice to keep MSDSs in a brightly colored binder, attached to a wall or stored on a special shelf in the work area. All MSDSs for the chemicals used in a specific work area should be kept in that area. Complete sets of all MSDSs for the workplace should be kept in at least on central location in the plant. Most MSDS information is broken into 9 to 12 sections on the data sheet. MSDS interpretation and familiarization training should address the contents of each section and the various types of MSDSs that exist in the specific workplace or work area.

If an inventory check reveals missing MSDSs, the employer should send a letter to the manufacturer, importer, or distributor, requesting a copy of the missing MSDS. By checking for MSDSs every time a chemical inventory is performed and every time a chemical is received in the workplace, the employer should always have the correct MSDS available for the workplace and work areas.

It is primarily the responsibility of the manufacturer, importer, or distributor to provide timely updates of MSDSs to the employer. However, the employer also should ensure that the MSDS master file, and all area files, contain the most recently updated MSDSs for the hazardous chemicals used in the workplace. This step can be done by sending an annual form letter to the manufacturer, distributor, or importer, requesting updated MSDSs for the chemicals used by the employer. These letters are part of the documentation showing that the employer is complying with the intent of the HCS, and copies should be kept permanently on file.

Several current computer programs and on-line services have a database containing the most common

MSDSs. However, these programs and services can be costly and may not cover all the mixtures and compounds an employer might use, store, or transfer. Whether MSDSs are supplied by the manufacturer, distributor, or importer, developed in-house; or accessed through a database or on-line service, the employer should write a policy that describes how MSDSs will be acquired, updated, integrated with inventory and receiving procedures, distributed, and accessed by employees.

The HCS does not specifically require that MSDSs be kept on site for a particular length of time after the chemical is no longer in use. According to OSHA 29 *CFR* 1910.120, access to the records, however, is required for 30 years, so they must be safely stored. It is suggested that a record of the identity of substances used be kept indefinitely to comply with other federal regulations regarding retention of employee exposure and medical records. An employer may discard an original data sheet and retain only new data sheets if a record of the original formulation is also kept.

Employee Information and Training

Given today's complex job demands and diverse workforce, employee information and training regarding hazardous materials has become even more critical. Employers must make sure that workers have the proper information and training, and that the information is provided in a way they can understand and interpret correctly.

HCS Requirements

Employers must give their employees information and training regarding hazardous chemicals in their work areas. This information and training must be provided with the employee's first work assignment, and whenever a new hazard (including a process change) is introduced.

All employees who are, or may potentially be, exposed to hazardous chemicals under normal work conditions or in foreseeable emergencies, must be provided with required information and training before beginning work in the area that contains hazardous chemicals.

Employees must be informed of:

- HCS requirements
- Any operations in their area where hazardous materials are present
- Location and availability of employer's written hazard communication program. The program must include required list(s) of hazardous chemicals and associated MSDSs.

Employers must provide the following information in the training sessions:

- All methods and observations that may be used by the employee to detect hazardous chemical presence or release in the work area
- Physical and health hazards of chemicals used in the work areas. Chemicals are rated, per the HCS and OSHA, by the type of hazard they present: health, physical, or both. It is the responsibility of the chemical manufacturer, importer, distributor, and employer, to make sure that each hazardous chemical is properly identified by its hazard type, and that appropriate warning labels are affixed to the chemical's container. The respective MSDSs may expand on the warning label information, which is why the MSDSs must be readily available to all employees who are, or who may be, exposed to hazardous chemicals. This information must be explained in training sessions so employees will learn how to read the warning labels and recognize the hazards the chemicals pose.
- Measures employees can take to protect themselves from chemical hazards in their work areas. This includes instruction in appropriate work practices; emergency procedures; and use, maintenance, and selection of PPE.

The employer is responsible for:

- Choosing correct types of PPE for chemicals being used
- Providing a written policy that addresses PPE use
- Providing employees with training and medical surveillance needed when PPE is used in the workplace.
- Details of the hazard communication program developed by the employer. This part of the training must include explanation of the labeling system, MSDS use and information content, how to correctly use the hazard information sheets and charts, and how the container labels are tied to the MSDSs.

Best Practices Suggestions. Product and employee liability concerns make recordkeeping an essential part of an employee training program. The record keeping documents can include:

- Notices posted for each training session (Figure 11–7)
- Topics discussed in each training session
- Minutes of each training session
- An attendance sheet that contains the session date, session topic, printed employee name, and employee signature (Figure 11–8). It is recommended that these attendance records be kept for employee's

(Text continues on page 172.)

HAZARD COMMUNICATION EMPLOYEE TRAINING MEMO

TO:_____DATE:_____

DEPARTMENT:_____ _____

CLOCK, EMPLOYEE OR SOC. SEC. #_____

All employees exposed or potentially exposed to hazardous chemicals under normal operating conditions are required to attend a training session at the time of initial assignment or when a new hazard is introduced into the workplace. If the date assigned conflicts with another important activity, please have your supervisor contact this office. Otherwise, please sign and return this form to:

_____

You have been scheduled to attend the next hazard communication training program:

        DATE:_____

        STARTING TIME:_____

        COMPLETION TIME:_____

        LOCATION:_____

☐ Yes, I will be able to attend.

        Employee's
        Signature_____

        Supervisor's
        Approval_____

☐ No, I cannot attend. My supervisor will explain.

The Hazard Communication Employee Training Memo can be used to notify the employee of a training session. It should be returned to the Compliance Manager and will become part of the permanent record of the employee's training.

Figure 11–7. Sample Hazard Communication Employee Training Memo

HAZARD COMMUNICATION TRAINING LOG

SUBJECT:_____ COMPANY: _____
A/V _____ ADDRESS: _____
Handouts _____ TRAINER'S NAME _____
TITLE _____
DATE OF TRAINING _____

EMPLOYEE NAME	EMPLOYEE NO	DEPARTMENT	EMPLOYEE SIGNATURE

Use a separate training log for each lesson. Employees attending the lesson should signify their presence with their signature. Note absentees, and schedule makeup classes.

Figure 11–8. Sample Hazard Communication Training Log

EMPLOYEE TRAINING EVALUATION SHEET

Date: _____

1. List the hazardous chemicals you use in your work area: _____
_____
_____
_____

2. Were all of these chemicals covered in the training program? ____ Yes ____ No
   If NO, list those chemicals not covered.
_____
_____
_____

3. Did the program explain the protective clothing/equipment needed to work with the hazardous chemicals you use? ____ Yes ____ No. If NO, which chemicals were omitted?
_____
_____
_____

4. Do you have and use the required protective clothing/equipment? ____ Yes ____ No
   If NO, explain.
_____
_____
_____

5. Did the training program cover all the questions you had about the safe and proper use of hazardous chemicals?
   ____ Yes ____ No. If NO, explain.
_____
_____
_____

6. Use the back of this sheet to offer any comments on how to improve this training program.

NAME: _____

DEPARTMENT: _____

In addition to training on my rights and obligations under the law, I have...

• been instructed in how to read a MSDS and a label...

• I have knowledge of where the company written hazard communication program and MSDSs are kept and that I have access to these.

• I understand the necessary precautions to be taken when dealing with hazardous chemicals in my work area...

• I know where medical supplies and safety equipment are kept.

DATE: _____

_____
(employee signature)

The Employee Training Evaluation Sheet is used for student opinion and is also documentary evidence of training attendance.

Figure 11–9. Sample Employee Training Evaluation Sheet

employment duration, and for at least five years after.

- A document signed by the employee stating that he or she understands the session content and that continued employment with the company depends on attending all training sessions and following all company policies and procedures (Figure 20–9)
- Documentation of qualifications of instructor(s) teaching the training sessions

Some companies may want to create a learning lab (usually an area that is set up for supervised, unsupervised audiovisual or computer-based instruction on company policies and procedures) for training employees. This type of training program is viewed as a time and personnel saver by many companies, but it has some major flaws. (See also the best practices suggestions in Elements of a Hazard Communication Program earlier in this chapter.)

For example, the HCS requires that all employees be trained before they begin to work with hazardous chemicals. Unless a supervisor or trainer is assigned to ensure that employees report to the learning lab before beginning to work with hazardous chemicals, the company is in violation of the HCS.

Another flaw common to an unsupervised learning lab, a home study program, or an interactive computer-based training program, is that often no one is available to answer questions, a practice that can lead to dangerous situations in the workplace. Similarly, because the HCS requires that all employees understand the company HazCom program and the dangers of the specific workplace chemicals, even the best interactive computer learning program is limited, because it cannot answer unprogrammed questions.

Language and literacy barriers also may pose problems for the company training and documentation program. OSHA requirements for employee training and understanding of information presented in formal training programs do not differentiate between employers with more or less formal education or those who speak English as their second language or do not understand it, and any other employees. All employees must meet the HCS requirements for training, and the employer must prove that all employees understand the policies and procedures of the company's HazCom program.

The training program for employees with little formal education, or with English as a second or foreign language, may require more class time, translation of English materials into another language, or the use of international pictograms.

Whatever is done to facilitate the employee's understanding of hazards associated with the workplace, the training should be shown to be effective and in compliance with regulations, through audits and other documentation processes.

Proving that the training program is effective and operational will be an ongoing job for supervisors, trainers, and management staff. The employer can conduct periodic walk-through inspections to assess workplace conditions, job performance, and adherence to the company's HazCom program.

Another way to assess success of the training program is to conduct OSHA-like audits during walk-throughs. This type of audit could entail random questioning of individual employees regarding the company's HazCom program, checking on availability of MSDSs, having employees find certain sections in the HazCom program and explaining their content to the "inspector," or asking employees if they have experienced problems performing their job tasks safely and within the HazCom program guidelines. Information obtained during walk-throughs can be used to correct and update the HazCom program, and can serve as the basis for a reward system (e.g., the department with the most correct answers and least deficiencies within a six-month period could receive a monetary or other award).

The success of the training program also depends on the completeness of the chemical inventory. Once the initial inventory is finished and all chemicals have been identified as hazardous or nonhazardous, the hazardous chemicals can be categorized further by the type(s) of hazard they pose to the employee and the workplace, and then incorporated into the training program. It is recommended that the chemical inventory be continually updated as needed.

It is important that employees understand that there are between 5 and 6 million currently-known chemicals, and that only some of them are considered to be hazardous. Chemicals considered to be hazardous are referenced in OSHA, EPA, NIOSH, DOT, and ACGIH standards, guidelines, and lists (reference sources for these lists and other information sources appear at the end of this chapter). Copies of chemical lists and information sources can be kept on file for reference and training purposes.

In addition, the employer should incorporate information into the HazCom program that addresses types of symptoms usually associated with acute and chronic exposure to hazardous chemicals. The employer may choose to have a medical professional, toxicologist, safety expert, or industrial hygienist conduct these information and training sessions.

There are generally four methods that the employer can adopt to reduce unnecessary employee exposure to hazardous chemicals:

1. **Product Substitution.** The employer should try to choose a different, less harmful chemical for a specific process or operation, or change the process

to avoid using any chemicals. This is the preferred method of employee protection. Care must be taken to completely remove the more harmful chemical from the work area to prevent employees from accidentally using it. The employer is responsible for constantly reviewing the types of chemicals used in work processes and in the workplace. Reduce the number and type of chemicals by consolidating the use of oils, cleaning agents, etc.

2. **Use of Engineering Controls.** These controls include proper equipment when handling containers of hazardous chemicals, process enclosures, worker isolation, and local exhaust and general ventilation. The employer should develop written policies and training programs for appropriate use of these types of engineering controls. Safety eyewash stations and deluge showers are also engineering controls employees can use.

3. **Use of Administrative Controls.** These controls include rotating employees through areas or jobs with high probability of exposure to chemicals. This method is not acceptable with highly harmful chemicals, such as carcinogens. Establishing proper work practices and housekeeping measures are other administrative controls.

4. **Use of Personal Protective Equipment.** When a less hazardous chemical cannot replace a more hazardous chemical, and engineering controls do not adequately reduce exposure, employees usually must wear PPE specifically chosen for use with the hazardous chemical.

The written hazard communication program should include procedures and requirements for documenting these methods and steps taken to control exposure. In addition, employees should receive training in basic first aid for chemical exposures.

The employer should post emergency phone numbers at all phone locations and develop a written policy that addresses what the employees are to do in case of accident, fire, or explosion. Employees should be trained in these procedures, and the employer should hold several fire drills, plus other drills each year, to test the emergency response policies.

A spill, fire, or explosion involving hazardous chemicals may require action from employees who have received special training in responding to and abating hazardous chemical emergency situations. (This training refers to the requirements of the OSHA standard [29 *CFR*, 1910.120] of March 6, 1989, for Hazardous Waste Operations known as Emergency Response and HAZWOPER.) The employer should then develop a policy that addresses these types of emergencies and supply the extra

training courses and equipment required to teach employees how to handle the dangers involved in a hazardous chemicals incident.

Hazardous waste site operations fall under many specific regulations and jurisdictions, including the HCS. The basics of the HCS apply to all hazardous waste site personnel, including use of MSDSs, training, PPE, hazard identification, and written policies and procedures. However, other regulations, standards, and agencies also affect the hazardous waste site employee (such as SARA Title III, RCRA, and DOT. These types of operations are discussed in Chapter 12, Environmental Management.

NONROUTINE ELEMENTS OF A HAZARD COMMUNICATION PROGRAM

Temporary and Contract Employees

In a general sense, the employer is responsible for the safety, health, and well-being of all people who enter the workplace or work area. To this end, the employer should establish procedures covering all employees and visitors as a part of the company's HazCom written program. These procedures should specifically address the type of hazardous chemical training and information that maintenance workers, outside contractors, office personnel, and cleaning crews should receive before they enter workplaces where they may be exposed to hazardous chemicals.

Best Practices Suggestions. These procedures should contain at least the following:

- List of hazardous chemicals located in assigned work areas and any areas through which workers must pass to reach their work area
- Floor plan showing areas where workers are not permitted
- Location of written HazCom program
- Location of MSDS file for the assigned work area
- PPE requirements to be used in work area
- Review of contractor's PPE program, and if necessary, training in use of specific PPE and engineering controls to be used in assigned work area
- Emergency procedure training for fires, spills, explosions, and other disasters
- Conditions to avoid in work area to reduce possibility of exposure to hazardous chemicals

To close the communication circle, the employer should request a list of the chemicals to be used in their workplace, a description of how they will be used, and copies of the MSDSs. Proper disposal information should also be shared with the contractors for the facility. Some companies schedule a prework meeting to discuss this information and approve/reject chemical usage at their facility.

All temporary and nonfacility employees, with no exceptions, should receive formal training before they begin work in the employer's workplace. The reasons for this rule are simple: the potential liability to the employer and the risk of a serious accident to an uninformed worker or "outsider" are too great. Even if outside crews have worked in similar situations, with the same or similar chemicals, they should be trained regarding specifics of each new employer's facility and HazCom program.

Visitor Orientation

Visitors can represent a serious liability risk to employers, particularly if supervisors do not provide visitors with appropriate orientation and protective equipment. Supervisors should always know where visitors are in their areas, and either assign someone to escort them or accompany the visitors themselves.

HCS Requirements. There are no specific HSC requirements for this group of people.

Best Practices Suggestions. If the visitor will be in areas where hazardous chemicals are used, stored, processed, or transferred, or must pass through these areas, it is suggested that appropriate training and PPE be supplied before the tour begins.

The employer should create a policy that addresses visitor restrictions, training, identification, access, and PPE and include it in the company's written HazCom program.

New Employee Orientation

Orientation is a critical part of new employee training. Supervisors must ensure that each new employee has been properly oriented to work requirements, safety procedures, and emergency measures before beginning their job assignment.

HCS Requirements. The OSHA HCS requires that all new employees receive initial training on workplace hazards, MSDSs, the company's written HazCom program, definitions, labeling and warning signs, chemical inventory, and the intent of the OSHA HCS.

Best Practices Suggestions. All new employees, even if they have worked previously with hazardous chemicals, must receive HazCom training before they begin to work with, or are potentially exposed to, hazardous chemicals used in the new employer's workplace. The employer can select the training methods to be used, provided they meet HCS requirements and intent, are in written form, and are part of the

company's HazCom program. It is suggested that the training should:

- Include hands-on and classroom sessions
- Include examples of actual labels, MSDSs, PPE and engineering controls used in the workplace
- Be performed in a location that promotes and stimulates learning (a boiler room, busy lunchroom, or other cramped or noisy area would not usually be a good place to hold the training session)
- Include an orientation tour of entire workplace
- Be conducted by a knowledgeable person
- Include provisions for immediate training if a new chemical or process is introduced to the workplace (mandatory under the HCS)
- Include a written test to audit training effectiveness and level of employee understanding
- Be scheduled so employees in training sessions are not interrupted by work problems that require them to leave
- Include annual refresher courses

The HCS helps the employer provide a safe, healthy work environment for all employees and give them the tools they need to perform assigned tasks safely and effectively. The best practices suggestions are offered to help employers meet the OSHA HCS intent. They are designed as examples and tools the reader can either use, modify, or discard when developing a company HazCom program.

SUMMARY OF KEY POINTS

Key points covered in this chapter include:

- The federal HCS, known as OSHA 29 *CFR*, 1910.1200, regulates an employer's duty to communicate chemical hazard information to employees who may be exposed on the job. Federal or state agencies and OSHA enforce this standard.
- Companies should understand the differences between HCS standards and best-practices suggestions. The best-practices suggestion are meant for use as guidelines for implementing HCS requirements or when no HCS requirements have been developed to cover specific situations. These suggestions must never replace HCS requirements.
- The HazCom program must include written information on hazardous chemicals, proper labeling, employee training, chemical inventory, and MSDS distribution. Each aspect of the hazard communication program must meet or exceed federal HCS requirements.
- A written hazard communication document provides employers and employees with specific policies

and procedures to follow when training employees to handle hazardous substances. A copy of the written program must be kept at the workplace and be available to employees and to OSHA compliance officers and NIOSH investigators. The policies and procedures developed must cover the exact chemicals and processes that exist at a company.

- All containers and drums of hazardous materials must be labeled accurately and the labels prominently displayed. Workers must be trained to understand the labels and interpret chemical hazards, first aid, or emergency instructions. Supervisors must know what substances are exempt from HCS requirements and what circumstances exempt employers from labeling hazardous chemical containers.

- MSDSs, prepared by chemical manufacturers and importers, provide information on hazardous chemicals, including chemical name, characteristics, physical and health hazards, exposure limits, precautions and control measures, medical treatment procedures, and name, address, and telephone number of a person who can provide additional information on the chemical. MSDSs should be kept in the workplace, be accessible to employees at all times, and be regularly updated and/or new ones must be requested from suppliers.

- Under HCS regulations, employers must provide employees with information about HCS requirements, hazardous operations in their work area, and location and availability of the written HC program. Training sessions should inform workers of methods they can use to detect the presence or release of a hazardous chemical, the physical and health hazards of chemicals, protective measures for safe use, and details of the HC program as developed by the employer.

- Employers are also responsible for the safety and health of visitors and of new, temporary, or outside employees working on a company property. These employees and visitors should receive orientation sessions and be provided with protective gear before they are allowed into work areas. The company's HCS program should include policies and procedures to cover these nonroutine elements of hazard communication.

RESOURCES

Annual Report on Carcinogens. National Toxicology Program (NTP).

Chemical Data Sheets. Chemical Cards and Water Information Sheets. Chemical Manufacturing Association, 1825 Connecticut Avenue, N.W., Washington, DC 20009.

Chemical Safety Slide Rule. National Safety Council, 1121 Spring Lake Drive, Itasca, IL 60143–3201.

Code of Federal Regulations. General Industry Safety and Health Standards: 29 *CFR* 1910; Toxic and Hazardous Substances. 29 *CFR*, Part 1910, Subpart Z, Transportation: 49 *CFR* Parts 100–199. Washington, DC 20402: Superintendent of Documents, U.S. Government Printing Office.

Computer Programs:
OHMTADS: Accessed through Chemical Information System, Inc., (800) 247–8737
CHRIS (Chemical Hazard Response Information System): Accessed through Chemical Information System.
*CAMEO (*Computer-Aided Management of Emergency Operations): Widely used database. U.S. Department of Commerce. NOAA Hazardous Materials Response Branch 7600 Sandy Point Way NE, Seattle, WA 98115, (206) 526–6317. Can also be purchased through the National Safety Council.
*TOXLINE (*Toxicology Information On-Line).
MEDLARS, Management Section, Specialized Information Systems, National Library of Medicine, 8600 Rockville Pike, Bethesda, MD 20814; (301) 496–6193; (800)638–8480.
*CHEMTREC (*Chemical Transportation Emergency Center): Access via telephone, (800) 424–9300; (202) 483–7616.

Guidelines to the Handling of Hazardous Materials. Source of Safety, Inc., 8303 East Kenyon Drive, Denver, CO 80237.

International Agency for Research on Cancer (IARC) Monographs

National Fire Codes. National Fire Protection Association, 1 Batterymarch Park, Quincy, MA 02269.

NIOSH Pocket Guide to Chemical Hazards. Washington, DC: U.S. Department of Health and Human Services.

NIOSH Registry of Toxic Effects of Chemical Substances. Washington, DC: U.S. Department of Health and Human Services.

Threshold Limit Value. American Conference of Governmental Industrial Hygienists, P.O. Box 1937, Cincinnati, OH 45201.

REFERENCES

Bronstein, AC, Currance, PL. *Emergency Care for Hazardous Materials Exposure, No. 2.* St. Louis, MO: C.V. Mosby, 1994.

Budavari, S, Ed. *The Merck Index of Chemicals, Drugs, and Biologicals.* 11th Edition. Rahway, NJ: Merck and Company, Inc., 1989.

Hazard Communication Guidelines for Compliance. OSHA Report No. 3111. U.S. Department of Labor, Occupational Safety and Health Administration. Washington, DC: U.S. Government Printing Office.

Lewis, RJ, Sr. *Hawley's Condensed Chemical Dictionary.* 12th Edition. New York, NY: Van Nostrand Reinhold Publishing Company. 1993.

Meyer, E. *Chemistry of Hazardous Materials.* Englewood Cliffs, NJ: Prentice-Hall, Inc., 1997.

12

ENVIRONMENTAL
MANAGEMENT

After reading this chapter, you will be able to:

- Recognize the major federal laws and state trends in environmental protection and understand their implications for the workplace

- List the principal elements in the management of hazardous chemicals

- List proper documentation required to show compliance with chemical management regulations

- Explain the elements of employee training to ensure company compliance with environmental regulations

- Describe proper storage of and cleanup procedures for hazardous chemicals and materials

- Describe federal requirements for chemical emergency response planning

- Establish waste minimization, source reduction, and other policies to reduce environmental pollution

The supervisor's role, never an easy one, has been complicated further by the ever-increasing maze of environmental laws and regulations enacted in recent years. When people think of environmental problems, they usually recall full-scale disasters such as a chemical explosion, a pesticide spill, or oil spilled from a large oil tanker. They think of trains carrying dangerous chemicals that derail, forcing the evacuation of entire towns. They think of major corporations paying millions of dollars in fines because they violated federal environmental laws. The environmental problems that supervisors encounter at their facility or facilities may never reach those extreme levels. Nevertheless, they have potential for truly devastating consequences, not only for supervisors, workers, and corporations, but for the surrounding community and the entire planet as well.

Thus, it is essential that management at all levels in all industries take appropriate steps to protect their workers and the public health and welfare from toxic chemicals and environmental threats. The first step toward good environmental management is understanding what the issues are, and knowing which problems are considered serious enough to be regulated by the federal government. The supervisor needs to know what to look for, what the dangers are, and what situations require immediate action. If an environmental problem arises, the supervisor needs to understand what responses are required and where to go for additional assistance. He or she must know also what information is available to help make decisions and where to find that information.

Supervisors should be familiar with the regulations issued for control of potential environmental problems, and understand the consequences of failing to comply with those regulations. They must realize the gravity of both the problems and the implications of noncompliance. Supervisors must recognize environmental hazards when they appear and alert safety and/or environmental managers to these hazardous situations. They must report any observed problems, violations, or potential hazards to management.

This chapter provides an overview of the environmental issues facing supervisors. It will help them make responsible decisions for the benefit of their company's employees and of their communities. (Figure 12–1 is a guide to acronyms used.)

U.S. REGULATORY AGENCIES

The U.S. Environmental Protection Agency

The U.S. Environmental Protection Agency (EPA) was created in 1970 to help protect the environment

ACRONYMS RELATED TO ENVIRONMENTAL MANAGEMENT	
AHERA	Asbestos Hazard Emergency Response Act
ASHAA	Asbestos School Hazard Abatement Act
BACT	Best Available Control Technology
CAA	Clean Air Act
CAS	Chemical Abstract Services
CERCLA	Comprehensive Environmental Response, Compensation, and Liability Act
CWA	Clean Water Act
DOT	U.S. Department of Transportation
EPA	U.S. Environmental Protection Agency
EPCRA	Emergency Planning and Community Right-to-Know Act
Form R	Toxic Chemical Release Inventory Form
FIFRA	Federal Insecticide, Fungicide, and Rodenticide Act
HMIS	Hazardous Materials Identification System
HMTUSA	Hazardous Materials Transportation Uniform Safety Act of 1990
HSC	Hazard Communication Standard
HSWA	Hazardous and Solid Waste Amendments
LAER	Lowest Achievable Emission Rate
LEPC	Local Emergency Planning Committee
MACT	Maximum Achievable Control Technology
MOA	Memorandum of Agreement
MSDS	Material Safety Data Sheet
NAAQS	National Ambient Air Quality Standards
NESHAP	National Emission Standards for Hazardous Air Pollutants
NFPA	National Fire Protection Association
NPDES	National Pollutant Discharge Elimination System
NSPS	New Source Performance Standards
OSHA	Occupational Safety and Health Administration
PCBs	Polychlorinated Biphenyls
POTW	Public Owned Treatment Works
PRP	Potentially Responsible Parties
RCRA	Resource Conservation and Recovery Act
SARA	Superfund Amendments and Reauthorization Act of 1986
SERC	State Emergency Response Commission
SIC	Standard Industrial Classification
TSCA	Toxic Substances Control Act
TRI	National Toxic Release Inventory

Figure 12–1. Commonly Used Acronyms

and the public health, and to control and abate environmental pollution under the laws enacted by Congress. The specific actions EPA takes to carry out these laws depend on, among other things, the best available scientific knowledge and control technology.

EPA staff in Washington, DC, and around the country work to develop environmental policies, set national standards, manage a research and development program, enforce laws and regulations, and develop regulations for pesticides, toxic substances, hazardous waste, air, and water. Most of EPA's dealings with the public and regulated industries are handled by the agency's 10 geographic regional offices. These offices work directly with state and local government agencies and with individuals to carry out environmental laws and regulations and provide technical assistance. They also review the environmental problems of these governments to make sure the programs are adequate and consistent with federal laws and regulations. Departments within EPA are the Office of the Administraton; International Activities; Administration and Resource Management; Enforcement; General Counsel; Policy, Planning and Evaluation; Inspector General; Water; Solid Waste and Emergency Response; Air and Radiation; Pesticides and Toxic Substances; and Research and Development.

Although some states are delegated responsibility for administering EPA programs, the EPA retains enforcement authority and oversight responsibilities. A Memorandum of Agreement (MOA) between the state and EPA's Regional Administrator outlines the nature of these responsibilities and oversight powers. The MOA specifies the level of coordination between the state and EPA in implementing the program.

No two MOAs are exactly alike, as they contain state-specific agreements. However, several provisions, required by rule, are common to all MOAs:

- Specifying the frequency and content of reports that the state must submit to EPA
- Coordinating compliance monitoring and enforcement activities between the state and EPA
- Joint processing of permits for those facilities that require a permit from both the state and EPA
- Specifying the types of permit applications to be sent to the EPA Regional Administrator for review and comment
- Transferring permitting responsibilities upon authorization.

Occupational Safety and Health Administration

The Occupational Safety and Health Administration (OSHA) was created within the Department of Labor to require employers and employees to reduce workplace hazards and implement new or improve existing safety and health programs. OSHA requires records of job-related injuries and illnesses to be kept. It develops and enforces mandatory job safety and health standards, and guarantees employers and workers the right to be fully informed and to appeal actions.

OSHA standards fall into four major categories: general industry, maritime, construction, and agriculture. These standards require adoption or use of many practices, means, methods, or processes reasonably necessary and appropriate to protect workers on the job and the environment around them. Employers and supervisors are responsible for becoming familiar with standards applicable to their workplaces. They must ensure that employees have a workplace that is free from recognized hazards.

Under OSHA regulations, employers of 10 or more employees must maintain records of occupational injuries and illnesses as they occur. OSHA requires inspection of workplaces by OSHA compliance health and safety officers (CHSO).

U.S. GOVERNMENT REGULATIONS AND LAWS

Over the past 30 years, the federal government has passed numerous landmark environmental laws. These range from broad-based guidelines to specific regulations protecting air, water, and soil quality and overseeing workplace health and safety. This section discusses the provisions of major federal environmental regulations and how they affect both supervisors' day-to-day responsibilities and the company's role as a corporate citizen in the community.

Clean Air Act

Air pollution has long been one of the greatest threats to human health and the environment. The first laws addressing air pollution were passed by cities as early as the 1880s. One of the country's landmark federal environmental laws was the Clean Air Act (CAA), passed by Congress in 1970 to give the EPA the power to establish air quality standards. The Clean Air Act of 1970 was amended in 1974, 1977, and 1990.

Under the CAA, the EPA sets and periodically reviews National Ambient Air Quality Standards (NAAQS) for "ambient" air pollutants. These are pollutants emitted by industrial operations and vehicles at levels that "may measurably be anticipated to endanger" public health and welfare. These criteria pollutants include ozone, carbon monoxide, particulate matter, sulfur dioxide, nitrogen oxides, and lead.

Primary standards for emissions are set for these pollutants to protect human health. Secondary standards are designed to protect crops, livestock, vegetation, buildings, and visibility. To control air emissions from cars and other vehicles, EPA must prescribe and revise emission standards for new vehicle engines in several categories and develop programs to test and certify these engines for compliance with national standards.

EPA also establishes New Source Performance Standards (NSPS) for stationary sources, such as factories, power facilities, and other buildings. These sources generate air pollutants mainly by burning fuel for energy and as by-products of industrial processes. Regulations are also in place for emission contaminants from existing sources. Electric utilities, factories, and commercial buildings that burn coal, oil, natural gas, wood, and other fuels are the principal sources of such pollutants as sulfur dioxide, nitrogen oxides, carbon monoxide, particulates, lead, and the volatile organic compounds that are precursors to ozone pollution and smog.

The CAA also regulates prevention of air deterioration in parts of the country that have met NAAQS by requiring new or modified sources to use Best Available Control Technology (BACT) to accomplish new source performance standards (NSPS).

In addition, the CAA requires EPA to set National Emission Standards for Hazardous Air Pollutants (NESHAP) that can contribute to an increase in mortality or serious illness in humans. Those standards have been issued for asbestos, beryllium, mercury, vinyl chloride, arsenic, radionuclides, benzene, and coke oven emissions.

Each state must prepare a State Implementation Plan describing how it will control emissions from mobile and stationary sources to meet the NAAQS. When a state develops a plan, it must be approved by municipal and state governments and by EPA. It then becomes part of federal and state laws and may be enforced by either the EPA or the state. If a state does not meet national air pollution standards, it must develop a new plan. Some states have completed modified plans.

Amendments to the CAA were passed in 1990. Congress added regulations for areas not meeting national air pollution standards—called nonattainment areas—and regulations for mobile sources of air pollution, air toxic chemicals, chlorofluorocarbons, and other chemicals that destroy the ozone layer. Those amendments identified 189 hazardous air pollutants that must be regulated. They set a schedule for identifying and regulating 250 categories of sources of toxic air pollutants, using Maximum Achievable Control Technology (MACT) and Lowest Achievable Emission Rate (LAER). They categorized areas not meeting national air pollution standards into four categories: extreme, severe, serious, and moderate/marginal. States were required to revise their implementation plans for all but the marginal nonattainment areas.

Summary. The CAA is one of the most complex federal regulatory laws. It may help supervisors to deal with clean air regulations by prompting questions such as: Do company processes require air permits? Where is the information stored about the company's history of air permits and/or violations? What pollution control equipment is available and what is being utilized to reduce air emissions? Which personnel in other departments can help supervisors understand specific regulations, order equipment, or identify potential problems? Are there state clean air regulations, in addition to the federal ones, that the company must follow?

Clean Water Act

The principal federal law dealing with water pollution is the Clean Water Act (CWA). It was first passed in 1972 as the Federal Water Pollution Control Act (FWPCA) and strengthened and renamed the Clean Water Act in 1977. Under this law, EPA has developed regulations and programs to reduce pollutants entering all surface waters, including lakes, rivers, estuaries, oceans, and wetlands. The 1977 amendments ensured continued support for municipal sewage treatment facilities, initiated a new state and federal program to control pollution from "nonpoint" sources such as city streets and farmland, and accelerated imposing tighter controls on toxic pollutants.

When the name of the act was changed in 1977, so was its regulatory focus. The objective of the act became that of controlling toxic water pollutants more effectively through use of technology-based standards. The Clean Water Act is composed of five major elements:

1. Permit program
2. National effluent standards
3. Water quality standards
4. Oil and hazardous substances
5. Stormwater discharges

EPA also develops uniform, nationally consistent limitations for pollutant discharges for industry and wastewater treatment facilities, with standards particular to different industries. These guidelines are used by EPA, the states, and local publicly owned treatment works to establish National Pollutant Discharge Elimination System (NPDES) permits. All industrial and municipal facilities that discharge wastewater directly into rivers and streams must have an NPDES permit. They are responsible for monitoring and reporting discharge levels. Industries with permits to discharge to Public Owned Treatment Works (POTW) also are responsible for monitoring and reporting. The local POTW, the states, and/or EPA inspect the dischargers to determine if they are complying with the law.

Under the CWA, states adopt water quality standards for every stream within their borders. These standards include a designated use for the body of water (such as fishing or swimming) and criteria to protect that use. They also prohibit degradation of existing water conditions. The criteria are specific to pollutants and represent permissible levels of substances in the water that still allow the designated use.

The discharge of oil and hazardous substances into navigable waters of the United States is prohibited. Specific hazardous substances are identified and spill quantities of these substances deemed too "harmful" are established. The Act requires that "harmful" spills into navigable waterways be reported. The amended Act requires that certain categories of industrial or municipal stormwater discharges also be regulated as NPDES discharges. For example, facility floor drains are discharge points. Procedures are required to ensure that such potential discharge points be blocked or diked in the event of a hazardous material spill. These water quality standards are the basis for nearly all water quality management decisions. They are reviewed periodically to ensure compliance with the CWA and its standards.

Summary. Questions to ask about clean water compliance include the following: Do the company's processes require water discharge permits? Into what bodies of water is the wastewater discharged? Where is the information about the company's history of water discharge permits and/or violations? What pollution control equipment is available and what controls have been implemented to reduce water discharges? Which personnel in other departments can help supervisors understand specific regulations, order equipment, or identify potential problems? Are there state/local in addition to federal clean water regulations that the company must follow?

Resource Conservation and Recovery Act/Hazardous and Solid Waste Amendments

When the potential environmental problems posed by chemical and industrial waste disposal became clear, Congress passed the Resource Conservation and Recovery Act (RCRA) in 1976. RCRA promotes "cradle-to-grave" management of hazardous wastes, from point of generation to final disposal location, depending on the hazardous waste determination made by

management. The program defines requirements for hazardous waste generators; transporters; and treatment, storage, and disposal facilities.

The Act requires industries to identify, quantify, and characterize their hazardous wastes as part of a comprehensive federal and state hazardous waste program. Current hazardous waste generators are subject to RCRA requirements to varying degrees, depending upon the amount of hazardous waste materials in their facilities in any one month. Generators are classified in three categories:

- Large-quantity generator—generating over 1,000 kg/month
- Small-quantity generator—generating 100 to 1,000 kg/month
- Conditionally exempt small-quantity generator—generating less that 100 kg/month of hazardous waste and no more than 1 kg/month of acutely hazardous waste.

Specific RCRA standards for hazardous waste generation include the need for:

- Waste identification
- Notification to EPA of hazardous waste activity
- Obtaining an EPA "generator" ID number
- Secure storage facilities
- Storage time and quantity limitations
- Container labeling
- Record keeping
- Reporting
- Emergency preparedness, prevention, and response plans
- Personnel training
- Waste minimization plans
- Transportation and manifest tracking
- Use of permitted treatment, storage, and disposal facilities

In addition to standards for hazardous waste generators, RCRA specifies performance standards and establishes a permit system for owners and operators of hazardous waste treatment, storage, and disposal facilities (TSDs).

RCRA was amended by the Hazardous and Solid Waste Amendments (HSWA) in 1984. The amendment addressed the protection of groundwater quality by mandating:

1. Technological standards for land disposal facilities, including double linings, spill collection systems, and groundwater monitoring
2. New requirements for managing and treating small quantities of hazardous wastes, such as those generated by auto repair shops and dry cleaners
3. Release detection, prevention, and correction

regulations for underground storage tanks (USTs) that store petroleum, petroleum products, and chemicals classified hazardous under CERCLA. These regulations also define the types of tanks permitted and EPA notification requirements.

4. Upgraded design and performance criteria for disposal of municipal solid waste in municipal and industrial landfills
5. Restrictions on land disposal of many untreated hazardous wastes (i.e., "land ban rule")

These regulations required EPA to focus on issuing permits for land disposal facilities and on eventually phasing out land disposal of some wastes. The law also required businesses that generate small amounts of hazardous wastes to be regulated by RCRA. Recycling and waste minimization provisions were included in the amendments to encourage overall reduction of waste.

Summary. Questions to ask about RCRA compliance include the following: How does the company dispose of its waste? Do these waste disposal methods require permits? Where is the information about the company's history of waste disposal permits and/or violations? What equipment or programs are available and what has been implemented to increase the company's recycling efforts and minimize waste? Is the company investigating the use of chemicals and processes to reduce hazardous waste? Which personnel in other departments can help supervisors understand specific regulations, order equipment, or identify potential problems? Are there state/local, as well as federal, waste disposal regulations that the company must follow? (See also the section on Waste Minimization later in this chapter.)

Toxic Substances Control Act

In 1976, Congress saw the need to control the risks that may be posed by the 65,000 commercial chemical substances and mixtures that are not regulated as either drugs, food additives, cosmetics, or pesticides. With the passage of the Toxic Substances Control Act (TSCA), EPA regulates the manufacture, use, and disposal of a chemical substance, and requires testing for carcinogenic and other harmful effects of the chemical.

The law also mandates that EPA be notified of any new chemical prior to its manufacture to determine if it presents an unreasonable risk to human health or to the environment. Once a chemical is on the market, its manufacture, processing, distribution, use, and disposal can be regulated by EPA if there are unreasonable public health or environmental risks associated with them. Regulatory tools available under TSCA range from labeling and use restrictions to outright bans on manufacture (Figure 12-2).

TSCA authorizes EPA to require that industry test a chemical when insufficient data exist to assess the risks and identify when people are likely to experience substantial exposure or unreasonable risk from the chemicals. EPA also requires that an industry maintain records of workers' allegations of significant adverse reactions to the substance and to report any new information that suggests that it may pose substantial risks.

One of the most significant elements of TSCA is its extensive regulation of polychlorinated biphenyls (PCBs). Covered by those regulations are manufacturing, processing, distribution in commerce, prohibitions, authorization, and use of PCBs (40 *CFR*, 761.60–79); exemptions for manufacturing, processing, and distribution in commerce (40 *CFR*, 761.80); PCB spill-cleanup policy (40 *CFR*, 120–135); and records and monitoring, certification program or records by importers, and maintenance of monitoriing records by persons who import, manufacture, process, distribute in commerce, or use chemicals containing inadvertently generated PCBs (40 *CFR*, 761.180–193).

Summary. Questions to ask about TSCA compliance include the following: Does the company manufacture, use, or dispose of chemicals that are covered by TSCA regulations? Does the company import or export chemicals regulated by TSCA? Does the company deal with any carcinogens? Who is notified if a new chemical is introduced at the facility? Which personnel in other departments can help supervisors understand specific regulations, order equipment, or identify potential problems? Who maintains records of reported significant adverse reactions? Are there state/local, as well as federal chemical regulations that the company must follow?

Asbestos School Hazard Abatement Act of 1984/Asbestos Hazard Emergency Response Act of 1985 (Title II of TSCA)

Asbestos is now known to be a hazardous substance and is recognized to have carcinogenic potential. It is commonly found in pipe insulation, furnace insulation, ceilings, and other surfaces where it was applied many years ago.

Under the Asbestos School Hazard Abatement Act (ASHAA) of 1984, interest-free loans have been made available to schools for asbestos control projects. Under the Asbestos Hazard Emergency Response Act (AHERA) of 1985, which amended TSCA, schools are required to identify and respond to their asbestos problems. By using EPA or state-certified professionals, schools must identify potentiial problem areas, submit plans for correcting the problem

STATE OF ILLINOIS

ENVIRONMENTAL PROTECTION AGENCY DIVISION OF LAND POLLUTION CONTROL

P.O. BOX 19276   SPRINGFIELD, ILLINOIS 62794-9276 (217) 782-6761

State Form   LPC 62 8/81   IL532-0610

EPA Form 8700-22 (Rev. 9-88)

FOR SHIPMENT OF HAZARDOUS AND SPECIAL WASTE

Form Approved. OMB No. 2050-0039, Expires 9-30-92

PLEASE TYPE   (Form designed for use on elite (12-pitch) typewriter.)

UNIFORM HAZARDOUS WASTE MANIFEST	1. Generator's US EPA ID No.	Manifest Document No.	2. Page 1 of	Information in the shaded areas is not required by Federal law, but is required by Illinois law.

3. Generator's Name and Mailing Address   Location If Different

A. Illinois Manifest Document Number

IL 3976600 FEE PAID IF APPLICABLE

4. *24 HOUR EMERGENCY AND SPILL ASSISTANCE NUMBERS*

B. Illinois Generator's ID

5. Transporter 1 Company Name   6.   US EPA ID Number

C. Illinois Transporter's ID

D. (   )   Transporter's Phone

7. Transporter 2 Company Name   8.   US EPA ID Number

E. Illinois Transporter's ID

F. (   )   Transporter's Phone

9. Designated Facility Name and Site Address   10.   US EPA ID Number

G. Illinois Facility's ID

H. Facility's Phone ( )

GENERATOR

11. US DOT Description (Including Proper Shipping Name, Hazard Class, and ID Number)	12. Containers		13. Total Quantity	14. Unit Wt/Vol	I. Waste No.
	No.	Type			
a.					EPA HW Number X X __ Authorization Number
b.					EPA HW Number X X __ Authorization Number
c.					EPA HW Number X X __ Authorization Number
d.					EPA HW Number X X __ Authorization Number

J. Additional Description for Materials Listed Above

K. Handling Codes for Wastes Listed Above In Item #14

G = Gallons   Y = Cubic Yards

15. Special Handling Instructions and Additional Information

16. **GENERATOR'S CERTIFICATION:** I hereby declare that the contents of this consignment are fully and accurately described above by proper shipping name and are classified, packed, marked, and labeled, and are in all respects in proper condition for transport by highway according to applicable international and national government regulations.

If I am a large quantity generator, I certify that I have a program in place to reduce the volume and toxicity of waste generated to the degree I have determined to be economically practicable and that I have selected the practicable method of treatment, storage, or disposal currently available to me which minimizes the present and future threat to human health and the environment; **OR,** if I am a small quantity generator, I have made a good faith effort to minimize my waste generation and select the best waste management method that is available to me and that I can afford.

Printed/Typed Name	Signature	Date Month Day Year

TRANSPORTER

17. Transporter 1 Acknowledgement of Receipt of Materials

Printed/Typed Name	Signature	Date Month Day Year

18. Transporter 2 Acknowledgement of Receipt of Materials

Printed/Typed Name	Signature	Date Month Day Year

FACILITY

19. Discrepancy Indication Space

20. Facility Owner or Operator: Certification of receipt of hazardous materials covered by this manifest except as noted in item 19.

Printed/Typed Name	Signature	Date Month Day Year

This Agency is authorized to require, pursuant to Illinois Revised Statute, 1989, Chapter 111 1/2, Section 1004 and 1021, that this information be submitted to the Agency. Failure to provide this information may result in a civil penalty against the owner or operator not to exceed $25,000 per day of violation. Falsification of this information may result in a fine up to $50,000 per day of violation and imprisonment up to 5 years. This form has been approved by the Forms Management Center.

In case of a spill call the Illinois Office of Emergency Response at 217 / 782-7860 and the National Response Center at 800 / 424-8802 or 202 / 426-2675.

COPY 1. TSD MAIL TO GENERATOR

COPY 2. TSD MAIL TO IEPA (RCRA AND PCB WASTES)

COPY 3. TSD COPY

COPY 4. TRANSPORTER 1 COPY

COPY 5. GENERATOR MAIL TO IEPA (RCRA AND PCB WASTES)

COPY 6. GENERATOR'S COPY

Figure 12–2. Sample Uniform Hazardous Waste Manifest

to their state governor, and begin to implement their plans.

In addition, EPA provides recommendations on how to handle asbestos problems in public and commercial buildings. OSHA also has issued standards to protect people in workplaces where asbestos is a problem and where it is being removed.

Summary. Questions to ask regarding asbestos include the following: Has a complete asbestos inspection been conducted in the facility? Is any asbestos present in the facility? Must it be removed? Who in the company can help to find a certified asbestos removal contractor? What departments should be notified and what procedures should be followed to protect their workers if asbestos must be removed? What should supervisors do to protect their workers?

Federal Insecticide, Fungicide, and Rodenticide Act

Another legislative tool EPA uses to control the risks of toxic substances is the Federal Insecticide, Fungicide, and Rodenticide Act (FIFRA), which encompasses all pesticides used in the United States. When it was enacted in 1947, FIFRA was administered by the U.S. Department of Agriculture to protect consumers against fraudulent pesticides; the health and environmental effects of pesticides were unknown at the time. In 1970, EPA became responsible for FIFRA, which was amended in 1972 to shift its emphasis to public health and environmental protection.

Under FIFRA, EPA is responsible for controlling the risks of pesticides through a registration process designed to ensure that when used properly, a pesticide presents no unreasonable health or environmental risks that do not outweigh its benefits to society. Before a new pesticide may be marketed, it must be registered with EPA. As part of the application, the manufacturer must submit test data that include information about the risk of cancer, birth defects, and any chronic effects, as well as risks to wildlife.

If test data show that using a pesticide may harm human health or the environment, EPA can either refuse to register it, restrict its use, or require that only certain applicators use the pesticide. The person applying the chemical must be licensed or registered as well. Once a pesticide is registered, it must be clearly labeled, indicating its approved uses.

Summary. Questions to ask about FIFRA compliance include the following: Does the company produce pesticides that require FIFRA registration? Where is information stored about the company's past history of FIFRA registration? Which personnel

in other departments can help supervisors understand specific regulations, order equipment, or identify potential problems? Are there state/local, as well as federal chemical registration regulations that the company must follow?

Comprehensive Environmental Response, Compensation, and Liability Act of 1980 (Superfund)

The Comprehensive Environmental Response, Compensation, and Liability Act (CERCLA)—commonly known as the Superfund Act—was enacted in 1980 after several incidents of uncontrolled, dangerous disposal of toxic chemicals made it clear that RCRA could not address all the cleanup needs of hazardous waste sites on land. The Superfund Act started a comprehensive program to address the worst abandoned or inactive waste sites in the country through a priority list. The funds for this program came from taxes on crude oil and 42 commercial chemicals.

A significant feature of this Act is the requirement to report to the National Response Center releases of hazardous substances into the environment.

CERCLA's scope is far broader than any other federal environmental statutes. CERCLA's jurisdiction applies to any type of industrial, commercial, or noncommercial facility, regardless of whether there are specific regulations affection the facility, and regardless of how the "hazardous substance, pollutant, or contaminant" is released into the environment.

Under the Superfund Act, potentially responsible parties (PRP) can clean up sites themselves with EPA or state oversight (Figure 12–3). EPA and the states also can start actions to clean up sites after attempts at negotiation with the companies that created the problem have failed, or if there is an emergency. The government can later sue the responsible companies to recover the costs of cleanup. In most cases, the EPA and PRPs negotiate a cleanup procedure and associated administrative costs.

Summary. Questions to ask about Superfund include the following: Where and how is a company's hazardous waste shipped? What should management do if they receive a letter naming the company as a potentially responsible party in a Superfund waste site? Where is the information stored about the company's past history of waste shipments and possible Superfund liability? Is the company taking proactive steps to minimize its potential liability? Which personnel in other departments can help supervisors understand specific regulations or identify potential problems? Are there state/local, as well as federal, hazardous waste disposal regulations that the company must follow?

Figure 12–3. A crew with full personal protective equipment cleans up a hazardous waste site.

Superfund Amendments and Reauthorization Act of 1986

The Superfund Amendments and Reauthorization Act of 1986 (SARA) authorized funds for both emergency response to chemical emergencies and accidents, and for longer term cleanup programs. Under SARA, Congress strengthened EPA's ability to focus on permanent cleanups at Superfund sites, to involve the public in decision processes at sites, and to encourage states and Native American tribes to participate actively with EPA to address these sites. The law also expanded EPA's research, development, and training responsibilities, and strengthened EPA's enforcement authority to compel others to clean up hazardous waste problems for which they are responsible.

Emergency Planning and Community Right-to-Know Act

EPCRA was enacted as a free standing provision of SARA. This act is known as SARA Title III. EPCRA is the congressional effort to compel state and local governments to develop plans for responding to unanticipated releases into the environment of a number of hazardous substances. EPCRA provisions are contained in four major parts:

- Sections 301–303: *Emergency Planning*—intended to develop state and local governments' emergency response and preparedness capabilities by establishing:
  - State emergency response commissions
  - Local emergency planning committees
  - Emergency response plans for chemical releases involving 400 + extremely hazardous substances designated by the EPA.

- Section 304: *Emergency Notification*—specifies that facilities using or storing extremely hazardous materials must immediately notify the local emergency planning committee and the state emergency response commission if there is a release of an extremely hazardous substance that exceeds the "reportable quantity" for the substance.
- Sections 311–312: *Community Right-to-Know*

Reporting Requirements—Requires that facilities maintaining MSDS for OSHA reasons (i.e. for hazardous chemicals used) must submit an MSDS for each hazardous chemical and an annual inventory (Tier I and Tier II reports) detailing types, amounts, and locations of these materials to:

– State emergency response committee (SERC)
– Local emergency planning committee (LEPC)
– Local fire department.

• Section 313: *Toxic Release Reporting*—requires that facilities using, manufacturing, or processing "threshold quantities" of toxic chemicals must submit to the EPA an annual toxic chemical release form summarizing releases from the preceding calendar year.

General provisions of EPCRA restrict the use of trade secrecy claims as reasons for not completely identifying chemicals in reporting, and establish penalties for violation of various provisions of the Act.

Hazardous Materials Transportation Uniform Safety Act

The Hazardous Materials Transportation Uniform Safety Act of 1990 (HMTUSA) requires the Secretary of Transportation to regulate the transportation of hazardous materials in intrastate, interstate, and foreign commerce. It defines the relationship of federal and nonfederal laws governing the transportation of hazardous materials. It defines the standards to be used should a conflict between federal and nonfederal laws arise.

HMTUSA requires certain persons engaged in the transportation of hazardous materials to register with the Secretary of Transportation, who may collect fees to cover costs of processing the registrations. Safety permits are required by motor carriers who transport a class A or B explosive, a liquefied natural gas, a hazardous material designated as extremely toxic by inhalation, or a certain quantity of radioactive materials will be required to obtain a safety permit.

The Secretary of Transportation sets standards for states and Native American Tribes to use in establishing and enforcing specific highway routes over which hazardous materials may be transported. This law gives grants for emergency response planning by states and training for public sector responders and hazardous materials employees. It cites the need for a central reporting system and computerized telecommunications data center.

HMTUSA includes requirements for training to be given by all hazard materials employers to their hazard materials employees regarding the safe loading, unloading, handling, storing, and transporting of these materials. The requirements include emergency preparedness for responding to accidents or incidents involving the transportation of hazardous materials.

Pollution Prevention Act of 1990

Citing "significant opportunities for industry to reduce or prevent pollution at the source," the Pollution Prevention Act of 1990 declares that "it is to be the national policy of the United States that pollution should be prevented or reduced at the source whenever feasible; pollution that cannot be prevented should be recycled in an environmentally safe manner, whenever feasible; pollution that cannot be prevented or recycled should be treated in an environmentally safe manner whenever feasible; and disposal or other release into the environment should be employed only as a last resort and should be conducted in an environmentally safe manner."

The act authorizes the EPA Administrator to facilitate the adoption of source reduction techniques by businesses, including the use of the Source Reduction Clearinghouse, to foster the exchange of information regarding source reduction techniques, the dissemination of such information to businesses. It establishes an annual award program to recognize companies that operate outstanding or innovative source reduction programs.

Matching grants are made to states to promote use of source reduction techniques by businesses and to provide training in source reduction techniques through local engineering schools and other appropriate means.

STATE TRENDS IN ENVIRONMENTAL LAWS

In recent years, many states have adopted environmental regulations that may either be stricter than federal laws or deal with an area not covered by federal regulations. Here are several examples of such state laws:

• **California.** This state has the nation's strictest air pollution laws. Beginning in 1998, state regulations will require automakers to sell some cars and trucks, such as solar and electric vehicles, that emit no pollution, or vehicles that run on alternative fuels. California also has passed legislation requiring all products containing substances known to cause cancer or birth defects to bear a warning label alerting the public that these products are hazardous.

• **Massachusetts.** The state leads the country in the trend toward preventing all types of pollution with legislation ordering reduction in use of toxic materials. The law's goal is to cut waste generation in half by 1997. The largest 2,400 chemical users in the state are required to file annual reports on how they use 1,000 identified toxic substances.

- **New Jersey.** This state has adopted the toughest water enforcement law in the country, with mandatory fines and penalties for violations of wastewater discharge permits. It has a strong Worker and Community Right-to-Know Act. The Right-to-Know Department prepares Hazardous Substance Fact Sheets, which are distributed by the EPA. New Jersey also has passed a pollution prevention act with the goal of cutting toxic waste use by certain industries by 50% by 1996.
- **Oregon.** The state passed the nation's first pollution prevention law for toxic substances, which requires companies using large amounts of toxic chemicals to analyze that use and develop detailed plans for reduction.
- **Wisconsin.** This state has passed a law establishing recycling standards for newspapers and plastic bottles, controls on toxic substances in packaging, funding for recycling market development, a policy hierarchy from source reduction to landfilling, a Pollution Prevention Board to promote pollution prevention source reduction, and a ban on out-of-state waste unless the exporter meets its own state's and Wisconsin's recycling requirements.

It is critical for supervisors to be familiar with their state and local environmental regulations. A company's environment, safety, or legal departments can help in these efforts.

CHEMICAL MANAGEMENT

Each year new chemicals are introduced into industry, adding to the already staggering number used by businesses, industries, and service institutions. Federal and state laws have been enacted to help identify, handle, transport, and dispose of hazardous chemicals to protect public health and the environment. Supervisors may be responsible for complying with many of these regulations. It is important to work with the company's environmental deprtment. They should be thoroughly familiar with standards and guidelines that apply to their companies' operations.

Identification of Chemicals

There are several ways to identify chemicals, especially hazardous materials. Each chemical has its own Chemical Abstract Services (CAS) registry number, a specific chemical name, and a chemical formula, and many chemicals have their own Chemical Abstract

Services (CAS) registry number. In identifying a chemical, its physical state (gas, liquid, or solid) must be noted, as well as whether it is a pure chemical or a mixture. Hazardous materials can be identified in three additional ways:

- The National Fire Protection Association (NFPA) system is used on storage tanks and smaller containers at fixed facilities. The NFPA system uses numbers and colors on a diamond-shaped sign to define the basic hazards of a specific material. Health, flammability, and reactivity are identified and rated on a scale of "0" to "4" depending on the degree of hazard presented, with "0" indicating no particular hazard (Figure 12–4).
- The U.S. Department of Transportation (DOT) system is used exclusively on containers and tanks transported in interstate commerce. The DOT's Hazardous Materials Transportation Administration regulates more than 1,400 hazardous materials. It requires labels on individual, nonbulk containers (such as drums and pails) and placards on tanks and trailers. The diamond-shaped DOT label indicates the nature of the hazard in the cargo with a hazard symbol, a four-digit DOT identification number for a United Nations (UN) hazard number which classifies the material in one of nine hazard classes (Figure 12–5).
- The Hazardous Materials Identification System (HMIS), developed by the National Paint & Coatings Association, uses ratings of "0" to "4" to warn employees of the health, flammability, and reactivity of chemicals. An alphabetical code designates equipment needed for personal protection.

Waste Disposal

In addition to the federal government's regulations for disposing of waste, each state has regulations for waste disposal. Many of these are as restrictive or more restrictive than federal regulations. There also may be local regulations in place. Supervisors must be familiar with state and local regulations governing solid waste disposal. Many states impose restrictions on the disposal of specific nonhazardous solid waste.

The first step for supervisors in managing and disposing of waste is to take an inventory of all raw materials, and all residues and by-products created by the production process. They must determine if the waste is categorized as a solid waste according to RCRA. The definitions are:

- A solid, liquid, semisolid, or containerized gas that is discarded, has served its intended purpose, or is a manufacturing or mining by-product
- A solid, liquid, semisolid, or containerized gas that is not either domestic sewage, a Clean Water Act

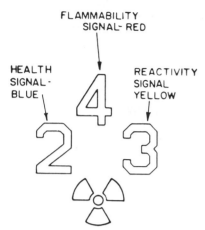

Fig. 1. For Use Where White Background is Not Necessary.

Fig. 2. For Use Where White Background is Used With Numerals Made From Adhesive-Backed Plastic

Fig. 3. For Use Where White Background is Used With Painted Numerals, or, For Use When Signal is in the Form of Sign or Placard

ARRANGEMENT AND ORDER OF SIGNALS — OPTIONAL FORM OF APPLICATION

Distance at Which Signals Must be Legible	Size of Signals Required
50 feet	1"
75 feet	2"
100 feet	3"
200 feet	4"
300 feet	6"

NOTE:
This shows the correct arrangement and order of signals used for identification of materials by hazard

Figure 12-4. NFPA 704 M Hazard Identification System

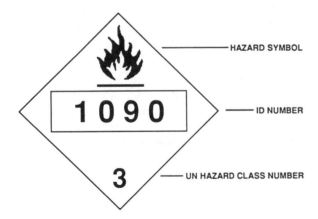

Figure 12-5. Modification of DOT Hazard Identification System

point source discharge, irrigation return flow, Atomic Energy Commission source, or mining waste

• Garbage, refuse, or sludge

Supervisors must then determine which of the solid wastes also are defined as hazardous waste. As much waste as possible should be reused or recycled. All waste must be disposed of properly, with the appropriate approvals, at properly permitted disposal sites. In selecting a disposal site for nonhazardous solid waste, supervisors should, to avoid those listed as Superfund sites, work with the facility or company environmental manager to review potential sites, have the sites inspected, or review the inspection reports of others. When selecting a transporter for hazardous materials, supervisors and/or environmental managers should ensure that the transporter's compliance program, training and testing programs, drivers' logs, and any known violations are reviewed.

Hazardous Materials

Although hazardous materials can be defined in many ways, the simplest definition is a material that has the capability of causing harm to life, property, or the environment, when it is released from its container or emitted during a production process. Hazardous materials usually are pure substances or uncontaminated mixtures designed to perform certain tasks. In industry, they are raw materials used in a manufacturer's product or process.

Hazardous wastes are usually an end result of a

process; they are the materials created, or used and contaminated. They can be spilled raw materials or materials left over after a manufacturing or other process is complete, to be discarded or recycled.

When a company is managing its waste disposal, it must determine which of the solid wastes are hazardous. If any hazardous waste is generated, the facility must be identified as one of four categories: conditionally exempt small-quantity generator, small-quantity generator, 90-day accumulation generator, and permitted storage facility, or large-quantity generator. EPA-assigned identification numbers are required by generators from both the federal EPA and state environmental management departments.

Supervisors and/or environmental managers must determine how to transport and dispose of the hazardous waste. It is critical for them to be familiar with federal, state, and local regulations governing disposal. It is important for supervisors and/or environmental managers to inspect each disposal site and transporter before they are used, and to refer to their company's files and those of other companies that have used the sites and transporters.

All facilities handling hazardous materials and hazardous wastes must develop a formal training program and emergency response procedures. Each facility should designate one or more individuals to be responsible for all phases of hazardous chemical regulatory reporting, training, and regulatory compliance. (These training and emergency procedures are discussed in more detail in the Emergency Planning section at the end of this chapter.)

The management of hazardous materials is a complicated, critical challenge. The EPA, OSHA, and individual companies must work together in this job under the terms of each state's MOAs.

CHEMICAL MANAGEMENT RECORDS

Proper handling of hazardous chemicals requires careful documentation and compliance with regulations; employee training; spill response; and transportation, storage, disposal, and tracking of materials. This section discusses the documentation that companies are required to generate.

Record Keeping

Many local, state, and federal environmental regulations require companies to create, file and maintain extensive records to document their hazardous materials management policies and procedures. These records include:

- Compliance management manual
- Documents detailing how the handling and disposal

of chemicals have been done
- Copies of permits
- Inspection reports
- Personnel and training information
- Reports on emissions, releases, discharges, and excessive amounts of pollutants and hazardous materials
- Details of all monitoring hat has been performed
- Copies of important correspondence relating to environmental and chemical management
- PCB inventory report
- Material Safety Data Sheets (MSDSs)
- SARA Title III reports
- Emergency response notification information
- Contacts with and reports to government agencies
- Pending applications and permits

During an agency inspection, these records may answer all of an inspector's questions because of their accessibility and their documentation of chemical management practices.

Material Safety Data Sheets

Under the OSHA standard on hazard communication, companies are required to keep MSDSs on file for all hazardous chemicals in the workplace (29 *CFR*, 1910.1200, Hazard Communication). They must also make this information available to employees, so workers will know about the chemical hazards to which they are exposed, and can take necessary precautions in handling the substances.

MSDSs contain information regarding a chemical's or mixture's identification (i.e., chemical name, other names, identification number, and the manufacturer). The MSDS also contains the chemical's ingredients, exposure limitations, physical characteristics, fire and explosion hazards, reactivity information, health hazard information, first aid recommendations, spill, leak, and disposal procedures, special protection information including personal protection equipment, and special comments and precautions for working with and storing the chemical.

An MSDS or chemical inventory must be submitted to community right-to-know organizations when new hazardous chemicals are introduced at a facility in quantities above the established threshold quantities. A revised MSDS must be provided by the manufacturer if significant new information is discovered about a chemical.

Environmental Auditing

EPA encourages each company to perform a voluntary environmental audit or an on-site inspection as "an investigation into the history and current status of a particular piece of property (site)." An environmental audit should accomplish the following:

- Identify the presence and extent of environmental contamination or hazardous materials from current or previous site activities
- Determine the level of compliance with current standards or regulations
- Establish a base for compliance
- Identify opportunities for source reduction
- Provide a general review of environmental risks associated with the site and its operations
- Define management's compliance programs and commitment to those programs
- Assess hazards to employees, community, and environment
- Reaffirm a facility's good environmental standing and record of achievement, if reported to the facility's community and the public

An audit is an important review of a facility's environmental risks and of its record of compliance with environmental laws. Companies should conduct internal audits on a consistent, company-wide basis, and request audits by outside consultants as often as needed.

Regulatory Inspections

Every facility should develop a policy for handling state and federal regulatory inspections, which can range from a walk-through to a full-fledged official audit. Such visits are often unannounced and unscheduled. Supervisors should be courteous, cooperative, and responsive to an inspector's questions without volunteering any additional information. Regardless of the circumstances, the inspector's goal is the same: to ensure compliance with the rules of her or his agency. The inspector usually will collect samples, review documents and permits, look for violations, request laboratory reports, and tour the facility.

Several people should be trained to accompany a government inspector and be knowledgeable of environmental issues and the company's environmental documents and records. They should keep notes of all questions and comments and keep a record of any documents and inspector reviews.

If an inspection results in a citation for a violation, the agency will issue a "Notice of Noncompliance/ Violation" or administrative order. The senior manager at a facility being inspected should be made aware of any inspection activity and whether such a notice is issued, so that appropriate members of management are informed and can participate. If the company is not in compliance, the firm's environment department should gather the relevant documents and discuss the situation with the agency.

Reporting Requirements

Companies must notify the U.S. National Response Center immediately after the release of a reportable quantity of any hazardous substance or pollutant, according to several state and federal regulations, including RCRA, CERCLA, and SARA. A reportable quantity is a specified amount of a hazardous substance that, if released, must be reported to an oversight agency; the exact quantity may vary by chemical. Supervisors should know the requirements for each section of the law.

These laws are strictly enforced, and the reporting times in many cases are almost immediate. Some air and waste water pollution releases must be reported within minutes of awareness of the situation, first by telephone and then by written communication.

SARA requires reporting of accidental chemical releases, annual inventories or chemicals, and annual release reporting of 325 toxic chemicals that are discharged into the air, water, and soil. Those releases are entered into the National Toxic Release Inventory (TRI) and made available to the public through a national database. (Toxic chemical release reporting is discussed in more detail in the Emergency Planning section at the end of this chapter.)

TRAINING AND PERSONNEL ISSUES

A well-trained, knowledgeable workforce is essential to a company's effective compliance with environmental regulations. Supervisors are largely responsible for ensuring that workers receive proper training and updates as new chemicals, processes, and regulations are introduced into the workplace. Employees must be well prepared to handle routine procedures, as well as emergency situations to minimize the risk to life and property.

Hazardous Waste Training

Both EPA and OSHA have regulations for hazardous waste training. EPA requires training under RCRA, within its regulations applicable to treatment, storage, and disposal facilities. This training requirement also applies to generators of hazardous waste. The purpose of the training is to reduce the potential for mistakes that might threaten human health and the environment. This is accomplished by ensuring that facility personnel acquire expertise in the areas to which they are assigned.

The requirements specify when facility personnel must be trained, what record must be maintained, and the minimum number of times the initial training employees receive must be updated. Both on-the-job training and in-house training programs may be used to meet the training requirements.

Under the Hazard Communication Standard (HSC), OSHA establishes uniform requirements to make sure that the hazards of all chemicals imported onto, produced, or used in U.S. workplaces are evaluated, and that this hazard information is transmitted to affected employers and exposed employees. In line with this standard, employers must develop, implement, and maintain at the workplace a written, comprehensive hazard communication program. The program must include provisions for container labeling, collection and availability of MSDSs, and an employee training program.

The program also must provide a list or inventory of hazardous chemicals in each work area, how the employer will inform employees of the hazards of nonroutine tasks, and the hazards associated with chemicals in unlabeled pipes. If the workplace has multiple employers on-site, (e.g., at a construction work area), employers must ensure that information regarding hazards and protective measures be made available to other employers on–site., where appropriate. Finally, OSHA requires employers to establish a training and information program for employees exposed to hazardous chemicals, at the time of initial assignment and whenever a new hazard is introduced into their work area.

The following topics must be covered in the training and information program:

Training.
- How the hazard communication program is implemented in the workplace, how to read and interpret information on labels and the MSDSs, and how employees can obtain and use the available hazard information
- The hazards of the chemicals used or stored in the work areas
- Measures employees can take to protect themselves from the hazards
- Specific procedures provided, such as engineering controls, work practices, and the use of personal protection equipment
- Methods and observations workers can use, such as sight or smell, to detect the presence of a hazardous chemical they may be exposed to
- Employee registration and recordkeeping of those who attended training

Information.
- The existence of the Hazard Communication Standard and the requirements of the standard
- The components of the hazard communication program in the workplace
- Operations in work areas where hazardous chemicals are present
- The location of the written hazard evaluation procedures, communications program, lists or inventories of hazardous chemicals, and the MSDS library

All facilities should designate an individual to be responsible for all phases of hazardous chemical regulatory reporting, training, and regulatory compliance. For information on response training, see the Emergency Planning: Worker and Community Safety section, specifically about HAZWOPER, later in this chapter.

Spill Response

Each facility must have written plans and instructions for employees to follow, whenever an accidental release or spill of a hazardous substance occurs. That plan includes the facility's procedures, the names of the facility's emergency responder and the local emergency planning contact, the telephone numbers of agencies that need to be notified, and the people with the authority to report spills. As part of their training, employees should be made aware that these instructions exist and should know several places where copies of the plan can be quickly found. In addition, training and procedures may be required to comply with OSHA HAZWOPER (29 *CFR* 1910.120) regulation.

Before employees get involved in containing or cleaning up a spill, they must check their company's contingency plan for what personal protective equipment is needed to guard against exposure to the spilled material. That choice depends on the size of the spill, its location, and the toxicity of the chemical. The equipment may include goggles, gloves, boots, respirators, and chemical-resistant clothing. (See Chapter 9, Personal Protective Equipment, for more details.)

The CHEMTREC system, a service run by the Chemical Manufacturers Association, keeps MSDSs on file for use in emergency response. Each company's environmental or safety department can give supervisors that phone number and help them to become familiar with the service. The DOT national response telephone number for land/plant spills is 1–800–424–9300. All emergency numbers should be posted at appropriate telephones in the facility. Refer to the company's contingency plan information for the numerous steps that need to be taken in spill response procedures.

Emergency Equipment

If respiratory equipment is necessary, the appropriate equipment and procedures are to be covered in the company's emergency response plan. See also Chapter 9, Personal Protective Equipment, in this manual, and Chapter 7 in the *Accident Prevention Manual for Business & Industry: Engineering & Technology,* 11th edition, National Safety Council, 1997. (See also 29 *CFR,* 1910.120.)

The location of fire extinguishing equipment must be shown in the facility layout in the company's emergency response contingency plan. For details about effective fire safety and protection, see Chapter 17, Fire Safety, in this manual. Employees should receive annual training on the use of emergency equipment.

SPECIAL STORAGE CONSIDERATIONS

Hazardous chemicals and materials must be stored carefully to prevent contamination of the environment and risks to public health and safety. All containers and pipelines should be labeled with the identity of the contents and applicable hazard warnings. This section discusses regulations covering proper storage of hazardous materials and cleanup procedures for spills and leakage of these substances into the environment.

Underground Storage Tanks

An estimated 5 to 6 million USTs in this country contain petroleum products or hazardous chemicals. Thousands of these tanks and their pipes may be leaking, exposing people and the environment to toxic materials.

The use of USTs is regulated under the Hazardous and Solid Waste Amendments to RCRA, and the Superfund law. Many states and local governments also have developed programs to regulate underground and aboveground storage tanks.

EPA's regulations pertain to USTs that store petroleum, including gasoline and crude oil, and other substances defined as hazardous under Superfund. The regulations define the types of tanks allowed and initiate a tank notification system. They are intended to be run by the states, with EPA issuing regulations and guidance. EPA was required to issue technical standards for release detection, prevention, and corrective action for all tanks. Amendments to RCRA require the following standards for USTs:

- New technical standards for land disposal facilities, including double liners, and groundwater monitoring
- New requirements for managing and treating smaller quantities of hazardous wastes, such as those generated by auto repair shops and dry cleaners
- New release detection, prevention, and correction and spill control and overfill protection regulations for USTs that contain liquid petroleum or chemical products.

- Restrictions on the land disposal of many untreated hazardous wastes.

SARA authorized EPA, or states in a cooperative agreement with EPA, to require an owner or operator of a UST to take corrective action to remove or repair a tank if the action can be taken promptly and properly, and if it is necessary to protect human health and the environment. It is recommended that companies remove and replace USTs with aboveground storage tanks (ASTs) whenever possible and where allowed by local ordinances.

Aboveground Storage Tanks

Regulations require ASTs to be in good condition. These tanks should be inspected daily, while drums and similar containers should be inspected weekly. Regulations require secondary containment, such as dikes and vaults, to prevent leaks or spills and for use in performing special tests before a tank is drained.

Drums and Containers

Prior to handling a drum or container, the company must determine if it meets EPA and Department of Transportation regulations. It must also be properly inspected and labeled. Damaged or leaking drums or containers must be emptied of their contents, using a device classified for the materials being transferred, and must be properly discarded.

Supervisors must ensure that safe work practices are instituted before opening a drum or container. Only nonsparking tools should be used. The drum or container should be moved to a safe place, or all employees not involved in the operation should be located at a safe distance from the operation. Standing on or working from drums or containers is prohibited under OSHA's and EPA's regulations.

If drums or containers bulge, swell, or show crystalline material on the outside, they must not be moved onto or from a site unless appropriate containment procedures have been implemented. Virgin raw materials or finished products should be stored in different storage areas than those areas used for waste materials.

Location, Quantity, and Storage of Chemicals

For more details on the safe handling and storage of chemicals and other materials, see Chapter 15, Materials Handling and Storage. State and local regulations also govern storage of chemicals, and supervisors must be familiar with these laws.

EMERGENCY PLANNING: WORKER AND COMMUNITY SAFETY

As already discussed, federal requirements have been established for chemical emergency response planning for both worker safety and public safety.

Worker Safety—SARA Title I and HAZWOPER

Emergency response planning requirements for worker safety are covered under Title I of SARA, which gives OSHA the authority to implement its HAZWOPER regulations. HAZWOPER regulations pertain to private sector employees working at hazardous substance cleanup sites and employees involved in facility emergency response. EPA has companion regulations that pertain to federal, state, and local government employees.

SARA Title I. Title I plans focus on helping emergency responders approach, recognize, and evaluate chemical releases. They require employers to incorporate detailed procedures set out in the work protection regulations. Employers must also critique their responses to all emergency response incidents.

A site-specific incident command system to coordinate all emergency responders and their communication must be established for each incident and emergency, and an appropriately designated company official put in charge of the system. That person will control site operations, identify what hazardous materials are present, ensure that personal protective clothing and equipment are available and used, minimize the number potentially exposed individuals, and designate a safety official.

The incident command system also must determine how many emergency response workers are necessary at the scene, and arrange for backup workers to be available. Title I plans must also:

- Identify potential conditions that would require an evacuation
- Determine evacuation distances
- Identify sensitive populations in the area
- Identify who will order the evacuation and safe return when the incident is over
- Have a training program for emergency responders
- Determine what equipment to use to remediate or respond to an emergency

HAZWOPER. OSHA's HAZWOPER regulation covers employees involved in cleanup operations required by any government at uncontrolled hazardous waste sites. It also covers cleanup operations at RCRA sites, voluntary cleanup operations at uncontrolled hazardous waste sites, and emergency response operations for release of hazardous substances. The HAZWOPER regulation also requires:

- Development by each hazardous waste site of a safety and health program designed to identify, evaluate, and control safety and health hazards and provide for emergency response
- Preliminary evaluation of the site's characteristics prior to entry, by a person trained to identify potential site hazards and to aid in the selection of appropriate employee protection methods
- Training of employees before they engage in hazardous waste operations or emergency response that could expose them to safety and health hazards. Training is required for cleanup personnel, equipment operators, general laborers, and supervisors
- Documentation of all training
- Medical surveillance at least annually and at the end of employment for all employees exposed to hazardous substances
- Engineering controls, work practices, and personal protective equipment to keep exposure below established exposure levels for the hazardous substance involved
- Periodic air monitoring to identify and quantify levels of hazardous substances to ensure that proper protective equipment is used when needed
- An informational program with the names of key personnel and their alternates responsible for site safety and health
- Implementation of a decontamination procedure before any employee or equipment is allowed to leave an area of potential hazardous exposure
- An emergency response plan to handle possible on-site emergencies prior to beginning hazardous waste operations
- An off-site emergency response plan to better coordinate emergency action by local services and to implement appropriate control action

Community Safety—The Emergency Planning and Community Right-to-Know Act of 1986 (SARA Title III)

The Emergency Planning and Community Right-to-Know Act (EPCRA), in EPA's own words, "makes citizens full partners in preparing for emergencies and managing chemical risks" (Figure 12–6). This partnership involves the supervisor in several ways: as an individual citizen, as a possible member of a local emergency planning committee, and as a company representative with a responsibility to inform line workers about toxic health hazards and emergencies in their community. To meet these challenges, a supervisor must understand the law, its key provisions, and its enormous potential for providing information needed for effective emergency preparedness.

EPCRA has two purposes: to encourage and support emergency planning for responding to chemical accidents and to provide local governments and the public with information about possible chemical hazards in communities. The law operates through provisions detailed in four major sections.

Emergency Planning (Sections 301–303). Emergency planning requires state and local efforts to develop emergency response and preparedness capabilities based on chemical information provided by industry. These sections are designed to help communities prepare for and respond to emergencies involving hazardous substances. Every community in the United States must be part of a comprehensive plan.

As of April 17, 1987, the governor of each state was required to appoint a State Emergency Response Commission (SERC), which can be one or more existing state agencies or may consist solely of individual citizens. These commissions already have been named in all 50 states and in all U.S. territories and possessions.

Each SERC, in turn, must supervise the activities of LEPCs. They and LEPCs must establish procedures for receiving and processing public requests for information collected under other sections of the new law. SERCs also must review local emergency plans annually to evaluate their effectiveness in such areas as coordination of emergency plans across the state.

LEPCs are the grassroots groups that do most of the implementation of the new law. To truly represent their communities, LEPCs must include:

- Representatives of elected state and local officials
- Law enforcement officials, civil defense workers, and firefighters
- First aid, health, hospital, environmental, and transportation workers
- Representatives of community groups and the news media
- Owners and operators of industrial facilities and other users of chemicals, such as hospitals, farms, and small businesses

LEPCs must analyze hazards and develop plans to prepare for and respond to chemical emergencies in its district. The plans should be based on the chemical information reported to the LEPC by local industries and other facilities dealing with chemicals. All local emergency plans must:

- Use the information provided by industry to identify the location of chemicals and transportation routes where hazardous substances are present
- Establish emergency response procedures, including evacuation plans, for dealing with accidental chemical releases
- Set up notification procedures for those who will respond to an emergency
- Establish methods for determining the occurrence and severity of a release and the areas and populations likely to be affected
- Establish ways to notify the public of a release
- Identify the emergency equipment available in the community, including equipment at facilities
- Contain a program and schedules for training local emergency response and medical workers to respond to chemical emergencies
- Establish methods and schedules for conducting "exercises" (simulations) to test elements of the emergency response plan
- Designate a community coordinator and facility coordinators to carry out the plan. A common sense approach to this, would be to invite your local fire department into your facility for a tour and review of your hazardous materials and flammable materials use and storage.

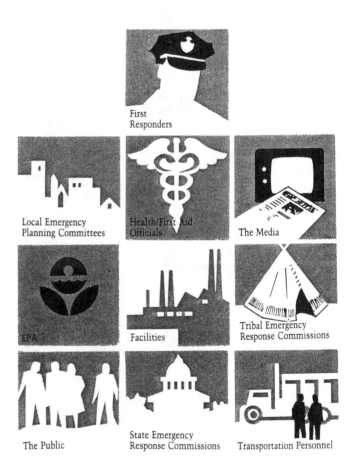

Figure 12–6. Participants in the Emergency Planning Community Right-to-Know (Title III) Program

Chemical Release Reporting. There are four groups of chemicals subject to reporting under EPCRA. Some chemicals appear in several groups:

1. Extremely Hazardous Substances (Sections 301 and 304): This list currently contains approximately 400 + chemicals. Because of their extremely toxic properties, these chemicals were chosen to provide an initial focus for chemical emergency planning. The list includes a "threshold planning quantity" (TPQ) for each substance. If at any time this amount or more of the chemical is present at any manufacturing facility, warehouse, hospital, farm, small business, municipal installation, or any other facility, the owner or operator must notify the SERC and the LEPC.

Emergency Release Notification requires immediate emergency notification to state and local authorities when any one of approximately 400 + chemicals specified under Superfund are released accidentally. Check Section 304 for reportable chemicals released on the company's property.

If a specified "hazardous" substance is released—in an accident at a facility or on a transportation route—in an amount that exceeds the reportable quantity for the substance, facilities must immediately notify the LEPCs and the SERCs likely to be affected. Other people who should be notified are the U.S. National Response Center, local emergency responders, and CHEMTREC.

Chemicals covered by this section of the law include not only the 400 + "extremely hazardous substances" but also approximately 720 hazardous substances subject to the emergency notification requirements of the Superfund law. See also CERLCA 302.4, which lists more than 10,000 materials that have MSDSs and might require reporting. Some chemicals are on both lists. Superfund requires immediate notification of substance releases to the National Response Center, which alerts federal responders. Part of the significance of Section 304 is that is requires immediate notification, not just of federal agencies, but also of local and state officials who must know if the spill affects the community and the environment.

Initial notification of a substance release can be made by telephone, shortwave radio, or in person. If the release resulted from a transportation accident, the transporter can dial 911, CHEMTREC, or the local authorities.

Notification will activate emergency plans. As soon as practical after the release, the facility coordinator must submit a written report to both the LEPC and the SERC. This report updates the original notification. It provides additional information about response actions taken, known or anticipated health risks, any advice regarding medical care needed by exposure victims, and corrective action to reduce possible recurrences.

Anyone who fails to notify authorities of a substance release is subject to civil penalties of up to $25,000 a day, for each day of noncompliance. Repeat offenders can be fined up to $75,000 a day. Criminal penalties also may be imposed on any person who knowingly and willfully fails to provide notice. Criminal violators face fines of up to $25,000 or prison sentences of up to two years. Repeat criminal offenders can be fined up to $50,000 and imprisoned for as long as five years.

2. Hazardous Substance Releases (Section 302): These are hazardous substances listed under previous Superfund hazardous waste cleanup regulations (40 CFR, 302.4). This list contains approximately 720 substances. Releases of these chemicals above certain amounts and conditions must be reported at once, because they may represent an immediate hazard to the community.

3. Community Right-to-Know Reporting Requirements (Sections 311–312): Although these chemicals are not on an official list, they are defined by OSHA regulations as chemicals that represent a physical or health hazard. They constitute those hazardous chemicals for which OSHA requires the user to maintain an MSDS. Material Safety Data Sheets for each of these chemicals and annual inventories (Tier i and Tier ii reports) detailing types, amounts and locations of these chemicals must be submitted by March 1 of each year.

4. Hazardous Chemical Reporting (Sections 311–312). Hazardous Chemical Reporting requires all businesses to submit information on chemicals broadly defined as "hazardous" to local emergency planners and local fire departments. Under Sections 311 and 312, facilities also must report the amounts, location, and potential effects of hazardous chemicals being used or stored in a facility at any one time in designated quantities in every community.

All companies—of any size, manufacturing or nonmanufacturing—potentially are subject to this requirement. It is critical to read those regulations and determine how they apply to each company. Companies must report this information to the relevant LEPCs, SERCs, and local fire departments. Facilities must report on the hazardous chemicals they use and store in two ways:

A. Material Safety Data Sheets (MSDSs)—Section 311: Under federal laws administered by OSHA, companies are required to keep MSDSs or lists on file for all hazardous chemicals used in the workplace. They also must make this information available to employees, so workers will know about the chemical hazards and can take necessary precautions in handling the substances.

The relevant chemicals are those defined as "hazardous chemicals" under OSHA's requirements—essentially, any chemical that poses physical or health hazards. As many as 500,000 products can be defined in this way. If they are present in the workplace or environment, they must be reported under the hazardous chemical reporting provisions. Once submitted to the LEPC, SERC, and local fire department, relevant MSDS information is available to the public upon request.

When EPCRA was passed in 1986, OSHA's regulations applied only to manufacturers—in approximately 350,000 facilities. In the wake of court decisions applying OSHA's Hazard Communication Standard beyond the manufacturing sector, these regulations now also apply to most facilities where workers are exposed to hazardous chemicals—about 4.5 million facilities.

B. Annual Chemical Inventories–Section 312:

Companies must report on hazardous chemicals by submitting annual chemical inventories to their LEPCs, SERCs, and local fire departments under a two-tier system. Under Tier I, a facility must report the amounts and general location of chemicals in certain hazard categories. (Example: A facility stores 10,000 pounds of substances that cause chronic health problems.) Threshold planning quantities of extremely hazardous substances must be reported as well. Specifically, Tier I forms must:

- Estimate (in ranges) the maximum amount of chemicals present at a facility at any time during the preceding calendar year
- Provide an estimate of the average daily amount of the chemicals present in each chemical category
- Provide the general location of hazardous chemicals within the facility

A Tier II report requires a brief description of how each chemical is stored, the specific storage location for each of the hazardous chemicals, and certain confidential information. (Example: A facility stores 500 pounds of benzene in the northwest corner storage room of the warehouse.)

Tier II information must be submitted to a local committee, the state commission, or a local fire department. This information is to include more specific details about each substance subject to a request:

- Chemical name or common name as indicated on the MSDS
- Estimate of the maximum amount of the chemical present at any time during the preceding calendar year
- A brief description of how the chemical is stored

- The location of the chemical in the facility, with a map showing those locations

Tier II reports also must indicate if the reporting facility has withheld location information from disclosure to the public for security reasons, such as protection against vandalism or arson.

Congress gave companies the choice of filing a Tier I or Tier II form, unless the SERC, LEPC, or fire department requests Tier II. The Tier I/II forms must be submitted annually by March 1 for the previous calendar year. The information reported under Sections 311 and 312 generally is to be available to the public upon request through local and state governments during normal working hours. The civil penalty for failing to submit MSDSs or lists of MSDS chemicals is up to $10,000 a day for each violation. For noncompliance with the annual inventory requirements, the penalty is $25,000 per violation.

Companies reporting under EPCRA can request, under limited conditions, that the identity of specific chemicals in their reports not be disclosed to the public. To protect a chemical's identity from disclosure, the company must be able to prove that the information has not been reported under any other environmental regulation and that it is a legitimate trade secret—that is, disclosure could damage the company's competitive position.

(See *Chemicals, the Press and the Public*, National Safety Council.)

5. Toxic Chemical Release Reporting and Inventory (Section 313). The Toxic Chemical Release Reporting and Inventory provision requires facilities using manufacturing, or processing "threshold quantities" of toxic chemicals to summarize and report any releases from the preceding calendar year to EPA. This provision applies to facilities with 10 or more employees. Manufacturing facilities with Standard Industrial Classification (SIC) codes 20–39 must, if they meet other criteria, submit Toxic Release Inventory forms. A small business exemption frees companies with nine or fewer employees from coverage.

Facilities must file annually, by July 1, a Toxic Chemical Release Inventory form (Form R) to estimate the total amount of each chemical that they release into the environment, either by accident or as a result of routine facility operations, or that they transport as waste to another location. A complete Form R must be submitted for each chemical. Releases covered include emissions to the air from fugitive and permitted stacks, liquid waste discharged into water, wastes disposed in landfills, and waste transported off-site to a public or private waste treatment or waste disposal facility.

Many of the chemicals covered by this section of the law, though not all, pose long-term (chronic) health and environmental hazards such as cancer, nervous system disorders, and reproductive disorders from ongoing routine exposure. Among the more commonly used substances required to be reported are ammonia, chlorine, copper, lead, methanol, nickel, saccharin, silver, and zinc. Specific information that must be gathered and reported includes:

- Which toxic chemicals were released into the environment during the preceding year
- How much of each chemical went into the air, water, and land
- How much of the chemical was transported away from the facility site for disposal
- How chemical wastes were treated on-site
- The efficiency of that treatment

These reports must be filed by July 1 of each year, covering releases in the previous calendar year. The are submitted to EPA headquarters in Washington, DC, and to the SERC.

EPA is required to compile the reports into a national computerized database called the Toxic Release Inventory (TRI), which must be accessible to the public through computer telecommunications or other means.

Companies that fail to file annual toxic chemical release reports are subject to civil penalties of up to $25,000 a day for each chemical they should be reporting.

WASTE MINIMIZATION, SOURCE REDUCTION, AND OTHER SOUND ENVIRONMENTAL POLICIES

In many areas, the emphasis has shifted to prevention of major environmental pollution and hazardous waste problems. This shift is further emphasized in state or local laws requiring pollution prevention actions. Pollution prevention can be accomplished by reducing the use and production of hazardous materials and recycling as much material as possible. This section discusses how the supervisor can help the company achieve its goals in these critical areas.

Waste Minimization

The ideal waste minimization program reduces the amount of waste generated at the source (or in the design stage) in terms of air emissions, effluents, and hazardous and nonhazardous waste. After waste minimization methods have been employed as much as possible, reuse and recycling becomes a goal. Recycling is one approach to waste minimization, but source

reduction and reuse are preferable approaches because they have a greater overall beneficial effect on the environment.

Waste minimization can save money by reducing operating costs such as waste treatment, disposal, and raw materials purchasing. It can protect the public health, worker health and safety, and the environment. Companies can minimize their wastes in several ways:

- Reduce emissions of chemicals into the air, water, and land.
- In a process review, track wastes from the acquisition of products to the actual product and to the disposal of the wastes. See where waste can be reduced along that route.
- Establish a waste inventory that distinguishes routinely produced operations wastes from those produced by nonrecurring events. Concentrate on reducing recurring waste streams.

Long-range waste minimization goals should include maximum reduction of manufacturing, mining, and processing discharges to air, water, and land; no emission, discharge, or disposal of any hazardous waste without treatment that minimizes or removes the hazards to the extent feasible; and regular review of all processes or operations to ensure minimization of potential employee or community exposure.

Source Reduction and Pollution Prevention

A viable and profitable trend among many companies is pollution prevention—the adoption of policies and processes that reduce and eventually prevent pollution from being generated—rather than cleaning it up or treating it once it has occurred. Under the Pollution Prevention Act of 1990, the EPA also is working on industrial pollution prevention by identifying measurable goals.

EPA has identified source reduction as its preferred option for the reduction of municipal waste nationwide. Many companies also have adopted source reduction policies to eliminate the amount of pollution at its source. This approach is more economical and more beneficial for the environment than treating higher amounts of emissions and/or waste. Some examples of pollution prevention policies adopted by companies recently include:

- Using water-based ink rather than solvent-based ink in packaging facilities, which cuts emissions and eliminates the need to buy more pollution control equipment
- Redesigning cleaning processes to eliminate use of chemicals that destroy the ozone layer, which also achieved considerable annual savings
- Developing a reating compound that makes fasteners

resistant to corrosion but needs fewer applications, thus reducing the effluent to be treated and saving research and development funds.

Effective Environment Strategies

Good environmental management strategies benefit everyone. They help to protect the environment, safeguard workers' health and safety, and make each facility a better "neighbor" to its community. They also can save a company money, especially in the long term.

The need to reduce hazardous and other wastes—to "zero levels" if possible—is compatible with every company's increasing emphasis on reducing operating costs. Polluters not only pay fines and go to jail but these occurrences also attract management's scrutiny as few other issues can. In recent cases, managers also have been subject to fines and jail sentences.

Effective environmental policies can have a positive global impact, just as poor environmental policies can inflict global harm. The chemicals that are destroying the earth's protective ozone layer may have come from facilities like those in many U.S. communities, as well as others around the world. The pollutants that rise into the air from any one facility may travel half-way around the world, where they pollute and harm crops, livestock, natural resources, and people.

Actions that protect the local environment also protect the larger environment of the entire planet. Whenever a company reduces its air pollution emissions or creates a responsible plan for responding to a chemical emergency that could cause the evacuation of the surrounding community, the benefits of these actions will multiply. They serve to safeguard not only the local community but also the global community in which everyone lives.

SUMMARY OF KEY POINTS

Key points covered in this chapter include:

- Environmental problems at facilities can have a major impact, not only on workers and the surrounding community, but on the rest of the world as well. As a result, management at all levels must take appropriate steps to protect their workers and the public health and welfare from a wide range of toxic materials and environmental threats.
- Supervisors must be familiar with federal and state regulations that address environmental problems, understand how to comply with these regulations, and realize the penalties for failure to comply. Supervisors must be able to recognize environmental

hazards when they appear. They must know the proper procedures for alerting management and taking initial action.

- The EPA was created in 1970 to protect the environment and public health and to control and abate environmental pollution under the laws enacted by Congress. The EPA has 10 regional offices that work directly with state and local authorities. MOAs outline the responsibilities and powers of state and EPA offices, and specify the level of coordination between the two government levels.
- Other landmark Congressional legislation that supervisors must know includes the Clean Air Act; Clean Water Act; Resource Conservation and Recovery Act; Hazardous and Solid Waste Amendments; Toxic Substances Control Act; Asbestos School Hazard Abatement Act of 1984; Asbestos Hazard Emergency Response Act of 1985; Federal Insecticide, Fungicide, and Rodenticide Act; Superfund and amendments; Hazardous Materials Transportation Uniform Safety Act; OSHA; and Pollution Prevention Act. These acts establish standards that most companies must meet and that supervisors may be responsible for enforcing on the job.
- State and local environmental protection legislation may be more stringent than federal laws. It is critical for supervisors to be familiar with state and local regulations. A company's environment, safety, or legal departments can help supervisors understand and apply these standards.
- Federal and state laws have been enacted to help identify, handle, transport, and dispose of hazardous chemicals used by businesses and industries. Supervisors must know these regulations and understand how to identify chemicals, establish procedures for disposing of solid and semisolid wastes, and store hazardous materials safely to prevent leakage into the environment.
- To ensure compliance with environmental protection regulations, supervisors and top management often are required to create, file, and maintain extensive records and keep pertinent MSDSs on hand. They must perform periodic environmental audits of company operations, cooperate with state and federal regulatory inspections, and establish reporting procedures for notifying officials of the release of any hazardous substance or pollutant.
- Supervisors are largely responsible for ensuring that workers receive proper training in environmental management. Topics include hazardous waste training, spill response, and emergency equipment. Workers must know how to comply with state and federal regulations and how to handle emergency situations.

- Hazardous chemicals and materials must be stored carefully to prevent contamination of the environment and risks to public health and safety. Supervisors and their workers must understand local, state, and federal legislation governing the proper use of USTs and ASTs, drums and containers, and the location and quantity of chemicals that a company is permitted to store.

- Companies must develop an emergency plan and contingencies to safeguard employees and the surrounding community in case of accidents, fires, or explosions of hazardous materials. Emergency response planning requirements are covered under Title I of SARA, which gives OSHA authority to implement its HAZWOPER regulations. EPCRA is designed to encourage and support emergency planning and to ensure that local governments and the public are informed about chemical hazards in communities.

- Emergency planning regulations also require state and local governments, and public safety agencies, to develop emergency response and preparedness capabilities appropriate to the chemical hazards in their immediate areas. States must appoint a State Emergency Response Commission, which in turn appoints a Local Emergency Planning Committee.

- Several state and federal regulations require companies to report on the amounts of various hazardous and toxic chemicals and substances they produce each year, and to report when these materials are released into the environment. Certain regulations, such as the Emergency Release Notification, mandate immediate emergency notice when an "extremely hazardous" chemical is released either intentionally or accidentally into the environment.

- All environmental protection and reporting laws mandate penalties in the form of jail terms and fines for private individuals or company officials who fail to comply with these laws. Supervisors should know the consequences for failing to meet state and federal environmental protection regulations.

- In many areas, companies seek to prevent environmental pollution and hazardous waste problems by recycling materials and reducing the amount of hazard waste produces, by eliminating the source of pollution, by substituting hazardous materials with nonhazardous materials, and by establishing sound environmental strategies. In today's global community, supervisors must realize that protecting the environment is everyone's business.

13

MACHINE
SAFEGUARDING

After reading this chapter, you will be able to:

- Explain the principles and benefits of safeguarding machines and equipment

- Describe the basic principles of effective safeguard design

- List the primary hazards of machinery and the safeguard options needed for each hazard

- Explain the safeguards needed for automated machines and equipment

- Plan, install, and maintain a guarding system

- Establish a maintenance and repair program for all safeguards in your department

The first step in any operation is to engineer the hazards out of a job as much as possible, to ensure continued production, employee safety and health, a good profit margin, and reduced equipment damage. At times, however, finding a better way is not feasible—or the technical knowledge is not sufficiently advanced. In these cases, you must then use safeguarding to protect employees from injury. If such precautions are not taken, the consequences can be serious. When an employee is injured while operating machines or equipment, production capacity and cost of operations are adversely affected. Poorly designed, improperly safeguarded, or unguarded machinery or equipment is an ongoing threat to the well-being of employees and to production capacity.

An injury can take an operator away from his or her job for a long time, sometimes permanently. The person must be replaced to keep the department going. Even if a qualified person is transferred from another department, eventually a new person must be hired. Finding, hiring, and training the right worker takes time and money.

Safeguards are a critical factor in controlling hazards and in preventing accidents. No matter how much training or experience a worker may have, no one can keep focused on work every minute. Also, some machines look so deceptively easy to run that the supervisor might allow semiskilled or unskilled persons to operate them. The results can be disastrous in terms of personal injury and machine or material damage.

The supervisor must have a complete understanding of machine safety principles and communicate these concerns to the workers.

PRINCIPLES OF SAFEGUARDING: GUARDS, DEVICES, AND SAFE DISTANCE

Safeguarding is frequently thought of as being concerned only with points of operation or the means of power transmission (Figure 13–1). Although safeguarding against these hazards is required, this step also can prevent injuries from other causes, both on and around machines and from equipment and damaged material.

The point of operation is that point where the work is actually being performed on the material (such as, cutting, boring, shaping, forming, shearing, ripping, or grinding). The point of operation can be safeguarded by guards (physical barriers that prevent access by the operator and others to the point of operation) or by devices (a mechanism or control designed for safeguarding at the point of operation). Guards can include die enclosures, fixed barriers,

Figure 13–1. The cam and cam shaft of this turret lathe are completely enclosed, but a door (in open position for demonstration only) is provided for maintenance.

Figure 13–2. A barrier guard covers the spindle of this turret lathe to prevent injuries from direct contact with moving parts.

adjustable barriers, and interlocked barriers. Devices can include two-hand trips; two-hand controls; pull-backs; and presence-sensing, movable-barrier, hold-out, or restraint devices.

Barrier guards can prevent injuries from the following sources:

1. Direct contact with exposed moving parts of a machine—either points of operation on machines (power presses, shears, milling machines, machine tools, or woodworking equipment), or power-transmitting parts of mechanisms (gears, pulleys and sheaves, slides, or couplings), as shown in Figure 13–2.
2. Work in process: for example, pieces of wood that kick back from a power ripsaw, or metal chips that fly from machining operations.
3. Machine failure, which usually results from lack of preventive maintenance, overloading, metal fatigue, or misuse.
4. Electrical failure, which may cause malfunctioning of the machine or electrical shocks or burns.
5. Operator error or failure caused by lack of knowledge, instructions, training, or skill, emotional distractions, misunderstandings, unsafe operation, illness, fatigue, and so on.

The experience of more than six decades of organized accident prevention proves that it is unwise to rely entirely on education and training of operators or their cooperation to avoid mishaps. Many factors affect a person's attitude, judgment, and ability to concentrate. For example, skilled workers with emotional or physical problems may not pay strict attention to their production responsibilities or give their best effort.

Safeguarding Hazards Before Accidents

Guarding the hazard is a fundamental principle of accident prevention, and is not limited to machinery. When you survey your department from the point of view of safeguarding against all hazards, you will list many potential sources of injury that should be protected by barricades, rails, toeboards, enclosures, or other means. As you list electric switches and equipment, motors, engines, fixed ladders, stairs, platforms, and pits, you might ask, "Can an accident occur here, or here . . . or here?" See Chapter 1, Safety Management, and Safety Analysis in Chapter 5, Promoting Safety and Health, for details on finding hazards.

Benefits of Safeguarding

When operators are preoccupied with being injured, afraid of their machines or of getting close to moving parts, they obviously cannot pay strict attention to production responsibilities. Once their fears have been removed, workers can concentrate on the operation and often are more productive. In addition, when moving parts are left exposed, other objects may fall into them, or get caught and damage the machine or material in process.

Well-designed and carefully maintained safeguards assure workers that management is committed to preventing accidents and ensuring safe production. When employees realize this fact, they are more inclined to contribute to safety efforts.

An added benefit of a proper safeguarding program is raising employee morale, especially when workers who must use the safeguards are consulted before the safeguards are made or bought. Employees often have good ideas that contribute to both safety and economy. Even if they have no suggestions, workers will usually feel satisfied at being consulted about their jobs. As a result, they may take a greater interest in the project, and be less likely to remove guards or barriers, or fail to replace them. Finally, a proper safeguarding program will help to reduce injury, damage, downtime, and workers' compensation resulting from injuries, and help to ensure compliance with state and federal regulations.

Types of Safeguards

Many standardized guards and devices, and specially designed safeguards are available from the manufacturer and others to protect machine operators, particularly their hands, at the point of operation. Table 13–A presents concise descriptions of the nature, action, advantages, and limitations of a variety of such guards and devices.

Sheet metal, perforated metal, expanded metal, heavy wire mesh or steel stock may be used as material for the construction of guards. The best practice is to follow the requirements of OSHA and ANSI standards when selecting material for the design and construction of guards. Figures 13–3 through 13–5 show examples.

If moving parts must be visible, transparent impact plastic or safety glass can be used where strength of metal is not required. Guards or barriers may be aluminum or other soft metal where resistance to rust is essential or where iron or steel guards can cause damage to machinery. Wood, plastic, and glass fiber barriers are usually inexpensive, but are low in strength compared with steel. However, these materials effectively resist effects of splashes, vapors, and fumes from corrosive substances that would react with iron or steel and reduce its strength and effectiveness.

(*Text continues on page 207.*)

Table 13-A. Point-of-Operation Protection

Type of Guarding Methods	Action of Guard	Advantages	Limitations	Typical Machines on Which Used
Enclosures or Barriers				
Complete, simple fixed enclosure	Barrier or enclosure admits stock but not hands into danger zone because of feed-opening size, remote location, or unusual shape	Provides complete enclosure if kept in place Both hands free Generally permits increased production Easy to install Ideal for blanking on power presses Can be combined with automatic or semiautomatic feeds	Limited to specific operations May require special tools to remove jammed stock May interfere with visibility	Bread slicers Embossing presses Meat grinders Metal square shears Nip points of inrunning rubber, paper, and textile rolls Paper corner cutters Power presses
Warning enclosures (usually adjustable to stock being fed)	Barrier or enclosure admits the operator's hand but gives warning before danger zone is reached	Makes "hard to guard" machines safer Generally does not interfere with production Easy to install Admits varying sizes of stock	Hands may enter danger zone— enclosure not complete at all times Danger of operator not using guard Often requires frequent adjustment and careful maintenance	Band saws Circular saws Cloth cutters Dough brakes Ice crushers Jointers Leather strippers Rock crushers Wood shapers
Barrier with electrical contact or mechanical stop activating mechanical or electrical brake	Barrier quickly stops machine or prevents application of injurious pressure when any part of the operator's body contacts it or approaches danger zone	Makes "hard to guard" machines safer Does not interfere with production	Requires careful adjustment and maintenance Possibility of minor injury before guard operates Operator can make guard inoperative	Dough brakes Flat roll ironers Paper box corner stayers Paper box enders Power presses Paper calenders Rubber mills
Enclosure with electrical or mechanical interlock	Enclosure or barrier shuts off or disengages power and prevents starting of machine when guard is open; prevents opening of guard while machine is under power or coasting (Interlocks should not prevent manual operation or "inching" by remote control)	Does not interfere with production Hands are free; operation of guard is automatic Provides complete and positive enclosure	Requires careful adjustment and maintenance Operator may be able to make guard inoperative Does not protect in event of mechanical repeat	Dough brakes and mixers Foundry tumblers Laundry extractors, driers, and tumblers Power presses Tanning drums Textile pickers, cards

(continued)

Table 13–A. (Continued)

Type of Guarding Methods	Action of Guard	Advantages	Limitations	Typical Machines on Which Used
Automatic or Semiautomatic Feed				
Nonmanual or partly manual loading of feed mechanism, with point of operation enclosed	Stock fed by chutes, hoppers, conveyors, movable dies, dial feed, rolls, etc. Enclosure will not admit any part of body	Generally increases production Operator cannot place hands in danger zone	Excessive installation cost for short run Requires skilled maintenance Not adaptable to variations in stock	Baking and candy machines Circular saws Power presses Textile pickers Wood planers Wood shapers
Hand-Removal Devices				
Hand restraints	A fixed bar and cord or strap with head attachments which, when worn and adjusted, do not permit an operator to reach into the point of operation	Operator cannot place hands in danger zone Permits maximum hand feeding; can be used on higher-speed machines No obstruction to feeding a variety of stock Easy to install	Requires frequent inspection, maintenance, and adjustment to each operator Limits movement of operator May obstruct work space around operator Does not permit blanking from hand-fed strip stock	Embossing presses Power presses
Hand pull-away device	A cable-operated attachment on slide, connected to the operator's hands or arms to pull the hands back only if they remain in the danger zone; otherwise it does not interfere with normal operation	Acts even in event of repeat Permits maximum hand feeding; can be used on higher-speed machines No obstruction to feeding a variety of stock Easy to install	Requires unusually good maintenance and adjustment to each operator Frequent inspection necessary Limits movement of operator May obstruct work space around operator Does not permit blanking from hand-fed strip stock	Embossing presses Power presses
Two-Hand Trip				
Electrical	Simultaneous pressure of two hands on switch buttons in series actuates machine	Can be adapted to multiple operation Operator's hands away from danger zone No obstructions to hand feeding Does not require adjustment Can be equipped with continuous pressure remote controls to permit "inching" Generally easy to install	Operator may try to reach into danger zone after tripping machine Some trips can be rendered unsafe by holding with the arm, blocking or tying down one control, thereby permitting one-hand operation Not used for some blanking operations	Dough mixers Embossing presses Paper cutters Pressing machines Power presses Washing tumblers
Mechanical	Simultanous pressure of two hands on air control valves, mechanical levers, controls interlocked with foot control, or removal of solid blocks or stops permits normal operation of machine			

(continued)

Table 13–A. (Continued)

Type of Guarding Methods	Action of Guard	Advantages	Limitations	Typical Machines on Which Used
		Miscellaneous		
Limited slide travel	Slide travel limited to $1/4$ in. or less; fingers cannot enter between pressure points	Provides positive protection Requires no maintenance or adjustment	Small opening limits size of stock	Foot power (kick) presses Power presses
Presence-sensing device	Sensing field and brake quickly stop machine or prevent its starting if hands are in danger zone	Does not interfere with normal feeding or production No obstruction on machine or around operator	Expensive to install Does not protect against mechanical repeat Generally limited to use on slow-speed machines with friction clutches or other means to stop the machine during the operating cycle Can be circumvented	Embossing presses Power presses Rubber mills Squaring shears Press brakes
Light curtains	Sense presence of person through electric beam	Provides positive protection Does not interfere with work	Must be closely maintained	
Special tools or handles on dies	Long-handled tongs, vacuum lifters, or hand die holders which avoid need for operator's putting hand in the danger zone	Inexpensive and adaptable to different types of stock Sometimes increases protection of other guards	Operator must keep hands out of danger zone Requires good employee training, close supervision Only one part of prevention needed Will not bring employer into OSHA compliance	Dough brakes Leather die cutters Power presses Forging hammers
Special jigs or feeding devices	Hand-operated feeding devices of metal or wood which keep the operator's hands at a safe distance from the danger zone	May speed production as well as safeguard machine Generally economical for long jobs	Machine itself not guarded; safe operation depends upon correct use of device Requires good employee training, close supervision Suitable for limited types of work	Circular saws Dough brakes Jointers Meat grinders Paper cutters Power presses Drill presses

Figure 13–3. A sheet metal guard with transparent plastic viewing window guards this lathe operation.

Figure 13–4. A sheet metal hinged guard (in open position for demonstration only) encloses the drive belt of this lathe.

SAFEGUARD DESIGN

It is easier to establish effective methods for safeguarding power transmissions than it is for safeguarding points of operation, because power transmissions are more standardized. OSHA and ANSI standards for mechanical power-transmission apparatus (ANSI/ASME B15.1 and 29 *CFR* 1910.219) apply to all moving parts of mechanical equipment. Point-of-operation guarding can be found in ANSI standards for specific machinery, such as in the following safety requirements:

- Bakery Equipment, Z50.1
- Construction, Care, and Use of Drilling, Milling, and Boring Machines, B11.8
- Construction, Care, and Use of Grinding Machines, B11.9
- Construction, Care, and Use of Lathes, B11.6
- Construction, Care, and Use of Mechanical Power Presses, B11.1
- Construction, Care, and Use of Metal Sawing Machines, B11.10
- Construction, Care, and Use of Power Press Brakes, B11.3
- Safety Standard for Stationary and Fixed Electric Tools, ANSI/UL 987
- Safety Standard for Mechanical Power Transmission Apparatus, ANSI/ASME B15.1

Figure 13–5. The transparent plastic splash guard on this index screw machine protects workers from operating oils.

OSHA standards for the safeguarding of mechanical equipment. 29 *CFR*, Part 1910.212–219 Subpart "O," cover safeguarding of machinery and machine guarding. It is a violation of law to ignore these regulations.

For safeguards to be effective, whether built by the user or purchased from manufacturers, they should meet certain performance or design standards. If a company does not have a safety professional, it should get the advice of the insurance company safety engineer, manufacturer, consulting engineer, or the NSC Consulting Department as to what designs are available in machinery and machine guarding. The National Safety Council offers guidance to members; the Council's *Accident Prevention Manual for Business & Industry, Engineering & Technology* volume, *Safeguarding Illustrated*, and *Power Press Safety Manual* give many ideas in addition to those presented here.

To be generally acceptable, a safeguard should:

1. Conform to or exceed applicable ANSI standards and requirements of OSHA and/or OSHA-approved state plans.
2. Be considered a permanent part of the machine or equipment.
3. Afford maximum protection for operators and those performing maintenance, repairs, servicing, and also for passersby.
4. Prevent access to danger zones and points of operation (hazards created by moving parts of the machine and equipment).
5. Not weaken machine structure.
6. Be convenient. Guards must not interfere with efficient machine operation, cause operator discomfort, or complicate light maintenance and/or cleaning the machine area.
7. Be designed for the specific job, piece part, and specific machine. (Provisions must be made for oiling, inspecting, adjusting, and repairing of machine parts.)
8. Be resistant to fire and corrosion. Be easy to repair.
9. Be strong enough to resist normal wear and shock, and durable enough to serve over a long period with minimal maintenance.
10. Not be a source of additional hazards, such as splinters, pinch points, sharp corners, rough edges, or other injury sources. If possible, a safeguard covering rotating parts should be interlocked with the machine itself so the machine cannot be operated unless the safeguard is in place.

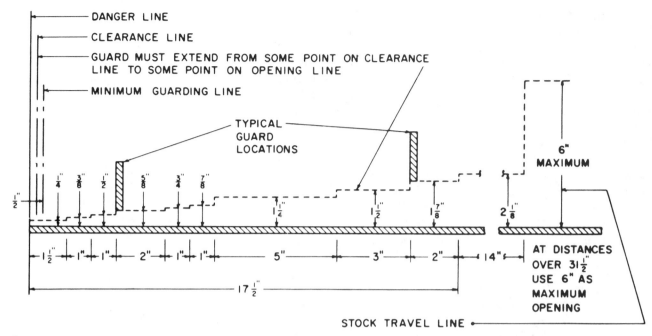

Figure 13–6. These point-of-operation guard locations show where barriers should be placed to prevent hand or forearm from contacting a hazard. The danger line is the point of operation. The clearance line marks the distance required to prevent contact between the guard and the moving parts. Minimum guarding line is ¹⁄₂ in. from the danger line. (From OSHA requirements 29 *CFR*, 1910.217.)

A Safeguarding Program

When installations of new machinery or the relocation of existing machinery or machine guarding and retrofitting are contemplated, persons in the organization who have a specific interest in them should be consulted and function as a team. Management should seek the opinions of facility engineers, facility layout, safety department, and purchasing department. They should also include: machine operators, supervisors, setup and maintenance personnel, millwrights, oilers, and electricians. Also, check with manufacturers. In regards to safeguarding, the manufacturers have safeguards available that are superior to any others that might be installed (see discussion in this chapter under Built-In Safeguards). Remember to consult manufacturers if proposed safeguarding changes or interferes with the original equipment design or safety function.

OSHA and ANSI standards should be studied before design, construction, and installation of safeguarding takes place. Every safeguard, whether installed when machinery is purchased or constructed or purchased later, must meet or exceed OSHA requirements. The team, supervisor, or person responsible for purchasing, constructing, or maintaining safeguards should be familiar with applicable OSHA 29 *CFR*, 1910, Subpart "O" (or other applicable portion) and ANSI standards.

A safeguarding program should include all equipment in the establishment. When several pieces of equipment are connected to form a manufacturing "cell"—they must be guarded as a cell.

A good safeguarding program can also help reinforce observance of other preventive measures. If hands and arms are protected by safeguarding, it makes sense to protect other parts of the body. Like providing eye and foot protection where applicable, a safeguard is a warning signal—a constant reminder to workers that enclosed is a potential hazard and danger.

Maximum-Sized Openings

An important factor to consider in safeguard design is the maximum size of openings. If a barrier is to provide complete protection, openings in it must be small enough to prevent the operator or others from reaching into the danger zone. If, for operational purposes, an opening must be larger than that specified by the standard (Figure 13–6), further protection must be provided.

Built-In Safeguards

Although sometimes a safeguard made in the facility can do a good job, often the best guard or barrier is the one provided by the manufacturer of a machine. Many standard machinery and safeguarding manufacturers design first-class, serviceable safeguards for equipment, available when contacted or specified on a purchase order. The manufacturer's safeguards are usually designed to be an integral part of the machine and are sometimes superior in appearance, placement, and function to those made in house.

Companies may sometimes believe partial or makeshift barriers can do the job well enough.

Disadvantages of makeshift safeguards are obvious. Because they give a false sense of security, they may be more harmful than no safeguard at all. They require operators to be constantly alert in order to make up for the guard's inadequacy and thus give little protection against human failure. Makeshift guards are often flimsy and are certain to become damaged and ineffective, sometimes shortly after installation. At times they may not be used by workers who feel they serve no useful purpose.

A completely effective safeguard is one that eliminates the hazard completely and permanently and can withstand handling and normal wear and tear.

Too often, machinery is purchased and installed without necessary safeguards to protect operators and other workers. Frequently, an excuse given is that purchase of a machine as a capital expense is closely budgeted, while in-house constructed safeguarding is categorized as operating cost and is more acceptable. As a result, some companies buy machinery and make safeguards after the machinery has been installed.

This tactic is poor economy. Machine safeguards can be provided by the manufacturer more cheaply because costs are spread over a large number of machines. In addition, the employer increases its liability.

Advantages of securing as much built-in protection as possible from the manufacturer are as follows:

1. The cost of safeguards designed and installed by the manufacturer is usually lower than the cost of installing safeguards after a machine has been purchased.
2. Built-in safeguards conform more closely to the machine contours.
3. A built-in safeguard can strengthen a machine, act as an exhaust duct or oil retainer, or serve some other functional purpose, thereby simplifying the design and reducing the cost of the machine.
4. Built-in safeguards generally improve production and prevent damage to equipment or material in process.

Substitution as a Safeguard

Substituting one type of machine for another sometimes can mean elimination or reduction of hazards. For example, substitution of direct-drive machines or individual motors for overhead line-shaft transmission decreases hazards inherent in transmission equipment (assuming that some older line-shaft installations still remain). Speed reducers can replace multicone pulleys. Remote-controlled automatic lubrication can eliminate need for employees to work dangerously close to moving parts.

Matching Machine or Equipment to Operator

This chapter discussion has emphasized safeguarding transmission parts or points of operation. Safe operation of machinery, however, involves more than eliminating or covering hazardous moving parts. Overall accident potential of the machine operation must be considered. These basic questions should be asked:

- Is there a materials-handling hazard?
- Are the limitations of a person's manual effort—lifting, reaching, bending, pushing, and pulling—recognized?
- Is the design of existing or proposed safeguards based on physiological factors and human body dimensions?

All physical or design features of a production machine and the workplace station should be evaluated as though the machine were an extension of a person's body and can do only what that person wants it to do. To match the machine or equipment to the operator, consider the following factors:

- **The workplace.** Machines and equipment should be arranged so the operator does minimal lifting and traveling. Conveyors and skids to feed raw stock, and chutes or gravity feeds to remove finished stock, should be considered.
- **The work height.** The workstation should be of optimal height in relation to stand-up or sit-down methods of operation. Proper height and type of chair or stool must be determined. Elbow height is the determining factor in minimizing worker fatigue. Effective work level is generally 41 in. (1 m) from floor to work surface, with chair height from 25 to 31 in. (0.6 to 0.8 m).
- **Controls.** Machine speed and ON-OFF controls should be readily accessible. Position and design of machine controls—such as dials, push buttons, and levels—are important. Controls should be standardized on similar machines. Operators then can be switched back and forth, as necessary, without having to use different controls. E-STOP locations must be available for both left and right hands.
- **Materials handling aids.** Aids should be provided to minimize manual handling of raw materials and in-process or finished parts, both to and from machines. Overhead chain hoists, belts or roller

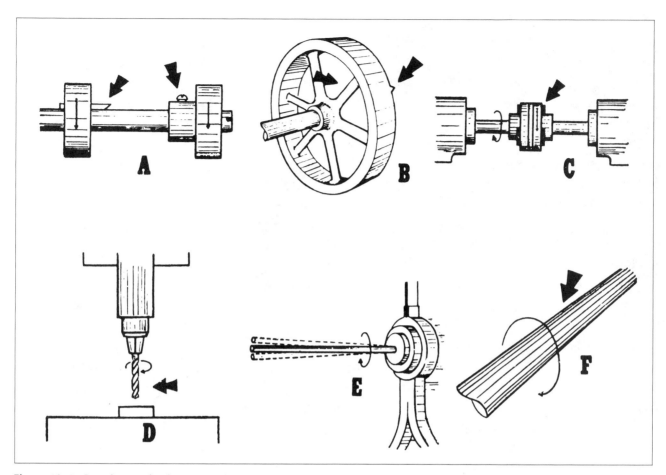

Figure 13–7. Rotating mechanisms can seize and wind up loose clothing, belts, hair, etc. They should, therefore, be guarded. Left to right, they are (A) projecting key and set screw, (B) spokes and burrs, (C) coupling bolts, (D) bit and chuck, (E) turning bar stock, and (F) rotating shaft.

conveyors, and work positioners are typical examples.

- **Operator fatigue.** Workers become fatigued at a machine workstation usually as the result of combined physical and mental activities, not simply from expending energy. Excessive speed-up, boredom from monotonous operations, and awkward work motion or operator position also contribute to fatigue.
- **Adequate lighting and other environmental considerations.**
- **Excessive noise.** More than an annoyance, excessive noise can be a real hazard because it can cause permanent hearing damage. Methods of protecting workers are given in Chapters 8 and 9 and are covered by OSHA regulations.

SAFEGUARDING MECHANISMS

Machines have certain basic mechanisms which, if exposed, always need safeguarding. These mechanisms,

which incorporate the primary hazards involved in machinery, can be grouped under the following headings:

- Rotating mechanisms
- Cutting or shearing mechanisms
- In-running nip points
- Screw or worm mechanisms
- Forming or bending mechanisms
- Impact mechanisms

A piece of equipment may involve more than one type of hazardous exposure. For instance, a belt-and-sheave drive is a hazardous rotating mechanism. It also has hazardous in-running nip points.

Rotating Mechanisms

A rotating part (Figure 13–7) is dangerous unless it is safeguarded. Mechanical power transmission apparatus represents most of this type of hazardous mechanism. Although relatively few injuries are caused

Figure 13–8. This hinged screw machine gearbox guard (shown in open position) is required even if the gears seem to be protected by location.

by such apparatus, the injuries often are permanently disabling.

Small burrs or projections on a shaft can easily snag hair or clothing, or catch a cleaning rag or apron and drag a person against and around the shafting. Vertical or horizontal transmission shafts, rod or bar stock projecting from lathes, set screws, flywheels and their cross members, drills, rotating couplings, and clutches are common hazardous rotating machine parts.

Shafting, flywheels, pulleys, gears, belts, clutches, prime movers, and other types of power transmission apparatus may seem safe by virtue of their location. However, many accidents happen where such apparatus is located—"where no one ever goes." The supervisor, oiler, and maintenance people go into these seldom-entered places. For their safety, hazards in such areas must also be safeguarded (Figure 13–8).

When a flywheel, shaft, or coupling is located so that any part of it is less than 8 ft (2.4 m) (ANSI/ASME B15.1; or 7 ft in OSHA 29 *CFR* 1910.212 and 219) above a floor or work platform, it must be safeguarded in one of several methods set forth in the federal standards. For example, exposed pulleys, belts, and shafting with rotating parts within 8 ft of the

floor should have substantial safeguards designed to protect employees from all possible contact. The underside of belts running over passageways or work areas should be protected by heavy screening so they can cause no harm if they break.

Enclosures should be removable, or provided with hinged panels and interlocked, to facilitate inspection, oiling, or repairing of the mechanism parts. However, enclosures should not be easily removed, and should be taken off only when the machine is stopped and all energy sources isolated and lockout procedures implemented, and then only by the supervisor's special order. Machines should never be operated until it is placed back in service by the person(s) performing the work and the enclosures have been properly replaced and secured.

Ends of rotating shafts projecting into passageways or work areas should be cut off or protected by nonrotating caps, safety sleeves, or guarded. Hangers should be securely fastened and well lubricated. Setscrews with slotted or hollow heads should be used instead of the projecting type. Couplings should be protected to prevent contact not only with coupling, but with the exposed rotating shafts also.

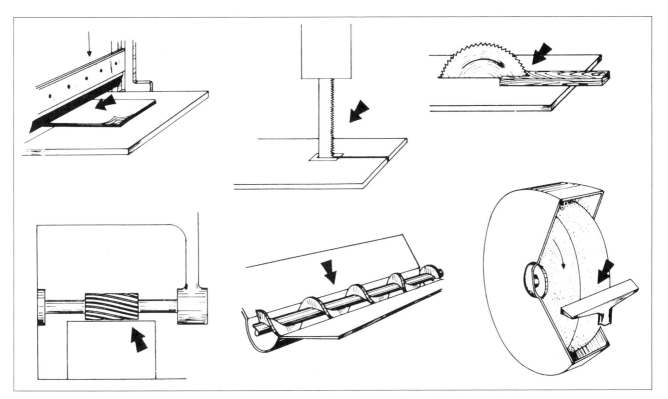

Figure 13–9. Protection should be provided for all variations of common cutting or shearing mechanisms.

Exposed gears should be entirely enclosed by substantial fixed barriers or protection in some other equally effective way. Sheet metal is preferable. The in-running side of the gearing should be treated with special care. Removable barriers should be equipped with interlocks and the power disconnect locked in the OFF position for inspection and maintenance. Other materials can be used where excessive exposure to water or chemicals might cause rapid deterioration of metal barriers.

It is essential to enclose the gearing flywheel in power presses because the large forces involved may cause overloading, which, in turn, can cause fatigue cracks in the shaft or gear. The fact that gears have split in half, and have fallen, indicates the need for strong enclosures.

Clutches, cutoff couplings, and clutch pulleys with projecting parts 7 ft (2 m) or less from the floor should also be enclosed by fixed barrier safeguards (29 *CFR* 1910.212 and 219). Some clutches within the machine may be considered "guarded by location," since they are out of normal reach. But if any possibility of contact exists, a complete enclosure should be provided.

Cutting or Shearing Mechanisms

The hazards of cutting or shearing mechanisms (Figure 13–9) lie at points where the work is being done, and where the movable parts of the machine approach or cross the fixed parts of the piece or machine. Guillotine cutters, shear presses, band and circular saws, milling machines, lathes, shapers, and abrasive wheels are typical of machines that present cutting or shearing hazards.

Saws. A circular saw must be shielded by a guard that covers the blade at all times to at least the depth of the teeth. The hood must adjust itself automatically to the thickness of the material being cut to remain in contact with it. The hood should be made so that it protects the operator from flying splinters or broken saw teeth. A table saw should be equipped with a spreader or splitter. It should also have an antikickback device to prevent materials from being thrown back at the operator.

All band saw blade parts must be enclosed or otherwise safeguarded except the working portion of the blade between the bottom of the guide rolls and the table. The enclosure for the blade part between the sliding guide and upper saw wheel guard should be self-adjusting, if possible. Band saw and band knife wheels should be completely enclosed. The barrier should be constructed of heavy material, preferably metal.

The upper half of a swing cutoff saw should be equipped with a complete enclosure. The lower half must be designed so it will ride over the fence and drop down on the table or the work being cut. A counterweight or other device should automatically return the saw to the back of the table when the saw is released at any point in its travel. Limit chains or other equally effective devices should be used to prevent the saw from swinging past the table toward the operator. Saw blades should be kept sharp at all times to prevent force feeding and development of cracks.

Cutters. On a milling machine, cutters should be shielded by an enclosure that provides positive protection for the operator's hands. A rotary cutter or slitter should have a barrier that completely encloses the cutting disks or knives. This makes it impossible for the operator to come into contact with the cutting edges while the machine is in motion.

Shears. The knife head on both hand- and power-operated shears should be equipped with a barrier to keep the operator's fingers away from the cutting edge. The barrier should extend across the full width of the table and in front of the hold-down. This barrier may be fixed or automatically adjusted to the thickness of the material to be cut.

If material being cut is narrower than the knife blade width, adjustable finger barriers should be installed. These will protect the open area at the sides of the material being sheared so the operators cannot get their hands caught at the sides under the blade. The barrier may be slotted or perforated to allow the operator to watch the knife, but the openings should not exceed 1/4 in. (6 mm), according to ANSI B11.1, and should preferably be slotted vertically to provide maximum visibility. A cover should be provided over the entire treadle length on the shears, leaving only enough room between the cover and treadle for the operator's foot. This device prevents accidental tripping of the machine if something should fall on the treadle.

Grinding Wheels. Since the abrasive grinding wheel is a common power tool and often used by untrained workers, it is the source of many injuries. Stands for grinding wheels should be heavy and rigid enough to prevent vibration. They should be securely mounted on a substantial foundation. Wheels should neither be forced on the spindle nor be too loose.

A person trained in correct and safe procedures should be in charge of wheel installations. Before a wheel is mounted, it should be carefully examined for cracks or other imperfections that might cause it to disintegrate. A "ring" test must be performed prior to mounting (29 *CFR* 1910.215). The work rest should be rigid and set no farther away than 1/4 in. (6 mm) from the wheel face so the material cannot be caught between the wheel and the rest. Wheels should be kept true and in balance.

Each wheel should be enclosed with a substantial hood made of steelplate to protect the operator in case the wheel breaks. The threaded ends of the spindle should be covered so that clothing cannot get caught in them. Hoods should be connected to effective exhaust systems to remove particles and prevent harmful dust from entering the grinding operation.

The manufacturer's recommended wheel speed should be adhered to at all times. Eye and face protection should be provided and properly fitted. Instruct workers that respirators should be used if considerable dust is generated, or if a hazardous or dangerous material is present and the air contaminants cannot be properly exhausted. Respirators should be of the type required to reduce the exposure to the acceptable level and cleaned in accordance with the manufacturer's instructions. Respirators should not be issued for use by more than one employee.

It is important that grinding wheels be properly stored and handled to prevent cracking or other damage. They should be stored in a dry place where they will not absorb moisture and should be kept in racks, preferably in a vertical position. The wheel manufacturers can supply additional information on proper storage and handling methods.

Buffing and Polishing Wheels. This equipment should be equipped with exhaust hoods to catch particles the wheels throw off. Not only do hoods protect the operator—they also prevent accumulation of particles on the floor and in the area. Eye or face protection should be provided for operators. They should not wear gloves or loose clothing. Protruding nuts or the ends of spindles, which might catch on operators' hands or clothing, should be covered by caps or sleeves.

Wire Brush Wheels. The same machine setup and conditions that apply to polishing and buffing wheels apply to brushes. The speed recommended by the manufacturer should be followed strictly.

The hood on scratch wheels should enclose the wheel as completely as the nature of the work allows. It should be adjustable so protection is maintained as the diameter of the wheel decreases. The hood should cover the spindle end, nut, and flange protection. Personal protective equipment is especially important in operating scratch wheels because wires can break off and fly into the operator's face.

The materials should be held at the horizontal center of the brush. The wire tips of the brush should do the work. Forcing the work into the brush only results in (1) merely wiping or dragging the wires across the material, with no increase in cutting action, (2) increasing wire breakage, and (3) snagging the work. Small pieces should be held in a jig or fixture. More details about abrasive wheels, buffers, and scratch brushes are given in Chapter 14, Hand Tools and Portable Power Tools.

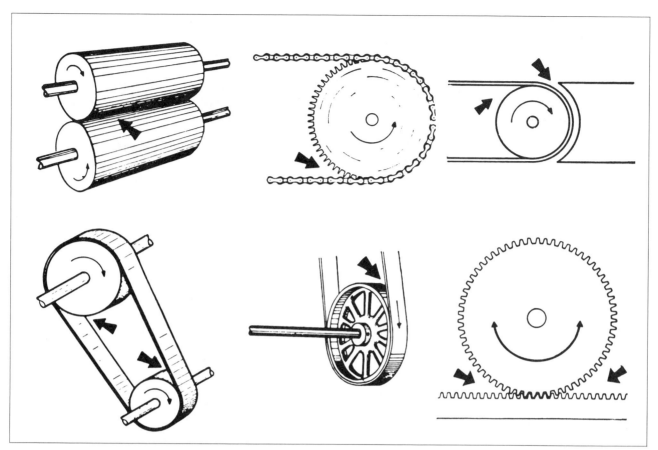

Figure 13–10. These typical in-running nip points require guarding.

In-Running Nip Points

Whenever two or more parallel shafts that are close together rotate in opposite directions (Figure 13–10), an in-running nip point is formed. Objects or parts of the body may be drawn into this nip point and be crushed or mangled. Typical examples of nip points are found on intermeshing gears, rolling mills, calenders, chains and sprockets, conveyors, belts and sheaves, racks and pinions, and at points of contact between any moving body and a corresponding stationary body.

Nip points should be made inaccessible by fixed barrier guards, or should be protected by instantaneous body-contact cutoff switches with automatic braking devices. The in-running side of rolls like those used for corrugating, crimping, embossing, printing, or metal graining should be protected. Arrange barriers so operators can feed material to the machine without catching their fingers between the rolls or between the barriers and the rolls.

Protect calenders and similar rolls with a device that the operator can use to shut off the rolls immediately at the feed point by means of a lever, rod, or treadle. The E-STOP must be fully accessible at front and back of machines (29 *CFR* 1910.218). Otherwise, the nip should be guarded by an automatic electronic device that will stop the rolls if anything but stock approaches the intake points. Enclosures are usually the best way to protect chains and sprockets, racks and pinions, belts and pulleys (or sheaves), and drive mechanisms for conveyors.

Screw or Worm Mechanisms

The hazard involved in the operation of screw or worm mechanisms is the shearing action set up between the moving screw and the fixed housing parts. Screw or worm mechanisms are generally used for conveying, mixing, or grinding materials. Examples are food mixers, meat grinders, screw conveyors, dry material mixers, and grinders of various types.

Screw conveyor covers should not be used as a walkway. If they must be walked on, additional protection should be provided. Corrosion and/or abrasive material in this conveyor type can erode the metal from the conveyor cover underside so that a person might, without thinking, step through it. When screw conveyors are constantly fed while in motion and use of an interlocked cover is impractical, install a heavy screen or mesh guard or grid. In this way, when the cover is removed for product inspection, the mesh will prevent anyone from falling through the opening.

Covers should also be provided for all mixers. The cover should be hinged to prevent removal and should have an interlock that will cut off the power source to stop action immediately when the cover is raised. Power switches for such machines should be locked out during maintenance or cleaning operations.

Rigid grids should cover openings on grinder hoppers. Such coverings should be large enough to permit materials to be fed to the grinder, but small enough to prevent any part of the operator's body from touching the cutting knives or the worm. Removable hoppers should be interlocked so the grinders cannot be operated when the hoppers are removed.

Other safeguards for screw or worm mechanisms may include guards and devices that require the operator to use both hands on the controls. Operating controls should be located so that they cannot be activated while any body part is in a position to be caught in the machine. Points of operation often can be guarded by regulating the size, shape, and location of the feed opening.

Metal Forming, Shearing, Punching, Bending Mechanisms

Use of metal-forming machines for stamping and forming pieces of metal and other materials has grown rapidly. Hand and finger injuries on these presses, as they are commonly operated, have become frequent. Misuse or abuse of these machines constitutes one of the most serious sources of mechanical problems in accident prevention.

Factors that make the problem difficult are variations in operations and operating conditions. Size, speed, and type of press; size, thickness, and kind of pieces to be worked; construction of dies; degree of accuracy required in the finished work; and length of run vary greatly. It is unwise to depend simply on operator skill for protection.

To make certain that operating methods are safe, insist that the die setter not only set the dies for a new run and test the machine for proper operation, but also set and adjust all other safeguards to the piece part and operator. Then ask the operator to run the machine a few strokes to make sure the safeguards are in place, and the adjustments are correct. See to it that die setters or setup person and operators know they must complete each of these steps before starting a production run. This procedure applies equally to shearing mechanisms: for example,

guillotine cutters, veneer chippers, paper-box corner cutters, leather dinking machines. As a group, these and similar machines present the worst kind of point-of-operation hazards, yet supervisors often believe that safeguarding them would slow down production. The right safeguards, properly designed and utilized, will not only diminish the hazard, but will usually increase production.

A procedure that omits any of these steps provides no guarantee that the dies are in alignment, that the kickout is working properly, that the clutch is in proper condition, or that the guard (if it is removable) is in place and operating. Supervisors with long experience in hazard control customarily require the machine setup crew to fill out a tag as written evidence that the procedure was thorough and complete for each setting. According to 29 *CFR* 1910.217 (e)(1):

> Inspection, maintenance, and modification of presses—inspection and maintenance records. It shall be the responsibility of the employer to establish and follow a program of periodic and regular inspections of his power presses to insure that all their parts, auxiliary equipment, and safeguards are in a safe operating condition and adjustment. The employer shall maintain records of these inspections and the maintenance work performed.

Each job poses questions. Should fixed guards, pull-back devices, two-hand control devices, or presence-sensing devices be used? The answer depends upon such factors as the type of press, full or part revolution, type of press, full or part revolution, kind of stock (piece, strip, roll); type of feed (hand, slide, dial, automatic); and type of knockout or ejection (mechanical or pneumatic) used to complete the operation.

Antirepeat Protection

Power presses sometimes repeat (an uninitiated or unexpected successive powered stroke of the press resulting from a malfunction). Safeguarding, therefore, must be designed to prevent injuries that could result from such occurrences. They include antirepeat mechanism, control component failure (control reliability), and brake monitor. Also workers have sustained injuries when they reached impulsively for misplaced stock after the press has been tripped. Details of how to prevent these accidents are given later in this chapter under the heading Automatic Protection Devices. Even the best training, experience, and supervision cannot substitute for well-designed safeguarding and the constant observance of safety practices.

Primary and Secondary Operations

Power-press operations consist of primary and secondary operations. Primary operations (blanking) are the easiest to safeguard at the point of operation. Because flat material is used, guards can be constructed or adjusted to allow only sufficient opening for the material to pass through the guard into the die. The trailing edge of strips can often be processed (1) by holding the scrap skeleton and pulling the stock through, (2) by pushing it through with the leading edge of the next skeleton or (3) by pushing it through with the leading edge of the next strip. When none of these is possible, inward shaping (a funnel shape) of a guard to meet the die at the point of stock entry to the die can help to fully use a manually advanced strip. If this customizing of the guard is not possible, the remaining tail end of the material should be discarded as scrap.

In secondary operations, parts or finished products are produced using blanks from primary operations. Safeguarding generally is more difficult than in primary operations, since manipulation of parts is required at or near the point of operation. The operator should be able to work efficiently without reaching into the point of operation. Ways to achieve safe, efficient operation include (1) providing a choice of operator-controlled feeding mechanisms, (2) using special tools, and (3) designing in positive part ejection and scrap removal.

Wherever possible, high-production or long-run dies should be guarded individually. This procedure saves setup time, and a permanent guard will not be detached or lost.

If die enclosures or fixed barriers cannot be used on secondary operations, use a presence-sensing device, a gate or movable barrier device, two-handed trip device or two-handed control device or pull-back device that removes the hands from the point of operation. Unless the press has automatic or semiautomatic feeding and ejection with a die-enclosure or a fixed-barrier guard, hand tools should be used. However, a hand-feeding tool is not a point-of-operation guard or protection device, and should never be used instead of it.

Safety devices must be maintained. Their use must be supervised strictly in accordance with the manufacturer's instructions. Otherwise they may fail, resulting in injuries and loss of workers' confidence. Usually your safety department, machine manufacturing, an insurance company safety engineering department is willing to assist in the design and installation of the best guard for a specific operation.

Feeding Methods

Automatic or semiautomatic feeds, if they are used in conjunction with a fixed enclosure guard. In most cases, they either increase production or reduce costs or both. With these feeds and safeguarding, operators do not need to place their hands into the point of operation during ordinary feeding. An air-ejection jet can help in removing parts from the die. The enclosure should be interlocked with the clutch brake mechanism.

The choice of feed will depend on the design of the die, the shape and quantity of parts being processed, and the type of equipment available. Feed types include gravity or chute, push, follow, magazine, automatic magazine, dial, roll, reciprocating, hitch, and transfer. Feeds can be automatic, semiautomatic, or manual.

Among the advantages gained with automatic feeding are (1) the operator does not have to reach into the point of operation to feed the press; (2) the feeding method usually makes it possible to enclose the die completely; and (3) the operator can load the feed mechanism, start the press, and then leave the vicinity of the press for a considerable number of strokes. Sometimes, the operator may be able to run several presses at once.

Figure 13–11. The slide feed allows loading of the die outside the danger zone. The permanent plastic barrier guard permits full visibility of the operation. The guard is separated at the top to allow die maintenance. The overlap of the guard at the separation eliminates shear hazard during travel of the slide. (Courtesy Allis-Chalmers Mfg. Co.)

With semiautomatic feeds, the operator does not have to reach into the point of operation, but usually must manually load the feed mechanism repeatedly or at frequent intervals. The feeding method usually makes it possible to enclose the die completely. Semiautomatic feeding is not adaptable for certain blanking operations or for nesting off-shaped pieces (Figure 13–11).

When manual feeding is required, eliminate the need for operators to place their hands or fingers within the point of operation. If this cannot be done, some method must be employed to protect the employee should the ram descend. Special tools have been developed and used successfully on operations where automatic feeds or enclosure guards are impractical. Such tools include pushers, pickers, pliers, tweezers, forks, magnets, and suction disks. These are usually made of soft metal to protect the die. Strict discipline is necessary to force operators to use them consistently. Such hand tools are not substitutes for guards, but should be used in conjunction with guard devices.

Bear in mind that few kinds of press guards provide complete protection. An automatic or semiautomatic feed may make it unnecessary for operators to place their hands in the danger zone, but may not prevent their doing so. If the operator attempts to straighten a part just before it passes under the slide, this method of safeguarding provides no real protection. Therefore, good practice combines such guards with a two-hand control device that requires constant pressure or control during the downward stroke of the press. (A two-handed control device can only be used when the press has a part revolution clutch or direct drive capable of being stopped during the stroke.)

Automatic or semiautomatic methods of feeding usually can be installed on jobs that are fed manually. Automatic feeds should be supplemented by a substantial enclosure at the point of operation, especially on slow-moving equipment, for complete protection. If possible, machine parts should be adjusted to reduce the hazard. For instance, the stroke on a press may be limited so that fingers cannot enter between the dies. More details are given later in this chapter.

Ejecting Material

Safe removal of material is as necessary as safe feeding. Since the way the finished piece is removed from the machine may influence the choice of feeding method, the removal method must be considered when an automatic or semiautomatic feed is selected.

Various methods of ejection may be used: compressed air, punch, knockouts, strippers, and gravity. Operators should not be required to remove finished parts or scrap material from the die manually because of the hazards involved. Air blowoff systems, crankshaft-operated scrap cutters, and other devices may be used.

Controls

Power presses should have actuating devices that prevent hands from getting under the slide or ram when the press is operated. Such devices include two-hand control devices or levers, treadle bars, pedals, and switches, located away from the point of operation. If two-hand controls or levers are used, relays or interlocks should be installed so that one switch or lever cannot easily be made inoperable and permit the press to be controlled with one hand, thus defeating the safeguard.

Press brakes are the source of many accidents. The gaging stops may be too low, and the piece being processed may slip beyond them. Because the ram motion is slow, some operators reach through the area between the ram and die to adjust the work. In doing so their hands can be caught. The proper type of starting device will make it impossible to reach under the ram after the press has begun to operate.

Foot controls should be covered by stirrup-type covers that extend over the entire length of the treadle

Figure 13–12. Foot pedals must be covered to prevent unintentional tripping of the press by the operator or by falling objects.

Figure 13–13. This expanded metal enclosure guard protects workers from protruding moving parts and power transmissions.

arm (an inverted U-shaped metal shield above the control) to prevent accidental tripping (Figure 13–12). When two or more operators run a press, foot controls or hand controls should be connected in series. Each person then is in the clear before the press can be operated.

On die-casting machines, two-hand tripping devices have been widely used. However, many consider it safer to install a sliding door that covers the die area. As the door closes, protecting the hazardous zone, it activates switches that set the machine in motion. Such a door virtually eliminates burns from splashing material.

It is prerequisite for safe die casting that no operator be allowed to place a hand or arm between the dies at any time. Long-handled pliers or tongs or similar tools should be used to place and remove stock; mechanical feeds and ejectors are even safer. Effective safeguards for these machines include two-hand tripping devices, sliding doors, treadle bars, and electrical or mechanical interlocking devices. To eliminate dangers involved in machine operation, enclosures may either be built and installed over hazardous areas or equipment may be redesigned to eliminate such exposed parts.

The modern lathe is a good example of machinery made safe through improved design. Its motor drive and gear box are enclosed so that line shafts, pulleys, and belts are dispensed with. The modern power press, in which all the working parts with the exception of the slide (ram) are enclosed, is another good example.

Safeguards used to make machinery safe include the fixed guard or enclosure, the interlocking guard or barrier, and the automatic protection device. Automatic or semiautomatic feeding and ejection methods are also ways of safeguarding machine operations.

Fixed Guards or Enclosures

The fixed guard or enclosure (Figure 13–13) is considered preferable to all other types of protection and should be used in every case, unless it has been definitely determined that this type is not at all practical. The principal advantage of the fixed guard is that it prevents access to the dangerous parts of the machine at all times. (If a fixed barrier is to provide complete protection, the openings in it must be small enough to prevent a person from getting into the danger zone. See Figure 13–6 earlier in this chapter.) This type of guard has another advantage. When the production job is finished, the guard remains with the die until it is needed for the next run.

Fixed safeguards may be adjustable to accommodate different sets of tools or various kinds of work. Once they have been adjusted, however, they should remain fixed. Under no circumstances should they be detached or moved.

Typical examples of the application of fixed safeguards are found on power presses, sheet-leveling or flattening machines, milling machines, gear trains, drilling machines, and guillotine cutters. Some remote feeding arrangements make it unnecessary for the operator to approach the danger point, and fixed barriers can be installed at a distance.

Interlocking Guards or Barriers

Where a fixed safeguard cannot be used, an interlocking guard or barrier should be fitted onto the machine as the first alternative. Interlocking may be mechanical, electrical, pneumatic, or a combination of types. The purpose of the interlock is to prevent operation of the control that sets the machine in motion until the guard or barrier is moved into position. Operators then cannot reach the point of operation, the point of danger.

When the safeguard is open, permitting access to dangerous parts, the starting mechanism is locked to prevent accidental starting. A locking pin or other safety device is used to prevent the basic mechanism from operating, for example, to prevent the main shaft from turning. When the machine is in motion, the enclosure cannot be opened. It can be opened only when the machine has stopped or has reached a fixed position in its travel.

To be effective, an interlocking safeguard must satisfy three requirements:

1. Guard the dangerous part before the machine can be operated.
2. Stay closed until the dangerous part is at rest.
3. Prevent operation of the machine, if the interlocking device fails.

Two-hand control devices are used in many types of interlocking controls on mechanical power presses they can be used only with part-revolution clutch equipment. These devices require simultaneous and sustained pressure of both hands on switch buttons, air-control valves, mechanical levers, or controls interlocked with a foot control, to name just a few. Two-hand operating attachments should be connected, so it is impossible to block, tie down, or hold down one button, handle, or lever, and still operate the machine.

When gate devices or hinged barriers are used with interlocks, they should be arranged so they completely enclose the pinch point or point of operation before the operating clutch can be engaged.

Interlocking controls are often installed on bakery machinery, guillotine cutters, power presses, dough mixers, some kinds of pressure vessels, centrifugal extractors, tumblers, and other machines on which covers or barricades must be in place before the starting control can be operated.

Automatic Protection Devices

An automatic protection device may be used, subject to certain restrictions, when neither a fixed barrier nor an interlocking safeguard is practicable. The device must prevent the operator from coming in contact with the dangerous part of the machine while it is in motion, or must be able to stop the machine in case of danger.

An automatic device functions independently of the operator. Its action is repeated as long as the machine is in motion. The advantage of this type of device is that tripping can occur only after the operator's hands, arms, and body have been removed from the danger zone.

An automatic protection device is usually operated by the machine itself through a system of linkage, through levers, or by electronic means. There are many variations. It can also be a hand-restraint or similar device, or a photoelectric relay.

Pull-back devices are attached to the operator's hands or arms and connected to the slide or ram plunger, or outer side of the press so that the operator's hands or fingers will be withdrawn from the danger zone as the slide or ram plunger, or outer slide descends. These devices should be installed and adjusted for each operator and readjusted for each job change, piece part, and press changes so they will properly pull the operator's hands clear of the danger zone.

All electronic safety devices (two-handed control devices) for power presses are made to perform the same function when energized. They interrupt the electrical current to the power press and shut off the machine (just as if the STOP button had been pushed). Electronic safety devices are effective only on power presses having air, hydraulic, or friction clutches. Such devices are not effective on power presses with full revolution clutch equipment. This is because once the operating cycle of this type of power press starts, nothing can prevent completion of the cycle.

To be effective, the electronic device should be operated from a closed electric circuit. Interruption of the current will automatically prevent the press from tripping. It is the supervisor's responsibility to make sure these devices are properly adjusted and maintained in peak operating condition. Many injuries have occurred because of improper adjustments and because parts have been allowed to become worn or need repair.

One advantage claimed for electronic devices (presence-sensing devices) is the absence of a mechanism in front of the operator. This is particularly advantageous on large presses. The electric-eye device should be installed far enough from the danger zone so that it will stop the slide or ram before the operator's hand can get underneath. Sufficient light beams should cover the bad-break area with a curtain of light.

Indexing is another press-shop term used to describe a mechanical method of feeding stock into press dies. One method is by a dial feeder. The dial feed is constructed, as its name implies, in the form of a dial having multiple stations that progress into the die by the indexing motion of the dial. The indexing of the dial should take place in conjunction with the up-stroke of the ram. When hand feeding, the index circuit should be controlled by dual-run buttons. Release of either button, during indexing, should stop the index cycle by releasing a safety clutch in the table. Safety guards should be connected to switches that stop the crank motion of the press whenever they are bumped. Both of these safety precautions will prevent accidents if the operator is tempted to reach into the die area to correct an improperly positioned part on the dial.

AUTOMATION

Automation, a somewhat misused term, is defined in this text as the mechanization of processes by the use of automatically controlled conveying equipment. Automation has minimized the hazards associated with manually moving stock in and out of machines and transferring it from one machine to another. It has also minimized exposure to the causes of hernias, back injuries, and foot injuries.

However, experience shows that those who work with automated equipment must have a thorough knowledge of its hazards and must be well trained in safe work practices to avoid accidents. Most finger and hand injuries result from operator exposure to the closing or working parts of a machine in the process of loading and unloading. The use of

indexing fingers, sliding dies, tongs, or similar hand tools reduces the hazard from such exposure. Nevertheless the supervisor must make sure that these devices are used consistently and correctly.

Like most innovations, automation has brought not only benefits, but hazards. Because each automatic operation is dependent on others, machine breakdown or failure must be corrected quickly. Also, because speed is highly important, maintenance employees may expose themselves, inadvertently perhaps, to working equipment parts. It is therefore imperative to have a mandatory policy that machines, equipment, and processes must be completely de-energized, isolated and locked out at the power source before servicing the machine. A well-defined lockout procedure should be written down and followed carefully.

Automated handling also has increased the use of stiles or crossovers. These should be constructed and installed in accordance with ANSI A12.1, *Safety Requirements for Floor and Well Openings, Railings, and Toeboards.* Automation eliminates or greatly reduces exposure to mechanical and handling hazards. In the single-operation process, however, the basic principles of safeguarding of equipment must still be applied. These principles are:

- Engineer the hazard out of the job as much as possible.
- Guard against the remaining hazards.
- Educate and test the performance of workers.
- Insist on use of safeguards provided.

Figure 13–14. This robot is guarded by a chain-link fence with gates. Clear spaces and areas for personnel who must maintain and repair the robot must be provided for in the design phase.

Robotics

Robots—machines specifically designed and programmed to perform certain operations—are rapidly becoming a part of the work environment. These machines can and have caused accidents to unsuspecting people working in the vicinity. Robot installation and safeguarding must conform to *Industrial Robots and Industrial Robot Systems—Safety Requirements*, ANSI/RIA 15.06–1992.

As early as possible, consult the engineer who designed or worked on the robot to learn about its capabilities, features, hazards, and operation. Allow for proper clearances when the robot is working with peripherals, such as machine tools, presses, transfer lines, palletizers, gaging stations, and so on. Even the robot with the best conceptual design may be efficient only if it has these clearances. Providing clearance for personnel, however, is an even more important factor. Although robots perform repetitive tasks for long periods of time, they are still machines and require periodic preventive maintenance. Eventually, the robots will need repairs to keep their devices, modules, and/or tooling working properly. Clear spaces and safe areas must be incorporated in the design stage (Figure 13–14).

A major hazard of robots involves their computer programming. When an operation is being performed, it may be virtually impossible to shut it off or reverse the cycle because of the automatic programming feature. Therefore, workers should be warned not to get near these machines. In some cases, it may be possible to totally enclose the robot, but in many instances, this may be impractical.

Control systems for robots usually work with a "priority interrupt" scheme. For example, higher priorities are interrupted by special hardware devices to signal that personnel are entering the work area, or that tooling needs to be changed or repaired. The supervisor should follow these basic rules:

1. Get as much information as possible on the operation of the robot.
2. Provide proper clearances and barrier guards around the robot (Figure 13–14).
3. Discuss all facets of the operation, including safety, with your people. All personnel should be trained.
4. Do not allow workers near robots in operation. Wait for the cycle to be completed and the machine de-energized before approaching.

Some jobs may require work to be done with power "on" (such as alignment and repair of servo systems). This must be realized in the design stage by the user's engineers. It cannot be assumed that work always will be performed in a power-off state. The simplest guarantee that equipment is safe to work on is to insist that repair personnel and the maintenance crew follow their company's lockout procedures. These are discussed later in this chapter.

Safe Practices

Safeguards are of primary importance in eliminating machine accidents, but they are not enough. Employees who work around mechanical equipment or operate a piece of machinery must have a respect for safeguards.

Before being permitted to run equipment, operators should be instructed in all the practices required for safe operation. Even experienced operators should be given refresher training, unless they know the hazards and the necessary precautions to be taken. In addition, employees who do not operate machinery, but who work in machine areas, should also receive instruction in basic safety practices.

Positive safety procedures should be established to prevent employees from misunderstanding instructions. You should enforce the following rules:

1. No guard, barrier, or enclosure should be adjusted or removed for any reason by anyone unless that person has permission from the supervisor, has been specifically trained to do this work, and machine adjustment is considered a normal part of the job.
2. Before safeguards or other guarding devices are removed so that repair or adjustments can be made, or equipment can be lubricated or otherwise serviced, the power for the equipment must be turned off and the main switch locked out and tagged.
3. No machine should be started unless the safeguards are in place and in good condition.
4. Defective or missing safeguards should be reported to the supervisor immediately.
5. Employees should not work on or around mechanical equipment while wearing neckties, loose clothing, watches, rings, or other jewelry.

MAINTENANCE OF SAFEGUARDS

The supervisor is responsible for scheduling inspection of machine safeguards as a regular part of machine inspection and maintenance. Such inspections are necessary, because employees tend to operate their machines without safeguards if they are not functioning properly, if they have been removed for repairs, or if they interfere in any manner with their operations. A guard or enclosure that is difficult to remove or to replace may never be replaced once it has been taken off. An inspection checklist can be developed for each machine to simplify the job and provide a convenient record for follow-up.

Maintenance and Repairs

Machines are subject to wear and deterioration, and can become unsafe to operate. Wear cannot be prevented, but it can be reduced, by controlling loads through proper manufacturing methods, alert supervision, employee attention, and good maintenance.

Lubrication is a basic maintenance function. Centralized lubrication will reduce hazards to oilers when climbing ladders to reach less accessible points. Some changes may be made so that most lubrication can be done at floor level. When oilers or repair crews must lubricate flywheel bearings, motors, and other parts at the tops of presses, good work practice suggests that permanent ladders with sturdy cages or wells should be installed.

Regardless of the type of lubrication necessary or the method used, it is the supervisor's responsibility to ensure that the machinery—including driving mechanisms, gears, motors, shafting hangers, and other parts—is being lubricated properly. Where automatic lubrication is not possible or feasible, extension grease or oil pipes should be attached to machines so that the oiler can avoid coming in contact with moving parts. Automatic lubrication should not be used, however, in cases where the oil or grease can congeal in the pipes.

Lockout/Tagout Procedures, the Control of Hazardous Energy Sources

Lockout or tagout procedures are designed to isolate or shut off machines, equipment, and processes from their power sources before employees perform any servicing or maintenance work. Employees must be trained in these procedures and instructed to replace safeguards after the work is completed. The supervisor, group leader, or authorized employee must enforce the company's lockout program.

Replacements

It is generally up to the supervisor whether to alert management whenever replacement of machinery and/or its safeguards is advisable. This step may seem like a production problem, but it is related to accident prevention. When machinery, safeguards, and safety devices become so worn that repairs cannot restore them to original operating efficiency, they constitute hazards. Often, unfortunately, these are concealed hazards. For example, if a nonrepeat device on a press becomes so worn that it does not operate or fails to operate properly, workers can be seriously injured.

This hazard can be eliminated only by replacing the machine part.

Several methods are used to determine when a machine needs maintenance or an overhaul to prevent a breakdown, or when replacements are in order. Some supervisors rely on reports of machine operators. Some have occasional inspections made by oilers, machine setters, or similar mechanics. Others set up inspections on a regular schedule and do it themselves, or have it done, routinely. The best plan is an established system of frequent, regularly spaced inspections.

SUMMARY OF KEY POINTS

Key points covered in this chapter include:

- The basic principles of safeguarding are: (1) engineer the hazard out of the job as much as possible, (2) guard against remaining hazards, (3) educate and test safety performance of workers, and (4) use safeguards provided. If such precautions are not taken, serious and property loss can result.
- Safeguarding is often thought of as being concerned only with the point of operation or with the means of power transmission. However, safeguards also prevent accidents from direct contact with exposed moving parts, work in process, machine failure, electrical failure, and operator error or human failure. The primary benefits of safeguarding are reducing the possibility of injury and improving production.
- Many safety devices, as well as barriers, enclosures, and tools, have been developed to protect machine operators. OSHA and ANSI standards require and present detailed guidelines for safeguarding such machines and equipment. Establishing effective methods of safeguarding power transmissions is easier than guarding points of operation because power transmissions are more standardized.
- An acceptable safeguard should conform to or exceed standards, be a permanent machine part, afford maximum protection, prevent access to the danger zone, be convenient, be designed for the specific machine and job, resist fire and corrosion, resist wear and shock, be long lasting, and not be a source of additional hazards. Built-in safeguards are generally more effective and less costly in the long run than "in-house" designed and constructed barriers.
- Safeguarding programs should adhere to pertinent regulatory codes and be instituted before machinery or equipment is installed. At times, substituting a less hazardous machine can solve a safeguarding problem. Regardless, supervisors should make

every effort to match equipment to operators in terms of workplace, work height, controls, materials handling aids, operator fatigue, adequate lighting, noise levels, and other environmental factors.

- Certain basic machine mechanisms, if exposed, always need safeguarding. These include rotating, cutting or shearing, screw or worm, forming or bending, impact mechanisms, and in-running nip points. These parts must have special safeguarding covers, barriers, or other equipment to shield operators and nearby workers from injury and to prevent equipment or product damage.

- Power presses consist of primary and secondary operations in which workers are exposed to potential injury. Wherever possible, high-production or long-run dies should be guarded individually. These devices should shut off the presses or prevent normal operation if workers' hands, clothing, or tools become caught in a press.

- Automatic or semiautomatic feeds, if they can be used, are generally successful safeguards. They can increase production, reduce costs, or both. The main advantages of these safeguards are that operators do not have to reach into the point of operation to feed the press, the die can be enclosed completely, and operators can leave the press untended for brief periods.

- Safe removal, or ejection, of material is as necessary as safe feeding. Various methods of ejection include compressed air, punch, knockouts, strippers, and gravity. Operators should not be required to remove finished parts or scrap material manually.

- Controls should be designed so machinery can be operated without endangering workers. Foot controls, for example, should be covered by stirrup-type covers to prevent accidental tripping. Dangers involved in machine operation can be eliminated either by enclosing hazardous machine parts, or by redesigning to eliminate exposed parts.

- Safeguards include fixed guards or enclosures, interlocking guards or barriers, and automatic protection devices. Fixed guards or enclosures are preferred to all other types because workers cannot remove them. When this safeguard is not practicable, interlocking guards or barriers can be fitted onto the machine to prevent operation of the controls until the guard or barrier is positioned.

When neither the fixed nor interlocking safeguard can be used, an automatic protection device can be applied. This safeguard prevents the operator from coming into contact with the dangerous part of the machine while it is in motion. It can also stop the machine in an emergency.

- Automation has minimized many hazards associated with manually moving stock in and out of machines but has brought its own hazards. Automated machinery must be completely deenergized (all energy sources isolated and locked out) before anyone is allowed to service or repair the machines. To apply effective safeguards on robots, supervisors must learn as much about the operation of the robot as possible, provide proper clearance and barriers around the robot, discuss all facets of operation and safety issues with workers, and not allow workers near robots in operation until the cycle is completed and the machine is deenergized.

- Supervisors must instruct workers in the use of effective safeguards and safe practices to prevent accidents and injuries. All workers, before operating machinery, should be thoroughly trained in how to use safeguards on equipment. Positive safety procedures should be established and vigorously enforced.

- Supervisors must set a schedule for inspecting and maintaining machine safeguards as a regular part of operations. Wear and tear on safeguards can be reduced by proper machine-handling methods, alert supervision, attention of employees, and regular maintenance. Before allowing workers to service machines and equipment, supervisors should be sure that lockout/tagout procedures have been implemented. Worn out, corroded, or damaged machine parts or safeguards should be replaced or repaired immediately.

14

HAND TOOLS AND
PORTABLE POWER TOOLS

After reading this chapter, you will be able to:

- Establish an accident control program based on four safe work practices: selecting the right tools for the job, using tools correctly and safely, keeping tools in good condition, and storing tools properly

- Describe the major hazards and safeguards for a wide variety of hand and portable power tools

- Explain the need for a centralized tool supply room or other way of controlling access to company tools

- Establish procedures for regular tool inspection, repair, and replacement

A major responsibility of the supervisor is to train people in safe handling of hand and power tools. This chapter discusses ways that hand and portable power tool injuries can be avoided. To pinpoint where major problems lie, first review department or company accident records. Identify all tool-related accidents/incidents, try to identify specific hazards, and check to see if any employees have had an unusual number of injuries. Such workers should have additional training and closer supervision. Make up written procedures for any tools used only rarely, to remind workers how to use the equipment properly and safely.

Training and close supervision are also needed when new tools, processes, and equipment are introduced in the operation or changes are made to an existing job. A job safety analysis of each task will serve to clarify tool needs, personal protective equipment (PPE) required, and safe work practices.

SAFE WORK PRACTICES

To be effective, a program to reduce tool-related injuries must include training in five basic safe practices. Your accident control program should include the following activities:

1. Evaluate and purchase the appropriate tools.
2. Train employees to select the right tools for each job. Good job safety analysis and job instruction training helps here. (See Chapter 4, Safety and Health Training.)
3. Train and supervise employees in the correct use of tools.
4. Set up regular tool inspection procedures, including inspection of employee-owned tools and provide good repair facilities as insurance that tools are being maintained in safe condition.
5. Set up a procedure for control of company tools. A check-out system at tool cribs is ideal. Provide proper storage facilities in the tool room and on the job (Figure 14–1).

Select the Right Tool

While many tools can be used to do a job, proper selection means choosing the right tool that can be most safely used by the employee.

Using improper tools can result in errors, injuries, and damage to products. For example, an adjustable wrench can tighten a nut, but a box end or a socket wrench of the proper size is better. Ergonomically designed tools, that meet job requirements, provide the best fit for the employee and the job.

Tool selection is related to other factors, such as job setup, workspace available, work height, and so forth, each related to selection of the safest, most efficient tool for the job. Cost and quality of tools are additional items to consider. Employers generally are expected to provide the correct tool(s) to do the job safely.

Use Tools Correctly and Safely

The supervisor shows employees how to use tools safely and correctly. Too often workers may select tools that are neither safe nor suited to the task, such as substituting a screwdriver for a crowbar, or using pliers instead of the proper wrench. In many situations the employee should study a tool's manual, which may suggest the use of proper PPE. Supervisors should ensure that workers understand the correct use of their tools and are aware of effects on other workers in the vicinity. Guards on tools must be checked to be sure they operate effectively.

Figure 14–1. Accident prevention begins with proper inspetion and storage of tools. They should be inspected when checked out and again when checked into the tool room. Only properly adjusted tools in good condition should be issued.

Keep Tools in Good Condition

Every tool should be examined before use. Employees in the tool crib and workers using tools on the job should be taught how to determine wear and damage on all tools used so that worn or damaged items can be repaired or discarded. Cutting edges must be kept sharp. Tools must be kept in good repair through a planned tool maintenance program. Good housekeeping is closely related to good, clean tools.

Tools that have deteriorated should not be used until repaired to meet manufacturers' specifications. Examples are wrenches with cracked or worn jaws, screwdrivers with broken bits (points) or broken handles, hammers with loose heads, dull saws, deteriorated extension or power cords on electrical tools, or broken plugs, and improper or removed grounding systems. Procedures should be established to check and remove from service any deteriorated tools.

Employees are usually not expected to do tool maintenance work. If they are, they should be trained and supervised.

Store Tools in a Safe Place

Tools need to be stored safely either at the work area or in a common tool crib. If employees own their own tools, safe storage is of equal importance.

Make a complete check of operations to determine need for special tools to do work more safely. Special tools may require unusual handling and storage. For example, tools like powder-actuated hand tools should be kept under lock and key and made available only to personnel trained and certified in their use.

USE OF HAND TOOLS

The misuse of common hand tools is a major source of injury to industrial workers. Injury often results because it is assumed that "anybody knows how to use" common hand tools. Observation and records of injuries show that this is not the case. The supervisor should study each job, train new workers, and retrain old employees on correct procedures for using tools. Where employees have the privilege of selecting or providing their own tools, you should advise them on using the safest equipment for each task. Regardless of who provides or owns the tools, they must be of the type specified by the employer and maintained as required by the employer. This requirement is so important that considerable attention must be given to establishing safe work practices with hand and portable power tools.

Be sure to enforce all established practices. Check the condition of tools frequently to see they are maintained and sharpened correctly, and that guards are not altered or removed. Make routine checks regarding the use and the condition of PPE used with these tools. Adopt specific practices for using hand tools and associated PPE for each operation. The required assessment of hazards should identify your protective equipment needs.

Cutting Tools

All edged hand-operated cutting tools should be used in such a way that, if a slip occurs, the direction of force will be away from any part of the user's body. For safety and efficiency, edged tools should be kept sharp, and ground to the proper angle. A dull tool does a poor job and may stick or bind. Any unexpected difference in tool performance may cause the user to lose control of the tool. Thus, all cuts should be made along the grain when possible.

Metal Chisels. Factors determining selection of a cold chisel are (1) materials to be cut, (2) size and shape requirements of the tool, and (3) depth of the cut to be made. The chisel should be heavy enough so that it does not buckle or spring when struck. For best results and maximum safety, flat and cape chisels should be ground so that the faces form an angle of 70 degrees for working on cast iron, 60 degrees for steel, 50 degrees for brass, and about 40 degrees for babbitt and other soft metals.

A chisel edge large enough for the job should be selected, so the blade rather than the point or corner is used. Also, a hammer heavy enough to do the job should be used. Some workers prefer to hold the chisel lightly in the hollow of their hands supporting the chisel by the thumb and first and second fingers. If the hammer glances from the chisel, it will strike the soft palm rather than the knuckles. Other workers think that a grip with the fist holds the chisel steadier and minimizes the chances of glancing blows. In some positions, this may be the only grip that is natural or even possible.

Hand protection can also be supplied by a rubber pad, forced down over the chisel to provide a hand cushion. Workers should chip in a direction away from the body. When shearing with a cold chisel, workers should hold the tool at the vertical angle that permits one cutting edge bevel to be flat against the shearing plane. Employees should wear eye protection when using chisels, and set up a shield or screen to prevent injury to other workers in the vicinity. If a shield does not afford enough protection to all exposed employees, then these workers should also wear eye protection.

Bull chisels held by one person and struck by another require tongs or a chisel holder to guide the tool so the worker is not exposed to injury. Both workers should wear eye, head, and foot protection.

Table 14–A. Selector for Hack Saw Blades

Pitch of Blade (Teeth per Inch)	Stock to Be Cut	Explanation
14	Machine Steel Cold Rolled Steel Structural Steel	The course pitch makes saw free and fast cutting
18	Aluminum Babbitt Tool Steel High-Speed Steel Cast Iron	Recommended for general use
24	Tubing Tin Brass Copper Channel Iron Sheet Metal of 18 gage or over	Thin stock will tear and strip teeth on a blade of coarser pitch
32	Sheet Metal of less than 18 gage (18 gage = 1.27 in.)	

The one who swings the sledge hammer should not wear gloves.

Dress all chisel heads at the first sign of mushrooming, because mushroomed heads often produce flying chips. It is suggested that when using a $\frac{1}{2}$ in. steel chisel, a hammer with $1\frac{1}{2}$ in. diameter face be used.

Tap and Die Work. This type of work requires certain safeguards. The work piece should be firmly mounted in a vise. Only a T-handle wrench or adjustable tap wrench should be used. Steady downward pressure should be applied on the taper tap. Excessive pressure causes the tap to enter the hole at an angle or bind the tap, causing it to break. A properly sized hole must be made for the tap, and the tap should be lubricated as necessary.

Keep hands away from broken tap ends. Broken taps should be removed with a tap extractor. If a broken tap is removed by using a prick punch or a chisel and hammer, the worker should wear eye protection. When threads are being cut with a hand die, hands and arms should be kept clear of the sharp threads coming through the die, and metal cuttings should be cleared away with a brush.

Hack Saws. These tools should be adjusted in the frame to prevent buckling and breaking, but should not be so tight that the pins that support the blade break. Install blades with teeth pointing forward.

The preferred blade to be used is shown in Table 14–A. A general rule is that at least two teeth should be in the cutting piece. It is advisable to release tension on the blade before storing hack saws and retension the blades before starting work.

Files. Selecting the right kind of file can prevent injuries, lengthen the life of the file, and increase production. A file-cleaning card or brush should be used to keep the file in peak condition. Files should not be hammered or used as prying tools. Such abuse frequently results in the file's chipping or breaking, causing injury to the user. A file should not be made into a center punch, chisel, or any other type of tool because the hardened steel can fracture. Use a vise, whenever possible, to hold the object being filed.

The correct way to hold a file is to grasp the handle firmly in one hand and use the thumb and forefinger of the other to guide the point, using smooth file strokes. This technique gives good control, and produces better work more safely.

A file should never be used without a smooth, crack-free handle; otherwise, if the file binds, the tang may puncture the palm of the hand, the wrist, or other part of the body. Under some conditions, a clamp-on, raised offset handle can provide extra clearance for the hands. Files should not be used on lathe stock turning at high speeds (faster than three turns per file stroke). The end of the file can strike the chuck, dog, or face plate and throw the file (or a metal chip) back at the operator, inflicting serious injury.

Tin Snips. This tool should be heavy enough to cut the material easily so the worker needs use only one hand on the snips and the other to hold the material. The material should be well supported before the last cut is made so the cut edges do not press against the hands. When cutting long sheet-metal pieces, push down the sharp edges next to the hand holding the snips.

The jaws of snips should be kept tight and well lubricated. When not in use, tin snips should be hung up or laid on a shelf. Burrs on the cut usually indicate a need for adjusting or sharpening the snips.

Workers should always wear gloves and use eye protection when trimming corners or slivers of sheet metal. Small particles can fly with considerable force into a worker's face.

Other cutters, such as those used on wire, reinforcing rods, or bolts, should have ample capacity for the stock. Otherwise, the jaws may spring or spread. Also, a chip can fly from the cutting edge and injure the user.

Cutters are designed to cut only at right angles to the plane of the material being cut. They should not be "rocked," hammered, or pushed against the floor to facilitate the cut because they are not designed to withstand excessive strain. This practice can nick the cutting edges, lessening the cutter's ability to perform.

A good rule is, "If it doesn't cut with ease, use a larger cutter or a cutting torch." Cutters require frequent lubrication. To keep cutting edges from becoming nicked or chipped, cutters should not be used as nail pullers or pry bars.

Cutter jaws should have the hardness specified by the manufacturer for the particular kind of material to be cut. By adjusting the bumper stop behind the jaws, workers can set the cutting edges to have a clearance of 0.003 in. (0.076 mm) when closed. Always store cutters in a safe place (consult manufacturer's recommendations).

Punches. Punches are used like chisels. They should be held firmly and securely, and be struck squarely. The tip should be kept shaped as the manufacturer has specified. Punches should be held at right angles to the work.

Wood Chisels. All employees should be instructed in the proper method of holding and using wood chisels. While molded plastic and metal handles are often available, wood handles are also found on wood chisels. If the handle is wood, it should be free of splinters and cracks. The wood chisel handle that is struck by a mallet should be protected with a metal or leather cap to prevent splitting and mushrooming. (NOTE: Do not use a metal hammer on wood chisels.)

Work to be cut should be free of nails to avoid damage to the blade or to prevent a chip from flying into the user's face or eye. (Eye protection is required.) The steel in a chisel is hard so that the cutting edge will hold, so it is also brittle enough to break if the chisel is used as a prying bar. When not in use, the chisel should be kept in a rack, on a workbench, or in a slotted section of the tool box so that the sharp edges will be protected. Sharp edges can be safeguarded, as shown in Figure 14–2.

Wood Saws. These tools should be carefully selected for the work they are to do. For fast crosscut work on green wood, a coarse saw (4 to 5 points per in.) is best. A fine saw (over 10 points per in.) is better for smooth, accurate cutting in dry wood. Saws should be kept sharp and well set to prevent binding. When saws are not in use, make sure they are placed in racks.

Saw set (the amount of angle or lean of a point from the blade) is needed to cut wood or other materials properly and cleanly. Inspect the material to be cut to avoid sawing into nails or other metal.

Manual sawing should be a one-hand operation. When starting a cut, workers should guide the saw with the thumb of their free hand held high on the side of the saw blade. They should not place a thumb on material being cut. Begin with a short, light stroke toward the body. After the cut begins, increase pressure and increase stroke length. Use light, shorter strokes as the cut is completed.

Axes. To use an axe safely, workers must be taught to check the axe head and handle, clear the area for an unobstructed swing, and swing correctly and accurately. Accuracy is attained by practice and proper handhold—and good training. All other workers must be kept a safe distance away from the direct line of the axe swing.

A narrow axe with a thin blade should be used for hard wood, and a wide axe with a thick blade for soft wood. A sharp, well-honed axe gives better chopping speed and is safer to use because it bites into the wood. A dull axe may glance off the wood being cut and strike the user in the foot or leg.

The person using the axe should make sure there is a clear circle in which to swing before starting to chop. Vines, brush, and shrubbery within the range, especially overhead vines that may catch or deflect the axe, should be removed.

Axe blades must be protected with a sheath or metal guard wherever possible. When the blade cannot be covered, it is best to carry the axe at one's side. The

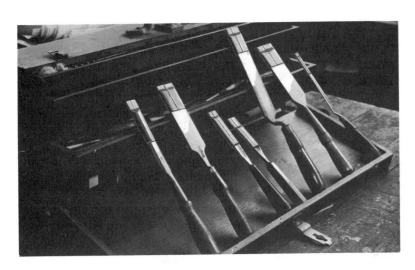

Figure 14–2. Guard sharp edges of hand tools with metal, fiber, or heavy cardboard sleeves that fit over them. Check tool boxes regularly to assure that sharp edges of tools are covered.

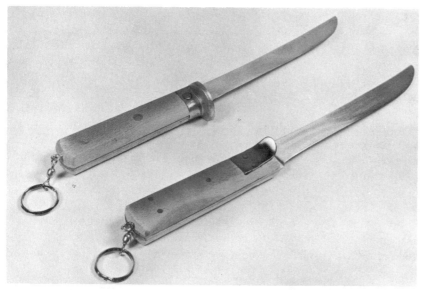

Figure 14–3. Knives equipped with ring-and-swivel guards and handle guards prevent the user's hand from sliding over the handle onto the blade in the event that the knife stabs against a solid object.

blade on a single-edged axe should be pointed down.

Hatchets. Hatchets, which are used for many purposes, frequently cause injury. For example, when workers attempt to split a small piece of wood while holding it in their hands, they may strike their fingers, hand, or wrist. Hatchets are dangerous tools in the hands of an inexperienced worker. To start the cut, it is good practice to strike the wood lightly with the hatchet, then force the blade through, by striking the wood against a heavier or larger block of wood.

Hatchets should not be used for striking metal because flying chips may injure the user or others nearby.

Knives. These tools cause more disabling injuries than any other hand tool. In the meat packing industry, for example, hand knives account for more than 15% of all disabling injuries. The principal hazard in using knives is that the hands may slip from the handle onto the blade, or the knife may strike the body or the free hand. A handle guard or a finger ring (and swivel) on the handle reduces these hazards (Figure 14–3).

The cutting stroke should be away from the body. If that is not possible, then the hands and body should be in the clear, a heavy leather apron, cut-resistant

gloves, or other protective clothing should be worn, and, where possible, a rack or holder should be used for the material to be cut. Jerky motions and the use of excessive force should be avoided to help maintain balance and control. Be sure employees are trained and supervised. Training in handling knives is very important in the food-service and similar industries.

Workers who must carry knives with them on the job should keep them in sheaths or holders. Never carry a sheathed knife on the front part of a belt. Always carry it over the right or left hip, toward the back, for added protection by locating it away from vital areas of the body.

Knives should never be left lying on benches or in other places where they could cause hand injuries. When not in use, they should be kept in racks with the edges guarded. Safe placement and storage are important to knife safety.

Ring knives—small, hooked knives attached to a finger ring—are used where string or twine must be cut frequently. Supervisors should make sure that the cutting edge is kept outside the hand, not pointed inside. A wall-mounted cutter or blunt-nose scissors would be safer.

Carton cutters are safer than hooked or pocket knives for opening cartons. They not only protect the user, but limit the depth of cuts that could damage carton contents. Frequently, damage to contents of soft, plastic bottles may not be detected immediately; subsequent leakage may cause chemical burns, damage other products, or start fires.

Make certain that employees who handle knives have ample room, so they are not in danger of being bumped by moving machinery or other employees. For instance, a left-handed worker should not stand to the right of a right-handed person. The left-handed person might be placed at the left end of the bench or otherwise given more room. Workers should be trained to cut away from or out of line with their bodies.

Be particularly careful about employees leaving knives hidden under product, scrap paper, or wiping rags, or among other tools in work boxes or drawers. Knives must be kept separate from other tools to protect the cutting edges and workers.

Work tables should be smooth and free of slivers. Floors and working platforms should have slip-resistant surfaces and be kept unobstructed and clean. If sanitary requirements permit mats or wooden duck boards, they should be in good repair, so workers do not trip or stumble. Conditions that cause slippery floors should be controlled as much as possible by good housekeeping.

Careful job analysis and accident investigation may suggest some changes in the work procedures that will make knives safer to use. For instance, on some jobs, special jigs, racks, or holders may be provided so it is not necessary for the operator to stand close to the piece being cut.

Wiping dirty or oily knives on aprons or clothing should be eliminated. The blade should be wiped with a towel or cloth with the sharp edge turned away from the wiping hand. Sharp knives should be washed separately and in a way that they will not be hidden under wash water.

Horseplay should be prohibited around knife operations. Throwing, "fencing," trying to cut objects into smaller and smaller pieces, and similar practices are dangerous.

The supervisor should make sure that nothing is cut that requires excessive pressure on the knife, such as frozen meat. Food should be thawed before it is cut or else frozen food should be sawed. In addition, knives should not be used as substitutes for openers, screwdrivers, or ice picks.

Miscellaneous Cutting Tools

Planes, scrapers, bits, and draw knives should be used only by experienced personnel. Keep these tools sharp, clean, dry, and in good condition. When not in use, they should be placed in a rack on the bench or in a tool box to protect the user and prevent damage to cutting edges.

Torsion Tools

Many tools used to grip materials are available. Proper tool selection and usage are important. Damaged tools should be removed from service.

Wrenches. All wrenches should be pulled, not pushed (Figure 14–4). Workers' footing should be secure. Ample workspace is needed near the wrench to provide clearance for fingers. Use a short, steady pull. If a nut does not loosen or tighten fully, a larger wrench may be needed. Open-end and box wrenches should be inspected to be sure that they fit properly. Do not mix metric with nonmetric sizes. Wrenches should not be hammered or struck unless they are designed for striking.

Figure 14–4. Adjustable (crescent) wrenches are probably the most often misused wrenches. They are made for use on odd-sized nuts and bolts that other wrenches might not fit. To use, the wrench should be placed on the nut so that when the handle is pulled, the moveable jaw is closer to the user's body. The pulling force will then be applied to the fixed jaw, which is the stronger of the two.

Socket wrenches give great flexibility. The use of special types should be encouraged on jobs where workers can be injured. In some situations, socket wrenches can be safer to use than adjustable or open-end wrenches, and they also protect the bolt head or nut.

Wrench jaws that fit prevent damage to nuts and bolt heads because they are less likely to slip.

Adjustable wrenches are not intended to take the place of open-end, box, or socket wrenches. They are used mainly for nuts and bolts that do not fit standard wrench sizes. The correct method is to apply the open side of the jaws toward you so that pressure is applied to the fixed jaw as the wrench is pulled toward the body (Figure 14–4).

Pipe Wrenches. Pipe wrenches, both straight and chain tong, should have sharp jaws and be kept clean to prevent slipping.

The adjusting nut of the wrench should be inspected frequently. If it is cracked, the wrench should not be used. A cracked nut may break under strain, causing failure of the wrench and possible injury to the user.

Workers on overhead jobs have been seriously injured when pipe wrenches slipped on pipes or fittings, causing them to lose their balance and fall. Using a wrench of the wrong length is another cause of accidents. A wrench handle too short for the job does not give proper grip or leverage. An oversized wrench handle may strip the threads or break the fitting or the pipe suddenly, causing a slip or fall.

A pipe wrench should never be used on nuts or bolts. The corners will break the teeth of the wrench, making it unsafe to use on pipe and fittings. Also, a pipe wrench used on nuts and bolts can damage their heads. The wrench should not be used on valves, struck with a hammer, or used as a hammer unless, as with specialized types, it is designed for such use.

Torque Wrenches. These have a scale to show the amount of force applied to the nut or bolt (usually foot pounds). It is important that a torque wrench be well cared for to ensure accurate measurements. Cleanliness of the bolt or nut threads is also important.

Tongs. Although tongs are usually purchased, some companies make their own to do specific jobs. To prevent pinching hands, the end of one handle should be upended toward the other handle, to act as a stop. It is also possible to braze, weld, or bolt bumpers on the handles a short distance behind the pivot point so that the handles cannot close against the fingers.

Pliers. Side-cutting pliers sometimes cause injuries when short ends of wire are cut. A guard over the cutting edge and the use of eye protection will help to prevent eye injuries.

The handles of electrician's pliers should be insulated. Because pliers do not hold the work securely, they should not be used as a substitute for a wrench.

Vise-grip pliers can be locked on the object, but they should be used with care because they can pinch the fingers or hand.

Special Cutters. These tools include those for cutting banding wire and strap. Claw hammers and pry bars should not be used to snap metal banding material. Only cutters designed for the work provide safe and effective results.

Nail Band Crimpers. These crimpers make it possible to keep the top band on kegs and wood barrels after nails or staples have been removed. Use of these tools eliminates injury caused by reaching into kegs or barrels that have projecting nails and staples.

Pipe Tongs. Such tongs should be placed on the pipe only after the pipe has been lined up and is ready to be fashioned. A 3–in. or 4–in. (7.5- or 10-cm) block of wood should be placed near the end of the travel of the tong handle. The block should be parallel to the pipe to prevent injury to the hands or feet if the tongs slip.

Workers should not stand or jump on the tongs, or place extensions on the handles to obtain more leverage. They should use larger tongs, if necessary, to do the job.

Screwdrivers. The screwdriver is probably the most commonly abused tool. Discourage workers from using screwdrivers as punches, wedges, pinch bars, or pries. If used in such ways, they become unfit for the work they are intended to do. A broken handle, bent shaft, or dull, broken, or twisted blade may cause a screwdriver to slip out of the slot and injure a worker's hand (Figure 14–5).

The tip must be kept clean and sharp to permit a good grip on the head of the screw. A screwdriver blade should fit the screw snugly. A sharp, square-edged blade will not slip as easily as a dull, rounded one, and requires less pressure. The part to be worked on should never be held in the hands. It should be laid on a bench or flat surface, or held in a vise. This practice will lessen the chance of hand injury if the screwdriver should slip. In addition, workers should observe the following precautions:

- Never use any screwdriver for electrical work unless it is insulated.
- In wood and sheet metal, make a pilot hole for the screw.
- Do not carry screwdrivers in your pockets.
- Keep screwdriver handles clean.
- Keep screwdriver blades properly dressed.

Screwdrivers used in a shop are best stored in a rack. This layout allows easy selection of the proper screwdriver.

Allen Wrenches. These wrenches are used like screwdrivers. The key factor is choosing the correct-size wrench for the job. It is easy to apply too much force on small Allen wrenches.

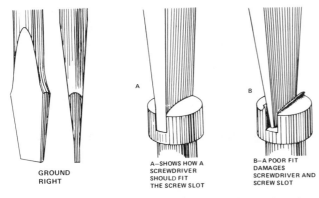

GROUND
RIGHT

A–SHOWS HOW A
SCREWDRIVER
SHOULD FIT
THE SCREW SLOT

B–A POOR FIT
DAMAGES
SCREWDRIVER AND
SCREW SLOT

Figure 14–5. Screwdriver blades must be ground flat in that portion of the tip that enters the screw slot, and then gradually taper out to the diameter of the shank (*left*). When selecting a screwdriver, make sure the blade makes a good fit in the screw slot (*right*).

Vises. Vises in a variety of sizes, jaws, and attachments are used to grip and hold objects being worked. Vises should be secured solidly to a bench or similar base. When sawing material held in a vise, make the cut as close to the vice jaws as possible. Lightly oil all moving parts of a vise.

Clamps. Many kinds of clamps are used in a variety of operations. Nearly all clamps can be used with pads to prevent marring the work. Over-tightening a clamp can break it or damage the product. If there is a swivel tip on the moving jaw of the clamp, it must operate or move freely. Lightly oil moving parts such as the threads. Clamps should be stored on a rack and not in a drawer.

Impact Tools

The term impact tools refers to a variety of hammers and related tools used to drive items such as nails into material by the use of manual or powered force. These tools are a common source of injuries, particularly to hands and arms. (See Chapter 10, Ergonomics, for more information on potential injuries.) It may be difficult to get employees to follow safe work practices because people often believe they know how to handle hammers. In routine safety inspections, make sure employees are not being careless with these tools.

Hammers. All hammers should have a securely fitting handle suited to the type of head used. The handle, whether reinforced plastic, wood, or metal, should be smooth, free of oil, shaped to fit the hand, and of the specified size and length. Warn employees against using a steel hammer on hardened steel surfaces. Instead, they should use a soft-head hammer, ballpeen hammer, or one with a plastic, wood, or rawhide head. Protective eyewear should be worn to protect against flying chips, nails, or scale.

Hammers may chip or spall, depending on four factors:

1. The more square the corners of the hammer, the easier it chips.
2. The harder the hammer is swung, the more likely it is to chip.
3. The harder the object struck, the more chipping increases.
4. The greater the angles between the surface of the object and the hammer face, the greater the chances of chipping.

Selecting the proper hammer is important. One that is too light is as unsafe and inefficient as one that is too heavy.

To drive a nail, hold the hammer close to the end of the handle. Use a light blow to start, and increase power after setting the nail. Hold the hammer so that the angle of the hammer face and the surface of the object being struck will be parallel. The nail will drive straighter, and there will be less chance for damage. Placing the hammer on the nail before drawing it up to swing may increase accuracy of the aim.

Sledge Hammers. These tools have two common defects that present a hazard to workers: split handles, and loose or chipped heads. Because these tools are used infrequently in some industries, a loose or chipped head may not be noticed. Some companies place a steel band around the head and bolt it to the handle to prevent the head from flying off. The heads should be dressed whenever they start to check or mushroom. A sledge hammer so light that it bounces off the work is hazardous. Similarly, one that is too heavy is hard to control and may cause body strain.

Riveting Hammers. Often used by sheet metal workers, riveting hammers should have the same kind of use and care as ballpeen hammers. Inspect them often for checked or chipped faces.

Carpenter's or Claw Hammers. These hammers are designed primarily for driving and drawing nails. When a nail is to be drawn from a piece of wood, a block of wood may be used under the hammer head to increase leverage and protect the wood surface.

The striking faces should be kept well dressed at all times to reduce the hazard of flying chips or nails. A checker-faced head is sometimes used to reduce this hazard.

Eye protection is recommended for all nailers and all employees working in the same area.

Spark-Resistant Tools

Spark-resistant tools of nonferrous materials generally should be used where flammable gases, vapors, or explosive materials might be present. These tools will not produce sparks when striking hard surfaces, thus reducing risk of fire or explosion.

PORTABLE POWER TOOLS

Portable power tools are divided into four primary groups according to power source: (1) electric, (2) pneumatic, (3) internal combustion, and (4) explosive (powder actuated). Several types of tools, such as saws, drills, and grinders, are common to the first three groups, whereas explosive tools are used exclusively for penetration work and cutting.

A portable power tool presents hazards similar to those of a stationary machine of the same kind. Typical injuries caused by portable power tools are burns, cuts, contusions, abrasions, and sprains. Sources of injury include electrical shock, particles in the eyes, fires, falls, explosion of gases, and falling tools. Because of the extreme mobility of power-driven tools, they can easily come in contact with the operator's body, and it is difficult to guard such equipment completely. The operator is also more closely exposed to the source of power (electricity, compressed air, liquid fuel, or explosive cartridge), which creates additional hazards.

When using powder-actuated tools (explosive cartridge equipment) for driving anchors into concrete, or when using air-driven hammers or jacks, it is imperative that hearing protection and eye and/or face protection be worn by operators, assistants, and nearby personnel when the tool is in use.

All companies and manufacturers of portable power tools attach to each tool a set of instructions for safely operating the equipment. These are meant to supplement thorough training that each power tool operator should receive. (More details on using these tools are given below.)

Power-driven tools should be kept in safe places and not left in areas where they can be struck or activated accidentally by a passerby. The power cord should always be disconnected before accessories on a portable power tool are changed. Guards should be correctly adjusted before the tool is used.

Selecting the Proper, Best-Designed Tool

Power tools can be more hazardous than hand tools. Check with your safety professional to make sure the tools you select meet not only current safety standards but best practices. Tool manufacturers can recommend the best tool to do a job. Describe the job you want accomplished, the material to be worked on, and the space available in your shop or facility work area. Tell the manufacturer if the operation is intermittent or continuous. Power tools are available in a variety of light to heavy-duty models, and the manufacturers should be able to give you a reliable recommendation.

Electric Tools

Electrical shock is the chief hazard from electrically powered tools. Injury categories are minor and lethal shock. Serious electrical shock is not entirely dependent on the voltage of the power input. Nearly all portable power tools are powered at 110 volts. As explained in Chapter 16, Electrical Safety, the ratio of voltage to resistance determines the current. The current is regulated by the body's resistance to the ground and by environmental conditions. As little as 100 milliamps can be lethal. It is possible for a tool to operate with a defect or short in the wiring. Use of a ground wire, however, protects the operator under most conditions.

Electrical tools used in damp areas or in metal tanks expose the operator to conditions favorable to current flow through the body, particularly if the worker perspires. Many electrical shocks from tools are caused by failure of insulation between current-carrying parts and metal frames of tools. Only tools in good repair that are listed by an authorized testing laboratory should be used.

Insulating platforms, rubber mats, and rubber gloves provide an additional safety factor when tools are used in damp locations, such as in tanks, boilers, or on wet floors. Safe low voltage of 6, 12, or 24 volts, obtained through portable step-down transformers, will reduce shock hazard in damp locations.

Double-Insulated Tools. Protection from electrical shock while using portable power tools depends on third-wire protective grounding. However, "double-insulated" tools are available and provide reliable shock protection without third-wire grounding or a ground fault circuit interrupter (GFCI) (Figure 14–6). The National Electrical Code permits double insulation for portable tools and appliances. Tools in this category are permanently marked by the words "double insulation" or "double insulated." Units designed in this category that have been tested and listed by Underwriters Laboratories also carry the UL mark:

Many manufacturers are using the symbol for a variety of tools to denote "double insulation." This double-insulated tool does not require separate ground connections. The third wire or ground wire is

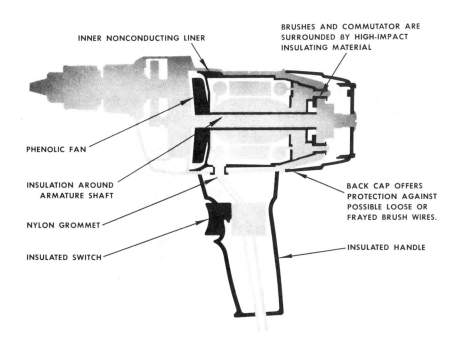

INNER NONCONDUCTING LINER

BRUSHES AND COMMUTATOR ARE
SURROUNDED BY HIGH-IMPACT
INSULATING MATERIAL

PHENOLIC FAN

INSULATION AROUND
ARMATURE SHAFT

NYLON GROMMET

INSULATED SWITCH

BACK CAP OFFERS
PROTECTION AGAINST
POSSIBLE LOOSE OR
FRAYED BRUSH WIRES.

INSULATED HANDLE

Figure 14–6. A double-insulated electrical tool has an internal layer of protective insulation that isolates the electrical components from the outer housing.

not needed and should not be used.

Failure of insulation is harder to detect than worn or broken external wiring, and points up the need for frequent inspection and thorough maintenance. Care in handling the tool, and frequent cleaning, will help prevent wear and tear that cause defects. See Chapter 16, Electrical Safety, for more details.

Grounding. The easiest way to safeguard the operator against electrical shock is to ground portable electrical tools. If any defect or short circuit inside the tool occurs, the current will flow harmlessly from the metal frame tool housing through a low-resistance ground wire rather than passing through the operator's body. All electrical power tools should be effectively grounded unless they are double insulated or of the cordless type.

The noncurrent-carrying metal parts of equipment may be grounded in two ways: first, by a grounding conductor, or second, by enclosing in metal the conductors feeding such equipment—provided an approved grounding-type attachment plug is used. The grounding conductor should be run with the power supply conductors in an insulated cable assembly or flexible cord. It must be properly terminated in an approved grounding-type attachment plug having a fixed grounding contacting member. If the grounding wire is separate, but insulated, it must be colored either in solid green or green with one or more yellow stripes.

A major safety concern is proper maintenance of the electrical power supply system, especially the ground. Periodic tests of the electrical system should be made by an electricalian to assure that it remains grounded.

A ground fault circuit interrupter, GFCI, is a device that in this case protects the tool operator should a ground fault occur. The GFCI can be installed anywhere in the circuit supplying the power tool. GFCIs come in a variety of overcurrent protection values and must be tested periodically. GFCIs are described in detail in Chapter 16, Electrical Safety.

Power tool electrical cords should be inspected frequently and kept in good condition. If the cord has been changed, have a qualified person check that the new cord is the right wire size. Use heavy-duty plugs that clamp to the cord insulation to prevent strain on current-carrying parts under manual use. Employees should be trained not to pull on cords and to protect them from sharp objects, heat, and oil or solvents that might damage or soften the insulation.

Extension Cords. Use only three-wire extension cords with three-prong, grounding-type plugs and three-pole receptacles that accept the tool's plug. Replace or repair damaged or worn cords immediately. Be sure to select the proper size cord because the flow of current (amperage) demanded by the electrical appliance will cause an undersize conductor to overheat. (See discussion in Chapter 16.) Table 14–B shows the proper gauge wire to use for various lengths of extension cord.

Electric Drills. These power tools can cause injuries in several ways: a part of the drill may be pushed into the hand, the leg, or other parts of the body. The drill may be dropped when the operator is not actually drilling. The eyes can be hit either by material being drilled or by parts of a broken drill bit. Avoid clearing chips using a finger—it can get pulled

Table 14–B. Minimum Gage Wire Size for Extension Cords

Tool Nameplate Amperes	Cord Length in Feet			
	25	50	100	150
0-6	18	16	16	14
6-10	18	16	14	12
10-12	16	16	14	12
12-16	14	12		

Source: Power Tool Institute, Inc.

into the drill and result in serious injury.

Oversized bit spindles should not be ground down to fit electric drills with smaller drill bit chucks. Instead, a reduction adapter can be used that will fit the large bit. However, this adapter is an indication of improper drill size and should be used sparingly to avoid placing excess strain on the smaller drill motor. When drills are used, the pieces of work should be clamped on or anchored to a sturdy base to prevent whipping.

Electric drills should always be of the proper size for the job. A 1/4 in. drill means a maximum 1/4 in. (diameter) drill bit that can be used for wood and light metal. A 1/2 in. drill bit would be needed to pierce steel and masonry. To operate, the chuck key should be attached to the cord, but removed from the chuck before starting the drill. Install the chuck key at the plug end so the employee must pull the plug before inserting the chuck key. If the drill has a side handle, it should be used. Make a punch mark to facilitate starting the drill and bit. Hold the drill firmly and at the proper angle for the job. Then start slowly. Gradually increase speed as needed.

Routers. The widespread use of routers is based on their ability to perform an extensive range of smooth finishing and decorative cuts.

Safety in using a router starts with an understanding that it operates at a very high speed, in the 20,000 RPM range, 25 times faster than a drill. It cuts quickly, is noisy, and the cutting blades are unguarded and out of sight when in use.

For work with routers, always wear eye protection against flying particles and airborne wood dust. Use a dust mask in dusty work conditions. Wear hearing protection during extended periods of operation.

Do not wear gloves, loose clothing, jewelry, or any dangling objects, including long hair, that may catch in rotating parts or accessories.

Install router bits securely, and according to the owner/operator's manual. Always use the wrenches provided with the tool.

Keep a firm grip with both hands on your router at all times. Failure to do so could result in loss of control,

possibly leading to serious injury. Read the operator's manual carefully regarding laminate trimmers and other small routers that are designed for use with one hand. Always face the cutter blade opening away from the body.

When the router is equipped with carbide-tipped bits, start by pointing the bit in a direction that will protect the operator and others against flying cutter pieces should the bit disintegrate.

Hold only those gripping surfaces of the router designated by the manufacturer. Check the owner/operator's manual. If a router is equipped with a chip shield, keep it properly installed. Keep your hands away from the bits or cutter area when the router is plugged in. Do not reach underneath the work while bits are rotating. Never attempt to remove debris while the router is operating.

Your desired cutting depth adjustments should be made only according to the tool manufacturer's recommended procedures for these adjustments. Tighten adjustment locks. Make certain the cutter shaft is engaged in the collet at least 1/2 in. Check the owner/operator's manual carefully.

Be certain to secure clamping devices on the work-piece before applying the router. The router switch should be in the OFF position before plugging into the power outlet. For greater control, always allow the motor to reach full speed before feeding the router into the work. Never force a router.

When removing a router from the workpiece, always be careful not to turn the base and bit toward the body. Unplug and store the router immediately after use and remove the cutter or cap before storage.

Electric Circular Saws. These saws are usually well guarded by the manufacturer, but employees must be trained to use the guard as intended (Figure 14–7). The guard should be checked often to be sure it operates freely and encloses the teeth completely when not cutting, and encloses the unused portion of the blade closest to the operator when it is cutting.

Circular saws should not be jammed or crowded into the work. The saw should be started and stopped outside the work. At the stroke beginning and end, or when the teeth are exposed, the operator must take extra care to keep the body extremities and the power cord away from the cutting line. All hand portable circular saws have a trigger switch to shut off power when pressure is released.

Injuries that occur when using portable circular saws are caused by contact with the blade; electrical shock or burns; tripping over the electrical cord, saws, or debris; losing balance; and kickbacks resulting from the blade being pinched in the cut.

Important requirements for safe operation of the hand portable circular saw are proper use, frequent inspection, and a rigid maintenance schedule. The manufacturer's recommendations for operation and maintenance must be followed faithfully and treated as standard procedure.

The following are specific safety "musts" when using any hand portable circular saw. Not doing so must be considered dangerous.

- Do not use a circular saw that is too heavy for workers to easily control.
- Be sure the trigger switch works properly. It should turn the tool on and return to the OFF position after release.
- Use sharp blades. Dull blades cause binding, stalling, and possible kickback; waste power; and reduce motor life.
- Use the correct blade for the application. Check these points carefully: Does it have the proper size and shape arbor hole? Is the speed marked on the blade at least as high as the no-load RPM on the saw's nameplate?
- The workpiece must be securely clamped. For maximum control, use both hands to properly and safely guide the saw.

Check blades carefully before each use for proper alignment and possible defects. Be sure blade washers (flanges) are correctly assembled on the shaft and that the blade is properly supported and tightened.

Is the blade guard working? Check for proper operation before each cut. Check the adjustment of antikickback dogs. Check often to be sure that guards return to their normal position quickly. If a guard seems slow to return or "hangs up," repair or adjust it

Figure 14–7. The lower movable guard of this portable circular saw always returns to the guarded position. The operator should keep fingers away from the trigger when the saw is not being used. (Courtesy Black & Decker Manufacturing Company.)

immediately. Never defeat the guard to expose the blade—for example, tying back or removing the guard.

Before starting a circular saw, be sure the power cord and any extension cord are out of the blade path and are sufficiently long to complete the cut. Stay constantly aware of the cord location. A sudden jerk or pulling on the cord can cause the operator to lose control of the saw.

For maximum control, hold the saw firmly with both hands after securing or clamping the workpiece. Check frequently to be sure clamps remain secure. Never hold a workpiece in your hand or across your leg when sawing. Avoid cutting small pieces of material that cannot be properly secured, or material on which the saw shoe cannot properly rest.

When making a "blind" cut (you can't see behind what is being cut), be sure that hidden electrical wiring, water pipes, or any mechanical hazards are not in the blade path. If there is any possibility of cutting into electrical conductors, disconnect power to these conductors. Contact with live wires could cause lethal shock or fire.

Set blade depth to no more than $1/8$ in. to $1/4$ in. greater than the thickness of the material being cut. Always hold the tools by the insulated grasping surfaces. When you start the saw, allow the blade to reach full speed before the workpiece is contacted. Be alert to the possibility of the blade binding and kickback occurring. If a fence or guide board is used, be certain the blade is kept parallel with it. Never over-reach.

When making a partial cut, or if power is interrupted, release the trigger immediately and don't remove the saw until the blade has come to a complete stop. Never reach under the saw or workpiece.

Portable circular saws are not designed for cutting logs or roots, or trimming trees or shrubs. To attempt this would be extremely hazardous.

Switch the tool off after a cut is completed, and keep the saw away from the body until the blade stops. Unplug, clean, and store the tool in a protected, dry place after use.

Kickback is a sudden reaction to a pinched blade, causing a combination of the saw being lifted up and out of the workpiece toward the operator and the workpiece being propelled in the direction of the blade rotations. Kickback is the result of tool misuse and/or incorrect operating procedures or conditions. Take specific precautions to help prevent kickback when using any type of circular saw:

- Keep saw blades sharp. A sharp blade will tend to cut its way out of a pinching condition.
- Make sure the blade has adequate set in the teeth. Tooth set provides clearance between the sides of the blade and the workpiece, thus minimizing

the probability of binding. Some saw blades have hollow-ground sides instead of tooth set to provide clearance.

- Keep saw blades clean. A buildup of pitch or sap on the saw blade surface increases the thickness of the blade and also increases the friction on the blade surface. These conditions cause an increase in the likelihood of a kickback.
- Don't cut wet wood. It produces high friction against the blade. The blade will tend to load up with wet sawdust and create a much greater probability of kickback.
- Be cautious of stock that is pitchy, knotty, or warped. These are most likely to create pinching conditions and possible kickback.
- Always hold the saw firmly with both hands.
- Release the switch immediately if the blade binds or the saw stalls.
- Support large panels so they will not pinch the blade. Use a straightedge as a guide for ripping.
- Never remove the saw from a cut while the blade is rotating.
- Never use a bent, broken, or warped saw blade. The probability of binding and resultant kickback is greatly increased by these conditions.

Overheating a saw blade can cause it to warp and result in a kickback. Buildup of sap on the blades, insufficient set, dullness, and unguided cuts can all cause an overheated blade. Never use more blade protrusion than is required to cut the workpiece—$1/8$ in. to $1/4$ in. greater than the thickness of the stock is sufficient. This minimizes the amount of saw blade surface exposed, and reduces the probability of kickback and severity if any kickback does occur. Minimize blade pinching by placing the saw shoe on the clamped, supported portion of the workpiece and allowing the cutoff piece to fall away freely.

When the saw is being used in a damp environment or outdoors, the operator should wear electrical insulating footwear and gloves. GFCI protection or a double-insulated saw is necessary for outside use.

Cutoff Saws. Do not use any cutoff saw beyond its rated speed. Check catalog RPM against safe saw speed. Never try to cut through thick material in one try; make a series of shallow cuts that gradually deepen to the cut desired. Operators who are exposed to harmful or nuisance dusts should wear approved respirators and eye protection.

Electric Reciprocating Saws. The common saber saw and reciprocating saw are two basic types of electric saws. Because the blade is almost fully exposed when in use and storage, the tool must be handled with extreme care. The blade should be removed when the saw is stored. The type of blade needed will be determined by the material being cut.

Hold the shoe as securely as possible against the work when cutting. Turn off the tool when the job is completed and do not remove the saw from the material until the motor has fully stopped.

The versatility of the reciprocating saw in cutting metal, plastic, wood, and other materials has made it a widely used tool. By design, it is a simple tool to handle. Its few demands for safe use, however, are very important.

- Without exception, use the blade specifically recommended for the job being done. Check your owner/operator's manual carefully concerning this.
- Position yourself to maintain full control of the tool. Avoid cutting above shoulder height.
- To minimize blade flexing and provide a smooth cut, use the shortest blade that will do the job.
- The workpiece must be clamped securely, and the shoe of the saw held firmly against the work to prevent operator injury and blade breakage.

Abrasive Wheels, Buffers, and Scratch Brushes. These tools should be guarded as completely as possible. Portable abrasive wheels must have proper guarding. For portable grinding, the maximum angular exposure of the periphery and sides should not exceed 180 degrees. The top portion of the wheel should always be enclosed. Guards should be adjustable so operators will be inclined to make the correct adjustment rather than remove the guard. However, the guard should be easily removable to facilitate replacement of the wheel. In addition to this mechanical guarding, the operator must use personal protective equiment (PPE) (including eye protection) in accordance with the hazard assessment.

A portable grinding wheel is exposed to more abuse than is a stationary grinder. The wheel should be kept away from water and oil; protected against dropping blows from other tools; and used carefully to avoid striking the sides of a wheel against objects. Cabinets or racks can help to protect the wheel against damage.

The speed and weight of a grinding wheel, particularly a large one, make it more difficult to handle than some other power tools. Since part of the wheel must necessarily be exposed, it is important that employees be trained in the correct way to hold and use the wheel so that it does not touch their clothes or body.

The wheels should be mounted by trained personnel only, with the wheels and safety guards conforming to ANSI standard B7.1, Safety Requirements for the Use, Care, and Protection of Abrasive Wheels. Grinders should be marked to show the maximum abrasive wheel size and speed. You must make sure the correct-speed wheel is used, for each job. Abrasive wheels should be sound-tested (ring-tested) before being mounted. A ring test involves tapping the wheel. For details, see the standard.

Sanders. Whether belt or disk type, sanders can cause serious skin abrasions when the rapidly moving abrasive touches the body. Because it is impossible to guard sanders completely, employees must be thoroughly trained to use them safely. The motion of the sander should be away from the body. All clothing should be kept clear of the moving parts.

A vacuum system attached to the sander or an exhaust system can be used to remove the dust from the work area. Eye protection should be worn. If harmful dusts are created, a respirator certified for the exposure by the National Institute for Occupational Safety and Health (NIOSH) or the Mine Safety and Health Administration (MSHA) should be worn. (See Chapter 9, Personal Protective Equipment.)

Safety precautions for using sanders include the following:

- Do not wear loose clothing, long hair, jewelry, or any dangling objects that may catch in rotating parts or accessories.
- Stay constantly aware of cord location.
- Never lock a portable sander in the ON position when the nature of a job may require stopping the sander quickly, such as using a disc sander on an automobile's fender well, where the rotating disc could get jammed in the well and result in an accident.
- With portable sanders, be careful not to expose the tool to liquids, or to use in damp, wet locations.
- When adjusting the tracking of the belt on a portable belt sander, be certain that you have the sander supported and positioned to avoid accidental contact with yourself or adjacent objects.
- Use jigs or fixtures to hold your workpiece whenever possible.
- Use appropriate personal protective equipment.

Sanders require especially careful cleaning because of the dusty nature of the work. If a sander is steadily used, it should be dismantled periodically, as well as thoroughly cleaned every day by being blown out with pressurized air. If compressed air is used, the operator should wear protective eyewear and use an air nozzle with sufficient side openings so that if "dead ended" against the skin, air pressure against the skin will not exceed 30 psig.

Because wood dust presents a fire and explosion hazard, keep dust to a minimum. Sanders can be equipped with a dust collection or vacuum bag. Electrical equipment should be designed to minimize the explosion hazard. Fire extinguishers approved for Class C (fires where there is an associated electrical exposure) should be readily available. Employees should be trained in what to do in case of fire.

Disc Grinders. Portable grinders basically remove material by contact with a rotating abrasive wheel or disc and buffing disc. There are safety precautions that apply to all grinders and other specific recommendations that apply to specialized grinding operations.

Before operating a grinder, compare the data on the nameplate with the voltage source. Be sure the voltage and frequency are compatible. Remove from the area material or debris that might be ignited by sparks. Be sure others are not in the path of the sparks or debris. Keep a properly charged fire extinguisher available.

Maintain proper footing and balance. Never attempt to grind from an awkward position. A portable grinder can kick, and glance off the work, if not tightly controlled.

Always disconnect the tool from the power source before installing or changing discs. Use only those discs marked with a rated speed that is at least as high as, or above, the speed rating on the nameplate of the tool. Do not use an unmarked wheel. Handle or store discs carefully to prevent damage or cracking.

Be careful not to over-tighten the spindle nut. Too much pressure will deform the flanges and stress the disc.

After mounting a wheel or disc and replacing the guard, stand to the side and allow a one-minute run-up at no load, to test the integrity of the wheel. Your grinder should come up to full speed each time before you contact the workpiece. Do not apply excessive pressure to the disc because that will stress the wheel, overheat the workpiece, and reduce your control.

Portable straight grinders should be used only with high-strength, bonded wheels. Position this type grinder away from the body and allow it to run for one minute before contacting work.

Tuck point grinders are a variation of straight grinders and are equipped with reinforced abrasive discs and the appropriate guard. Because using a tuck point grinder is very dusty work, a dust mask, face shield, and eye protection with side shields are recommended. Maintain firm control of the tool. Never over-reach. Carefully maintain balance. Do not allow the grinding wheel to bend, pinch, or twist in the cut, because kickback may result.

Angle grinders are used primarily with reinforced abrasive discs or wire cup brushes for the removal of metal or masonry. Use of the proper wheel and guard combination is critical. The user must follow the manufacturer's recommendations given in the owner/operator's manual.

Many angle grinders are equipped with guards that can be mounted with the opening in a variety of positions. Take care to position the guard to provide maximum protection. Many angle grinders can be converted for use as sanders. When guards are removed for a sanding operation, it is essential that they be replaced before the tool is again used for grinding.

Soldering Irons. Such tools are often the source of burns and of illness resulting from inhalation of fumes. Insulated, noncombustible holders will practically eliminate the fire hazard and the danger of burns from accidental contact. Ordinary metal covering on wood tables is not sufficient because the metal conducts heat and can ignite the wood.

Holders should be designed so that employees cannot accidentally touch the hot irons if they should reach for them without looking. The best holder completely encloses the heated surface. It is inclined so the weight of the iron prevents it from falling out. Such holders must be well ventilated to allow the heat to dissipate, otherwise the life of the tip will be reduced, and the wiring or printed circuit may be damaged. Also see the National Safety Council Occupational Safety & Health Data Sheet 445–1986, *Soldering and Brazing*.

Local exhaust ventilation is required to prevent inhalation exposure to solder and base metal fumes in excess of allowable levels established for each material. Sample the air to verify that the amount of contaminant in the air is not harmful. Food and beverages should not be consumed around soldering operations.

Particles should not be allowed to accumulate on the floor or work tables. Employees should wear protective eyewear, face shields, and respirators to reduce exposures not reduced through engineering.

Electrically Heated Glue Guns. These tools are being used increasingly in industry. Although all guns have insulated handles, there is danger in the high temperature of the glue and tip of the gun. Proper holders and storage can minimize the exposure. Electrical shock is also a risk and all tools should be properly grounded and insulated.

Percussion Tools. This family of tools is primarily associated with masonry applications as varied as chipping, drilling, anchor setting, and breaking of pavement. They range from pistol grip types to large demolition hammers. Normal operating modes include hammering, hammering with rotary motion, and rotation or drilling only. Many models incorporate a varied combination of the above types.

Always wear protective eyewear with side shields, and a full face shield. Use a respirator in dusty work conditions. Wear hearing protection during extended periods of operation so that the time-weighted noise exposure does not exceed the established levels. Do not wear loose clothing, long hair, jewelry, or dangling objects that may catch in rotating parts or accessories.

For maximum control, use the auxiliary handles provided with the tool. Do not tamper with clutches on those models that provide them. Have the clutch settings checked at the manufacturer's designated service facility at the intervals recommended in the owner/operator's manual.

Check for subsurface hazards such as electrical conductors or water lines before drilling or breaking blindly into a surface. If wires are present, they must be disconnected at the power source by a qualified person; or be certain they are avoided, to prevent the possibility of lethal shock or fire. Water pipes must be drained and capped.

Always hold the tool by the insulated grasping surfaces. Do not force the tool. Percussion tools are designed to hit with predetermined force. Added pressure by the operator only causes operator fatigue, excessive bit wear, and reduced control.

When cutting, drilling, or driving into walls, floors, or wherever "live" electrical wires may be encountered, workers should never touch any metal parts of the tool. Instead, they should hold the tool only by the insulated gripping surfaces to prevent electrical shock if contact is made with an energized conductor.

Air-Powered Tools

Operators of air tools should be instructed to:

- Keep hands and clothing away from the working end of the tool.
- Follow safety requirements applicable to the tool being used and the nature of the work being performed.
- Inspect and test the tool, air hose, and coupling before each use.

Most air-powered tools are difficult to guard. Therefore, the operator must be careful to prevent injury to hands, feet, or body if the machine slips or the tool breaks. Pistol or doughnut hand grips, or flanges in front of the hand grip, provide protection for the hands and should be used on all twist or percussion tools.

Air Hose. An air hose presents the same tripping hazard as do power cords on electrical tools. Anything accidentally hitting the hose may throw the operator off balance or cause the tool to fall. To prevent a tripping hazard, an air hose on the floor should be protected against trucks and pedestrians by placing two planks on either side or by building a runway over it.

Workers should be warned against disconnecting the air hose from the tool and using it for cleaning machines. Air hoses should not be used to remove dust from clothing.

Accidents sometimes occur when the air hose becomes disconnected and whips about. A short chain attached to the hose coupling and to the tool housing will keep the hose from whipping about if the coupling should break. Other couplings in the air line should also be safely pinned or chained. Air should be cut off before attempting to disconnect the air hose from the air line. Air pressure inside the line should be released before disconnecting.

A safety check valve installed in the air line at the manifold will shut off the air supply automatically if a fracture occurs anywhere in the line. If kinking or excessive hose wear is a problem, the hose can be protected by a wrapping of strip metal or wire.

Air Powered Grinders. These tools require the same kind of guarding as electric grinders. Be sure the speed regulator or governor on these machines is carefully maintained to avoid overspeeding the wheel (runaway). Hold grinder properly. Regular inspection by qualified personnel at each wheel change is ecommended.

Pneumatic Impact (Percussion) Tools. Percussion tools, such as riveting guns and jackhammers, are essentially the same, in that the tool is fitted into the gun and receives its impact from a rapidly moving reciprocating piston driven by compressed air at about 90 psig (625 kPa) pressure.

Two safety devices are needed to protect workers who operate these tools. First, the ON/OFF trigger should be located inside the handle where it is reasonably safe from accidental operation. The machine should operate only when the trigger is held in depressed position. Second, make sure the tool has a device holding it in place so that it cannot be shot accidentally from the barrel (Figure 14–8). On small air hammers not designed to use this device, use a spring clip to prevent the tool and piston from falling from the hammer.

It is essential that employees be thoroughly trained in proper use of pneumatic impact tools. Impress on all operators of small air hammers the following safety rule: Do not squeeze the trigger until the tool is on the work.

Air percussion tools either produce heavy blows or, due to rapid pulsating, vibrate strongly. Because tools such as rotary drills, saws, or grinders—especially those rotating at high speeds—produce excessive vibration, continued use of this equipment may cause damage to the body. Instruct workers to reduce transmission of this vibration by using rubber hand grips, air-cushion devices, and other vibration dampeners wherever possible. Prolonged use should be avoided. (See Chapter 10, Ergonomics, for more information on potential injuries.)

Jackhammers. Handling of heavy jackhammers causes fatigue and strain. Jackhammer handles should be provided with heavy rubber grips to reduce vibration and fatigue. Operators should wear protective footwear covering toes and metatarsal areas to reduce the possibility of injury, should the hammer fall.

Two chippers should work turned away from each other, back to back, to prevent face cuts from flying chips. Workers should not point a pneumatic hammer at anyone. They should never stand in front of operators handling pneumatic hammers.

Many accidents are caused by the steel drill breaking, causing the operator to lose balance and fall. If the steel is too hard, a particle of metal may break off and strike the operator. The manufacturer's instructions for sharpening and tempering the steel should be followed.

Air-Operated Nailers and Staplers. The principal hazard from these tools is accidental discharge of the fastener. In such instances, the fastener can become a dangerous projectile that can inflict serious injury at a considerable distance. Operators should be trained in use of these tools and must follow the manufacturer's operating instructions.

Personal Protective Equipment. As with all pneumatic impact tools, there is a hazard from flying chips. Operators should wear protective eyewear. If other employees must be in the vicinity, they should be similarly protected. Where possible, screens should be set up to shield persons near chippers, riveting guns, or air drills in use. Operators and others exposed should also wear hearing protection so that then exposure does not exceed established federal regulations. The sound of pneumatic tools can be extremely loud—similar to firecrackers or gunfire.

Eye and hearing protection should also be provided when using powered saws, grinding wheels, screwdrivers, buffers, scratch brushes, and sanders. The noise levels from some of these tools can reach or exceed 90 dBA (the federal standard time-weighted average for an eight hour work day).

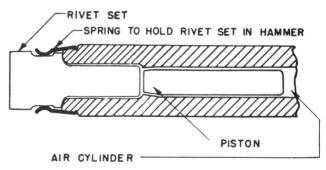

Figure 14–8. A spring clip, like the one shown here, should always be used to prevent a rivet set from falling from the air hammer.

Special Power Tools

Flexible Shaft Tools. These tools require the same type of personal protective equipment as do direct power tools of the same type. Abrasive wheels should be installed and operated in conformance with ANSI B7.1, discussed earlier. The flexible shaft must be protected against denting and kinking, that can damage the inner core and shaft.

It is important that the power be shut off whenever the tool is not in use. When the motor is being started, the tool end should be held with a firm grip to prevent injury from sudden whipping.

The abrasive wheel or buffer of the tool is difficult to guard. Because it is more exposed than the wheel or buffer on a stationary grinder, workers should use extra care to avoid damaging the equipment. When not in use, wheels should be placed on the machine or put on a rack, not on the floor.

Gasoline Power Tools. These tools are commonly used in logging, construction, and other heavy industry. The most well known is the chain saw. (See 29 *CFR* 1910.266(c)(5), Chain Saw Operations, and National Safety Council Occupational Safety & Health Data Sheet 320, *Portable Power Chain Saws*.)

Operators of gasoline power tools must be trained in their proper operation and follow the manufacturer's instructions. They must also be familiar with the fuel exhaust, heated surface, and noise hazards (see the discussion in Chapter 15, Materials Handling and Storage).

Powder-Actuated Fastening Tools. These tools are used for fastening fixtures and materials to metal, precast or prestressed concrete, masonry block, and to brick, stone, and wood surfaces, tightening rivets, and punching holes. Blank cartridges provide the energy and are ignited by means of a conventional percussion primer.

The hazards encountered in the use of these tools are similar to those encountered with firearms. The handling, storing, and control of explosive cartridges present additional hazards. Therefore, rules for the use, handling, and storage of both tools and cartridges must be just as rigid as those governing blasting caps and firearms (see OSHA regulations 1910.243(d) and National Safety Council Data Sheet 236, *Powder-Actuated Hand Tools*).

Additional specific hazards associated with these tools are accidental discharge, ricochets, ignition of explosive or combustible atmospheres, projectiles penetrating through the work, and flying dirt, scale, and other particles. In case of misfire, the operator should hold the tool in operating position for at least 30 seconds and follow the manufacturer's instructions for removing the load.

Powder-actuated tools can be used safely if workers are given special training and proper supervision. Manufacturers of the tools will aid in this training. Only properly qualified personnel should ever be permitted to operate or handle the tools. Workers can become qualified after a few hours of instruction (Figure 14–9).

QUALIFIED OPERATOR OF POWDER-ACTUATED TOOLS

Make(s)_____ Model(s)_____

This certifies that _____
(NAME OF OPERATOR)

Card No._____ Soc. Sec. No._____

Has received the prescribed training in the operation of powder actuated
tools manufactured by

_____
(NAME OF MANUFACTURER)

Trained and issued by _____
(SIGNATURE OF AUTHORIZED INSTRUCTOR)

I have received instruction in the safe operation and maintenance of powder actuated fastening tools of the makes and models specified and agree to conform to all rules and regulations governing their use. Failure to comply shall be cause for immediate revocation of this card.

_____   _____
(SIGNATURE)   (DATE)

Figure 14–9. A wallet card for qualified powder-actuated tool operators is available from tool manufacturers. A list of instructors should be maintained by each manufacturer.

Powder-assisted, hammer-driven tools are used for the same purposes as powder-actuated tools, and generally the same precautions should be followed. Workers should be trained to use powder charges of the correct size to drive studs into specific surfaces. They should be made responsible for safe handling and storage of cartridges and tools.

Powder-actuated tools should not be used on concrete with a thickness less than three times the fastener shank penetration depth, or on very hard or brittle materials, including cast iron, glazed tile, hardened steel, glass block, natural rock, hollow tile, or smooth brick. Fasteners should not be driven closer than 3 in. (7.5 cm) from an unsupported edge or corner.

Operators must wear adequate eye and face protection when firing the tool (Figure 14–10). Where the standard shield cannot be used for a particular operation, order special shields from the tool manufacturer. Hearing protection should also be worn to prevent damage from short, powerful bursts of sound. See Chapter 9, Personal Protective Equipment, for more details.

Propane Torches. Commonly used in many industries, propane torches are dangerous because of their open flame. Make sure the work area is clear of flammable and combustible materials. Proper storage of propane cylinders is important to prevent leakage of this highly flammable gas. Operating instructions are available from the manufacturer.

Figure 14–10. This powder-acuated tool drives studs into concrete slab. The operator should wear eye, ear, and head pro-tection. Note the holster (*left*) for carrying the tool. (Courtesy of Hilti, Inc.)

SUPERVISORY CONSIDERATIONS

As supervisor, you are responsible for controlling access to tools, for seeing that employee-owned tools meet company standards for ensuring that workers follow safe practices in carrying tools from place to place. You must be aware of these aspects of safety within your department, in addition to the hazards associated with particular tools.

Centralized Tool Control

Centralized tool control helps to ensure uniform inspection and maintenance of tools by a trained person. A toolroom attendant can promote tool safety by recommending or issuing the right type of tool, by encouraging employees to turn in defective or worn tools, and by stressing the safe use of tools. The attendant can be trained to recommend the correct protective equipment, such as welder's safety goggles or ear protection type to be used when the tool is issued. Centralized control and careful records of tool failure and other accidents will pinpoint hazardous conditions and unsafe practices.

Some companies issue each employee a set of numbered checks that are exchanged for tools from the toolroom. With this system, the attendant knows where each tool is and can recall it for inspection at regular intervals.

If you have a toolroom attendant you may want to set up a procedure so that the toolroom attendant can send tools in need of repair to a department or to the manufacturer for a thorough reconditioning.

Companies performing work at scattered locations may find it impractical to maintain a central toolroom. In such cases, the job supervisor should inspect all tools frequently and remove defective items from service. Many companies have supervisors check all tools weekly, using a preprinted checklist.

Personal Tools

In trades or operations where employees are required or prefer to have their own personal tools, you may as supervisor encounter a serious problem regarding the safety of such items. Personal tools are usually well maintained, but some people may purchase inferior or nonindustrial-rated tools, make inadequate repairs, or attempt to use unsafe tools (such as hammers with broken and taped handles, or electrician's pliers with cracked insulation).

To ensure that personal tools meet company standards, the company should spell out the general requirements that tools must meet before they can be used. First, where applicable, they should meet the appropriate ANSI standard(s) and/or be listed by a nationally recognized testing firm. Second, you should arrange a thorough inventory and initial inspection of personal tools. Inspect and list additional tools employees buy as though the items had been purchased by the company. Do not permit workers to use inferior or unsafe tools.

Carrying Tools

When climbing a ladder or any structure, workers should never carry tools in any way that might interfere with using both hands freely. A strong bag, bucket, or similar container should be used to hoist tools from the ground to the job. Tools should be returned in the same manner, not brought down by hand, carried in pockets, or dropped to the ground.

Mislaid and loose tools cause a substantial portion of hand-tool injuries. Tools are put down on scaffolds, on overhead piping, on top of stepladders, or in other locations from which they can fall on people below. Leaving tools overhead is especially hazardous in areas of heavy vibration or where people are moving about or walking.

Chisels, screwdrivers, and pointed tools should never be carried in a worker's pocket. They should be carried in a toolbox or cart, in a carrying belt (sharp or pointed end down) like that used by electricians and steel workers, in a pocket tool pouch, or in the hand, with points and cutting edges held away from the body.

Tools should be handed from one worker to another, never thrown.

Edged or pointed tools should be passed, preferably in their carrying cases, with the handles toward the receiver. Workers carrying tools on their shoulders should pay close attention to clearances when turning around. They should handle the tools so that they will not strike other people.

MAINTENANCE AND REPAIR

A toolroom attendant or tool inspector should be qualified by training and experience to pass judgment on the condition of tools for further use. Dull or damaged tools should not be returned to stock. Keep enough tools of each kind on hand so that when a defective or worn tool is removed from service, you can replace it immediately.

Efficient tool control requires periodic inspections of all tool operations. These inspections should cover housekeeping in the tool supply room, tool maintenance, service, inventory, handling routine, and condition of the tools. Responsibility for such periodic inspections is usually given to you as the department supervisor. This job should not be delegated. However, all employees need to recognize signs indicating when a tool is damaged, dull, or unsafe so they can report it at once.

Hand tools receiving the heaviest wear—chisels, wrenches, sledges, drills and bits, cold cutters, and screwdrivers—will require regular maintenance. Proper maintenance and repair of tools require adequate facilities: workbenches, vises, a forge or furnace for hardening and tempering, a variety of protective eyewear, repair tools, grinders, and good lighting.

Tempering Tools

Such tools as chisels, stamps, punches, cutters, hammers, sledges, and rock drills should be made of steel and be heat-treated. This procedure makes them hard enough to withstand blows without excessive mushrooming, and yet not so hard that they chip or crack. Hardening and tempering of tools require special skills in manufacture.

Metal Fatigue

Metal fatigue is a common problem in cold weather. Metals exposed to subzero temperatures undergo cold-soaking. A molecular change takes place that makes steel brittle and easily broken. This phenomenon shortens the useful life of such equipment as pneumatic impact tools, drill bits, and dies for threading conduit.

Ferrous metals, including carbon steel, alloy steel, and cast iron, are characterized by decreased toughness and corresponding brittleness at low temperatures. Nonferrous metals, on the other hand, such as aluminum and aluminum alloys, nickel, copper and copper alloys, chromium, zinc and zinc alloys, magnesium alloys, and lead, are more resistant to low temperatures and are often used in areas of extreme cold.

Safe-Ending Tools

Such tools as chisels, rock drills, flatters, wedges, punches, cold cutters, and stamping dies should have their heads properly hardened by a qualified worker. Short sections of tight-fitting rubber hose can be set flush with the striking ends of chisels, hand drills, mauls, and blacksmith's tools to keep chips from flying, since they usually embed themselves in the rubber sleeve. Chisels, drift punches, cutters, and marking tools that manufacturers claim do not spall or mushroom are available. This feature is achieved by using a combination of alloys and scientific heat treatment to temper the metal.

Dressing Tools

Tools should have regular maintenance of their cutting edges or striking surfaces. In most cases, once the cutting or striking surfaces have been properly hardened and tempered, only an emery wheel, grindstone, or oilstone is needed to keep the tool in good condition. Be sure to grind in easy stages. Keep the tool as cool as possible with water or other cooling medium.

Tools that require a soft or medium-soft head should be dressed as soon as they begin to mushroom. A slight radius ground on the edge of the head, when it is dressed, will enable the tool to stand up better under pounding and will reduce the danger of flying chips. A file or oilstone, rather than an abrasive wheel, is recommended for sharpening pike poles and axes.

A wood-cutting tool, because of its fine cutting edge, can be initially dressed on a grinder with a wheel recommended by the manufacturer. An oilstone set securely in a wood block placed on a bench should be used to obtain a fine, sharp cutting edge. The oilstone should never be held in the hand; a slip off the face of the stone could cause a severe hand injury. Often a few finishing strokes on a leather strop will produce a keener edge.

Metal-cutting tools, because they generally have greater body, can be dressed or sharpened (or both) on an abrasive wheel. Do not use too much pressure against the wheel or the tool will overheat.

Make sure workers follow the manufacturer's recommendations for choosing the proper kind of abrasive wheel to sharpen wood or metal-cutting tools. Each cutting edge should have the correct angle according to its use.

Handles

The wooden handles of hand tools should be $3/4$ in. to $1\frac{1}{2}$ in. of the best straight-grained material, preferably hickory, ash, or maple. They should be free from slivers. Make sure that they are properly attached. Poorly fitted or loose handles are a hazard. They can damage the work material, and make it difficult for the worker to control the tool.

No matter how tightly a handle may be wedged at the factory, both use and shrinkage will loosen it. Inspect tools regularly for damaged and loose handles. These should be removed from service and repaired, if feasible.

SUMMARY OF KEY POINTS

Key points covered in this chapter include:

- One of a supervisor's major responsibilities is to train and supervise people in the safe handling of hand and portable power tools. Supervisors must pinpoint where major problems in tool handling lie, and devise a program to reduce tool injuries. Four basic safe work practices include (1) select the right tool designed for the job and user, (2) use tools properly and safely, (3) keep tools in good condition, and (4) store tools in a safe place.

- Supervisors must establish procedures for regular inspection, repair, and replacement of tools and set up a central tool supply room or other method of controlling access to company tools. Special tools may need to be kept under lock and key.

- The misuse of common hand tools is a major source of injury to industrial workers. These tools include cutting, torsion, shock, and spark-resistant equipment. Supervisors must make sure that employees are trained in the proper selection and use of these tools. They must see that employees follow all safety guidelines and procedures. Personal protective equipment (PPE) such as gloves, protective eyewear and helmets, hearing protection, aprons, and protective footwear should be worn.

- Cutting tools include metal and wood equipment, such as chisels, tap and die work, hack saws, files, tin snips, punches, saws, axes, hatchets, knives, and miscellaneous tools. All edged tools should be used so that the direction of force is away from the body. Cuts should be made with the grain when possible. Edged cutting tools should be kept sharp and ground to the proper angle. Saws must be sharpened and oiled regularly or they may stick in the cut. Workers should wear personal protection equipment to shield them from flying chips or metal slivers and from cuts. Cutting tools should never be used frivolously in mock fights or contests of skill. They can inflict serious injuries in a moment of carelessness.

- Torsion tools are used to fasten or grip materials and include various types of wrenches, tongs, pliers, special cutters, nail band crimpers, screwdrivers, Allen wrenches, vises, and clamps. These tools must fit the job precisely to prevent slippage, the jaws or blades should be kept sharp and clean, and pressure should be applied away from the body. Workers must protect their hands, arms, and faces with special care when using these tools.

- Shock tools include regular hammers, sledge hammers, riveting hammers, and carpenter's or claw hammers. All hammers should have secure handles and properly dressed heads. Workers must be sure to select the proper weight and size hammer to perform work safely.

- Spark-resistant tools of nonferrous materials are used where flammable or explosive materials are stored. These tools will not create sparks when striking metal or other hard surfaces.

- Portable power tools are divided into four groups according to the power source: electric, pneumatic, internal combustion, and explosive. Typical injuries caused by portable power tools include burns, cuts, and sprains. These tools must be carefully handled and stored so that they cannot be activated accidently. Workers must be thoroughly trained in proper use. In addition to the hazards associated with the tool itself, the power source represents a risk to workers. All employees must wear hearing protection and eye and/or face protection when operating portable power tools or working near them.

- Electrical shock is a significant hazard from electrically powered tools. Workers can suffer minor or lethal shock. Insulating platforms, rubber mats, rubber gloves, ground fault circuit interrupters, double-insulated tools, grounding wires, and grounded extension cords are all ways of protecting workers who use electrical tools. Workers should wear appropriate personal protective gear to minimize risk of injury and hearing loss. All damaged or worn cords should be replaced immediately. Cords should be inspected frequently to be certain they can carry the required electrical current safely.

- Important requirements for safe operation of electrically powered drills, routers, saws, wheels, sanders, soldering irons, percussion, and other portable power tools are proper use, frequent inspection, and a rigid maintenance schedule. The manufacturer's recommendations for operation and maintenance must be followed strictly.

- Air-powered tools such as grinders and pneumatic tools (jackhammers, nailers, and staplers) present serious hazards to workers. Operators must be instructed to keep hands and clothing away from the working end of the tool, protect the air hose and check the safety value, follow safety requirements for the tool being used and work performed, and inspect and test the tool, air hose, and coupling before each use. Hands, feet, and torso should be protected in case the machine slips or the tool breaks, and hearing and eye protection should be used. Pistol or doughnut hand grips, or flanges in front of the hand grip, provide protection and should be used on all twist or percussion tools.

- Special power tools include flexible shaft tools, gasoline power tools, powder-actuated fastening tools, and propane torches. This equipment requires the same type of personal protection gear as do electric or air-powered tools. In addition, workers must be trained in fuel hazards and the dangers associated with explosive cartridges used in powder-actuated tools. Workers generally are given special instruction to operate these tools safely. They must know the manufacturer's recommendations for using the tools under various conditions.

- Supervisors must establish routine procedures for controlling access to company tools. For example, centralized tool control helps to ensure uniform inspection and maintenance by a trained person. The attendant can recommend proper safety equipment and issue the right tool for the right job.

- When employees purchase their own tools, supervisors must ensure that such tools meet established company standards, are regularly inspected and repaired or replaced, and fit the work to be done. Supervisors should also instruct employees on the proper way to carry tools. Workers should never carry tools while climbing ladders, should always carry pointed tools with the sharp end covered or held away from the body, and should always hand, never throw, tools to another worker.

- Supervisors must assume responsibility for inspection and maintenance and repair of tools in their department. Tools should be checked for quality construction and manufacture, for signs of metal fatigue in extremely cold environments, for properly hardened heads, for appropriately dressed cutting edges or striking surfaces, and for properly fitting handles.

15

MATERIALS
HANDLING AND STORAGE

After reading this chapter, you will be able to:

- Describe common materials handling problems and precautions

- Explain safe manual handling methods for lifting and carrying loads and handling specific container shapes

- Describe hazards, safeguards, and operating guidelines for common materials handling equipment

- Establish a training program for workers to ensure that they follow safe practices while moving materials

- Establish a program of regular inspection, repair, and replacement of materials handling equipment and tools

- List the principles and guidelines for safe storage of materials

Materials handling, whether done manually or with mechanical equipment, can be a major source of occupational injuries. Safe materials handling off the job is also important.

This chapter presents basic materials handling hazards and safeguards to help prevent injuries and destruction of property. The topics covered include materials handling problems; manual handling methods; equipment used for materials handling; inspection of ropes, chains, and slings; and safe storage of materials. The supervisor must know how to instruct workers to move materials safely.

MATERIALS HANDLING PROBLEMS

Manual handling of materials accounts for an estimated 25% of all occupational injuries. These injuries are not limited to the shipping department or warehouse, but come from all operations, because it is impossible to run a business without moving or handling materials. Moreover, such injuries are not limited to back strains, but can affect legs, feet, arms, fingers, hands, and neck. Given the cost of injuries, the importance of proper materials handling techniques becomes increasingly apparent.

Common injuries workers suffer include strains and sprains, cuts, fractures, and bruises. These are caused primarily by unsafe conditions and practices—improper lifting techniques, carrying too heavy a load, incorrect gripping, back and leg positioning, etc.

Another major cause of materials handling accidents can be traced to poor job design. Ask the following questions about your present operating practices:

- Can the job be re-engineered to eliminate manual handling of materials or minimize reaching, bending and twisting (Figure 15–1)
- Can employees be given mechanical material handling aids—or properly sized boxes, containers, powered industrial trucks, or hoists—that will make their jobs safer (Figure 15–2)
- Will personal protective clothing, or other equipment, help to prevent injuries?

These are not the only questions that might be asked, but they serve as a start for an overall appraisal. It is beneficial for everyone to adhere to safe lifting and handling techniques. These methods are discussed in the next section.

MANUAL HANDLING METHODS

Since most injuries occur to backs, fingers, and hands, people need to be taught how to pick up and

Figure 15–1. Manual handling can be engineered out of many operations by using the appropriate mechanical equipment.

Figure 15–2. A drum hand truck minimizes strain in handling and moving drums. (Courtesy Liftomatic Material Handling, Inc.)

put down heavy, bulky, or long objects. Some general precautions are in order.

1. Inspect materials for slivers, jagged edges, burrs, rough or slippery surfaces.
2. Get a firm grip on the object.
3. Keep fingers away from pinch points, especially when putting materials down.
4. When handling lumber, pipe, or other long objects, keep hands away from the ends to prevent them from being pinched.
5. Wipe off greasy, wet, slippery, or dirty objects before trying to handle them.
6. Keep hands free of oil and grease.

In most cases, gloves, hand leathers, or other hand protectors can be worn to prevent hand injuries. In other cases, handles, holders, or other material handling-assisted aids can be attached to the objects themselves, for example, handles or straps for moving auto batteries, wheeled carts, trays, or baskets for carrying laboratory samples.

Workers receive a large share of injuries to feet, legs, and toes while handling materials. One of the best ways workers can avoid such injuries is to wear footwear with integral protective toes and metatarsal guards.

The eyes, head, and trunk of the body are also vulnerable to injury. Especially when removing steel or plastic strapping from boxes, containers, piping and bales, make sure to follow instructions on their safe removal and at the same time employees should wear eye protection as well as stout gloves. Care should be taken to prevent strapping ends from flying loose and striking the face or body. The same precaution applies to coiled steel, wire, piping, lumber, or cable. If a material is in bags and the likelihood of airborne dust or material is created or present, the person handling it should consider the use of a respirator or other suitable personal protective equipment. Workers also need training in handling heavy objects. See the directions below for lifting and carrying.

Lifting and Carrying

Obviously, the best means to reduce back injuries is to try to eliminate manual lifting. If this cannot be done, another way is to reduce exposure. This can be achieved by cutting weight loads, using mechanical handling aids, rearranging or redesigning the workplace or asking coworkers for help. Basic rules for manual lifting include:

1. Never let workers overexert themselves when lifting. If the load is thought to be more than one person can handle, assign another person to the job.
2. Lift gradually, without jerking, to minimize the effects of acceleration.

3. Keep the load close to the body.
4. Lift without twisting the body.
5. Follow the safe lifting procedures described below.

In reference to the safe lifting procedure, remember that some researchers working in the area of safe lifting now feel that it is better to let workers choose the lifting position most comfortable for them.

Training for Safe Lifting Practices

Numerous attempts have been made to train materials handlers to do their work, particularly lifting, in a safe manner. Unfortunately, hopes for significant and lasting reductions of overexertion injuries through the use of training have been generally disappointing. There are several reasons for the disappointing results:

- If the job requirements are stressful, behavioral modification alone will not eliminate the inherent hazard. Designing a safe job is still necessary.
- People tend to revert to previous habits and customs if practices to replace previous ones are not reinforced and refreshed periodically.
- Emergency situations, the unusual case, the sudden, quick movement, increased body weight, or impaired physical well-being/nutrition may overly strain the body, since training does not include these conditions.

Training for safe materials handling (which is not limited to lifting) should not be expected to solve the problem of job overload, lack of proper equipment, etc. If properly applied and periodically reinforced, training and behavior modification can and do help to alleviate some aspects of the basic problem.

Originally, the straight back/bent knees lifting posture was recommended. However, frequency and intensity of back injuries was not reduced during the last 40 years while this lifting method has been taught. Biomechanical and physiological research has shown that leg muscles used in this lifting technique do not always have the needed strength. Also, awkward and stressful postures may be assumed if this technique is applied when the object is bulky, for example. Hence, the straight back/bent knees action evolved into the "kinetic" lift. The back is kept mostly straight and the knees are bent, but the positions of the feet, chin, arms, hands, and torso are prescribed. Another variant is the "free style" lift, which may be better for male (but not female) workers than the straight back/bent knee technique. It appears that no single lifting method is best for all situations.

Training of proper lifting techniques is an unsettled issue. It is unclear what exactly should be taught, who should be taught, how and how often a technique should be taught. This uncertainty concerns both objectives and methods, as well as expected results. Claims about effectiveness of one technique or another are frequent but are usually unsupported by convincing evidence.

A thorough review of existing literature indicates that the issue of training for prevention of back injuries in manual materials handling is confused at best. In fact, training may not be effective in injury prevention, or its effect may be so uncertain and inconsistent that the money and effort paid for training programs might be better spent on research and implementation of techniques for worker selection and ergonomic job design. Nevertheless, according to the National Institute for Occupational Safety and Health (NIOSH):

> The importance of training in manual materials handling in reducing hazard is generally accepted. The lacking ingredient is largely a definition of what the training should be and how this early experience can be given to a new worker without harm. The value of any training program is open to question as there appear to have been no controlled studies showing a consequent drop in the manual material handling (MMH) accident rate or the back injury rate. Yet so long as it is a legal duty for employers to provide such training or for as long as the employer is liable to a claim of negligence for failing to train workers in safe methods of MMH, the practice is likely to continue despite the lack of evidence to support it. Meanwhile, it may be worth considering what improvements can be made to existing training techniques. (NIOSH, 1981, p. 99.)

Rules for Lifting. There are no comprehensive and sure-fire rules for "safe" lifting. Manual materials handling is a very complex combination of moving body segments, changing joint angles, tightening muscles, and loading the spinal column. The following DOs and DO NOTs apply, however:

- DO design out manual lifting and lowering in tasks and workplace. If it must be done by a worker, perform it between knuckle and shoulder height.
- DO be in good physical shape. If you are not accustomed to lifting and vigorous exercise, do not attempt to do difficult lifting or lowering tasks.
- DO think before acting. Place material conveniently within reach. Have handling aids available. Make sure sufficient space is cleared.
- DO get a good grip on the load. Test the weight and balance before trying to move it. If it is too bulky or heavy, get a mechanical lifting aid or somebody

else to help, or both.

- DO get the load close to the body. Place feet close to the load. Stand in a stable position with feet pointing in the direction of movement. Lift mostly by straightening the legs.
- DO NOT twist the back or bend sideways.
- DO NOT lift or lower awkwardly.
- DO NOT hesitate to get mechanical help or help from another person.
- DO NOT lift or lower with the arms extended.
- DO NOT continue lifting when the load is too heavy.
- DO try to avoid lifting above your shoulders and below your knees.

Handling Specific Shapes

Boxes, Cartons, and Crates. These are best handled by grasping at alternate top and bottom corners and drawing one corner between the legs. Any box, carton, or crate that appears either too large or too heavy for one person to handle should be handled by two people or with mechanical handling equipment, if possible.

When two people handle a crate, box, or carton, it is preferable that they be nearly the same size so both carry an equal portion of the load. They must be able to see in the direction they are walking and be aware of pedestrians, industrial trucks, machines, walls, and equipment.

Bags or sacks are grasped in the same manner as boxes. If a sack is to be raised to shoulder height, it should be raised to waist height first, and rested against the belly or hip before it is lifted to the shoulder. It should rest on its side. If two people are to lift a large bag or sack, they should be nearly the same size. This makes the job easier and keeps the load balanced. The two should lift at the same time, on an agreed signal.

Barrels and Drums. Workers need special training to handle barrels and drums safely. If two people are assigned to upend a full drum, they should use the following procedure:

- Stand on opposite sides of the drum and face one another.
- Grasp both edges near their high points. Lift one end; press down on the other.
- As the drum is upended and brought to balance on the bottom edge, release the grip on the bottom chime and straighten it up with the drum.

When two people are to overturn a full drum, they should use this procedure:

- Make sure there is enough room. Cramped quarters

can result in badly injured hands.

- Stand near one another, facing the drum. Grip the closest point of the top edge with both hands. Rest palms against the side of the drum, and push until the drum balances on the lower edge.
- Step forward a short distance, and release one hand from the top chime in order to grip the bottom chime. Ease the drum down to a horizontal position until it rests solidly on its side.

If one person is to overturn a drum, he or she should:

- Make sure there is enough room.
- Stand in front of the drum, reach over it, and grasp the far side of the top chime with both hands. (A short person can grasp the near side of the chime, if this is easier.) If the drum is tight against a wall or against other drums, pull on the edge with one hand and push against the wall (or other drum) with the other hand for additional control.
- Pull the top of the drum toward you, until it is balanced on the edge of the lower chime.
- Transfer both hands to the near side of the top chime. Keep the hands far enough apart to avoid being pinched when the drum touches the floor.
- Lower the drum. Keep the back straight, inclined as necessary. Bend the legs so that the leg muscles take the strain.

If one person is to upend a drum, reverse this procedure. Actually, if one person must handle a drum, he or she should have a lifter bar that hooks over the chime and gives powerful leverage and excellent control. Barrel and drum lifters are commercially available. To roll a barrel or drum, push against the sides with the hands. To change direction of the roll, grip the chime. No one should kick a drum with the feet.

To lower a drum or barrel down a skid, turn it and slide it on end. Do not roll it. To raise a drum or barrel up a skid, two workers should stand on opposite sides of the skid (outside the rails, not inside, and not below the object being raised). Then, roll the object up the incline.

Handling drums and barrels can be hazardous, even when using utmost care. Special handling equipment and tools (approved by the safety professional or safety department) is recommended and should be available to make this job safer and easier (Figure 15–2).

Long objects, like ladders, lumber, or pipe, should be carried over the shoulder. The front end should be held as high as possible to prevent its striking other employees, especially when turning corners. When two or more people carry an object, they should place it on the same shoulder, respectively, and walk in step.

MATERIALS HANDLING EQUIPMENT

Hand Tools

Many hand tools are available for specific jobs and for specific materials. They should be used only for their designated purpose. Further, hand tools should always be kept in good condition for best results and for safety reasons. For instance, if a chisel is used, the blade should be sharp and free of burrs. A mushroom-headed chisel should be repaired or discarded. (Refer to Chapter 14, Hand Tools and Portable Power Tools, for more details.)

Crowbars are probably the most common hand tools used in materials handling. Select the proper kind and size for the job. Because the bar can slip, people should never work astride it. They should position themselves to avoid being pinched or crushed if the bar slips or the object moves suddenly. The bar should have a good "bite" and not be dull or broken. When not in use, crowbars should be hung on a rack or placed so they cannot fall on or trip workers.

Special rollers are designed for moving heavy or bulky objects. Workers should be careful not to crush fingers or toes between the rollers and the floor. They should use a sledge or bar, not their hands and feet, to change the direction of the object. They should avoid unnecessary turns by making sure the rollers are properly placed under the equipment before moving.

Handles of tongs and pliers should be offset so hands and fingers will not be pinched.

Hand hooks used for handling materials should be kept sharp so they will not slip when applied to a box or other object. The handle should be strong, securely attached, and shaped to fit the hand. The handle and the point of long hooks should be bent on the same plane so the bar will lie flat and not be a tripping hazard. The hook point should be shielded when not in use.

Dipping a shovel into a pail of water occasionally will help to keep it free from sticky material. Greasing or waxing the shovel blade will also prevent some kinds of material from sticking. When not in use, shovels should be secured upright against a wall or kept in racks or boxes.

Nonpowered Hand Trucks

Two-Wheeled Manual Hand Trucks. These hand trucks look as though they would be easy to handle, but workers must follow several safe procedures:

1. Keep the center of gravity of the load as low as possible. Place heavy objects below lighter objects.
2. Balance the load so the weight will be carried by the axle, not by the handles.
3. Position the load so it will not slip, shift, or fall. Load only to a height that will allow a clear view ahead.
4. Let the truck carry the load. The operator should only balance and push.
5. Never walk backwards with a hand truck.
6. When going up or down an incline, keep the truck in front of you.
7. Move the truck at a safe walking pace. Do not run.
8. Keep the truck under control at all times.

A truck that is designed for a specific purpose should be used only for that purpose. A curved bed truck should be used for handling drums or other circular materials, where a horizontal platform truck should be used. Except for handling acetylene or compressed gas cylinders, foot brakes can be installed on wheels of two-wheeled trucks so operators need not place their feet on the wheel or axle to hold the truck. Handles should have knuckle guards, unless they are narrower than the distance between the outside of the wheel axle or the chisel, whichever is greater.

Manually Operated Four-Wheeled Hand Trucks. Operating rules for these vehicles are similar to those for two-wheeled trucks. Extra emphasis should be placed on proper loading, however. Four-wheeled trucks should be evenly loaded to prevent tipping. These trucks should be pushed rather than pulled, except for those with a fifth wheel and a handle for pulling.

Trucks should not be loaded so high that operators cannot see where they are going. If there are high racks on the truck, two persons should move the vehicle. One should guide the front end, the other the back end. Handles should be placed at protected places on the racks or truck body so that passing traffic, walls, or other objects will not crush or scrape the operator's hands. Truck contents should be arranged so they will not fall or be damaged if the truck or the load is bumped.

General Precautions. Truckers should be warned of four major hazards: (1) running wheels off bridge/dock plates or dock platforms; (2) colliding with other trucks or obstructions; (3) jamming hands between the truck and other objects; and (4) running wheels over feet.

When not in use, trucks should be parked in a designated area, not in aisles or other places where they constitute tripping hazards or traffic obstruction. Trucks with drawbar handles should be parked with handles up and out of the way. Two-wheeled trucks should be stored on the chisel with handles leaning against a wall or the next truck.

Powered Hand Trucks

Only trained and authorized operators shall be permitted to operate powered industrial trucks. Use of powered hand trucks, controlled by a walking operator, is increasing. Principal hazards are (1) operator being caught between the truck and another object, and (2) possible collisions with objects or people.

The truck should be equipped with a "dead-man" control, wheel guards, and ignition key that can be taken out when the operator leaves the truck. Personnel must be trained not to use trucks unless authorized. Training should follow instructions included in the truck manufacturer's manual. General instructions include:

1. Do not operate the truck with wet or greasy hands.
2. Lead the truck from right or left of the handle. Face the direction of travel and keep one hand on the handle.
3. Powered hand trucks must enter an elevator or other confined areas with load end forward.
4. Always give pedestrians the right of way.
5. Stop and proceed with caution at blind corners, doorways, and aisle intersections to prevent collisions.
6. Never operate the truck faster than normal walking pace.
7. Handle flammable or corrosive liquids only when they are in approved containers.
8. Never ride on the truck, unless it is specifically designed for the driver to ride.
9. Never permit others to ride on the truck.
10. Do not indulge in horseplay.
11. Report immediately to your supervisor damage or any unsafe conditions.

Powered Industrial Trucks

To comply with OSHA 29 *CFR* 1910.178, only trained and authorized operators are permitted to operate powered industrial trucks (Figure 15–3). Training programs must include safe operating and material handling practices as outlined in the standard. The supervisor must emphasize workers' safety awareness. Trained and authorized drivers should have badges or other clear authorization to drive, which they should display at all times. These steps will help meet OSHA requirements.

Powered trucks have either a battery-powered motor or an internal combustion engine. Trucks should be maintained according to the manufacturers' recommendations. All trucks acquired on or after February 15, 1972, must meet the design and construction requirements established in ANSI/ASME B56.1–1988, *Low Lift and High Lift Trucks*. Modifications and additions that affect capacity and safe operation should not be made by the customer or user without the manufacturer's prior written approval. All nameplates and markings must be accurate, in place, and legible.

Eleven different designations of trucks or tractors are authorized for use in various locations (*Powered Industrial Trucks, Including Type Designations, Areas of Use, Maintenance, and Operation*, ANSI/NFPA 505–1987).

Battery-Charging Installations. This equipment must be located in areas designated for that specific purpose. The company must provide facilities for flushing and neutralizing spilled electrolyte, fire protection, and barriers to prevent trucks from damaging the charging apparatus. There must also be adequate ventilation for dispersal of gases from gassing batteries. Eyewash stations are necessary if workers are adding water to cells. Racks used to support batteries must be made of spark-resistant materials or be coated or covered to prevent spark generation.

A conveyor, overhead hoist, or equivalent equipment must be used for handling batteries. Reinstalled batteries must be properly positioned and secured in the truck. Use a carboy tilter or siphon for handling

Figure 15–3. This truck enables the operator to ride the vehicle while transporting a load over long distances. (Courtesy Clark Equipment Company.)

electrolyte. Teach workers that acid must always be poured into water; water must never be poured into acid—it overheats and splatters. During charging operations, vent caps must be kept in place to avoid electrolyte spray. Make sure vent caps are functioning. Battery or compartment cover or covers must be open to dissipate heat.

Take all precautions to prevent open flames, sparks, or electric arcs in battery-charging areas. Tools and other metallic objects must be kept away from uncovered battery terminals. Employees charging or changing batteries must be authorized to do this work, trained in proper handling techniques, and required to wear protective clothing, including face shields, long sleeves, rubber boots, aprons, and gloves. Smoking is prohibited in the charging area.

Refueling. All internal combustion engines must be turned off before refueling. Refueling should be in the open or in specifically designated areas where ventilation is adequate to carry away fuel vapors. Liquid fuels, such as gasoline and diesel fuel, must be handled and stored in accordance with the National Fire Protection Association's *Flammable and Combustible Liquids Code*, NFPA 30. Liquefied petroleum gas fuel (or LP-gas) must be stored in accordance with *Storage and Handling of Liquefied Petroleum Gases*, NFPA 58. LP-gas tanks must be secured with both straps while on the truck so as not to shake loose and possibly catch fire. Install NO SMOKING signs in prominent places. Smoking must be strictly prohibited in the service areas.

Train employees to follow general rules for driving powered industrial trucks in your operations. For each job, you should draw up a specific set of rules (referenced from OSHA 29 *CFR* 1910.178).

1. All traffic regulations must be observed, especially plant speed limits.
2. Safe distances must be maintained approximately three truck lengths from the truck in front. Trucks must be kept under the operator's control at all times.
3. The right of way must be yielded to ambulances, fire trucks, and other vehicles in emergency situations.
4. Other trucks traveling in the same direction must not be passed at intersections, blind spots, or other dangerous locations.
5. Drivers are required to slow down and proceed with caution, sound horn at crosswalks, aisles and other locations where vision is obstructed or when approaching pedestrians. If the forward view is obstructed because of the load being carried, the load should be adjusted to provide clear view of travel.
6. Railroad tracks must be crossed diagonally and

drivers must never park within 8 ft (2.5 m) of the center of the nearest railroad track bed.

7. Drivers are required to keep their eyes on the direction of travel and to have a clear view of the path ahead at all times. They should never back up without looking. Be especially careful on loading docks (Figure 15–4).
8. Grades shall be ascended or descended slowly. For grades in excess of 10%, loaded trucks will be driven with the load upgrade. On all grades, load and load-engaging means are to be tilted back if applicable and raised only as far as necessary to clear the floor or road surface.
9. Trucks must always be operated at a speed that will permit them to stop safely. Drivers are required to slow down on wet and slippery floors. Stunt driving and horseplay are not tolerated.
10. Dockboards or bridge plates are to be driven over carefully and slowly and only after they have been properly secured. Their rated weight capacity must never be exceeded.
11. Elevators shall be approached slowly and then entered squarely after the elevator car is properly leveled. Once on the elevator, the controls should be neutralized, power shut off, and the brakes set.
12. Never run over loose objects on the roadway surface.
13. While negotiating turns, reduce speed to a safe level and turn the hand steering wheel in a smooth, sweeping motion. When maneuvering at a low speed, the hand steering wheel must be turned at a moderate and even rate.
14. Only stable or safely balanced loads shall be handled, and extra caution must be exercised when handling off-center loads.
15. Only loads within the rated load capacity of the truck shall be handled, and long or high (including multiple-tiered) loads that may affect capacity must be adjusted.
16. Load-engaging means must be placed under the load as far as possible; the mast is then carefully tilted backward to stabilize the load.
17. Extreme care must be used when tilting the load forward or backward, particularly when placing items on high tiers. Tilting forward with the load-engaging means elevated is not permitted except to pick up a load. Elevated loads must not be tilted forward, except when the load is in a deposit position over a rack or stack. When stacking or tiering, use only enough backward tilt to stabilize the load.
18. When operating in close quarters, keep hands out of the way so they cannot be pinched between steering controls and projecting stationary objects. Keep legs and feet inside the guard or the operating stations of the truck.

Figure 15–4. Forklift truck operators should be well trained in safe operating procedures, including keeping eyes in the direction of travel, having a clear view of the path ahead, and assuring that the dock plate is secured in place. (Courtesy Bader Rutter.)

19. Do not use the reverse control on electric trucks for braking.
20. Park trucks only in designated areas—never in an aisle or doorway or where they will obstruct equipment or material. Fully lower the load-engaging means flat on the floor surface, neutralize the controls, shut off the power, and set the brake and wheel block when parking on an incline.

Research on safety belts in power trucks conducted during 1982 and 1983 convinced some manufacturers that safety belts should be installed for the protection of the drivers. Seatbelts should be installed and used when sit-down type trucks are operated outdoors, on uneven surfaces, near loading docks, and where there is a possibility of tipping. In tests, it was found that if a truck tipped over, the driver was safer sitting on the seat rather than attempting to jump clear. A driver jumping clear could be struck by part of the truck body or overhead guard, and be seriously injured or killed.

Depending on its use, a power truck should have a dead-man control and be equipped with guards, such as a hand-enclosure guard or overhead guard. A lift truck should also have upper and lower limit switches to prevent overtravel. No powered industrial truck should be used for any purpose other than the one for which it was designed.

A forklift should not be used to elevate employees (for example, in servicing light fixtures or when stacking materials) unless a safety platform meeting OSHA requirements, with standard railing and toeboards, is fastened securely to the forks (Figure 15–5). Failure to use safety platforms has resulted in many serious accidents.

Combustion By-Products. When internal-combustion engine trucks are operated in enclosures, the concentration of carbon monoxide should not exceed limits specified by local or state laws. In no case should the time-weighted average concentration exceed 35 parts per million (ppm) for an 8-hour exposure. The atmosphere must contain a minimum of 19% oxygen, by volume—air usually contains 20.8% oxygen, by volume. Many companies equip their fuel-powered trucks with catalytic exhaust purifiers to

Figure 15–5. When a forklift is used to elevate employees, a safety platform enclosed with standard railing and toeboards must be securely fastened to the forklift.

burn the carbon monoxide before it reaches the atmosphere. Some power trucks are approved and marked only for use in specific locations. These include areas where flammable gases or vapors are present in quantities sufficient to produce explosive or ignitable mixtures, or areas where combustible dust or easily ignitable fibers or flyings are present. (Refer to ANSI/NFPA 505–1987, *Powered Industrial Trucks*, for classification of types.)

Loads. Operators should not carry unstable loads. If material of irregular shape must be transported, make sure it is securely placed and blocked, tied, or otherwise strapped down. If possible, crosstie the load after neatly stacking it. Never place an extra counterbalance on a forklift to help handle overloads because it puts added strain on the truck and endangers the

operator and nearby employees. It is difficult to operate an overloaded truck safely.

Many stand-up-type electric trucks are being used in industry. They vary in design and operation, and workers must be trained in the operating procedures for each one they use.

Operators of powered industrial trucks should not drive up to anyone standing in front of a wall or other fixed object, or allow anyone to stand or pass under the elevated portion of any truck, whether loaded or empty. It is the operator's responsibility to protect pedestrians. Operators should not expect pedestrians to see them. Unauthorized personnel are not permitted to ride on trucks; even if authorized, they are permitted only when they are in the space provided for a passenger. No one should be allowed to place arms or legs between uprights of the mast or outside the running lanes of the truck.

Raise or lower loads at the point of loading or unloading, not during travel. Operators should be trained to check the area around the truck before raising a load, so they will not strike structural members of the building, electric wiring or cables, or piping (especially ones carrying gases or flammable fluids).

Operators must make sure there is sufficient headroom under overhead installations, such as lights, pipes, and sprinkler systems, before driving underneath them. Overhead guards must be used on lift trucks as protection against falling objects. An overhead guard is intended to offer protection from the impact of small packages, boxes, bagged materials, and so forth, but not to withstand the impact of a falling capacity load. Load backrest extensions must be used whenever necessary to minimize the possibility of the load or part of the load falling toward the rear.

Operators must make sure they check decking of semi trucks before entering with a lift truck. Floors on which power trucks travel should be kept in peak condition. They should have a load capacity great enough to support both truck and load. The combined weight of the heaviest truck and load should not exceed one-fifth of the design strength of the floor. For example, a floor originally designed to bear 1,000 pounds per square foot (psf)—or 5,000 kg/m²—should not be allowed to bear more than 200 pounds per square foot. Maximum floor loadings should be prominently displayed (ANSI standard A58.1–1982, *Minimum Design Loads for Buildings and Other Structures*).

The 5 to 1 ratio, although not excessively conservative, is important in establishing a margin of safety as protection against possible miscalculations.

Dock Plates

Dock plates (bridge plates) that are not built in should be secured to prevent them from "walking" or sliding when used.

Secure by drilling holes and dropping pins or bolts through them.

Plates should be large and strong enough to hold the heaviest load, including the equipment. A safety factor of five or six should be used. Plates can have the edges turned up, or have angle-iron welded to the edges, to keep trucks from running over their edges. Keep plates in good condition. Make sure they do not have curled corners or bends. They should be kept clean and dry—clear of snow and ice, and free of oil and grease. Canopies can be used to protect them from the weather.

Dock plates should be positioned with a forklift or other mechanical equipment to avoid back injuries

and pinched fingers. The dock plate must be kept under control and not allowed to bounce. Handholds or other effective means must be provided on portable dock plates to permit safe handling.

Permanent, adjustable dock levelers are the safest and most efficient method of traversing between the dock and a vehicle. They should be of sufficient capacity to handle present and future loads, and be long enough to reduce operating grades. Dock levelers should be inspected and maintained on a regular basis (Figure 15–6).

Trucks or trailers backed up to the dock must have their brakes set and rear wheels chocked to prevent movement when power trucks are loading or unloading. This would also apply to railroad cars. This should be the lift-truck operator's responsibility. Fixed jacks must be used when the trailer is not coupled to a tractor. When not in use, dock plates should be stored safely to avoid acting as tripping hazards. They should not be placed on edge so they can fall over. For details of loading highway trucks and cars, see later sections in this chapter.

Conveyors

The primary rule regarding conveyors is simple: no one is allowed to ride or climb on or over them. Post prominent warning signs making this rule clear. If people must cross over conveyors, the company should provide bridges or stiles so that they do not have to crawl across or under the conveyor.

In the case of certain roller conveyors, it is possible to provide a hinged section that opens, and permits people to walk through without having to climb over. The hinged section should be returned to its original position as soon as the individual has reached the other side. Tell workers to make sure that no merchandise is on the roller conveyor when they use the hinged section.

Gravity conveyors, usually the chute or roller type, should be equipped with warning or protective devices to prevent workers' hands from being caught in descending material or from being jammed between material and a receiving table. If the conveyor jams up, first try to free it from the top side. If someone must enter the chute to free a jam, confined space entry and lockout procedures should be implemented. Adhere to OSHA's confined space requirements (29 *CFR* 1910.146). Inform your employees of these hazards and train them to handle such hazards safely.

Pinch points on conveyors—gears, chain drives, and revolving shafts—should be guarded. Emergency stop systems, such as tripwire cables, can be installed. Tell workers not to remove these guards. Should the conveyor need servicing, all energy sources should be isolated and locked out.

Figure 15–6. This hydraulic dock leveler prevents forklifts and other vehicles from rolling or being driven off the dock when the truck or trailer is not present. (Courtesy Rite Hite Corp.)

Pneumatic or blower conveyors must be shut off and locked out before inspections or repairs are made. Inspection ports, equipped with transparent coverings made of durable, nonflammable, shatter-resistant material, permit inspections without stopping the operation.

Screw conveyors should be completely enclosed unless otherwise effectively guarded. Their principal danger is that workers may try to dislodge material or free a jam with their hands or feet, and be caught in the conveyor. Screw conveyors should be repaired only when all energy sources have been isolated and locked out and OSHA's 29 *CFR* 1910.147 lockout/ tagout requirements are met. Any exposed sections should be covered with a metal grate with openings no larger than $1/2$ in. (13 mm), or with a solid cover— ANSI/ASME standard B20.1–1987, *Safety Standards for Conveyors and Related Equipment.* The cover should be strong enough to withstand abuse. Before removing any cover, grating, or guard, implement lockout procedures. Do not rely on interlocks. Never step or walk on covers, gratings, or guards.

Cranes

All overhead and gantry cranes must meet the design specifications of the ANSI/ASME B30.17–1985, *Overhead and Gantry Cranes,* and OSHA 29 *CFR* 1910.129; all mobile, locomotive, and truck cranes must meet ANSI/ASME B30.5–1989, *Mobile and Locomotive Truck Cranes* (OSHA 29 *CFR* 1910.180).

Although many types of cranes are used in industry, safe operating procedures are much the same. Standard signals (Figures 15–7 and 15–8) should be thoroughly understood by both operator and signaler. Assign only one signaler for each crane, and tell the crane operator to obey only this person's instructions. People working with or near a crane should keep out from under the load, be alert at all times, and watch/ listen to warning signals, or combination of flashing lights, etc., closely. When a warning signal sounds, they must move to a safe place immediately.

No crane should be loaded beyond its rated load capacity. The weight of all auxiliary handling devices, such as hoist blocks, hooks, and slings, or other attachments must be considered as part of the load.

Substantial and durable rating charts with clearly legible letters and figures are to be fixed to the crane cab in a location clearly visible to the operator while seated at the control station.

Hoist chain or hoist rope must be free of kinks and twists and must not be wrapped around the load. Loads must be attached to the load line hook by means of a sling or other approved device. Care must be taken to make certain that the sling clears all obstacles. The load must be well secured, and properly balanced in the sling or lifting device, before it is lifted more than a few inches. Before starting to hoist, make sure that multiple-part lines are not twisted around each other.

Hooks should be brought over the load slowly to prevent swinging. During hoisting operations, there must be no sudden acceleration or deceleration of the moving load. The load must not come in contact with any obstruction.

Cranes must not be used for side pulls, unless specifically authorized by a responsible person who can determine that the stability and integrity of the crane is not endangered, and that the various parts of the crane will not be overstressed. There must be no hoisting, lowering, or traveling while any employee is on the load or hook. The operator must not carry loads over people's heads. On overhead cranes, a warning signal must be given when starting the bridge and when the load or hook comes close to, or is about to pass over, any person's head so that personnel can get out of the way.

Brakes must be tested each time a load approaching the rated load is handled. Loads must not be lowered below the point where less than two full wraps of rope remain on the hoisting drum. When two or more cranes are used to lift a load, only one qualified, responsible person must be in charge of the operation. He or she must analyze the operation, and instruct all personnel involved about the proper positioning and rigging of the load, and all other movements. An operator must not leave the controls while the load is suspended. All necessary clothing and personal belongings that are stored in an operator's cab must be stored so as not to interfere with access or operation. Tools, oil cans, waste, extra fuses, and other necessary articles must be kept in a toolbox and not carelessly left in or around the cab. A portable fire extinguisher must be kept in the cab. Access to the cab must be by fixed ladder, stairs, or platform, but no steps over gaps exceeding 12 in. (30 cm) are allowed [OSHA *CFR* 1910.179, (c) Cabs, (2) Access to crane].

The upper load limit switch of each hoist must be tried out under no-load conditions, at the beginning of each operator's shift. If the switch does not operate properly, a qualified person must be notified immediately. The hoist limit switch that controls the upper limit of travel of the load block must never be used as an operating control.

Inspections. Make sure that cranes in regular service are inspected periodically. The intervals depend on the nature of the crane's critical components and the degree of its exposure to deterioration or malfunction. Daily inspections must be made of all control mechanisms for maladjustments, including deterioration or leakage in air or hydraulic systems.

Frequent inspections at daily-to-monthly intervals, or as specifically recommended by the manufacturer, must be conducted of the following items:

- All control mechanisms for excessive wear of components or contamination by lubricants or other foreign matter
- All safety devices for malfunction
- Crane hooks for deformations, cracks, or more than 15 % in excess of normal throat opening or more than a 10-degree twist from the plane of the unbent hook, and spring-loaded safety latch.
- Rope reeving not in compliance with manufacturer's recommendations
- Electrical apparatus for malfunctioning and excessive deterioration, dirt, and moisture accumulation

Periodic inspections, from one- to twelve-month intervals, or as recommended by the manufacturer, cover the complete crane. These inspections include the items listed under frequent inspections as well as:

- Deformed, cracked, or corroded members in crane structure and hook, and spring-loaded safety latch
- Loose bolts or rivets
- Cracked or worn sheaves and drums, OSHA 1910.179J, Inspections, (2) Frequent inspections (iii)
- Worn, cracked, or distorted parts, such as pins, bearings, shafts, gears, rollers, and locking devices
- Excessive wear on brake and clutch system parts linings, pawls, and ratchets
- Load, boom, angle, and other indicators over their full range to detect significant inaccuracies
- Gasoline, diesel, electric, or other power plants for improper performance or noncompliance with safety requirements, excessive wear of chain-drive sprockets, and excessive chain stretch
- Travel steering, braking, and locking devices for malfunction and excessively worn or damaged tires
- Determine if the above items constitute a hazard and, if so, correct them. Accurate records must be kept; see OSHA requirements.

Prior to initial use, all new and altered cranes must be checked periodically and frequently. Cranes not in regular service and idle for one month but less than six months, must meet standards of items outlined for frequent inspections. Cranes idle for six months or

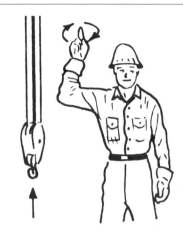

HOIST. With forearm vertical, forefinger pointing up, move hand in small horizontal circle.

LOWER. With arm extended downward, forefinger pointing down, move hand in small horizontal circles.

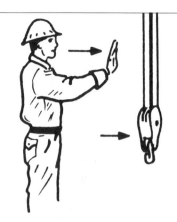

BRIDGE TRAVEL. Arm extended forward, hand open and slightly raised, make pushing motion in direction of travel.

TROLLEY TRAVEL. Palm up, fingers closed, thumb pointing in direction of motion, jerk hand horizontally.

STOP. Arm extended, palm down, move arm back and forth.

EMERGENCY STOP. Both arms extended, palms down, move arms back and forth.

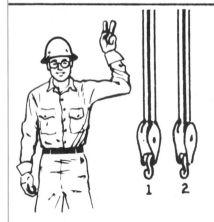

MULTIPLE TROLLEYS. Hold up one finger for block marked ''1'' and two fingers for block marked ''2''. Regular signals follow.

MOVE SLOWLY. Use one hand to give any motion signal and place other hand motionless in front of hand giving the motion signal. (Hoist slowly shown as example.)

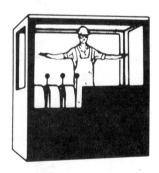

MAGNET IS DISCONNECTED. Crane operator spreads both hands apart palms up.

Figure 15–7. Standard hand signals for controlling operation of overhead and gantry cranes. (Reprinted with permission from ANSI/ASME B30.5-1989.)

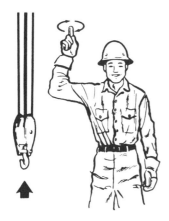

HOIST. With forearm vertical, forefinger pointing up, move hand in small horizontal circle.

LOWER. With arm extended downward, forefinger pointing down, move hand in small horizontal circles.

USE MAIN HOIST. Tap fist on head; then use regular signals.

USE WHIP LINE. (Auxiliary Hoist) Tap elbow with one hand; then use regular signals.

RAISE BOOM. Arm extended, fingers closed, thumb pointing upward.

LOWER BOOM. Arm extended fingers closed, thumb pointing downward.

MOVE SLOWLY. Use one hand to give any motion signal and place other hand motionless in front of hand giving the motion signal. (Hoist Slowly shown as example)

RAISE THE BOOM AND LOWER THE LOAD. With arm extended thumb pointing up, flex fingers in and out as long as load movement is desired.

LOWER THE BOOM AND RAISE THE LOAD. With arm extended, thumb pointing down, flex fingers in and out as long as load movement is desired.

Figure 15–8. Standard hand signals suitable for crawler, locomotive, wheel, and truck-mounted cranes. (Reprinted with permission from ANSI/ASME B30 series.)

more must be inspected in accordance with OSHA requirements (29 *CFR* 1910.179 and 180). Standby cranes must be inspected at least semiannually. Cranes exposed to adverse environmental conditions must be inspected more often.

All ropes must be inspected thoroughly once a month. A full written, dated, and signed report of rope condition must be kept on file where it is readily available to appointed personnel. Any deterioration, resulting in appreciable loss of original strength, must be carefully noted. The rope may constitute a hazard if there is a reduction of rope diameter below normal dimensions caused by loss of core support, internal or external corrosion, or wear or breakage of outside wires; corroded or broken wires at end connections; or severe kinking, crushing, cutting, or unstranding.

All rope not used for a month or more, due to shutdown or storage of a crane on which it is installed, must be given a thorough inspection before being placed in service. This inspection should identify all types of deterioration. It should be performed by a responsible person whose approval is required before the rope is further used. A written, dated report of the rope condition must be available for inspection.

On limited-travel ropes, heavy wear and/or broken wires may occur in sections that come in contact with (1) equalizer sheaves or other sheaves where rope travel is limited, or (2) saddles. Particular care must be taken to inspect all ropes at these locations, with special attention paid to all nonrotating ropes.

Safety Rules. These safety rules for electric overhead cranes are suggested by one crane manufacturer:

1. Only regularly authorized operators should use cranes.
2. When on duty, operators should remain in the crane cabs ready for prompt service.
3. Before traveling, make certain the hook is high enough to clear all obstacles.
4. Under no circumstances should one crane bump another.
5. Examine the crane at every shift for loose or defective gears, keys, runways, railings, warning bells, signs, switches, sweep brushes, ropes, cables, and other parts. Report defects. Keep the crane clean and well lubricated.
6. No one should go to the top of a crane without first opening the main switch, placing a warning sign on it, and locking it out. Operators should unlock the switch and remove the sign promptly when they come down.
7. After completion of a repair job, make sure that bolts, tools, and other materials have been removed to prevent damaging the machinery when the crane is restarted, and to be sure nothing falls off the crane. Keep tools, oil cans, and other loose objects in a box provided for that purpose.
8. Do not carry a load over workers' heads on the floor. Sound a gong or siren when necessary. No one should ride the load or the crane hook.
9. If the power goes off, move controller to OFF position until power is again available.
10. Make sure the fire extinguisher on the crane is in good condition; and if used, that it is refilled immediately.
11. Never let an operator who is ill or not physically fit run a crane.
12. Do not drag slings, chains, or ropes. After the load is taken off, do not move the crane until the hook is lowered and the hook-on person has hooked the chain or rope.
13. If the operator feels it is unsafe to move a load, a check with the supervisor should be made.
14. Be sure that when an operator leaves the cage, he or she leaves the main switch open. A magnetic crane must be empty and its controller turned off.
15. When an outside crane is parked at the end of a shift, the brake should be set or the crane should be chained to the track.
16. Operators should know that, if their crane fails to respond correctly, they should call the supervisor. Attempting to get out of difficulty by repeated operation may make the condition worse, not better.

Railroad Cars

Spotting of railroad cars is usually done by switch-engine crews. When employees must move railroad cars, the supervisor or crew leader should consider the following factors:

- Facilities or equipment available to move the cars
- Distance the cars must be moved
- Number of cars to be moved
- Whether cars are full or empty. If full, the type of load carried
- Whether the track is level or on a grade

After weighing these factors, you can decide whether to get power equipment, move the car by hand, or take material to (or from) the car's location. If possible, use power equipment to move the car.

If a car must be spotted, brakes must be set and wheels blocked when loading and unloading or moved without power equipment, a car mover (operated by only one person) can be used. If it takes two workers to move a car, then each one should have a car mover and they should work on opposite wheels. All hand-actuated car movers should be equipped with knuckle guards. A stationary car puller, if available, should be used.

When opening doors of boxcars, all workers should be alert to the danger of fingers being caught or of freight falling out. A bar or ratchet hoist or door puller should be used. Do not use lift trucks to open or close car doors.

Metal transfer plates must be securely anchored with bolts. Gangplanks and skids must be securely placed before using. Operators of mechanized equipment should avoid bumping doors or door posts. Metal bands should be removed or cut to avoid dangling ends. Stack special equipment, such as DF bars, at the ends of a car when empty, and close and latch doors before moving the car. Specially equipped cars with movable bulkheads have instructions in the cars that must be followed. When bulkhead partitions are difficult to move, examine rollers to determine the reason for binding, to avoid having the bulkhead fall on those working in the car.

Highway Trucks

When loading or unloading highway trucks, set the brakes and place wheel chocks in front of the rear wheels or use dock-mounted vehicle restraints to prevent the trucks from rolling away from the dock (Figure 15–9). With vehicle restraints, trailers are held in place by their underride protection bar, or ICC bar. Unlike wheel chocks, vehicle restraints hold the truck safely to the dock if the landing gear or jack collapses, or if the driver tries to pull away from the dock before loading or unloading is completed. They are also effective in icy and wet environments.

If a powered industrial truck is used inside a trailer that is not coupled to its tractor, place fixed jacks as needed to support the trailer. Make sure jacks are snug against the trailer bottom. Where automated systems are used visual inspection should be conducted to make sure the trailer is secured. Check decking before entering in a lift truck. For more details of truck and truck terminal safety, see the National Safety Council's Motor Fleet Safety Manual.

Motorized Off-Road Equipment

Heavy-duty trucks, mobile cranes, tractors, bulldozers, and other motorized equipment used in the production of stone, ore, and similar materials and also in construction work have been involved in frequent, and often serious, accidents. In general, prevention of these accidents requires:

- Safe equipment
- Systematic maintenance and repairs
- Safety training for operators
- Safety training for repair personnel

- Strict observance of safety procedures, including on-site traffic
- Control of visitors, vendors, contractors and their training

Operation. Manufacturers' manuals contain detailed information on the operation of equipment. Recommended driving practices are similar to those necessary for the safe operation of highway vehicles. Off-road driving involves certain hazards that require special training and safety measures. The modern heavy-duty truck or off-road vehicle is a carefully engineered and expensive piece of equipment and warrants operation only by drivers who are qualified physically and mentally. For safety, driving standards should be especially high.

The time required for prospective drivers or operators to become thoroughly familiar with the mechanical/hydraulic/electrical features of the equipment, safety rules, driver reports, and emergency conditions varies. In no case should even an experienced driver be permitted to operate equipment until the instructor or supervisor is satisfied with her or his abilities. After workers and supervisors have been trained, supervisors have the important responsibility of seeing that drivers continue to operate vehicles as they were instructed.

Equipment. Safe operation of heavy equipment begins with the purchase specifications. A good policy is to specify safeguards over power transfer components, safe oiling devices, handholds, guard rails, and other devices when the order is placed with the manufacturer. Before equipment is put into operation, it should be thoroughly inspected and necessary safety devices installed and checked.

Both operators and repair personnel, whether experienced or not, should know and follow the manufacturer recommendations pertaining to lubrication, adjustments, repairs, and operating practices for equipment. A preventive maintenance program is essential to ensure worker safety and efficiency. Frequent and regular inspections and prompt repairs provide effective preventive maintenance.

In general, the operator is responsible for inspecting such mechanical conditions as hold-down bolts, brakes, clutches, clamps, hooks, and similar vital parts. Wire ropes should be lubricated (see manufacturer's instructions). For inspection guidelines, see the section that follows.

Many of the basic safety measures recommended for trucks also apply to motor graders and other earth-moving equipment. All machines should be inspected regularly by the operator, who should report any defects promptly. Equipment safety and efficiency of equipment are increased by adherence to periodic maintenance.

Figure 15–9. These are two types of vehicle restraints to secure a truck. (Courtesy Rite Hite Corp.)

ROPES, CHAINS, AND SLINGS

Special safety precautions apply to using ropes, rope slings, wire rope, chains and chain slings, and storing chains. You should know thoroughly the properties of these various types and the precautions pertaining to their use and maintenance. For additional help regarding ropes, chains, or slings, consult the supplier.

Fiber Ropes

Fiber rope is used extensively in handling and moving materials. The rope is generally made from manila (abaca), sisal, or nylon. Manila or nylon ropes give the best uniform strength and service. Other types of rope include those made from polyester or polypropylene or a composite of both that is adapable to special uses.

Sisal is not as satisfactory as manila because the strength varies in different grades. Sisal rope is about 80% as strong as manila. Manila rope is yellowish with a somewhat silvery or pearly luster. Sisal rope is also yellowish, but often has a green tinge and lacks the luster of manila. Its fibers tend to splinter, and it is rather stiff. The safe working loads and breaking strengths of the rope that you are using should be known and the limits must be observed.

Maintenance Tips. Precautions should be taken to keep ropes in good condition. Kinking, for example, strains the rope and may overstress the fibers.

It may be difficult to detect a weak spot made by a kink. To prevent a new rope from kinking while it is being uncoiled, first lay the rope coil on the floor with the bottom end down. Then pull the bottom end up through the coil and unwind the rope counterclockwise. If it uncoils in the other direction, turn the coil of rope over and pull the end out on the other side.

Rope should not be stored unless it has first been cleaned. Dirty rope can be hung in loops over a bar or beam and then sprayed with water to remove the dirt. The spray should not be so powerful that it forces the dirt into the fibers. After washing, the ropes should be allowed to dry and then shaken to remove any remaining dirt. Rope must be thoroughly dried out immediately after it becomes wet; otherwise it will deteriorate quickly. A wet rope should be hung up or laid in a loose coil until thoroughly dried. Rope will deteriorate more rapidly if it is frequently alternately wet and dry, than if it remains wet. Rope should not be allowed to freeze.

Store rope in a dry place where air circulates freely. Small ropes can be hung up. Larger ropes can be laid on gratings so air can get underneath and around them. Rope should not be stored or used in an area where the atmosphere contains acid or acid fumes, because it will quickly deteriorate. Signs of deterioration are dark brown or black spots.

Sharp bends should be avoided whenever possible because they create extreme tension in the fibers.

When cinching or tying a rope, be sure the object has a large-enough diameter to prevent the rope from bending sharply. A pad can be placed around sharp corners.

If at all possible, ropes should not be dragged. Dragging abrades the outer fibers. If ropes pick up dirt and sand, abrasion within the lay of the rope will rapidly wear them out. Lengths of rope should be joined by splicing. A properly made short splice will retain approximately 80% of the rope's strength. A knot will retain only about 50. (Details appear in Chapter 9, under Lifelines.)

Ropes should be inspected at least every 30 days; every day if they are used to support scaffolding on which people work. The following procedure will assist in inspection of natural fiber rope:

1. Check for broken fibers and abrasions on the outside.
2. Inspect fibers by untwisting the rope in several places. If the inner yarns are bright, clear, and unspotted, the strength of the rope has been preserved to a large degree.
3. Unwind a piece of yarn 8 in. (20 cm) long and $^{1}/_{4}$ in. (0.6 cm) diameter from the rope. Try to break this with your hands. If the yarn breaks with little effort, the rope is not safe.
4. Inspect rope used around acid or caustic materials daily. If black or rusty brown spots are noted, test the fibers as described in steps 1, 2, and 3, and discard all rope that fails the tests.
5. As a general rule, rope that has lost its pliability or stretch, or in which the fibers have lost their luster and appear dry and brittle, should be viewed with suspicion and replaced. This is particularly important if the rope is used on scaffolding or for hoisting where workers could be seriously injured or property damaged, by a break.

Rope Slings

Because of high tensile strength needed for a sling, fiber rope slings should be made of manila only. The following are some precautions to be observed in using manila-rope slings:

1. Make sure that the sling is in good condition and of sufficient strength. Take into account the factors of the splice and leg angles of the sling. Hooks, rings, and other fittings must be properly spliced.
2. Reduce the load by one-half after the sling has been in use for six months, even though the sling shows no sign of wear.
3. If the sling shows evidence of cuts, excessive wear, or other damage, destroy it.

Whether fiber or wire rope slings are used, certain general precautions should be taken. (These were given earlier under Fiber Ropes—Maintenance Tips.) In addition, it is best to keep the legs of slings as nearly vertical as possible (Figure 15–10).

Wire Ropes

Wire rope is used widely instead of fiber because it has greater strength and durability for the same diameter and weight. Its strength is constant either wet or dry and it maintains a constant length under varying weather conditions. Wire rope is made according to its intended use. A large number of wires give flexibility, while fewer strands with fewer wires are less flexible.

A wire rope used for general hoisting should not be subjected to a working load greater than one-eighth its breaking strength—a safety factor of five. Greater factors of safety (six or seven, for example) are always desirable for extra security. Wire ropes should be thoroughly inspected at least once a month; and written, dated, and signed reports kept on file. OSHA regulations (29 *CFR* 1910.184) spell out the details of this inspection. A tag attached to a sling can identify it and also indicate load capacity.

Wire rope should be lubricated at regular intervals to prevent rust and excessive wear. Sheaves for rope should be as large as possible. The less flexible the rope, the larger the sheave or drum diameter must be, otherwise the rope will be bent too sharply (Table 15–A). Rope should be wound in one layer only. Using several layers will mash or jam the rope and shorten its life.

Sheaves and drums should be aligned as much as possible to prevent excessive wear of rope. Reverse bending of rope, in which rope is bent first in one direction and then in the other, should be avoided, as this wears out wire rope faster than any other abuse. Fittings should be properly selected and attached if they are to be safe. Some of the principal wire rope attachments are illustrated in Figure 15–11, and include:

- Babbitt- or zinc-coated connections
- Wedge sockets
- Swagged attachments
- Thimble-with-clip connections
- Three-bolt clamps
- Spliced eye-and-thimble connections
- Crosby clips

The Crosby clip is probably the most commonly used fitting. It is important that these clips be installed properly. Wedge sockets are dependable and will develop approximately 75 to 90 percent of the strength of the rope. Swagged attachments are satisfactory for small-diameter ($^{1}/_{4}$ to 1 in. or 0.6 to 2.5 cm) ropes.

Wire rope should be spliced only by workers who have been trained to do it properly. Splices should be tested to twice the load they are expected to carry. Again, any questions should be referred to the supplier or manufacturer.

Chains and Chain Slings

Alloy steel chain, approximately twice as strong—size for size—as wrought-iron chain, has become standard material for chain slings. One advantage is that it is suitable for high-temperature operations. Continuous operation at temperatures of 800 F (425 C) (the highest temperature for which continuous operation is recommended) requires a reduction of 30% in the regular working load limit. For intermittent service, these chains can be used at temperatures as high as 1,000 F (540 C), but at only 50% of the regular working load limit (Table 15–B).

An alloy steel chain should never be annealed when the chain is being serviced or repaired. The process reduces the hardness of the steel; and this, in turn, reduces the strength of the chain. Wrought-iron chain, on the other hand, should be annealed periodically by the manufacturer (or persons specially trained in the operation) if it is used where failure might endanger human life or property.

Impact loads, caused by faulty hitches, bumpy crane tracks, and slipping hookups, can add significant stress in the chain. The impact resistance of heat-treated alloy steel chain does not increase in proportion to the strength of the chain. Under a full working load, it will fail before a fully loaded wrought-iron or heat-treated carbon steel chain will.

Chain slings preferably should be purchased complete from the manufacturer. When they need repair, they should be returned to the manufacturer. Query suppliers and manufacturers about details of strength and specifications of the slings.

Table 15–A. Tread Diameters for Sheaves and Drums

Rope Classification	Average Recommended	Minimum
6 x 7	72 times rope diameter	42 times rope diameter
6 x 19	45 times rope diameter	30 times rope diameter
6 x 37	27 times rope diameter	18 times rope diameter
8 x 19	31 times rope diameter	21 times rope diameter

Source: ANSI/ASME A17.1–1987, *Elevators and Escalators.*

Load and Wear. Chains should not be overloaded. The working load limit and break test limit for alloy steel chains are given in Table 15–B. Some of the principal causes of chain failure are:

1. Brittleness caused by cold working of the metal surface
2. Failure of the weld
3. Repeated and severe bending or deformation of the links
4. Metal fatigue
5. Brittleness caused by defects in the metal
6. Tensile failure—full elongation of the links

How much wear a chain can stand is determined by its use, the original factor of safety, and the loads the chain is expected to carry. A regular schedule of checking and calipering the links should be set up for each chain, depending on usage (Table 15–C).

To check elongation, chain should be calipered in gage lengths of 1 to 3 (preferably 5) ft when new, and a record of that test should be kept. The chain (or sling) should be discarded if it shows a stretch (includes link wear) of more than 5%. Some states

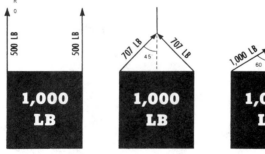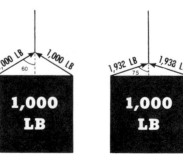

Figure 15–10. Increasing the sling angle (the angle of the sling leg to the vertical) increases the stress on each leg of the sling, even though the load remains constant. The stress on each leg can be calculated by dividing the load weight by the cosine of the angle.

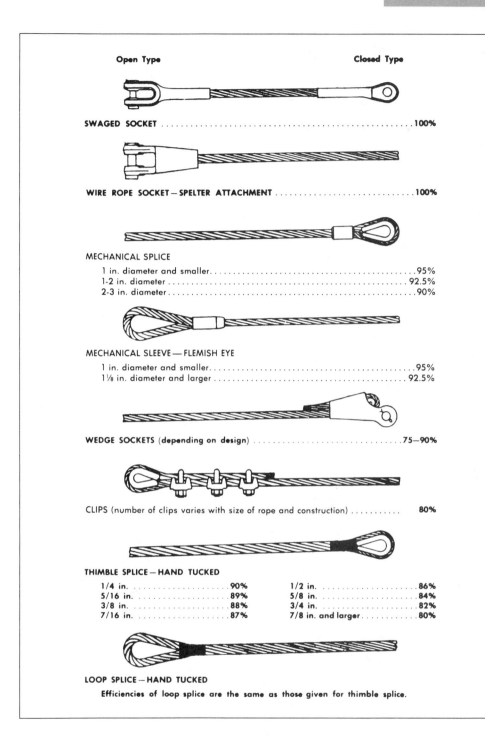

Open Type   Closed Type

SWAGED SOCKET 100%

WIRE ROPE SOCKET — SPELTER ATTACHMENT 100%

MECHANICAL SPLICE
    1 in. diameter and smaller. 95%
    1-2 in. diameter 92.5%
    2-3 in. diameter 90%

MECHANICAL SLEEVE — FLEMISH EYE
    1 in. diameter and smaller. 95%
    1⅛ in. diameter and larger 92.5%

WEDGE SOCKETS (depending on design) . . . . .75—90%

CLIPS (number of clips varies with size of rope and construction) **80%**

THIMBLE SPLICE — HAND TUCKED
    1/4 in. . . . . .90%   1/2 in. 86%
    5/16 in. . . . . .89%   5/8 in. 84%
    3/8 in. . . . . .88%   3/4 in. 82%
    7/16 in. . . . . .87%   7/8 in. and larger 80%

LOOP SPLICE — HAND TUCKED
    Efficiencies of loop splice are the same as those given for thimble splice.

specify that a chain must be destroyed or discarded when any 3-ft (0.9-m) length is found to have been stretched one-third of a link length.

A broken chain should never be spliced by inserting a bolt between two links. Nor should one link be passed through another and a nail or bolt inserted to hold them.

Use. Before making a lift, workers should check the chain to be sure it has no kinks, knots, or twists.

Workers should lift the load gradually and uniformly and make certain it is not merely attached to the top of the hook. They should never hammer a link over the hook as this will stretch the hook and the links of the chain. When lowering a crane load, brakes should be applied gradually to bring the load to a smooth stop.

Tagging. Each chain should be tagged to indicate its load capacity, date of last inspection, date of

Table 15–B. Working Load Limits and Break Test Limits for Alloy Steel Chain

Nominal Size of Chain Bar in Inches	Working Load Limits in Pounds	Minimum Break Test in Pounds
¼	3,250	10,000
³⁄₈	6,600	19,000
½	11,250	32,500
⁵⁄₈	16,500	50,000
¾	23,000	69,500
⁷⁄₈	28,750	93,500
1	38,750	122,000
1¹⁄₈	44,500	143,000
1¼	57,500	180,000
1³⁄₈	67,000	207,000
1½	80,000	244,000
1¾	100,000	325,000

1 in. = 2.54 cm; 1 kg = 2.5 lb

Source: *Specification for Alloy Steel Chain*, American Society for Testing and Materials, A391–1975. The strengths for wrought-iron chain are slightly less than half these figures.

Table 15–C. Correction Table for the Reduction of Working Load Limits of Chain Due to Wear

Reduce working load limits of chain by the following percent when diameter of stock at worn section is as follows:

Nominal Original Chain Size, Inches		5 Percent	10 Percent	Remove from Service
¼	0.250.........	0.244	0.237	0.233
⁵⁄₈	0.375.........	0.366	0.356	0.335
½	0.500.........	0.487	0.474	0.448
⁵⁄₈	0.625.........	0.609	0.593	0.559
¾	0.750.........	0.731	0.711	0.671
⁷⁄₈	0.875.........	0.853	0.830	0.783
1	1.000.........	0.975	0.949	0.895
1¹⁄₈	1.125.........	1.100	1.070	1.010
1¼	1.250.........	1.220	1.190	1.120
1³⁄₈	1.375.........	1.340	1.310	1.230
1½	1.500.........	1.460	1.430	1.340
1⁵⁄₈	1.625.........	1.590	1.540	1.450
1¾	1.750.........	1.710	1.660	1.570
1⁷⁄₈	1.875.........	1.830	1.780	1.680
2	2.000.........	1.950	1.900	1.790

1 in. = 2.54 cm

Source: ANSI/ASME B30.17–1985, *Overhead and Gantry Cranes (Top Running Bridge, Single Girder, Underhung Hoist)*.

purchase, and type of material. This information must not be stamped on the lifting links, however, as stamping can cause stress points that weaken the chain (Figure 15–12).

Chain attachments (rings, shackles, couplings, and end links) should be made of the same material to which they are fastened. Hooks should be made of forged or laminated steel and should be equipped with safety latches. Replace hooks that have been overloaded, loaded on the tips, or have a permanent set greater than 15% of the normal throat opening.

Storing Chains. Proper chain storage serves two purposes. It preserves the chain and promotes safe housekeeping. Chains are best stored on racks inside dry buildings that have a fairly constant temperature. Chains should be stored where they will not be run over by trucks or other mobile equipment, nor exposed to corrosive chemicals.

Storage racks and bins should be constructed in such a way that air circulates freely around the chain and keeps it dry. Use power equipment to lift large and heavy chains. Attempting to lift a chain manually can cause an injury.

The supplier or manufacturer can arrange a time to give you on-location instructions on the care, use, inspection, and storage of chains. By instructing maintenance workers on chains, you will be in a better position to evaluate your needs.

MATERIALS STORAGE

Both temporary and permanent storage of all materials must be neat and orderly. Materials piled haphazardly, or strewn about, increase the possibility of accidents. The warehouse supervisor must direct the storage of raw materials (and sometimes processed stock) kept in quantity lots for a length of time. The production supervisor is usually responsible for storing of limited amounts of materials and stock, for short periods, close to the processing operations. Planned materials storage minimizes (1) handling required to bring materials into production, and (2) removing finished products from production to shipping.

When employees plan materials storage, they must make sure that materials do not obstruct fire-alarm boxes, sprinkler system controls, fire extinguishers, first-aid equipment, lights, electric switches, and fuse boxes. Exits and aisles must be kept clear at all times.

There should be at least 18 in. (0.5 m) clearance below sprinkler heads to reduce interference with water distribution. This clearance should be increased to 36 in. (0.9 m), if the material being stored is flammable.

Aisles designed for one-way traffic should be at least 3 ft wider than the widest vehicle, when loaded. If materials are to be handled from the aisles, the turning radius of the power truck must be considered. Employees should keep materials out of the aisles and loading and unloading areas, which should be marked with white lines.

Storage is facilitated and hazards are reduced by using bins and racks. Material stored on racks, pallets, or skids can be moved easily and quickly from one workstation to another. When possible, crosstie material piled on skids or pallets.

In an area where the same type of material is stored continuously, it is a good idea to paint a horizontal line on the wall to indicate the maximum height to which material may be piled. This will help

Figure 15–12. In a typical double-chain sling, all components, such as the oblong master link, the body chain, and the hook, are carefully matched for compatibility. Note the permanent identification tag. (Courtesy Columbus McKinnon Corporation.)

to keep the floor load within proper limits and sprinkler heads in the clear.

Containers and Other Packing Materials

Because packing containers vary considerably in size and shape, height limitations for stacking these materials vary. The weight of the container itself must also be figured when keeping within the proper floor-loading limit.

Sheets of heavy wrapping paper placed between layers of cartons will help to prevent the pile from shifting. Better still, cross-tying will prevent shifting and sagging, and will permit higher stacking. Wire-bound containers should be placed so that sharp ends do not stick into the passageway. When stacking corrugated paper cartons on top of each other, remember that increased humidity may cause them to slump.

Bagged materials can be cross-tied with the mouths of the bags toward the inside of the pile, so contents will not spill if the closure breaks. Piles over 5 ft (1.5 m) high should be stepped back one row. Step back an additional row for each additional 3 ft (0.9 m) of height. Sacks should be removed from the top of the pile, not halfway up or from a corner. Remove sacks from an entire layer before the next layer is started. This keeps the pile from collapsing and possibly injuring people and damaging material.

Small-diameter bar stock and pipe are usually stored in special racks, located so that when stock is removed, passersby are not endangered. The front of the racks should not face the main aisle, and material should not protrude into the aisles. When the storage area is set up, the proper floor load should be considered.

Large-diameter pipe and bar stock should be piled in layers separated by strips of wood or by iron bars. If wood is used, it should have blocks at both ends; if iron bars are used, the ends should be turned up. Again, floor load must be considered.

Lumber and pipe have a tendency to roll or slide. When making a pyramid pile, operators should (1) set the pieces down rather than dropping them, and (2) not use either hands or feet to stop rolling or sliding material; this can result in serious injury.

Sheet metal usually has sharp edges and should be handled with care, using leathers, leather gloves, or gloves with metal inserts. Large quantities of sheet metal should be handled in bundles by power equipment. If sheet-metal piles are separated by strips of wood, they will be easier to handle when the material is needed, and will be less likely to shift or slide.

Tinplate strip stock is heavy and razor sharp. Should a load or partial load fall, it could badly injure anyone in its way. Two measures can be taken to prevent spillage and injuries: (1) band the stock after shearing, and (2) use wooden or metal stakes around the stock tables and pallets that hold the loads. It is everyone's responsibility who handle the bundles to band loads properly and make sure the stakes are in place when the load is on the table.

Packing materials, such as straw, excelsior, or shredded paper, should be kept in a fire-resistant room equipped with sprinklers and dustproof electric equipment. Materials received in bales should be kept baled until used. Remove only enough material for immediate use or for one day's supply. Packing materials should be taken to the packing room and placed in metal (or metal-lined wood) bins. Bin covers should have fusible links that close automatically in case of fire. To prevent injury in case a counterweight rope should break, the weight should be boxed in.

Steel and plastic strapping, which may be flat or round, requires that the worker be trained in both application and removal. In all cases, the worker should wear eye or face protection or combination of both, and leather-palm gloves. Heavy strap may require the wearing of steel-studded gloves.

Barrels or kegs stored on their sides should be placed in pyramids, with the bottom rows securely blocked to prevent rolling. If piled on end, the layers should be separated by planks. If pallets are used, they should be large enough to prevent overhang.

Hazardous Liquids

Acid Carboys. These are best handled with special equipment such as carboy trucks. Boxed carboys should generally be stacked no higher than two tiers and never higher than three. Not more than two tiers should be used for carboys of strong oxidizing agents, such as concentrated nitric acid or concentrated hydrogen peroxide. The best storage method is specially designed racks.

Before handling carboy boxes, inspect them thoroughly to make sure that nails have not rusted or that the wood has not been weakened by acid. Empty carboys should be completely drained and their caps replaced.

The safest way to draw off liquid from a carboy is to use suction from a vacuum pump or a siphon started either by a rubber bulb or an ejector. Another method is to use a carboy inclinator that holds the carboy the top, sides, and bottom, and returns the carboy to upright position automatically when released. Pouring by hand or starting pipettes or siphons by mouth suction to draw off the contents of a carboy should never be permitted.

Portable Containers. When liquids are handled in portable containers, contents should be clearly labeled. Substances should be stored separately in safety storage cabinets that meet pertinent OSHA 29 *CFR* 1910.106 and NFPA 30 requirements (Figure 15–13).

Figure 15–13. If the flammable liquid storage cabinets meet the requirements of NFPA 30, *Flammable and Combustible Liquids Code*, they are suitable for storing individual flammable liquids containers of up to five-gallon capacity. (Courtesy Eagle Manufacturing Company.)

Purchasing agents should specify to suppliers that all chemicals must be properly labeled showing contents, hazards, and precautions. Labels are discussed in Chapter 8, Industrial Hygiene.

Drums. Storage drums containing hazardous liquids should be placed on a rack, not stacked. It is best to have a separate rack for each type of material. Racks permit good housekeeping, easy access to the drums for inspection, and easy handling. Racks allow drums or barrels to be emptied through the bung. Self-closing spigots should be used, particularly where individuals are allowed to draw off their own supplies. If there is a chance that spigots might be hit by material or equipment, drums should be stored on end and a pump used to withdraw the contents. Drums containing flammable liquid, and the racks that hold them, should be grounded (Figure 15–14). In addition, the smaller container into which the drum is being emptied should be bonded to the edge of the drum by a flexible wire with a C-clamp at each end.

Storage areas for liquid chemicals should be naturally well ventilated. Natural ventilation is better than mechanical because the mechanical system may fail. Nevertheless, some rooms are required to have mechanical ventilation, especially near the floor, with an alarm if ventilation shuts off (NFPA 30). The floor should be made either of concrete or of some other material treated to reduce absorption of the liquids. The floor should be pitched toward one or more corrosion-resistant drains that can be cleaned easily.

Where hazardous chemicals are stored, handled, or used, emergency flood showers or eyewash fountains should be available. Operators should wear chemical goggles, rubber aprons, boots, gloves, and other protective equipment necessary to handle the particular liquid (Chapter 9, Personal Protective Equipment). Each type of hazardous material requires either special handling techniques or special protective clothing or both. In all cases, the supervisor or the purchasing agent should consult the safety professional or the safety department for handling precautions.

Additional Measures. When handling hazardous liquids, take these extra precautions:

1. Wear proper protective equipment.
2. Keep floors clean; do not allow them to become slippery.
3. Drain all siphons, ejectors, and other emptying devices completely before removing them from carboys or drums.
4. To dilute acid, always add acid to water: never add water to acid. Add the acid slowly and stir constantly with a glass implement.
5. In case of an accident, give first aid. Flush burned areas with lots of water for 15 minutes. Call a doctor immediately.
6. Do not force hazardous chemicals from containers by injecting compressed air. The container may burst or the contents may ignite.
7. Do not try to wash or clean a container unless its label specifically requires that it be cleaned before it is returned.

8. Do not store hazardous chemicals near heat or steam pipes, or where strong sunlight will strike them. The contents may expand and may cause a fire or explosion. Bottles may focus sunlight to ignite a fire in nearby combustibles.
9. Do not stir acids with metal implements.
10. Do not pour corrosives from a carboy by hand.
11. Do not move a carboy unless it is securely capped.

Pipelines to outside storage tanks are safer for large quantities of liquid chemicals or solvents, because they reduce and localize any spillage. When hazardous liquids are used in small quantities, only enough for one shift should be kept on the job and then is a closed/approved container. In many cases, local ordinances limit the amount of hazardous materials that can be stored or processed inside a building. The main supply should be stored in tanks located in an isolated place.

Pipes/Piping. These should be color-coded and labeled to identify their contents. Make sure workers understand the company's use of identifying colors and labels. Should outside contractors work in the area, they should be adequately informed of piping system contents and any hazards.

When a valve is to be worked on, it should be closed. The section of pipe in which it is located should be thoroughly drained before the flanges are loosened. Appropriate personal protective equipment (PPE) must be worn by workers.

Workers should loosen the flange bolts farthest away first, so that any remaining pipe contents will be directed away from them. A blind (or blank piece) should be fastened over the flanges on the open pipe

Figure 15–14. A storage rack reduces harzards, aids identification of the drum contents, and permits easy grounding.

ends as soon as they are parted. If a line is opened often, a two-ended metal blank can be permanently installed on a flange bolt so that the blank can be pivoted on the bolt when the flanges are parted. One end of the blank is shaped in the form of a gasket, to permit flow of liquid. The other end is blind and is swung to cover the pipe opening when the flanges are unbolted and parted.

Tank Cars. These storage cars should be protected on sidings by derailers and by blue stop flags or blue lights before they are loaded or unloaded. Hand brakes should be set and wheels chocked. Before the car is opened, it should be bonded to the loading line. The track and the loading or unloading rack should be grounded, and all connections checked regularly. Chemical tank cars should be unloaded through the dome rather than through the bottom connection.

Gas Cylinders

There are many regulations regarding the storage and handling of gas cylinders. Several government agencies and private organizations and their standards should be consulted—OSHA, National Fire Protection Association, Compressed Gas Association, and others. Care should be taken to comply with all local government codes as well.

Compressed gas cylinders should be transported and stored in the upright position. All cylinders should have protective caps in place and should be chained or otherwise secured firmly in place against a wall, post, or other solid object to prevent them from being knocked over. Different kinds of gases should either be separated by aisles or stored in separate sections of the building or storage yard. Fuel gas cylinders should be stored separately from oxygen cylinders. Empty cylinders must be stored apart from full cylinders. Oxygen must be separated from flammable gases by 20 ft or half-hour rated wall.

Set up storage areas away from heavy traffic. Containers should be stored where there is minimal exposure to excessive temperature, physical damage, or tampering. Never store cylinders of flammable gases near flammable or combustible substances.

Before cylinders are moved, check to make certain all valves are closed and that protective valve caps are in place. Always close the valves on empty cylinders. Never let anyone use a hammer or wrench to open valves. As stated before, cylinders should never be used as rollers to move heavy equipment. Handle all cylinders with care—a cylinder marked "empty" are never completely empty.

To transport cylinders, use a carrier that does not allow excessive movement, sudden or violent contacts, and upsets. When a two-wheeled truck with rounded back is used, chain the cylinder upright.

Never use a magnet to lift a cylinder. For short-distance moving, a cylinder may be rolled on its bottom edge, but never dragged. Cylinders should never be dropped or permitted to strike one another. Protective caps must be kept on cylinder valves when cylinders are not being used. Oxy-acetylene welding assemblies should be fitted with check valves to avoid inadvertent mixing of gases in the system. When in doubt about how to handle a compressed gas cylinder, or how to control a particular type of gas once it is released, ask the safety professional or the safety department for information and instructions.

Combustible Solids

Bulk storage of grains and granular or powdered organic chemicals and certain other materials presents fire and explosion hazards. Many materials that are not considered hazardous in solid form often become combustible when finely divided. Some of these are carbon, fertilizers, food products and by-products, metal powders, resins, waxes and soaps, spices, drugs and insecticides, wood, paper, tanning materials, chemical products, hard rubber, sulfur, starch, and tobacco. This list is by no means complete, mentioning only the general categories of materials that may generate explosive dusts.

Dust. To avoid dust explosions, prevent formation of an explosive mixture or eliminate all sources of ignition. Either the dust must be kept down, or enough air supplied to keep the mixture below the combustible limit. Good housekeeping and dust-collection equipment will go far toward preventing disaster. Make sure dust does not accumulate on floors and other structural parts of a building. A dust explosion is a series or progression of explosions. The initial one may be small, but it shakes up dust that has collected on rafters and equipment so that a second explosion is created. This chain reaction usually continues until the result is the same as if a single large explosion had taken place.

Wearing protective equipment is important when dusts are toxic. Such equipment may range from a respirator to a complete set of protective clothing. The Council, or a state or local safety agency, can recommend the proper type of protective clothing for specific jobs. Recommendations are detailed in Chapter 9, Personal Protective Equipment.

Bins. Bins in which solids are stored should have sloping bottoms to allow material to run out freely and to help prevent arching. For some materials, especially if they are in process, bins should have vibrators or agitators in the bottoms to keep material flowing.

If large rectangular bins, open at one side, are entered with power shovels or other power equipment, workers should take care not to undercut the material and endanger themselves if arched material should suddenly give way. Where practical, tops of bins should be covered with a 2-in. (5–cm) mesh screen, or at least by a 6–in. (15–cm) grating (or parallel bars on 6–in. centers) to keep people from falling into the bins.

Tests for oxygen content and for presence of toxic materials should be made before anyone enters a bin or tank for any purpose. If such tests indicate the need for protection, people should use air-supplied masks, in addition to a safety belt and lifeline, before entering. Filling, agitating and emptying equipment should be deenergized, tagged, and locked out so it cannot be started again, except by the worker after leaving the bin or by the immediate supervisor after checking the worker out of the bin. A person working in a bin should have a companion employee, equipped to act in case of emergency, stationed on top of the bin. Bins should be entered from the top only. All aspects of OSHA's confined space programs also need to be met.

If flammable vapors, dusts, or gases are found, workers should use only spark-resistant equipment and electrical equipment approved for use in such atmospheres by a recognized testing laboratory.

SUMMARY OF KEY POINTS

Key points covered in this chapter include:

- Manual handling of materials accounts for an estimated 25% of all occupational injuries. These injuries are caused by improper lifting, carrying too heavy a load, incorrect gripping, failing to observe proper foot or hand clearances, failing to use or wear proper safety equipment, and poor job design. Employees should assess their operations to see if manual handling can be eliminated, or powered equipment can be used to minimize the hazards, or work stations and areas can be redesigned using ergonomic principles.
- The largest number of injuries occur to fingers and hands. Workers must be taught how to pick up and put down heavy, bulky, or long objects. Protective gloves, eyewear, and safety shoes should be worn to prevent injuries. If manual lifting cannot be eliminated, workers must be taught correct lifting practices. Researchers have found that no single lifting method is best for everyone. Workers should choose the method that is most comfortable for them. Workers must be taught how to handle specific shapes—box, carton, crate, drum, or barrel—to minimize risk of injury to themselves and damage to containers. Special equipment or tools can make the job easier and safer.
- Materials handling equipment include hand tools, nonpowered and powered hand trucks, powered industrial trucks, dock plates, conveyors, cranes, railroad cars, highway trucks, and motorized equipment. Hand tools, such as chisels, crowbars, rollers, hooks, and shovels, should be in good condition and used safely to prevent injuries.
- Nonpowered hand trucks include two-wheel and four-wheel trucks. Four major hazards associated with these trucks are (1) running wheels off bridge plates or platforms, (2) colliding with other trucks or obstructions, (3) jamming hands between the truck and other objects, and (4) running wheels over feet. Only qualified personnel should be allowed to handle these trucks. They should be used only for the specific purpose they were designed for. Loads should be carefully placed so they do not shift, slide or obstruct the handler's view of the path ahead.
- Major hazards associated with powered hand trucks are (1) catching the operator between the truck and another object, and (2) collisions with objects or people. The truck should be equipped with a dead-man control, wheel guards, and an ignition key that can be removed when the operator leaves the truck. Only qualified workers should be allowed to use these vehicles.
- Powered industrial trucks are either battery operated or run by internal combustion engines. These trucks must meet stringent standards and be operated only by qualified drivers. Battery-charging installations must be in special areas, with adequate precautions for handling spilled electrolyte and dealing with fire hazards. Refueling areas must be adequately ventilated and secured against fire. Loads must be carefully stacked and secured to avoid spilling the load or overturning the truck. All employees must be trained in use and maintenance of this equipment.
- Dock plates not built in must be fastened down to prevent them from moving or sliding when used. Plates should be large and strong enough to hold the heaviest load—a safety factor of five or six should be adequate. They should be inspected frequently and kept in good condition.
- The primary rule in using conveyors is simple: never allow anyone to ride on them. Workers should know the specific safety rules for working on or around gravity, pneumatic or blower, or screw conveyors. Supervisors should train employees in how to maintain guards and other safety equipment, free conveyor jams, inspect conveyors for hazards, and cross safely from one side of a conveyor to the other.
- Although many types of cranes are used in industry,

safe operating procedures are much the same. Supervisors should assign only one signalman for each crane and tell crane operators to obey only this person's instructions. Loads must be properly balanced and carefully lifted before being moved from one location to another. Cranes and their chains and ropes must be inspected regularly for signs of wear, corrosion, or deterioration. Supervisors should pay particular attention to all control mechanisms, safety devices, crane hooks, ropes, and electrical apparatus.

- Moving any railroad car should be done after considering the facilities or equipment available to move the car, number of cars and distance they are to be moved, and whether the track is level or on a grade. The supervisor may need power equipment, or may have enough workers to move the car by hand. Workers should be alert to the dangers involved in opening and closing doors, transferring material in and out of cars, and securing car doors.

- When loading or unloading highway trucks, workers should set brakes and place wheel chocks under the rear wheels or use a vehicle restraint. They should use a fixed jack to support a semitrailer not attached to a tractor.

- All motorized equipment requires special handling to prevent accidents. Workers must be taught to operate this equipment according to detailed instructions in manufacturers' manuals. Equipment must be thoroughly inspected and given safety clearance before it is put into service. A preventive maintenance program is necessary to ensure worker safety and efficiency.

- Ropes, chains, and slings are common materials handling equipment that require careful inspection and maintenance to remain in good working order. Supervisors should become familiar with the properties of various types of ropes, chains, and slings, and the precautions for their use and maintenance. This equipment should never be used to lift a load if it has a kink, knot, or twist in the rope or chain. Fiber rope must be kept clean and dry, and stored in a dry, warm place. If the rope becomes frayed or worn, it should be replaced immediately. Wire ropes have greater durability and strength, and are used to handle heavier loads. They must be carefully lubricated to prevent rust and excessive wear.

- Chains and chain slings must be inspected frequently for wear, failure of a weld, metal fatigue, deformation, or stress in the links. They should not be overloaded or forced onto a hook as this weakens the links. Each chain should be tagged to indicate its load capacity. Chains should be stored in dry buildings with fairly constant temperatures.

- Both temporary and permanent storage of all

materials must be neat and orderly. Supervisors should make sure materials do not obstruct fire-alarm boxes, sprinkler system controls, fire extinguishers, first-aid equipment, lights, electric switches, fuse boxes, and exits and aisles. Use of bins and racks, pallets, and skids can reduce hazards associated with materials storage.

- Materials storage includes careful stacking and placement of containers and other packing materials, hazardous liquids, gas cylinders, and combustible solids. Supervisors must be familiar with hazards and precautions associated with each of these materials. Containers and other packing materials can be stacked safely by taking into consideration height and weight limitations to keep within safe floor-loading limits. Materials should be stacked to prevent loads from sliding or shifting.

- Hazardous liquids are best handled with special equipment. Acid carboys, portable containers, and drums must be moved only by trained personnel who are equipped with the proper tools and protective gear. Each type of hazardous material requires special clothing or handling techniques.

- Pipelines, tank cars, and gas cylinders also require special handling when used for storage. Pipelines should be color-coded and labeled to identify contents. Tank cars should be protected on sidings by derails and by blue stop flags or blue lights before they are loaded or unloaded. Gas cylinders must be carefully chained or fixed to a stationary object. Empty cylinders should be stored apart from full cylinders and kept away from flammable or spark-generating materials. Workers must be trained thoroughly in safe work practices to deal with these materials.

- Bulk storage of grains and granular or powdered chemicals or materials presents fire and explosion hazards. To avoid dust explosions, dust must be kept down or enough air supplied to keep the concentration below the combustible limit. Good housekeeping and dust-collection equipment will go far toward preventing disaster. Workers should wear protective equipment when working near toxic dust.

- If bins are used for storage, they should have a sloping bottom and, in some cases, agitators or vibrators to keep material moving. Supervisors must always check for oxygen content and for presence of toxic materials before allowing workers to enter a confined space such as a bin or tank for any reason. Procedures for permit-required entry into confined spaces should be followed. Workers should wear protective clothing and have a companion worker stationed outside in case of emergency.

16

ELECTRICAL SAFETY

After reading this chapter, you will be able to:

- Explain the action of electricity on the human body and how to use grounding to prevent shock hazards

- Describe how to use plug- and cord-connected equipment and extension cords safely

- Test branch circuits and electrical equipment

- Explain the uses and limitations of ground-fault circuit interrupters

- List the classes and divisions of hazardous locations and the proper electrical safeguards to use in each classified location

- Instruct workers in the proper procedures for handling electrical equipment safely on or off the job

Electrical safety is one of an employee's most important responsibilities in ensuring safe production. Although everyone knows high voltage electricity is dangerous, low voltage alternating-current (AC) levels can also cause fatalities by affecting the heart. An estimated 31% of the known electrocution fatalities occur in the home, 24% in general industry, and the remaining in the electrical power generation and distribution industries. This means that approximately 55% of electrocutions are due to contact with so-called low-voltage circuits (up to 600 volts AC or 750 volts DC).

Electricity usually does not *look* hazardous. Many people have been electrocuted when they *thought* they knew what they were doing. Unless you are trained and experienced in the electrical work to be done, call an electrical expert. A key concept to remember with any electrical exposure is: *Do not become a part of any electrical circuit.*

Electricity can also claim lives from fire-related accidents. The U.S. Product Safety Commission estimates that there are 169,000 house fires of electrical origin each year; claiming 1,100 lives and injuring 5,600. Property losses are estimated at $1.1 billion a year.

MYTHS AND MISCONCEPTIONS ABOUT ELECTRICITY

Misunderstandings about electricity can lead to serious injuries and property damage. Make sure your workers know the facts. Some of the more common myths and misconceptions are as follows:

1. Myth: "Electricity takes the path of least resistance." Actually, current will take any conductive paths, high or low resistance, in order to return to the source that provides it power. Small amounts of current will flow through paths of high resistance; but even small amounts can kill.
2. "Electricity tends to go to ground." A person is led to believe that electricity wants to go to ground and simply disappear. In reality, ground serves as just one of the electrical loops that misdirected current can use to get back to the grounded power source.
3. "If an electrical appliance or tool falls into a sink or tub of water, the item will short out and trip the circuit breaker." This is not reliable information, because the sink or tub may be non-conductive and therefore not part of the loop to ground.

    Persons should be warned not to reach into the water to retrieve the appliance or tool. The water serves as an electrical conducting path for the electricity to flow out of the electrical equipment. Any person reaching into the water with one hand when some other part of the body is touching a grounded object could receive a severe or even fatal shock.

4. "AC reverse polarity is not hazardous." The National Electrical Code (NEC) requires that power tools, attachment plugs, and receptacles be properly wired so as not to reverse the designated polarity. Many tools have switches in only one of the two conductors serving the item. The switch is supposed to be on the "hot" conductor supplying the power. If the tool (e.g., a drill) is plugged into a receptacle with reverse polarity and accidentally dropped into water, all of the internal wiring will be energized.

    The increased energized surface area in the water will allow more fault current to flow and increase the hazard of serious or fatal shock. If the tool and the receptacle are properly polarized, this would minimize the shock hazard. Only the terminal on the switch would be energized, not the entire tool, as in the case of reverse polarity.

5. Myth: "It takes high voltage to kill; 120 volts AC is not dangerous." Current is the culprit that kills. Voltage is a factor in determining how much current will flow. Five milliamps is human level of sensation; 100 milliamps can kill. Under the right conditions, AC voltage as low as 60 volts can kill. Respect all voltages as having the potential to kill.
6. Myth: "Double-insulated power tools are doubly safe and will always provide safety." Read the manufacturer's operating instructions carefully, but never place your trust completely in any electrical safety device. Double-insulated power tools can be hazardous if dropped into water. Electrical current can flow out of the power tool into the water.

Every piece of equipment is a potential source of electric shock. Even an electric shock too small to cause injury can trigger an involuntary reaction that results in physical injury; for example, after touching a live wire, a person may lose his or her balance and fall off a ladder. Design, layout, installation, and preventive maintenance programs can minimize exposure to live or "hot" electrical conductors. An effective electrical safety policy, employee training, and hazard awareness program can further reduce incidents of electric shock. It is also recommended that an electrical inspection program be implemented and inspections conducted periodically, as conditions warrant.

This chapter will help you to prevent electric shock injuries. Make sure that your employees do not believe in, or cling to, any of the myths and misconceptions about electricity.

ELECTRICAL FUNDAMENTALS REVIEW

Ohm's Law

Before discussing electrical hazards, a brief discussion of Ohm's Law is necessary. Important terms are voltage (v), current, amperes (amp), milliamperes (ma), watts (w), ohms, resistance (R), impedance (A), and power.

Ohm's Law simply states that one volt will cause a current of one ampere to flow through a conductor having the resistance of one ohm. As a formula, the relationship is represented by I(amps) = E/R (E = volts, R = resistance). An easy way to remember this formula, and the different ways it can be expressed, is to put the symbols in a circle (Figure 16–1).

1. Put your finger on I;I = E/R.
2. Put your finger on R;R = E/I.
3. Put your finger on E;E = I x R.

With this basic formula, you can better understand and explain the amount of electric current through a conductive body.

The above applies to direct current (DC); however, in alternating current (AC) usage, Ohm's Law is E = I x Z, where Z is called the impedance.

To demonstrate an understanding of Ohm's Law, try to solve this problem. Assume a person is working and perspiring, and has a hand-to-hand resistance of 1,000 ohms. This person contacts 100 volts with one hand and touches a ground surface with the other, completing a loop to the voltage source. What would be the current going through the body?

$$I = E/R = 100/1,000 = 0.1 \text{ amp (100 ma)}$$

Another important formula is the relationship of voltage, current, and power utilization (watts). Power is measured in watts and is equal to E (volts) x I

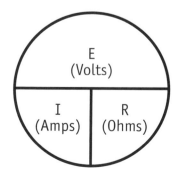

Figure 16–1. Ohm's Law.

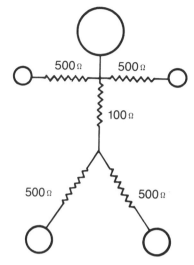

Figure 16–2. Human body resistance model.

(current). In other words, one watt would be equal to one ampere of current flowing through a resistor with one volt of potential difference. Another way to put this relationship to practical use would be to consider a 6-watt electric bulb, i.e., one requiring 6 watts of power to operate (about the size of a night light or panel-board light). Determine the current flowing through the filament of a 6-watt bulb that is being used in a 120-volt light socket.

$$P/E = 6/120 = 0.05 \text{ amp} = 50 \text{ ma}$$

This current is enough to affect the human heart and induce a fatal irregular heart beat. These examples of current levels from 50 ma to 20,000 ma are used to illustrate and clarify misconceptions about electricity and its effect on the human body.

Body Resistance Model

The human body impedance, or resistance to electric current, has three distinct parts: internal body resistance, and the two skin impedances associated with contact points on surfaces of different voltage potential. If the voltage applied is high enough, the skin offers virtually no impedance to the electric current (approximately 600 volts AC).

A body resistance model (Figure 16–2) would indicate approximately 1,000 ohms from hand to hand, or about 1,100 ohms from hand to foot. It is believed that the greatest number of injurious shocks involve a current pathway that is either hand-to-hand or hand-to-feet. Figure 16–2 illustrates that wet contact with an energy source (such as 120 volts AC) in one hand and a contact with a grounded object with the other produces current through the chest area equal to:

Table 16-A. Current and Its Effect on the Human Body

	Current in Milliamperes					
	Direct		Alternating			
			60 Hz		10,000 Hz	
Effect	Men	Women	Men	Women	Men	Women
Slight sensation on hand	1	0.6	0.4	0.3	7	5
Perception threshold	5.2	3.5	1.1	0.7	12	8
Shock—not painful, muscular control not lost	9	6	1.8	1.2	17	11
Shock—painful, muscular control not lost	69	41	9	6	55	37
Shock—painful, let-go threshold	76	51	16	10.5	75	50
Shock—painful and severe, muscular contractions, breathing difficult	90	60	23	15	94	63
Shock—possible ventricular fibrillation effect from 3-second shocks	500	500	100	100		
Short shocks lasting t seconds			$165/\sqrt{t}$	$165/\sqrt{t}$		
High voltage surges	50*	50*	13.6*	13.6*		

*Energy in watt-seconds or joules
Note: Data are based on limited experimental tests, and are not intended to indicate absolute values.

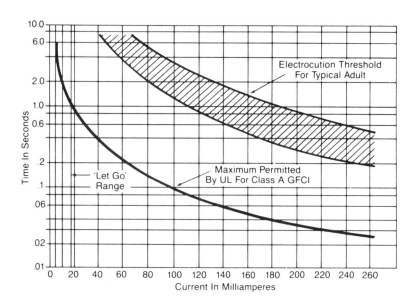

Figure 16-3. This chart shows the effects of electrical current intensity versus the time it flows through the human body.

$$I = E/R = 120/1,000 = 0.12 \text{ amp or } 120 \text{ ma}$$

This current level can be fatal. (Table 16-A)

It is easy to analyze electrical shock hazards today. A small amount of current (50 ma) can cause the heart to start beating rapidly and irregularly—the potentially fatal condition known as ventricular fibrillation. The example given previously, that a 6-watt bulb used 50 ma of current when energized by 120 volts, means that the small amount of current required to light a 6-watt bulb is capable of causing death.

For instance, once ventricular fibrillation occurs, death will ensue in a few minutes. Resuscitation techniques, if applied immediately, can save the victim. Thus, contrary to many other beliefs, even low AC current is dangerous; and it does not take high voltage to kill, as shown in Table 16-A.

Electrical Current Danger Levels

The danger threshold for ventricular fibrillation was established for typical adult workers as a result of research into the effects of electricity on the human body.

Figure 16-3 plots the range of electrical values that affect humans. Also shown is the requirement of Underwriters Laboratories for a Class A ground-fault circuit interrupter (GFCI) intended for protection. Note the time-current relationship at much lower currents and shorter time periods. The electrical current range below that, which causes muscles to freeze, (called the "let go" range) is also illustrated.

From Figure 16–3, one can determine that a 100 ma current flowing for 2 seconds through a human adult body will cause death by electrocution. This doesn't seem like much current when you consider that a small, light-duty portable electric drill draws 30 times that much. But, because a current as small as 3 ma (3/1,000 amp) can cause painful shock, it is imperative that all electrical plugs, cords, and extension cords be kept in good condition and connected to a properly wired grounded circuit.

Factors Enhancing Electrical Shock

The route that electrical currents take through the human body also affects the degree of injury. Voltage determines how much current flows. In cases where individuals come in contact with distribution lines, high voltage and high current can cause the moisture in the body to heat so rapidly that body parts literally explode. This extreme expansion is the result of body fluids changing to steam. This could result in a person's being injured severely, but not electrocuted.

Any report of a shock hazard should be investigated. When investigating shock incidents—whether "near misses" or those that have caused injury—the factors of voltage, current, and body resistance can be used as investigative tools. The combination of one or more of these factors can be analyzed and action taken to eliminate or control the hazard. Other factors that determine the degree of electric shock are as follows:

- Wet and/or damp locations
- Ground/grounded objects
- Current loop from power source back to power source
- Path of current through body and duration of contact
- Area of body contact and pressure of contact
- Physical size, condition, age of person
- Type and/or amount of voltage
- Personal protective equipment, gloves, shoes
- Metal objects such as watches, necklaces, rings
- Miscellaneous
  - Poor workplace illumination
  - Color blindness
  - Lack of safety training and knowledge of electricity
  - No safe work procedures
  - Noninsulated tools

Basic Rules of Electrical Action

There are four basic rules of electrical action that everyone should know. They are as follows:

1. Electricity isn't "live" until current flows.

2. Electric current won't flow until there is a complete loop, out from, and back to, the power source.
3. Electrical current always returns to its source, that is, the transformer or other source that supplied it.
4. When current flows, work (measured in watts) can be accomplished.

Rule 1. This rule explains why faulty power tools with live metal cases (due to a defective equipment-grounding conductor) can be used without causing a shock. Only when a person comes in contact with a ground loop with low enough resistance will current flow through the body and produce a shock.

Rule 2. When a current loop is formed, the electric current now has a path to return to its source. If the loop is broken, the current will stop flowing. To minimize electric shock hazards, look for ways to prevent ground loops. Insulated mats, insulated footwear insulated gloves, and insulated tools are some methods that can be used.

Rule 3. Current will find any available loop with sufficiently low resistance or path to return to its source. In a grounded system the *NEC* requires that the power source be electrically attached to an earth-driven ground rod as well as to the water piping. All other noncurrent-carrying metal parts must also be grounded. This creates many potential ground loops for current to get back to its source. Water pipes, metal ventilation ducts, metal door frames, T-bars holding suspended ceiling panels, and metal ridge rolls on a roof with a grounded lightning system are but a few of the many ground loops available.

Rule 4. If current flows through the body, it can do harm depending on the amount and duration of current flow. As noted in Table 16–A, the effect on the human body can be anything from a mild shock to complete tissue destruction.

These rules should help you analyze how and why someone received an electrical shock. Use the loop concept and Ohm's Law to investigate actual or potential electric shock situations and to determine appropriate safeguards for electrical hazards.

BRANCH CIRCUITS AND GROUNDING CONCEPTS

Single-Phase Service

A typical pole-mounted transformer (single-phase) is shown in Figure 16–4. Industrial plants may have their transformer vaults, feeder lines, and branch circuits distributed throughout the building in conduit systems. This system is used in residential, commercial,

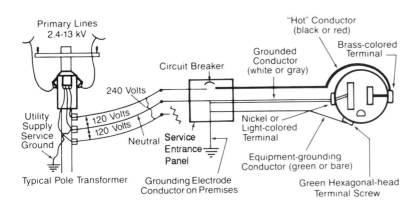

Figure 16–4. Note how the elements are identified in this single-phase service-branch circuit wiring.

and industrial operations to provide 120 volts for lighting and small appliances. The 240-volt, single-phase power is used for heavy appliances and small motor loads. It is important that the function of the simple, single-phase, three-element, 120-volt electrical system be understood, and that misconceptions about grounding, bonding, and reverse polarity be clarified for your workers, especially your electricians.

The *NEC* requires that electrical wiring design ensure continuity of the system. In Figure 16–4, the ungrounded conductor from the service entrance panel (SEP) is wired to the small receptacle slot. The terminal screw and/or the metal connecting to the small slot is required to be a brass color. Insulation on the ungrounded conductor is generally black or red. Remember the phrase "black to brass" or the initials "B & B." Insulation on the grounded conductor is either white or gray. This conductor fastens to the large slot side of the receptacle. It is defined by the code as the "grounded conductor." An easy way to remember this wiring method is to think "white to light."

It is important that the receptacle be wired in this way. If it is wired in reverse, that is, the white insulated conductor to the brass screw and the black insulated wire to the silver screw, it produces reverse polarity, mentioned previously in this chapter.

Caution: Do not consider yourself an electrical expert. Wiring and electrical repair should be done only by qualified personnel.

Reverse Polarity (AC)

Some individuals think that reverse polarity is not hazardous. An example of one hazardous situation could be an electric hand lamp. Figure 16–5 illustrates a hand lamp improperly wired and powered.

When the switch is turned off, the lamp socket shell is still energized. If a person accidentally touched the shell with one hand and a ground loop

back to the transformer, a shock could result. If the lamp had no switch and was plugged in as shown, the lamp shell would be energized when the plug was inserted into the receptacle. Many two-prong plugs have blades that are the same size and the situation described is just a matter of chance. If the plug is reversed, Figure 16–6, the voltage is applied to the bulb center terminal and the shell is at ground potential. Contact with the shell and ground would not create a shock hazard in this situation.

A second example: A person who works on a fluorescent lamp fixture deenergizes by turning off the wall switch. If the circuit has been wired backward, there will be 120 volts at the fixture, even though the switch is off (single-pole, single-throw switch).

Another example is the case of electric hair dryers or other plastic-covered electrical appliances. Figure 16–7 illustrates a hair dryer properly plugged into a receptacle with the correct polarity. When the appliance is plugged in with the switch in the hot or 120-volt leg, the voltage stops at the switch when it is in the OFF position. If the hair dryer were accidentally dropped into water, current could flow out of the plastic housing using the water as the conducting medium.

Caution: A person should never reach into water to retrieve an energized electrical appliance.

This fault current might be in the 100-ma range (sufficient to kill but not sufficient to trip a 20-amp circuit breaker). When the appliance is polarized or plugged in, as shown in Figure 16–7, the switch connection is the only surface area where the hot conductor is fastened. It is so small that high resistance allows only a small current to be available $(I = E/R)$. In this configuration, the fault current would be extremely low (unless the switch is in the ON position).

If the appliance is plugged in so that its polarity is reversed, as shown in Figure 16–8, when the switch is off, *voltage will be present throughout all the internal*

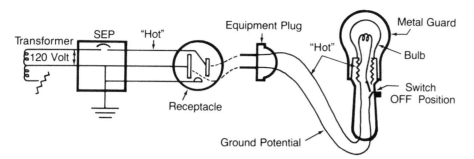

Figure 16–5. When this hand lamp is plugged in so that the polarity is reversed, the shell is still energized even when the switch is turned off. (SEP is the service entrance panel.)

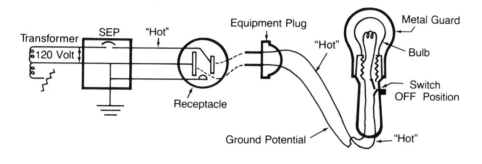

Figure 16–6. When this hand lamp is plugged in with the plug in correct polarity, current flows to the bulb center terminal and the shell is at ground potential.

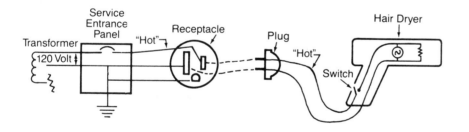

Figure 16–7. This hair dryer is plugged in correctly so that the switch is on the hot conductor.

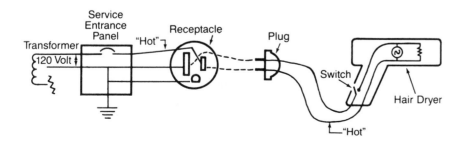

Figure 16–8. If the hair dryer is plugged in so that the polarity is reversed, voltage will be present throughout the internal wiring of the appliance even when the switch is turned off.

wiring of the appliance. Now if it is dropped into water, there would be increased current flowing through the water. The heating element has a large surface area, which decreases resistance and increases current. A person who accidentally tries to retrieve the dryer is now in a very hazardous position because the voltage in the water could cause current to flow through the body. This illustrates the concept that reverse polarity is a problem whenever appliances with plastic housings are used in areas near sinks, or where water can wet the appliance. Remember, most motorized appliances have air passages for cooling. Wherever air can go, so can moisture and water.

If the appliance had a double-pole double-throw switch (DPDT), then it would make no difference how the plug was positioned in the outlet. The hazard would be minimized, since the energized contact surface would be extremely small, resulting in a high resistance contact in the water and a resulting low available fault current. Later, we will see how ground-fault circuit interrupters (GFCIs) can be used to protect against shock hazards when using appliances with nonconductive housing around water.

Grounding Concepts

The terms "grounded conductor," "equipment-grounding conductor," "neutral," and "ground" are a cause of many difficulties in understanding basic electrical concepts. Grounding falls into two safety categories. One of these is system safety grounding, and the other is equipment safety grounding.

System grounding protects power company equipment during emergencies, such as lightning strikes. *Equipment safety grounding* provides a system where all noncurrent-carrying metal parts (such as in a metal drill press frame or an electric refrigerator) are bonded together and kept at the same potential. Grounding concept definitions are as follows.

Grounded Conductor. The *NEC* defines "grounded conductor" as "a system or circuit that is intentionally grounded." Figure 16–9 shows the conductor from the service entrance panel (SEP) to the outlet as the grounded conductor. As shown in the illustration, for 120-volt AC, 3-conductor outlets, the grounded conductor (white or gray insulation) is fastened to the silver screw, which is the large parallel slot.

The grounded conductor is a current-carrying conductor, but also a connected ground at the SEP. From the SEP back to the power source, the grounded conductor is called neutral. On 120-volt circuits, refer to the grounded conductor from the SEP to the receptacle as just that, the grounded conductor, not the neutral.

Neutral. The term "neutral" refers to a neutral conductor that carries only unbalanced current from the other conductors in the case of normally balanced circuits. From the SEP to the power source, the term "neutral" is used. In circuits from the SEP to large 240-volt AC appliance receptacles, the grounded conductor would also be referred to as the neutral.

Equipment-Grounding Conductor. Figure 16–9 shows the equipment-grounding conductor going from the SEP to the appliance outlet. This path is sometimes also called the "grounding wire." This grounding conductor, if insulated, is colored green, or green with a yellow stripe. This conductor may also be a bare wire like those used in plastic sheathed cable. The *NEC* allows the conduit system to be the grounding conductor path also.

The grounding conductor normally does not carry a current. Only when an equipment malfunction occurs—allowing current to flow from the motor or electrical device to the equipment case—does the grounding conductor carry the current back to the power source and keeps the equipment case at ground potential. (More on this in the discussion on grounded power tools and equipment.)

Portable equipment and portable power tools can be grounded (equipment grounding) by use of a separate grounding wire from the equipment housing to a *known* ground. This method is satisfactory for stationary equipment as long as conditions do not change. It is less than satisfactory for portable equipment where the integrity of the grounding wire could be questionable and the quality of the ground is unknown. However, it is better than no grounding provision at all. If used, the grounding wire should be at least 1 in. longer than the power cord, so the plug can be removed from the receptacle before the ground is disconnected.

Ground. The *NEC* defines this term as a "conducting connection, whether intentional or accidental, between an electrical circuit or equipment and earth, or to some conducting body that serves in place of earth." Ground, then, means an electrical connection to a water pipe or a driven ground rod, or, where pipe is not available, to the rod alone. To "ground" something means to connect it electrically to the ground. Since the power source and the SEP are both connected to ground, the earth or ground loop becomes an electrical loop hazard.

Figure 16–10 illustrates that accidental contact with the hot conductor, while a person is standing on the ground, could allow a current to return to the power source through the ground loop formed by the person and the earth. This can cause a shock.

Remember that grounded objects are many. The metal framing studs used in new construction are usually grounded from the electrical conduit system. The T-bar hangers for suspended ceilings can be a ground loop since the electric lighting fixtures are

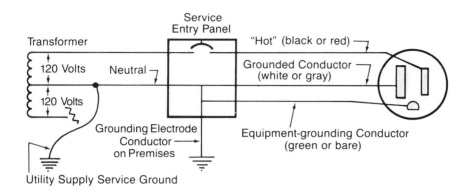

grounded and electrically may cause the T-bars to be ground potential. Ventilation duct work may be ground potential since electric switches, dampers, and other devices may be mounted on the duct work.

PLUG- AND CORD-CONNECTED EQUIPMENT AND EXTENSION CORDS

Grounded Power Tools and Equipment

The earlier discussion of two-conductor appliances with plastic housings revealed the hazard of dropping ungrounded electrical devices in water. Now it is important to explain how proper grounding of appliances with metal cases can provide safety if the appliances are accidentally immersed in water.

Virtually all portable power tools, except double-insulated tools, used in industry and on construction sites are equipped with a 3-prong plug. Unfortunately, not all receptacles are equipped with provisions for a 3-prong plug.

This fact can lead a worker, who is unwilling to make the effort to use an adapter, to break off the U-blade, or ground, and defeat the protection of an equipment-grounding circuit. This is a highly danger-ous practice on construction sites and in other areas that are often damp or wet. Cutting off the equip-ment-grounding conductor U-blade not only destroys the grounding path, but also may allow the plug to be inserted in the receptacle in a reverse polarity mode (Figure 16–11).

As shown in this figure, the grounding pin is cut off. In this situation, if a fault developed when the power tool was plugged into an outlet, the case would be energized with leakage voltage. The power tool would operate and the person using the power tool would be safe—but only until a ground loop was encountered. Then, current could flow through the person, through the ground loop, and back to the

power source. This would cause an electric shock to the person. The loop resistance would determine the seriousness of the shock.

Double-Insulated Power Tools

Double-insulated tools and equipment generally do not have an equipment-grounding conductor. Protection from shock depends upon the dielectric properties of internal protective insulation and the external housing to insulate users from the electrical parts. External metal parts, such as chucks and saw blades, are internally insulated from the electrical system.

Today, many electrical power tool manufacturers produce a line of double-insulated tools. As a result, many of their advertisements lead us to believe that double-insulated power tools are "doubly safe." However, you should take the following precautions when using double-insulated power tools:

- Double-insulated tools are designed so the inner electrical parts are isolated physically and electrical-ly from the outer housing. The housing is noncon-ductive. Dirt particles and other foreign matter from drilling and grinding operations may enter the housing through cooling vents. It can become lodged between the two shells, possibly voiding the required insulation properties.

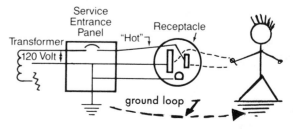

Figure 16–10. A shock is caused when a person contacts the hot wire while standing on the ground.

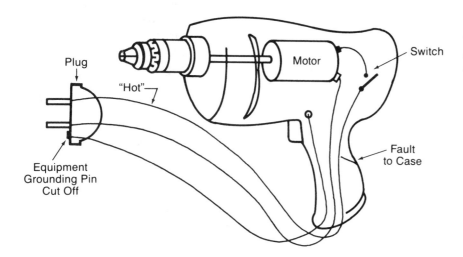

Figure 16–11. Removing the grounding pin destroys the grounding path and can allow the plug to be inserted in the receptacle in reverse polarity.

Plug

"Hot"

Motor

Switch

Equipment Grounding Pin Cut Off

Fault to Case

- Double insulation does not protect against defects in cords, plugs, and receptacles. Continuous inspection and maintenance are required.
- A product with a dielectric housing—for example, plastic—protects users from shock if the interior wiring contacts the housing. Immersion in water, however, can allow a leakage path that may be either high or low resistance. Handling the product with wet hands, in high humidity, or outdoors after a rain storm can be hazardous. The best indication of the safety of a double-insulated tool is the Underwriters Laboratories (UL) (or other recognized testing laboratory's) label attached to the housing. The UL listing is evidence that the tool design and construction technique meet minimum standards. A tool marked "double insulated," but without a UL label, may not be safe.

Double-insulated tools and equipment should be inspected and tested, like all other electrical equipment. Remember, do not use double-insulated power tools or equipment where water and a ground loop are present. They should not be used in any situation where dampness, steam, or potential wetness can occur, unless protected by a GFCI.

Attachment Plugs and Receptacles

Attachment plugs are devices fastened to the end of a cord so electrical contact can be made between the conductors in the equipment cord and conductors in the receptacle. The plugs and receptacles are uniquely designed for different voltages and currents, so only matching plugs will fit into the correct receptacle. In this way, a piece of equipment rated for one voltage/current combination cannot be plugged into a power system that is of a different voltage or current capacity (Figure 16–12).

The polarized 3-prong plug is also designed with the equipment-grounding blade slightly longer than the two parallel blades. This provides equipment grounding before the equipment is energized. Conversely, when the plug is removed from the receptacle, the equipment-grounding pin is the last to leave, assuring a grounded case until after power is removed. The parallel line blades may be the same width on some appliances since the 3-prong plug can only be inserted in one way.

Figure 16–12 illustrates the National Electrical Manufacturers Association (NEMA) standard plug and receptacle connector blade configurations. Each has been developed to standardize the use of plugs and receptacles for different voltages, amperes, and phases from 115 through 600 volts; from 10 through 60 amps; and for single– and three-phase systems.

You should watch for workers who jury-rig adapters. For example, trying to match a 20-amp attachment plug to a 50-amp receptacle configuration. The danger is getting the voltage and current ratings mixed up and causing fire and/or shock hazards. Equipment attachment plugs and receptacles should match in voltage and current ratings to provide safe power to meet the equipment ratings.

Electrical Extension Cord Deficiencies

Electrical extension cords are used in manufacturing, maintenance, construction, and in office operations. Defective, worn, or damaged extension cords have been the cause of death, injury, fire, and electrical equipment damage. Often the rush to meet schedules on time or lack of funds result in jury-rigged cords being fabricated and used. Purchasing staff might buy the cheapest units available, and workers end up with inferior extension cord sets that fail the contact tension test after a few cycles of use. Many times neither the receiving department nor the workers test the

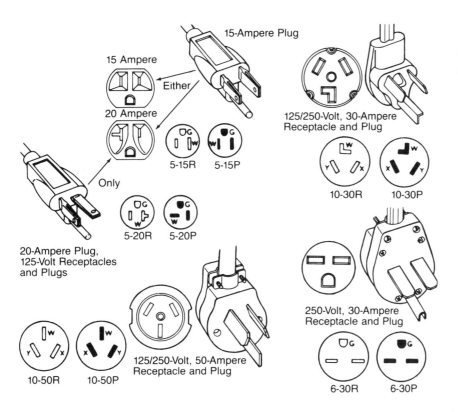

Figure 16–12. Some NEMA plug and receptacle blade configurations. (From OSHA Publication 3073, *An Illustrated Guide to Electrical Safety.*)

extension cords prior to using. This results in the possibility of brand-new, defective units being put into service.

Other problems resulting from extension cord use include overloading, cutting off the grounding blade, and incorrect maintenance and repair. The following safety criteria should be strictly observed and used for training and inspection purposes:

- Before new or repaired extension cords are put into use, they must be tested. In addition, always purchase UL-listed extension cords. One supplier purchased and distributed several thousand units that were incorrectly wired (reverse polarity). These defective units were not UL listed. If extension cords are repaired by your company's electrical shop personnel, be sure they test these units prior to returning them to service. Follow the Testing Extension Cords procedures described later in this chapter.
- A visual and electrical inspection should be made of the extension cord each time it is used. Cords with cracked, worn, or otherwise damaged insulation or damaged attachment plugs should be removed from service immediately. Cords with the grounding prong missing or cut off must also be removed from service.
- If extension cords are fabricated, only qualified electricians should do the work. The use of jury-rigged receptacles fabricated from junction boxes

should be prohibited. Only UL-listed attachment plugs, cords, and receptacle ends should be used when fabricating extension cords.

- Use of 2-conductor extension cords should be prohibited. Only grounding-type extension cord sets should be used.
- When cords must cross passageways, whether vehicular and/or personnel, the cords should be protected against physical damage and moisture and identified with appropriate warnings. Cords should be used in this manner only for temporary or emergency use. They must never be used where they pass through exit doorways, hazardous storage areas, smoke barriers, or fire barriers.
- Extension cords must be of continuous length without splices or taps. Personnel should be instructed to remove from service any cord sets that have exposed wires, cut insulation, loose connections at the plug or receptacle ends, and jury-rigged splices.
- Extension cords must not be connected or disconnected until all electrical load has been removed from the extension cord receptacle end.
- Extension cords used in hazardous locations must be approved for the applicable hazards (see the Hazardous Location section in this chapter).
- When the extension cord is not in use, it should be disconnected and neatly stored. Cords should be stored in a dry, room-temperature area and not exposed to excessively cold or hot temperatures.

- Never overload an extension cord electrically. If it is warm or hot to the touch, have it and/or the appliance being used checked by a qualified person to determine the problem. The wire size of the extension cord may be too small for the amperage of the appliance being used.
- Extension cords must not be draped over, or touch, hot surfaces such as steam lines, space heaters, radiators, or other heat sources.
- Do not plug in or pull out extension cords if you have wet hands.
- Extension cords must not be run through standing water, across wet floors, or other wet areas. If the extension cord is to be used to provide electrical power to equipment used in damp or wet locations, use Ground Fault Circuit Interrupter (GFCI) protection.
- Extension cords used outdoors should be listed and marked for outdoor use and should be protected by GFCIs. Do not allow damaged cord sets to be used outside.
- Extension cords cannot be used in place of permanent wiring.

BRANCH CIRCUIT AND EQUIPMENT TESTING METHODS

Testing Branch Circuit Wiring

Branch circuit receptacles should be tested periodically. The frequency of testing should be established based

Figure 16–13. A receptacle circuit tester checks the receptacle for proper connection of ground wire, correct polarity, and faults in any of the three wires.

on receptacle usage. In shop areas, quarterly testing may be necessary, while office areas may need only annual testing. A preventive maintenance program should be established, particularly because it is not unusual to find receptacles as old as the facility. A popular belief exists that receptacles never wear out or break. This is obviously false. For example, receptacles take severe abuse from employees disconnecting the plug from the receptacle by yanking on the cord. This can put severe strain on the contacts inside the receptacle and on the plastic face.

The electrical receptacle is a critical electrical system component. It must act as a strong mechanical connection for the appliance plug and provide a continuous electrical circuit for each of the prongs. The receptacle must be electrically correct or serious accidents can result. For this reason, a three-step testing procedure is recommended.

The first two steps can easily be done by most collateral duty safety inspectors. The third step is one that should be performed by electrical maintenance personnel prior to accepting contractor wiring jobs. Also, the third step should be done on any circuit where electrical maintenance has been performed. Some inspectors use the third step as a spot check of ground-loop quality.

Step 1. Plug in a 3-prong receptacle circuit tester and note the combination of indicator lights (Figure 16–13). (The circuit tester has three lights; the flashing light indicates which of three problems is present in the wires.) The tester checks the receptacle for proper connection of ground wire, correct polarity, or fault in any of the three wires. If these are normal, proceed to the next test. If the tester indicates a wiring problem, have it corrected as soon as possible. Remove the receptacle from all use until the problem is corrected by disengaging the circuit, if possible, or covering and tagging the receptacle "out of service." Retest after the problem is corrected.

Step 2. Once you are sure the receptacle is wired correctly, test the receptacle contact tension. A typical tension tester is shown in Figure 16–14. The tension should be 8 oz or more. If it is less than 8 oz, have it replaced. The first receptacle function that loses its contact tension is usually the grounding contact circuit. Because this is a critical circuit, it illustrates the importance of checking for receptacle contact tension.

The frequency of testing must be based on receptacle usage. All electrical maintenance personnel should be equipped with receptacle testers and tension testers. Other maintenance employees could also be equipped with testers and taught how to use them. In this manner, the receptacles can be tested before maintenance workers use them.

Figure 16–14. Use a receptacle tension tester to be sure receptacle contact tension is 8 oz or more.

Step 3. The ground loop impedance tester (GLIT) should be used next. This step may be conducted by the electrical maintenance person whenever any electrical circuit is reworked or repaired. An electrical preventive maintenance program should be established to perform regular (at least annual) ground loop testing in high usage areas. The safety inspector can use the GLIT for spot-checking (Figure 16–15).

To perform this test, simply insert the three-prong plug of the GLIT into the receptacle. Press the grooved portion of the red bar and observe the meter reading. A reading of 1 ohm or less is required. Any reading of more than 1 ohm should be reported to electrical maintenance for analysis and corrective action.

WARNING: The GLIT must never be used to test a branch circuit when patient monitoring devices or computer equipment of any kind is in use.

The GLIT should not be used until steps 1 and 2 have been completed successfully. The tester will indicate the wiring fault. Prior to making the circuit operational, all wiring faults must be corrected and the circuit retested.

Testing Extension Cords

Extension cord testing and maintenance is critical to worker safety. The extension cord conducts electrical energy from a fixed outlet or source to a remote location. It must be wired correctly, or the cord can

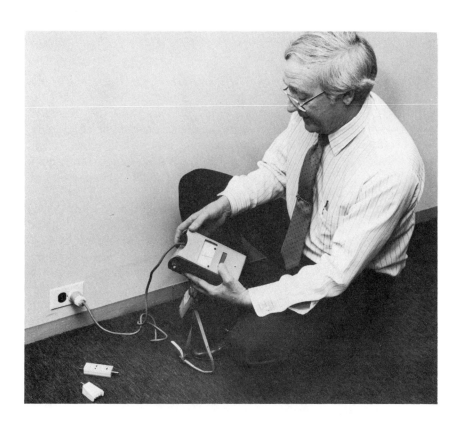

Figure 16–15. Check the electrical circuit with a ground loop impedance tester (GLIT). Note the tension tester and circuit tester at the lower left.

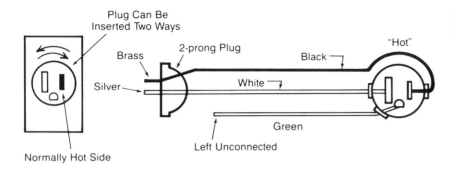

Figure 16–16. A three-conductor extension cord repaired with a two-prong plug.

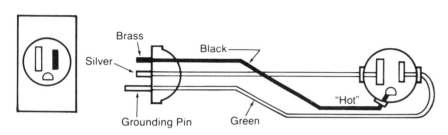

Figure 16–17. A three-conductor extension cord with hot and ground wires interchanged.

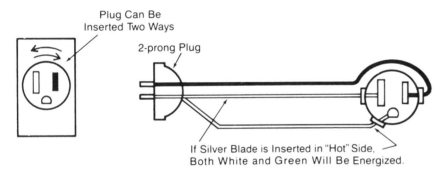

Figure 16–18. This three-conductor extension cord is repaired with a two-prong plug that has the white and green wires tied together.

become the critical fault path.

Many electrical accidents have been caused by faulty or incorrectly repaired extension cords. Many times the male plug on a three-conductor extension cord becomes damaged and the repair person installs a two-prong plug (Figure 16–16). This obviously means the receptacle end has no grounding path. It also means the plug can be inserted with correct polarity or reverse polarity.

Sometimes the repair may be made in such a way that the hot and ground are interchanged, as shown in Figure 16–17. This could happen on repair of the plug or the receptacle. In this configuration, when a grounded power tool is plugged into the extension cord, voltage will be applied to the tool casing. This could cause a dangerous or fatal electrical shock.

The same dangerous condition is illustrated in Figure 16–18. A repair was made replacing the broken three-prong plug with a two-prong plug. The repair person connected the white- and green-insulated conductors on one prong, then connected the black-insulated conductor to the other prong. Depending on how the plug was inserted into the receptacle, in one mode, the case would be grounded; and in the other mode, the case would be energized.

With a "hot" case, the user would have to find only a ground loop to the source and a fatal shock could occur. Interestingly enough, the power tool would operate in either mode.

Figure 16–19 illustrates a repaired or a jury-rigged extension cord fabricated with a three-prong plug and receptacle, but the connecting cord is only a two-conductor. This configuration would obviously show an open ground when tested. Because at times the cord may look like a three-conductor cord, the tester should be used to make sure the extension cord is safe.

New extension cords must be tested before being put into service. Many inspectors have found that new extension cords have open ground or reverse polarity. Don't assume that a new extension cord is good—test it!

Testing extension cords must follow the same

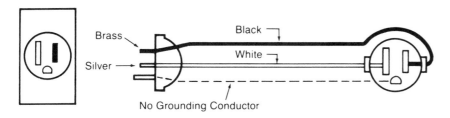

Figure 16–19. This extension cord with a three-prong plug and receptacle is connected to a two-conductor cord.

three steps used for electrical receptacle testing. These three steps are:

Step 1. Plug the extension cord into an electrical receptacle that has successfully passed the three-step receptacle testing procedure. Plug any 3-prong receptacle circuit tester into the extension receptacle. Note the combination of indicator lights. If the tester shows the extension cord is safe, proceed to the next step. If the tester indicates a faulty condition, return the extension cord to electrical maintenance for repair. Once the cord is correctly repaired and passes the 3-prong circuit tester test, it must be tested according to step 2.

Step 2. Plug a reliable tension tester into the receptacle end. The parallel receptacle contact tension and the grounding contact tension should check out at 8 oz or more. If the tension is less than 8 oz, the receptacle end of the extension cord must be repaired. As with fixed electrical receptacles, the receptacle end of an extension cord loses its grounding contact tension first. This path is the critical human protection path and must be both electrically and mechanically in good condition.

Step 3. Once an extension cord has passed steps 1 and 2, the ground-loop impedance tester may be used. The extension cord must be plugged into a properly connected receptacle (as noted in step 1). Plug the GLIT into the receptacle end of the extension cord. A reading of 1 ohm or less is required. If you obtain a reading of more than 1 ohm, return the cord to maintenance for analysis and repair, or replacement.

If a GLIT is not available, use an ohmmeter to test the continuity from the U-shaped grounding pin to the grounding outlet hole in the receptacle end of the extension cord. This test is done with the extension cord deenergized. *The extension cord should not be plugged into an electrical receptacle when using an ohmmeter. Damage to the ohmmeter could result.*

Testing Plug- and Cord-Connected Equipment

The last element of the electrical systems test is the cord- and plug-connected equipment. The electrical inspector should be alert for jury-rigged repairs made on electric power tools and appliances. Often a visual inspection will disclose three-prong plugs with the grounding prong broken or cut off. The grounding

path to the equipment case has been destroyed. The plug can now be plugged into the receptacle in the reverse polarity configuration.

Double-insulated equipment generally has a nonconductive case and will not be tested using the procedures in this section. Some manufacturers who have UL-listed double-insulation ratings may also provide a three-prong plug to ground any exposed noncurrent-carrying metal using the grounding path conductor. In these cases, the grounding path continuity can be tested.

A common error from a maintenance standpoint is installation of a three-prong plug on a two-conductor cord to the appliance. Obviously, there will be no grounding path if there are only two conductors in the cord. Anything can be jury-rigged, but normally the maintenance error is simply to hook up the black- and white-insulated conductors to the brass and silver screws, respectively, and leave the third screw unconnected. Some hospital-grade plugs have transparent bases that allow visual inspection of the electrical connection to each prong. Even in this situation, you should still perform an electrical continuity test.

Field Testing with an Ohmmeter. Many electrical power tool testers require availability of electric power. The ohmmeter can be used in the field and locations where electric power is not available or is not obtained easily. An additional safety feature is that the plug- and cord-connected equipment tests are made on deenergized equipment. Testing of deenergized equipment in wet and damp locations also can be done safely.

The plug- and cord-connected equipment test using a self-contained battery-powered ohmmeter is simple and straightforward. The two-step testing sequence can be performed on three-prong plug grounded equipment.

Voltage Detector Testing

When conducting an electrical inspection, a voltage detector (Figure 16–20) should be used in conjunction with the circuit tester, ohmmeter, and tension tester. There are several types of these devices on the market. The unit shown in Figure 16–20 is battery-powered and constructed of nonconducting plastic.

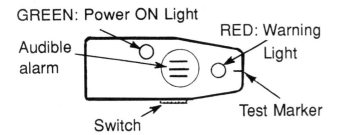

GREEN: Power ON Light

Audible alarm

RED: Warning Light

Switch

Test Marker

Figure 16–20. Use a voltage detector during an electrical inspection to find the "hot" conductor in a cord.

Being lightweight and self-contained, this tester makes an ideal inspection tool.

The voltage detector works like a radio receiver. It can receive or detect the 60-hertz electromagnetic signal from the voltage waveform surrounding an ungrounded ("hot") conductor. Figure 16–21 illustrates the detector being used to detect the hot conductor in a cord connected to a portable hand lamp. Whenever the front of the detector is placed near an energized (hot or ungrounded) conductor, the tester will provide an audible as well as visual warning. In Figure 16–21, if the plug was disconnected from the receptacle, the detector would not give an alarm since there would not be any voltage waveform present.

The detector can also be used to test for properly grounded equipment. Whenever the tester is positioned on a properly grounded power tool (e.g., the electric drill shown in Figure 16–22), the tester will not give a warning indication. The reason is that the electromagnetic field is shielded from the detector so that no signal is picked up. If the grounding prong had been removed and the drill was not grounded, the electromagnetic waveform would radiate from the drill. The detector would receive the signal and give an alarm.

As illustrated in Figure 16–22, the detector can test for many things during an inspection. Receptacles can be checked for proper AC polarity.

Circuit breakers can be checked to determine if they are ON or OFF. All fixed equipment can be checked for proper grounding. If the detector gives a warning indication on any metal-enclosed equipment, have it tagged out and repaired quickly. Ungrounded equipment can be just that or, in addition, there may be current leakage to the enclosure, making it hot with respect to ground. An experienced inspector could use the voltmeter to determine if there is voltage on the enclosure. The voltage detector should be used as an indicating tester, while qualitative testing should be done using other testing devices such as a multimeter. The use of the detector can speed up the inspection process by allowing you to check equipment grounding quickly and safely.

Another unique feature of the voltage detector shown in Figure 16–20 is that it is voltage sensitive. Table 16–B lists distances from a nonshielded hot conductor, at which the detector red light turns on, as compared to showing the strength of the voltage waveform. As an example, conductors energized with 120 volts AC can be detected from 0 to 1 in. from the conductor. On the other hand, power source wiring on a furnace automatic ignition system rated at 10,000 volts can be detected from 6 to 7 ft (1.8–2.1 m) away. You can use this feature to assist in making judgments regarding degree of hazard and urgency in obtaining corrective action on any appliance or power tool.

When using the voltage detector, you must understand how it operates. If you put the tester on an energized, double-insulated power tool, you will see the red warning light turn on. In this situation, the voltage waveform is detected because there is no metal enclosure to shield the waveform. The same would occur if you were to test many of the typewriters used in offices. This does not mean that the plastic- or non-metallic-enclosed equipment is unsafe, only that it is energized and not a grounding-type piece of equipment. By understanding the operational features of a voltage detector, you can see the benefits it offers and the many electrical hazards it can detect that other equipment may overlook.

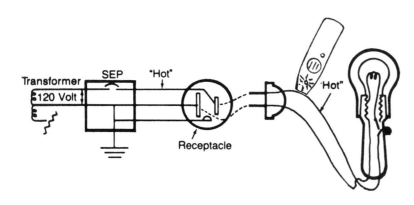

Transformer SEP "Hot"

120 Volt

Receptacle

Hot

Figure 16–21. To test for a "hot" conductor, place the front of the voltage detector near an energized conductor; the tester gives a visible and audible warning.

Safely and quickly check
for voltage at AC outlets

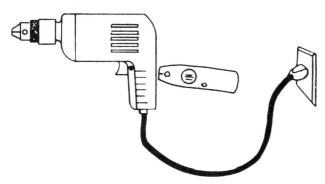

Easily check power tools
for proper grounding

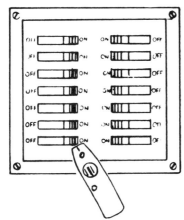

Quickly check circuit breakers

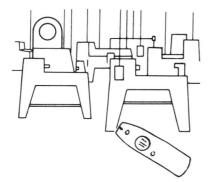

Check fixed equipment grounding

Figure 16–22. A voltage detector can be used to check receptacle voltage, equipment grounding, and circuit breaker condition.

Table 16–B. Voltage Versus Detection Distance

To Determine Approximate Voltage: Slowly approach the circuitry being tested with the front sensor end of the unit and watch at what distance the red LED light glows along with the "beep" sound. Use the chart below to determine the approximate voltage in the circuitry.

Voltage (V)	100	200	600	1K	5K	9K
Distance (inch)	0-1	1-2	3-5	15	5 ft	6-7 ft and up

These figures may vary due to conditions governing the testing, i.e., static created by standing on grounded material, carpets, etc.

GROUND-FAULT CIRCUIT INTERRUPTERS

Operational Theory

The ground-fault circuit interrupter (GFCI) is a fast-acting device that monitors current flow to a protected load. It can sense any leakage current returning to the power supply by any electrical loop other than through the white (grounded) and the black (hot) conductors.

When any leakage current 5 ma or over is sensed, the GFCI, in a fraction of a second, shuts off the current on both the hot and grounded conductors, thereby interrupting the fault current to the appliance and the fault loop. This action is illustrated in Figure 16–23.

As long as I1 is equal to I2 (normal appliance operation with no ground fault leakage), the GFCI switching system remains closed (N.C.). If a fault occurs between the metal case of an appliance and the hot conductor, fault current I3 will cause an imbalance (5 ma or greater for personnel protection), allowing the GFCI switching system to open (as illustrated) and remove power from both the white and hot conductors.

Another type of ground fault can occur when a person contacts a hot conductor directly or touches an appliance with no (or faulty) equipment-grounding conductor. In this case, I4 represents the fault current loop back to the transformer. This ground fault is generally the type to which personnel are exposed.

The GFCI is intended to provide worker protection by deenergizing a circuit, or portion of a circuit, in approximately $1/5140$ of a second when the ground fault current exceeds 5 ma. The GFCI should not be confused with ground fault protection (GFP) devices that protect equipment from damaging line-to-line fault currents. Protection provided by GFCIs is independent of the condition of the equipment-grounding conductor.

GFCI Protection Device

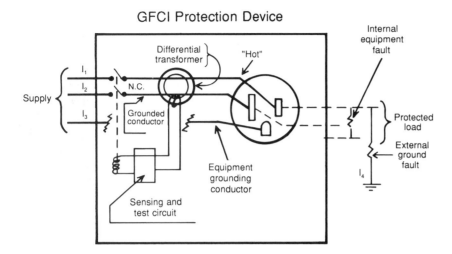

Figure 16–23. A circuit diagram for a GFCI.

Figure 16–24. Three types of GFCI: *(top)* a circuit breaker with a GFCI, *(middle)* a receptacle GFCI, *(bottom)* a portable GFCI.

The GFCI can provide protection to personnel even when the equipment-grounding conductor is accidentally damaged and made inoperative. When replacing receptacles in older homes or business buildings (equipped with nongrounding-type receptacles), an exception in the *NEC* allows GFCIs to be used when a grounding means does not exist for installation of a grounding-type receptacle. The existing non-grounding receptacles are permitted to be replaced with grounding-type receptacles, when power is provided to them through a GFCI-style receptacle or circuit breaker.

It is important to remember that a fuse or circuit breaker cannot provide hot-to-ground loop protection at the 5-ma level. The fuse or circuit breaker is designed to trip or open the circuit if a line-to-line or line-to-ground fault occurs that exceeds the circuit protection device rating. For a 15-amp circuit breaker, a short in excess of 15 amps or 15,000 ma would be required. The GFCI will trip if 0.005 amp or 5 ma starts to flow through a ground fault in a circuit it is protecting. This small amount of current (5 ma) flowing for the extremely short time required to trip the GFCI will not electrocute a person but will shock the person in the magnitude noted in Table 16–A.

GFCIs are available in several different types. Figure 16–24 illustrates three of the types available. Circuit breakers can be purchased with built-in GFCI protection. This combination unit can be secured in the circuit-breaker panel to prevent unauthorized persons from having access to the GFCI. The receptacle GFCI is the most convenient type. It can be tested and/or reset at the location where it is being used. It can also be installed so it is closest to the circuit-breaker panel and will provide GFCI protection to all receptacles on the load side of the GFCI. The portable type can be carried in a maintenance toolbox for instant use at a location where the available AC power is not GFCI-protected.

Other versions of GFCIs are available, including multiple outlet boxes for use by several power tools and for outdoor applications. The obvious difference between a regular electrical receptacle and a GFCI receptacle is the presence of the TEST and RESET buttons. The user should follow the manufacturer's instructions regarding the testing of the GFCI. Some recommend pushing the TEST button and resetting monthly. Defective units should be replaced immediately.

Be aware that a GFCI will not protect the user from line-to-line electrical contact. Suppose a person is standing on a surface insulated from ground (e.g., a dry, insulated floor mat), while holding a faulty appliance with a hot case in one hand. The worker then reaches with the other hand to unplug the appliance and contacts an exposed grounded (white) conductor. The worker has made a line-to-line contact. However, the GFCI will not protect her or him because there is no ground loop. To prevent this type of accident, it is essential to have an equipment-grounding conductor inspection program in addition to a GFCI program. All personnel should train and encourage employees to follow the electrical safety policy described later in this chapter.

GFCIs do not replace an ongoing electrical equipment inspection program. Rather, GFCIs should be considered an additional personnel safety factor for protection against the most common form of electrical shock and electrocution, the line (hot)-to-ground fault. It is recommended that GFCI protection be provided in any working area where damp or wet conditions can or do exist.

Off-the-Job Electrical Safety. The *NEC* requires GFCIs on 15-amp, 20-amp, 120-volt outside receptacles, garage circuits, and bathrooms. The *NEC* also requires that GFCIs be installed in crawl spaces at or below grade level and in unfinished basements. Receptacles installed within 6 ft of a kitchen sink to serve countertop surfaces must also be GFCI protected. Other locations where the *NEC* requires GFCI protection include pools, fountains, commercial garages, pipeline heating, marinas and boat yards, mobile homes, and recreational vehicles. It is recommended that you consult the NEC for the specific requirements applicable to any of these occupancies.

Nuisance GFCI Tripping

When GFCIs are used in construction activities, the GFCI should be located as close as possible to the electrical equipment it protects. Excessive lengths of temporary wiring or long extension cords can cause ground fault leakage current to flow by capacitive and inductive coupling. The combined leakage current can exceed 5 ma, causing the GFCI to trip.

Other nuisance tripping may be caused by one or several of the following items:

- Wet electrical extension cord to tool connections
- Wet power tools
- Outdoor GFCIs not protected from rain or water sprays
- Bad electrical equipment with case-to-hot-conductor fault
- Too many power tools on one GFCI branch
- Resistive heaters
- Coiled extension cords (long lengths)
- Poorly installed GFCI
- Defective or damaged GFCI
- Electromagnetic-induced current near high-voltage lines
- Portable GFCI plugged into a GFCI-protected branch circuit

Remember that a GFCI does not prevent shock. It simply limits the duration so the heart is not affected. The shock lasts about $1/5140$ second (0.025 second) and can be intense enough to knock a person off a ladder or otherwise cause an accidental injury. Emphasize to your workers that they should not consider a GFCI adequate or general protection against electrical hazards, and that they must follow all safety steps rigorously.

HAZARDOUS LOCATIONS

Overview of Classes and Divisions

This section provides only a summary of electrical equipment requirements as a guide for inspection or system analysis of various hazardous locations. For more in-depth design and engineering requirements, refer to subpart S, Electrical, in the OHSA Standards (29 *CFR* 1910) and Chapter 5 of the *NEC*. Hazardous locations are areas where flammable liquids, gases, or vapors, combustible dusts, or other easily ignitable materials exist, or can exist accidentally, in sufficient quantities to produce an explosion or fire. In hazardous locations, specially designed equipment and special installation techniques must be used to protect against the explosive and flammable potential of these substances.

Hazardous locations are classified as Class I, Class II, or Class III, depending on what type of hazardous substance is or may be present. In general, Class I locations are those in which flammable vapors and gases may be present. Class II locations are those in which combustible dusts may be found. Class III locations are those in which there are ignitable fibers and flyings.

Each of these classes is divided into two hazard categories, Division 1 and Division 2, depending on the likelihood of a flammable or ignitable concentration of a substance. Division 1 locations are designated as such because a flammable gas, vapor, dust, or easily ignitable material is normally present in hazardous quantities. In Division 2 locations, the existence of hazardous quantities of these materials is not normal, but they occasionally exist either accidentally or when material in storage is handled. In general, the installation requirements for Division 1 locations are more stringent than for Division 2 locations.

Additionally, Class I and Class II locations are also subdivided into groups of gases, vapors, and dusts having similar properties. Table 16–C summarizes the various hazardous (classified) locations.

Equipment Requirements

General-purpose electrical equipment can cause explosions and fires in areas where flammable vapors, liquids, and gases, and combustible dusts or fibers are present. Hazardous areas require special electrical equipment designed for the specific hazards involved. This includes explosion-proof equipment for flammable vapor, liquid, and gas hazards and dust-ignition-proof equipment for combustible dust. Other kinds of equipment are non-sparking equipment, intrinsically safe equipment, and purged and pressurized equipment.

Many pieces of electrical equipment have parts, such as circuit controls, switches, or motors, that arc, spark, or produce heat under normal operating conditions. These energy sources can produce temperatures high enough to cause ignition. When electrical equipment must be installed in hazardous areas, the sparking, arcing, and heating nature of the equipment must be controlled.

Installations in hazardous locations must be (1) intrinsically safe; (2) approved for the hazardous location; or (3) of a type and design that provides protection from the hazards arising from the combustibility and flammability of the vapors, liquids, gases, dusts, or fibers present. Installations can have one or more of these options. Each option is described in the following paragraphs.

Intrinsically Safe. Equipment and wiring approved as intrinsically safe are acceptable in any hazardous (classified) location for which they are designed. Intrinsically safe equipment cannot release sufficient electrical or thermal energy under normal or abnormal conditions to ignite specific flammable or combustible materials present in the location.

Make sure that flammable gases and vapors do not come into contact with or pass through the equipment. Additionally, evaluate all interconnections between circuits to be sure that an unexpected source of ignition is not introduced through other nonintrinsically safe equipment. Separation of intrinsically safe and nonintrinsically safe wiring may be necessary to make sure that the circuits in hazardous (classified) locations remain safe.

Approved for the Hazardous (Classified) Location. Under this option, equipment must be approved for the class, division, and group of locations. Two types of equipment are designed specifically for hazardous (classified) locations: explosion proof and dust-ignition proof. Explosion-proof apparatus is intended for Class I locations, while dust-ignition-proof equipment is primarily intended for Class II and Class III locations. Equipment listed specifically for hazardous locations should have a recognized testing laboratory's label or mark, indicating the class, division, and group of locations where it may be installed. Equipment approved for use in a Division 1 location may be installed in a Division 2 location of the same class and group.

Explosion-Proof. Generally, equipment installed in Class I locations must be approved as explosion-proof. Since it is impractical to keep flammable gases outside of enclosures, arcing equipment must be installed in enclosures that are designed to withstand an explosion. This minimizes risk of an external explosion that occurs when flammable gas enters the enclosure and is ignited by the arcs.

Not only must the equipment be strong enough to withstand an internal explosion, but the enclosures must be designed to vent the resulting explosive gases. This venting must ensure that the gases are cooled to a temperature below that of ignition temperature of the hazardous substance involved, before being released into the hazardous atmosphere.

There are two common enclosure designs: threaded-joint enclosures (Figure 16–25) and ground-joint enclosures (Figure 16–26). When hot gases travel through small openings in either of these joints, they are cooled before reaching the surrounding hazardous atmosphere.

Other design requirements (for example, sealing), prevent the gases, vapors, or fumes from passing into one portion of an electrical system from another. Motors, which typically contain sparking brushes or commutators and tend to heat up, must also be designed to prevent internal explosions.

Dust-Ignition-Proof. In Class II locations, equipment must generally be dust-ignition-proof. Section 502–1 of the *NEC* defines dust-ignition-proof as equipment,

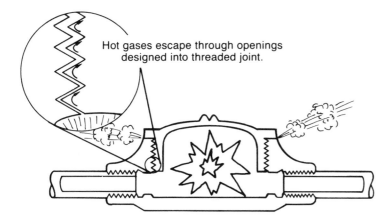

Hot gases escape through openings designed into threaded joint.

Figure 16-25. This threaded-joint explosion-proof exclosure is made of cast metal strong enough to withstand the maximum explosion pressure of a specific group of hazardous gases or vapors. Small openings designed into the threaded joint cool the hot gases as they escape. (From OSHA Publication 3073, *An Illustrated Guide to Electrical Safety*.)

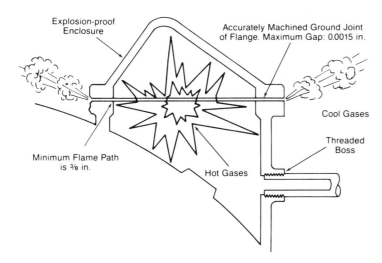

Explosion-proof Enclosure

Accurately Machined Ground Joint of Flange. Maximum Gap: 0.0015 in.

Cool Gases

Threaded Boss

Minimum Flame Path is ³/₈ in.

Hot Gases

Figure 16-26. This ground-joint enclosure cools gases as they escape through the ground joint of the flanges. (From OSHA Publication 3073, *An Illustrated Guide to Electrical Safety*.)

. . .enclosed in a manner that will exclude ignitable amounts of dust or amounts that might affect performance or rating and that, where installed and protected in accordance with the Code, will not permit arcs, sparks, or heat otherwise generated or liberated inside the enclosure to cause ignition of exterior accumulations or atmospheric suspensions of a specified dust on or in the vicinity of the enclosure.

Dust-ignition-proof equipment is designed to keep ignitable amounts of dust from entering the enclosure. In addition, dust can accumulate on electrical equipment, causing overheating as well as dehydration or gradual carbonization of organic dust deposits. Overheated equipment may malfunction and cause a fire. Dust that has carbonized is susceptible to spontaneous ignition or smoldering. Equipment must be designed to operate below the ignition temperature of the specific dust involved, even when blanketed. The shape of the enclosure must be designed to minimize dust accumulation when fixtures are out of reach of normal housekeeping activities, e.g., lighting fixture canopies.

Class II hazardous locations are comprised of three groups—E, F, and G. (See NFPA 497M for an explanation of hazardous atmosphere groups A–G.) Special equipment designs are required for these locations to prevent dust from entering the electrical equipment enclosure. For example, assembly joints and motor shaft openings must be tight enough to prevent dust from entering the enclosure. The design must take into account the insulating effects of dust layers on equipment and make sure the equipment will operate below the ignition temperature of the dust involved. If conductive combustible dusts are present, the design of equipment must take the special nature of these dusts into account.

In general, explosion-proof equipment is not designed for, and is not acceptable for use in Class II locations, unless specifically approved. For example, since grain dust has a lower ignition temperature than that of many flammable vapors, equipment approved for Class I locations may operate at a temperature that is too high for Class II locations. In

Table 16–C. Summary of Class I, II, and III Hazardous Locations

Classes	Groups	Divisions	
		1	2
I Gases, Vapors, and Liquids (Art. 501)	A: Acetylene B: Hydrogen, etc. C: Ether, etc. D: Hydrocarbons, Fuels, Solvents, etc.	Normally explosive and hazardous	Not normally present in an explosive concentration (but may accidentally exist)
II Dusts (Art. 502)	E: Metal dusts (conductive* and explosive) F: Carbon dusts some are conductive,* and all are explosive) G: Flour, Starch, Grain, Combustible Plastic or Chemical Dust (explosive)	Ignitable quantities of dust normally is or may be in suspension, or conductive dust may be present	Dust not normally suspended in an ignitable concentration (but may accidentally exist). Dust layers are present.
III Fibers and Flyings (Art. 503)	H: Textiles, woodworkings, etc. (easily ignitable, but not likely to be explosive)	Handled or used in manufacturing	Stored or handled in storage (exclusive of manufacturing)

*Note: Electrically conductive dusts are dusts with a resistivity less than 10^6 ohm-centimeters.
Source: *An Illustrated Guide to Electrical Safety*, OSHA Pub. No. 3073.

contrast, equipment that is dust-ignition-proof is generally acceptable for use in Class III locations, since the same design considerations are involved.

Safe Design for Hazardous Locations. Under this option, equipment installed in hazardous locations must be of a type and design that provide protection from the hazards arising from combustibility and flammability of vapors, liquids, gases, dusts, or fibers. The employer has the responsibility of demonstrating that the installation meets this requirement. Guidelines for installing equipment under this option are contained in the *NEC* in effect at the time of installation of the equipment. Compliance with these guidelines is not the only means of complying with this option; however, the employer must demonstrate that installation is safe for the hazardous (classified) location.

Equipment Marking

Approved equipment must be marked to indicate the class, group, and operating temperature range (based on 40 C ambient temperature) for which it is designed to be used. Furthermore, the temperature marked on the equipment must not be greater than the ignition temperature of the specific gases or vapors in the area. There are, however, four exceptions to this marking requirement:

- First, equipment that produces heat but that has a maximum surface temperature of less than 100 C (or 212 F) is not required to be marked with operating temperature range. The heat normally released from this equipment cannot ignite gases, liquids, vapors or dust.
- Second, any permanent lighting fixtures that are approved and marked for use in Class I, Division 2 locations do not need to be marked to show a specific group. This is because these fixtures are acceptable for use with all the chemical groups for Class I, that is, for Groups A, B, C, and D.
- Third, general-purpose equipment, other than lighting fixtures, considered acceptable for use in Division 2 locations, does not have to be labeled according to class, group, division, or operating temperature. This type of equipment does not contain any devices that might produce arcs or sparks. It is therefore, not a potential ignition source. For example, squirrel-cage induction motors without brushes, switching mechanisms,

Table 16–D. Most Often Violated Electrical Standards

NEC-NFPA 70-1990 Reference	Subject	OSHA Standard 29 *CFR* 1910
110-3	Suitability for safe installation and use in accordance with listing or labeling	.303(b)(1)
110-12(a)	Unused openings in cabinets, boxes, and fittings	.305(b)(1)
110-13(a)	Secure mounting of equipment	—
110-16(a)	Electrical terminal connections	—
110-16(b)	Electrical splices	.303(c)
110-16	Working space about electric equipment	.303(g)(1)
110-17	Guarding of live parts	.303(g)(2)
110-22	Disconnect and circuit identification	.303(f)
200-11	Reverse polarity	.304(a)(2)
210-7 & 410-58	Grounding-type receptacles, cord connectors, and attachment plugs	.305(j)(2)(i)
210-8	Ground-fault circuit interrupters	—
210-63	Heating, air-conditioning, and refrigeration equipment outlet	—
250-42	Grounding fixed equipment—general	.304(f)(5)(iv)
250-45	Founding of cord- and plug-connected equipment	.304(f)(5)(v)
250-51	Effective grounding path	.304(f)(4)
250-59	Methods of grounding cord-connected equipment	—
400-7	Flexible cord & cable uses permitted	.305(g)(1)(i)
400-8	Flexible cord & cable uses not permitted	.305(g)(1)(iii)
400-9	Flexible cord & cable splices	.305(g)(2)(ii)
400-10	Pull at joints and terminals	.305(g)(2)(iii)

or similar arc-producing devices are permitted in Class I, Division 2 locations (See *NEC*, Section 501-8(b)). Therefore, they need no marking.

- Fourth, for Class II, Division 2, and Class III locations, dust-tight equipment (other than lighting fixtures) is not required to be marked. In these locations, dust-tight equipment does not present a hazard so it need not be identified.

ELECTRICAL STANDARDS MOST OFTEN VIOLATED

To aid the new or even experienced inspector, a listing of the most often-cited electrical standard violations is provided (Table 16–D). This table is a compilation of federal and private sector inspection experience. It primarily concerns electrical personnel safety. The average inspector should become adept at recognizing the items in this listing. These items should be included in any electrical inspection checklist you may be using.

Table 16–D also provides a cross reference to either the *NEC*, 1990 edition, and/or the OSHA standard paragraph. Each electrical standard noted in Table 16–D is discussed below. Consult the *NEC* or the OSHA standard for additional details and interpretations.

- **110–3.** *Examination, Identification, Installation, and Use of Equipment.* This standard provides operational characteristics for practical employee and facility safeguarding. You should evaluate suitability, mechanical strength of enclosures, insulation, heating and wiring effects, proper use of listed or labeled equipment, and any other factor that would provide personnel safeguarding. Fabricating and using extension cords with junction box receptacle ends would be a violation of this standard.

- **110–12(a).** *Unused Openings.* This reference requires that electrical equipment be installed in a neat and professional manner. All openings in junction boxes and electrical equipment must be effectively closed to prevent metal objects from entering the enclosure and causing arcing or shorting of the supply conductors. Personnel protection is also provided by preventing contact with electrically live parts.

- **110–13(a).** *Secure Mounting of Equipment.* All fixed electrical equipment must be firmly secured to the surface on which it is installed. Many times you may discover a conduit hanging loose or equipment boxes not secured to the wall. These are examples of violations of this standard.

- **110–14(a).** *Electrical Terminal Connections.* Loose or improperly tightened terminal connections have been determined as the cause of many electrical fires and equipment burnouts. It is important to be alert to this problem when intermittent equipment operation or flickering lights are observed.

- **110–14(b).** *Splices.* Splices are required to be joined by suitable splicing devices or methods. The three elements of a proper splice are (a) mechanical strength, (b) electrical conductivity, and (c) insulation quality. These factors must at least be equivalent to the conductors being spliced.

- **110–14.** *Working Space About Electrical Equipment* (600 volts, nominal, or less). This paragraph requires that sufficient access and working space be provided and maintained around all electrical equipment. It is important that the electrical equipment clearance space does not become storage space. It may be necessary to make the clearance space obvious by using floor stripes or other methods.

- **110–17.** *Guarding of Live Parts* (600 volts, nominal, or less). This reference requires that the live parts of electrical equipment operating at 50 volts or more must be guarded against accidental contact. Approved enclosures are recommended. When the alternatives of location (e.g., 8-ft elevation above the floor) are used, employee safeguarding should be carefully evaluated.

- **110–22.** *Disconnect and Circuit Identification.* This reference requires that the disconnecting means for motors and appliances, and each service, feeder, or branch circuit at the point where it originates, be legibly marked to indicate its purpose unless located and arranged so the purpose is evident. Many times a contractor will install new wiring and leave circuit-breaker panels with blank circuit identification cards. Whenever electrical modifications are completed, the circuit identifications cards should be updated.

- **200–11.** *Polarity of Connections.* No grounded conductor should be attached to any terminal or lead so as to reverse designated polarity. This usually can be determined by using the 3-prong circuit tester. Be sure to test wall receptacles and also extension cord receptacles.

- **210–7 & 410–58.** *Grounding-Type Receptacles, Cord Connectors, and Attachment Plugs.* These references require that receptacles installed on 15- and 20-amp branch circuits be of the grounding type. Grounding-type attachment plugs and mating cord connectors must also be used on electrical equipment where grounding-type receptacles are provided. An exception would be the use of UL-listed double-insulated tools or appliances.

- **210–8.** *Ground-Fault Circuit-Interrupter Protection (GFCI) for Personnel.* This section applies to dwelling units, hotels, and motels, but GFCI protection may be provided for other circuits, locations, and occupancies where personnel can be protected against line-to-ground shock hazards. GFCI-protection requirements for other specific applications are as follows: 305–6 Construction Sites; 426–6 Fixed Heating for Pipelines; 426–31 Fixed Outdoor De-Icing Equipment; 511–10 Commercial Garages; 551–41 (c) Recreational Vehicles; and 680

Swimming Pools, Tubs, and Fountains. The various codes and regulations are minimal requirements and in many cases not retroactive. From an accident-prevention standpoint, you should look for areas where potential line-to-ground shock hazards could occur and recommend GFCI protection regardless of the lack of retroactivity.

- **210–63.** *Heating, Air-Conditioning, and Refrigeration (HVAC) Equipment Servicing Outlet.* It is now a requirement that a 125-volt, 15- or 20-amp receptacle be installed on rooftops and in attic spaces for use by maintenance personnel for servicing HVAC equipment. These receptacles must be on the same level and within 75 ft (22.9 m) of the HVAC equipment. Although the Code does not require it, you should consider providing GFCI-protected receptacles if maintenance work must be done during wet weather or in roof areas where water may be present.

- **250–42.** *Grounding Fixed Equipment—General.* This paragraph requires exposed noncurrent-carrying metal parts of fixed equipment to be grounded. By using the noncontact voltage detector, you can quickly determine whether metal equipment housing is properly grounded.

- **250–45.** *Grounding of Cord- and Plug-Connected Equipment.* This paragraph requires that exposed noncurrent-carrying metal parts of cord- and plug-connected equipment (listed) shall be grounded. Exceptions to this requirement are tools and appliances listed as double-insulated, or that are supplied through an isolating transformer with an ungrounded secondary of not more than 50 volts.

- **250–51.** *Effective Grounding Path.* The path to ground from circuits, equipment, and conductor enclosures must (a) be permanent and continuous, (b) have capacity to conduct safely any fault current likely to be imposed on it, and (c) have sufficiently low impedance to limit the voltage to ground and to facilitate the operation of the circuit-protection devices. If a portable power tool has the equipment-grounding prong broken off, the 250–51 (1) requirement would not be met since the grounding path is not continuous. Extension cords being used with defects in the equipment-grounding conductor path would also not meet this requirement. Finally, if the grounding path tested out at more than 10 ohms, 250–51 (3) would not be met. It is recommended that the equipment-grounding conductor loop impedance be less than 1 ohm. This can be determined only by using a ground-loop impedance tester.

- **250–59.** *Cord- and Plug-Connected Equipment.* This section describes three methods that can be used to ground equipment effectively. One method is to use the electrical conduit in conjunction with approved connectors and receptacles. The second method is to run the equipment-grounding conductor along with the power-supply conductors in a cable assembly or flexible cord. The third method is to use a separate flexible wire or cable. The main purpose of alternative methods is to ensure that all metal enclosures and frames are at the same ground potential.

- **400–7.** *Flexible Cord and Cable Uses Permitted.* Specific applications where flexible cords and cables are permitted include pendants, wiring of fixtures, portable lamps and appliances, and stationary equipment that may have to be moved frequently for maintenance or other purposes. Another important use is to prevent the transmission of noise and vibration. Special attention should be given to proper installation and protection of flexible cords and cables.

- **400–8.** *Flexible Cord and Cable Uses Not Permitted.* Flexible cords must not be used as a substitute for fixed wiring. Additionally, they must not be run through holes in walls, ceilings, or floors nor through doorways or windows.

- **400–9.** *Splices.* This section prohibits use of flexible cords that have been spliced or tapped. Using splices to repair hard-service cord No. 14 or larger is permitted if done in accordance with the previously covered Section 110–14(b) Splices.

- **400–10.** *Pull at Joints and Terminals.* This reference requires that flexible cords be connected to devices and to fittings so that tension will not be transmitted to joints or terminal screws. This requirement applies to the points where the cord connects to the appliance and to the attachment plug.

The above list is not intended to be comprehensive, but rather identifies some of the most common electrical safety problem areas. To determine the exact requirements, consult the *NEC* or the OSHA standard noted. Using the testing methods noted in the Branch Circuit and Equipment Testing Methods section of this chapter will help you to find and document many of these hazards.

INSPECTION GUIDELINES AND CHECKLIST

Before you conduct an electrical inspection, check your test equipment to be sure it works properly. As a minimum, you should have a circuit tester, a GFCI tester (combination circuit/GFCI testers are available), a contact tension tester, and a volt-ohmmeter. Avoid wearing metal jewelry or loose hanging necklaces. Wear protective footwear that is electrical hazard-rated per ANSI 241, 1991. Your footwear

should have synthetic soles (do not wear leather soles). A clipboard, writing material, and the Table 16–E checklist should also be included. A camera is optional; if used, remember that before-and-after photos or slides make valuable visual aids for training use.

Table 16–E provides general guidelines that will assist you in checking for electrical hazards from the service entrance panel(s) to the equipment using the power. Further explanation of these guidelines is as follows:

1. Service Entrance Panel—Check the branch circuit identification. It should be up to date and posted on the panel door. Be sure the panel and cable or conduit connectors are secure. There should be no storage within 3 ft (1 m) of the panel. No flammable materials of any kind should be stored in the same area or room. Look for corrosion and water in or around the area. Missing knockouts, covers, or openings must be covered properly to eliminate workers' exposure to live parts.

2. System Grounding—Check connection of the grounding electrode conductor to the metal cold-water pipe and to any driven ground rod. Also check any bonding jumper connections and any supplemental grounding electrode fittings. These items should not be exposed to corrosion but should be accessible for maintenance and visual inspection.

3. Wiring (General)—Temporary wiring used for an extended time should be replaced with fixed wiring. Conduit and/or cable systems must be protected from damage by vehicles or other mobile equipment. All fittings and connections to junction boxes and other equipment must be secure. No exposed wiring can be allowed. Check for missing knockouts and cover plates. Jury-rigged splices on flexible cords and cables must be repaired correctly. Electrical equipment should be installed in a neat and professional manner. Check for damaged insulation on flexible cords and pendant drops.

4. Electrical Equipment/Machinery—The most important item is to test for proper grounding. All electrical equipment and machinery must be grounded effectively so that no potential difference exists between the metal enclosures. Use the voltage detector to find discrepancies and use other test equipment to determine the corrective action required.

   Disconnects should be identified clearly regarding the specific machinery they shut off, and should be placed near the machinery for emergency use. The disconnects should be exercised periodically to be sure they are operable. All electrical connections to equipment must be secure so that no wiring tension is transmitted to the

Table 16–E. Inspection Guidelines

1. **Service Entrance Panel**—Circuit I.D., Secure Mounting, Knockouts, Connectors, Clearances, Live Parts, Ratings.
2. **System Grounding**—Secure Connections, Corrosion, Access, Protection, Proper Size.
3. **Wiring (General)**—Temporary, Splices, Protected, J Box Covers, Insulation, Knockouts, Fittings, Workmanship.
4. **Electrical Equipment/Machinery**—Grounding, Wiring Size, Overcurrent/Disconnect Devices, Installation, Protection.
5. **Small Power Tools**—Attachment Plugs, Cords, Cord Clamps, Leakage, Grounding.
6. **Receptacles**—Proper Polarity, Adequate Number, Mounting, Covers, Connections, Protection, Adapters.
7. **Lighting**—Grounding, Connections, Attachment Plugs and Cords, Cord Clamps, Live Parts.
8. **GFCI Protection**—Bathrooms, Wet Locations, Fountains, Outdoor Circuits, Testing.

electrical terminals within the equipment. The wiring installation should be protected from damage at all times while machinery is in use.

5. Small Power Tools—Attachment plugs should be checked for defective cord clamps and broken or missing blades. Connection of the cord to the power tool should be secure. Use your ohmmeter to check for leakage and an effective equipment-grounding conductor.

6. Receptacles—The receptacles should be tested for proper wiring configuration. Be sure to install enough receptacles to eliminate use of extension cords as much as possible. Covers should be in place and not broken. Multiple outlet adapters on a single outlet should be discouraged to prevent overloading. Surface-mounted receptacle boxes should be protected from damage by housekeeping or other mobile equipment.

7. Lighting—Cord- and plug-connected metal lamps and fixtures should be tested for grounding. Use the ohmmeter to do this. Check all cord clamps for secure connections. Frayed or old cords should be replaced.

8. GFCI Protection—Generally, GFCI protection is not required by the NEC on a retroactive basis. However, where employees are exposed to potential line-to-ground shock hazards, GFCI protection is recommended. This is especially important in workplaces where portable electrical equipment is being used in wet or damp areas in contact with earth or grounded conductive surfaces.

Typically, areas where GFCIs should be installed include bathrooms, outdoor fountains, tubs, countertop sinks, crawl spaces under buildings, outdoor

circuits, swimming pools, and other wet locations. Use your GFCI tester to be sure that the GFCI is operable. After years of use, they can become defective and will need to be replaced. Check to make sure that receptacles being provided with GFCI protection are labeled as such.

SAFEGUARDS FOR PORTABLE HOME ELECTRICAL APPLIANCES

This section is primarily provided for use in off-the-job safety training. There are instances where home appliances are used on the job, and the following general precautions would be applicable to those situations. It is recommended that any electrical training include discussions of safeguards for small, portable, electrical home appliances. You should recommend installation of GFCIs in bathrooms and other locations required by the *NEC* for new dwelling construction.

When using electrical appliances, especially when children are present, basic safety precautions should always be followed. Appliance manufacturers provide written procedures for safe use of their appliances. Their instructions should be read by adults and children alike. Safety precautions should be understood. Strong emphasis should be given on the danger of using small electrical appliances in or near water. The following list established a common sense baseline for the safe handling, use, and storage of electrical appliances in the home. To reduce risk of electrocution:

- Install GFCI-protected receptacles in bathrooms, near kitchen sinks, outdoors, and in other areas near water or where wet or damp conditions may exist.
- Always unplug and store appliances after using.
- Do not use any electrical appliance while bathing. For examples of warning labels, see Figure 16–27.
- Do not place or store appliances where they can fall or be pulled into a sink or tub (such as on towel bar on tub door).
- Do not place or drop any appliance into water or other liquid.
- Do not reach for an appliance that has fallen into water. Unplug the electrical cord immediately.

To reduce the risk of burns, electrocution, fire, or injury:

- Never leave an appliance unattended when it is plugged in.
- Closely supervise an appliance used by, on, or near children or invalids.
- Use the appliance only for its intended use as described in the manufacturer's manual. Do

not use attachments not recommended by the manufacturer.

- Never operate an appliance if it has a damaged cord or plug, is not working properly, has been dropped or damaged, or has been immersed in water. Return the appliance to dealer or repair center for examination and electrical or mechanical adjustment and repair.
- Keep the cord away from heated surfaces.
- Never block the air openings of an appliance. Keep the air openings free of lint, hair, paper, or other materials.
- Never use appliances when you are drowsy.
- Never drop or insert any object into any appliance opening.
- Do not use an appliance where flammable liquid vapors or aerosol (spray) products are being used or where oxygen is being administered.
- Do not attempt to repair appliances yourself. Most small electrical appliances do not lend themselves to normal repair procedures.
- Keep your appliance in a cool, dry place and out of the reach of children. When handling or storing your unit, do not put any stress on the line cord and the point where it enters the unit. Excessive stress could cause the cord to fray and to break.
- Never plug an appliance into a damaged receptacle or one that has a missing face plate.
- Keep all electrical appliances away from hot tubs, bath tubs, circulating water tubs, sinks, and swimming pools.

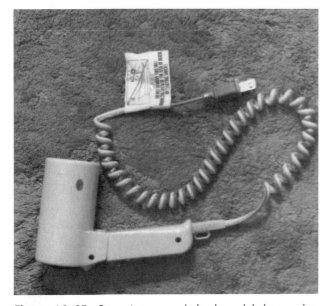

Figure 16–27. Do not remove hair dryer labels warning against using them in bathtubs.

SAFETY PROGRAM POLICY AND PROCEDURES

Safety Policy for Electrical Equipment

Each facility should have an electrical safety program policy. The policy should cover the responsibilities of all employees, including supervisors, workers, and the specialists who inspect, install, and maintain the electrical systems and equipment. The policy should stress management's concern and support of safety issues. Individuals responsible for applying and enforcing the electrical policy should have standards of performance which include periodic assessment of their electrical safety performance.

In addition to policy and implementation procedures, an electrical safety program should also include four basic areas of concern: training and education, hazardous condition identification and reporting, work practices, and housekeeping. The National Safety Council Occupational Safety and Health Data Sheets (Table 16–F) available through the NSC library provide an excellent source of technical and training information to support any facility electrical safety program. In addition, the professional electrician responsible for installing, repairing, and maintaining electrical equipment and systems should be familiar with NFPA 70E, *Electrical Safety Requirements of Employee Workplaces.*

Management should support an effective, preventive maintenance program. Use NFPA 70B, *Electrical Equipment Maintenance,* as a guide to implement or fine-tune this type of program. All employees should be made responsible for detecting and reporting unsafe electrical equipment. The following discussion provides suggestions for ensuring that these responsibilities are carried out effectively.

Supervisory Responsibilities

Training and education. Supervisors should have training courses to help them carry out the electrical safety program responsibilities for their specific areas. All employees exposed to electrical hazards should receive electrical safety training. In addition to electrical workers, this may include janitors, machine operators, etc. See list of job types in OSHA 29 *CFR* 1910.332. You must also monitor employees and assess performance against established facility safety program procedures.

Hazardous Condition Reporting. A written procedure promoting observation and reporting of electrical hazards should be implemented. You might also institute an employee recognition program in conjunction with hazard reporting. This will reward employees who identify and locate electrical hazards,

Table 16–F. National Safety Council Occupational Safety & Health Data Sheets that Pertain to Electrical Safety Programs (available in NSC Library)

Data Sheet No.	Title
I-240	Cathode-Ray Tubes
I-248	Auxiliary Electrical Systems and Emergency Lighting
I-316	Low-Voltage Extension Light Cords and Systems
I-385	Electric Cords and Fittings
I-498	Live Line Tools
I-515	Temporary Electric Wiring for Construction Sites
I-544	Electrical Switching Practices
I-546	Maintenance of Electric Motors for Hazardous Locations
I-547	Static Electricity
I-579	Applications of Electric Plug and Receptacle Configurations
I-581	Silicon Diode Grounding Devices
I-598	Flexible Insulating Protective Equipment for Electrical Workers
I-607	Direct Buried Utility Cables
I-620	Care, Fitting, and Maintenance of Lineman Climbers
I-624	Electrical Controls for Mechanical Power Presses
I-635	Lead-Acid Storage Batteries
I-636	GFCIs for Personnel Protection
I-641	Electrical Testing Installations
I-644	Treatment of Extraneous Electricity in Electric Blasting
I-657	Underground Residential Distribution of Electricity
I-660	Electrical Safety in Health Care Facilities
I-675	Electric Hand Saws, Circular Blade Type
I-684	Equipment Grounding

so they can be controlled.

Work Practices. A sample of suggested work practices is included in this section under Employee Responsibilities. Employees should be rated on performance in following safe work practices.

You should also be familiar with OSHA standards as they apply to your responsibility.

Housekeeping. Housekeeping must be monitored closely by the supervisor. Floor-area problems always present challenges. Areas around electrical equipment such as circuit-breaker panels, disconnects, and fixed power tools should be kept free from stored items, debris, and any liquids or material that could create slippery floors or obstruct access to the equipment for maintenance or emergencies. When hazards in these areas are reported, they should be logged in, and work orders issued to correct them.

Employee Responsibilities

All employees have the responsibility of following safety procedures that they have been taught. They are also responsible for reasonable alertness and cooperation in maintaining safety procedures on the job.

Training and Education. Employees must receive training in electrical safety work practices, equipment operation, and required PPE. New job duties also require new safety training. Many accidents are caused by employees' lack of knowledge about equipment or its operation. Sometimes blame for accidents is placed on employees when, in reality, specific training was not provided. OSHA requires it under 29 *CFR* 1910.330–335.

Hazardous Condition Reporting. Employees should always report unsafe equipment, conditions, or procedures. Team effort is required by employees and supervisors. Getting equipment repaired should receive priority, even if it requires rescheduling a process or project. Under no condition should defective electrical equipment causing electrical shocks to employees be used to get a job done. The Electrical Safety Policy should be followed and deviations reported immediately.

Work Practices. Employees are responsible for following safe work practices, procedures, and safety policy as established by the employer. Workers should also be familiar with OSHA regulations as they apply to workplace safety.

Housekeeping. In the process of routine housekeeping, employees should be observant and report conditions that could cause electrical shock hazards. For example, the use of improperly grounded electrical equipment in areas that have water on the floor can expose workers to risk of electrical shock. Storing tools or other materials around electrical panels or equipment disconnects can create hazards as well as prevent immediate disconnection in an emergency. Cleaning electrical equipment with solvents can create health and physical safety problems.

Electrical Safety Policy

Supervisors must know all facets of their employers' electrical safety procedures and make sure workers understand what is expected of them. You should stress that, as a minimum, the following items must be observed:

- Plug power equipment into wall receptacles with power switches in the Off position.
- Unplug electrical equipment by grasping the plug and pulling. Do not pull or jerk the cord to unplug the equipment.
- Do not drape power cords over hot pipes, radiators, or sharp objects.

- Check the receptacle for missing or damaged parts. Do not plug equipment into defective receptacles.
- Check for frayed, cracked, or exposed wiring on equipment cords. Check for defective cord clamps at locations where power cords enter equipment or attachment plugs.
- Generally speaking, extension cords should not be used in office areas. Extension cords should be limited to use by maintenance personnel.
- "Cheater plugs," extension cords with junction box receptacle ends, or other jury-rigged equipment should not be used.
- Consumer electrical equipment or appliances should not be used if not properly grounded.
- Personnel should know the location of electrical circuit-breaker panels that control equipment and lighting in their areas. Circuits and equipment disconnects must be identified.
- Temporary or permanent storage of materials must not be allowed within 3 ft (1 m) of any electrical panel or electrical equipment.
- When defective electrical equipment is identified by personnel, it should be tagged immediately and removed from service for repair or replacement.
- Any electrical equipment causing shocks or with high leakage potential must be tagged with a DANGER—DO NOT USE label or equivalent.

ELECTRICAL DISTRIBUTION SYSTEM REVIEW

The electrical distribution system should be designed, installed, operated, and maintained to safely and reliably provide electrical power for all required operations. The facility should be analyzed to determine that the following requirements are documented and observed regarding overall electrical system operations and maintenance. This questionnaire should be used to determine compliance to OSHA and *NEC* requirements. If there are any No answers, obtain professional electrical services to correct the situation and eliminate hazards.

- There is a program of preventive maintenance and periodic inspection to ensure that the electrical distribution system operates safely and reliably. Inspections and corrective actions are documented.
- The facility identifies components of the electrical distribution system to be included in the program. Consideration is given to reliability of receptacles, wiring, fuse and/or breaker panels, transformers, switch gear, and service equipment.
- There is a current set of documents, which indicate the distribution of and controls for partial or complete shutdown of each electrical system.

- Electrical maintenance and operating personnel are provided with appropriate job training, authorization to perform specific work, and written records of this training are maintained.
- The capacity of electrical feeds and transformers is adequate for the electrical demands of the facility.
- Any special-purpose electrical subsystems or installed devices are maintained as required.
- The electrical distribution system is designed with enough receptacles and circuits to power devices used in each area without the use of extension cords.

SUMMARY OF KEY POINTS

This chapter covered the following key points:

- Electrical safety is one of a everyone's most important responsibilities. Supervisors must know the facts about electricity and be able to counter myths and misconceptions that workers may hold.
- The four basic rules of electrical action are (1) electricity isn't "live" until current flows, (2) current won't flow until there is a complete loop out from and back to the power source, (3) current always returns to its source, and (4) when current flows, work measured in watts is performed. These four rules help supervisors to analyze electrical shock hazards or investigate how and why electrical injuries occurred in their departments.
- According to Ohm's Law, it is current that can injure or kill a worker, while voltage determines how—either by burns or electrocution. The human body presents resistance to current. Should a worker establish an electrical loop between herself or himself and a ground, the current will pass through the body on the way to the ground. Even a small amount of current can cause the heart to go into ventricular fibrillation, which can produce sudden death if the heart is not resuscitated within a few minutes.

  Factors that can enhance shock potential include wet or damp locations, nearby grounded objects that the worker or electrical tool may touch, lack of worker training, and type and amount of voltage and current.
- The keys to preventing electrical shock hazards are (1) to train workers thoroughly in safe work practices; (2) to wire, insulate, and ground all electrical equipment properly so there are no ground loops between an electrical tool or machine and the worker; and (3) routinely inspect, maintain, and replace such equipment, along with all cords, outlets, plugs, attachments, and power sources.
- Employees should understand the single-phase,

three-element, 120-volt electrical system that powers industrial and commercial operations in the United States. The 240-volt, single-phase power is used for heavy-duty appliances and small motor loads. If the system is not wired properly, or if a tool or appliance is not plugged in properly, it can create reverse polarity. In reverse polarity the entire casing of a tool or machine is energized and creates a shock hazard, particularly if accidentally immersed in water.

- A ground is a conducting connection, whether intentional or accidental, between an electrical circuit or equipment and earth, or to some conducting body that serves in the place of earth. A ground loop can become an electrical hazard if workers accidentally form part of the loop. Grounding falls into two safety categories—system grounding and equipment safety grounding.

  System grounding provides protection of the power company equipment and the consumer. Equipment safety grounding provides a system where all non-current-carrying metal parts are bonded together and kept at the same potential. Grounding concepts include the grounded conductor, neutral conductor, and equipment-grounding conductor.

- Power tools and equipment can be designed to protect workers against shock hazards by being double insulated or by being equipped with 3-pronged plugs in which one prong acts as a ground. However, these safeguards are not foolproof, and workers must remember to inspect and test their equipment regularly. Double-insulated equipment should not be used in damp or wet conditions nor immersed in water.

- Attachment plugs and receptacles are designed for specific voltage/current combinations and should not be used for any other combination. To do so can cause fires or shock hazards to workers. Electrical extension cords must be the proper gage to carry the current load required. These cords should be UL-listed, tested before being put into use, and inspected regularly.

- Employees can use several testing procedures for branch circuits and electrical equipment. Branch circuit receptacles should be tested periodically to ensure that electrical outlets are safe and in good working order. Inspectors often use a three-step process using a 3-prong receptacle circuit tester, a receptacle contact tension tester, and a ground-loop impedance tester. Extension cords should be tested using the same three-step procedure, to make sure cords are in good repair and can carry the required current safely.

  Finally, cord- and plug-connected equipment should be tested using a two-step procedure of ground-to-pin testing and an appliance leakage test. Inspectors should be alert for jury-rigged repairs made on electric power tools and appliances, and for signs of wear or deterioration. An ohmmeter can be used where electric power is not available for testing equipment. A voltage detector can be used in conjunction with the circuit tester, ohmmeter, and tension tester to identify electrical hazards. It can detect voltage waveforms surrounding an ungrounded, unshielded, or hot, conductor. It is also voltage-sensitive and can detect voltage from a distance.

- The ground-fault circuit interrupter (GFCI) is a fast-acting device that monitors current flow to a protected load. If it senses a current leakage, it will shut off the current on grounded and hot conductors. The GFCI protects workers against electrical hazards caused by ground faults, but it will not protect workers from line-to-line electrical contact. GFCIs should be used where workers are exposed to wet or humid conditions and where they may come in contact with ground or grounded equipment. GFCIs do not prevent shock but limit the duration so the heart is not affected.

- Hazardous locations are classified as Class I (flammable vapors and gases present), Class II (combustible dusts present), and Class III (ignitable fibers and flyings present). Each class is divided into two hazard categories: Division 1—flammable gas, vapor, dust, or easily ignitable material is present in hazardous quantities, and Division 2—hazardous quantities of these materials in hazardous locations must consist of equipment and wiring that is (1) intrinsically safe, (2) approved for hazardous locations, or (3) of a type or design that provides protection from the hazards present. Generally, equipment in Class I locations must be approved as explosion proof, while Class II equipment must be dust-ignition proof. With some exceptions, equipment must be clearly marked to indicate the class, group, and operating temperature range for which it is designed.

- Inspectors must be familiar with most often violated electrical standards. This knowledge will help them identify and correct electrical hazards frequently occurring in the workplace. Guidelines established for inspecting electrical equipment and wiring include the following items: service entrance panel, system grounding, electrical equipment machinery, small power tools, receptacles, lighting, and GFCI protection.

- Workers should also instruct workers in off-the-job safety training to guard against electrical hazards in the home. These precautions can prevent burns, electrocution, fire, or injury to workers and their family members.

- An electrical safety program should include training and education, hazardous condition reporting, safe work practices, and housekeeping. Management personnel and employees should know clearly what their responsibilities are and what is expected of them. Part of the safety program involves setting up an electrical distribution system review to provide safe, reliable electric power for all required operations.

17

FIRE SAFETY

After reading this chapter, you will be able to:

- Explain the basic principles of fire safety, including the chemistry of fires

- Identify fire hazards, causes of fires, and safeguards required to prevent fires

- Conduct regular, periodic inspections to ensure that work areas remain in a fire-safe condition

- Describe the use and operation of fire protection equipment and systems

- Instruct employees in procedures for reporting fires and evacuating work areas

- Develop a effective fire protection education program for both on and off the job

An effective fire protection program depends on everyone. Although the overall program may be developed and administered by the director of safety, fire protection, security, engineering, or maintenance, each supervisor has a direct interest in, and responsibility for, the program's success.

Fire protection is a science in itself. This manual cannot cover all aspects of fire prevention and protection, but this chapter presents condensed, basic information to help everyone conduct fire-safe operations. More specialized information is available, and sources are cited throughout this chapter.

BASIC PRINCIPLES

Fire is an oxidation process that emits light, heat, and other products of combustion. In order to explain or understand fire development, experts have devised various fire models. One of the earliest is shown as a tetrahedron in Figure 17–1. To sustain most fires, four elements must be available at the same time: elevated temperature, oxygen, fuel, and an uninhibited chain reaction. Fire extinguishing agents act by removing one or more of the tetrahedron's sides. However, as Figure 17–1 shows, an uninhibited chain reaction is not necessary for a deep-seated, surface-glowing type of fire to continue. Under the proper conditions, many materials will burn. A good example is steel, which in the form of steel wool burns quite readily.

Fire begins with an ignition source in the presence of a fuel source and can spread to other fuel sources by conduction, convection, and radiation. Conduction transfers heat through contact with solid material. Convection transfers heat through heated air. Radiation heat transfers through electromagnetic waves given off by flames.

Employees, because of knowledge of day-to-day operations, are in an excellent position to determine necessary fire prevention measures. You should recognize the need for specific fire protection equipment and see to it that such equipment is provided. Take time to become thoroughly familiar with the fire equipment best suited to your particular operations.

Departmental housekeeping is also under the everyone's control. Make sure your employees follow safe housekeeping practices to prevent fires. Continuous training in fire-safe work procedures,

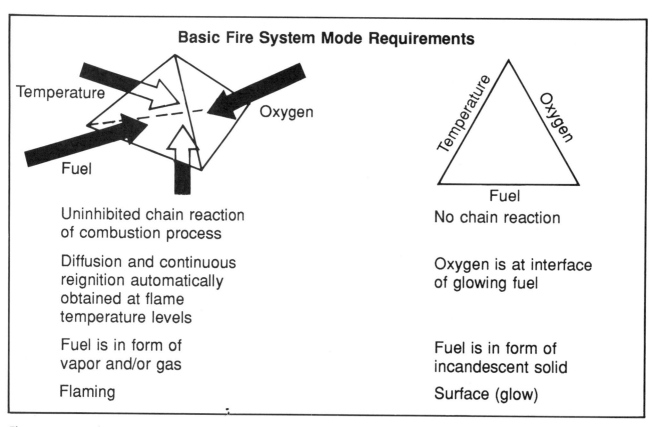

Figure 17–1. In the presence of oxygen, fuel, and elevated temperature, an uninhibited chain reaction can produce a flaming fire, but without a chain reaction you can still produce a deep-seated, surface-glowing fire.

regular inspections of work areas, and close supervision of employee job performance are primary requirements of a successful fire prevention program. Although fire protection equipment may be maintained by others, you have ultimate responsibility for the safety of yourself, materials, and equipment.

Understanding Fire Chemistry

Expertise, with regard to fire prevention and control problems, begins with an understanding of basic fire chemistry. Every ordinary fire (one that does not produce its own oxygen supply) results when a substance (fuel) in the presence of air (oxygen) is heated to a critical temperature, called its "ignition temperature."

This important concept is best illustrated by the fire tetrahedron. By looking at this model, you can understand how most fires are extinguished. The methods used include:

- **Oxygen Removal.** Removing or lowering oxygen levels is difficult because a fire needs about the same amount for burning (percentage of oxygen in the air) that humans need for breathing. In some cases, oxygen levels can be reduced below the minimum percentage needed for combustion by purging and rendering the atmosphere inert in closed containers or processing systems. Firefighting foam extinguishes fires by smothering (and cooling) action.
- **Fuel Removal.** In many cases, it is neither possible nor practical to remove all fuels (solids and liquids). However, try to keep the quantity of stored flammable and combustible materials at a minimum.
- **Heat Source Control.** Eliminating and controlling heat sources also are elementary steps in fire prevention. Conscientious workers can control use of welding and cutting equipment, torches, heating equipment, spark-producing equipment, electricity, and smoking materials. Keep heat and ignition sources away from fuel.

Table 17–A lists common sources of ignition that cause industrial fires, gives examples in each case, and suggests preventive measures. A following section, Causes of Fire, discusses each in detail.

For many years, the principle of extinguishing fires focused on the fire triangle and the removal of any of its three sides, representative of three components. Remodeling the fire triangle into a fire tetrahedron (Figure 17–1) presents a more realistic concept of extinguishment. This is because it also takes into consideration the chemical chain reaction needed to sustain combustion. The tetrahedron has four sides or faces, one for each of the ways to extinguish a fire. Because each face is directly adjacent to and connected

to the others, a tetrahedron accurately represents their interdependency. Removing one or more of the faces will put out the fire.

Consequently, to extinguish a fire, the following steps can be taken:

1. Reduce or remove the oxygen by smothering (for example, by shutting the lid over a tank of burning solvent, or by covering it with foam) or by dilution (replacing the air with an inert gas such as carbon dioxide).
2. Remove or seal off the fuel by mechanical means, or divert or shut off the flow of liquids or gases fueling the fire.
3. Cool the burning material below its ignition point with a suitable cooling agent (hose streams or water extinguishers).
4. Interrupt the chemical chain reaction of the fire (using dry chemical agents).

Once people understand the fire tetrahedron and its practical application, they will be more alert to, and aware of, fire prevention and control methods.

Determining Fire Hazards

You can contribute to a fire protection program in two ways: first, identify existing fire hazards and, second, take action to resolve them. To do this, seek the best technical advice available from fire protection experts.

Self-Inspections. Develop an inspection checklist (in conjunction with fire protection specialists) that specifies as many places, materials procedures, classes of equipment, conditions, and circumstances where fire hazards are likely to exist as possible. Use these category names as headings on the list, and under each one write down the precise fire-safe practices that must be followed. Specify which personnel—for example, machine operator, electrician, maintenance crew, or porter—have specific responsibilities (see discussion of inspections in Chapter 6, Safety, Health, and Environmental Auditing).

When, after inspection, you have made brief notations beside each of the described safe practices, you will have a detailed picture of how many or how few fire protection measures are actually in effect. You will be able to spot neglected precautions readily. The sample fire prevention checklist shown in Figure 17–2 can serve as a guide. You should make your own list and include special points for conditions that are not listed in this broad example.

When conducting an inspection, you may discover that some points previously escaped your notice. Unless you have had considerable experience in fire prevention, you should ask your superior and your company's fire safety personnel, or the fire insurance

Table 17–A. How to Control Sources of Ignition in Industrial Fires

Sources of Ignition	Examples	Preventive Measures
Electrical equipment	Electrical defects, generally due to poor maintenance, mostly in wiring, motors, switches, lamps, and hot elements.	Use only approved equipment. Follow *National Electrical Code.* Establish regular maintenance.
Friction	Hot bearings, misaligned or broken machine parts, choking or jamming of materials, poor adjustment.	Follow a regular schedule of inspection, maintenance, and lubrication.
Foreign substances	Tramp metal, which produces sparks when struck by rapidly revolving machinery (a common cause in textile industry).	Keep foreign material from stock. Use magnetic or other separators to remove tramp metal.
Open flames	Cutting and welding torches (chief offenders). Gas and oil burners. Misuse of gasoline torches.	Follow established welding precautions. Keep burners clean and properly adjusted. Do not use open flames near combustibles.
Smoking and matches	Dangerous near flammable liquids and in areas where combustibles are used or stored.	Smoke only in permitted areas. Use prescribed receptacles. Make sure matches are out.
Spontaneous ignition	Deposits in ducts and flues. Low-grade storage. Industrial wastes. Oily wastes and rubbish.	Clean ducts and flues frequently. Remove waste daily. Isolate stored materials likely to heat spontaneously.
Hot surfaces	Exposure of combustibles to furnaces, hot ducts or flues, electric lamps or irons, hot metal being processed.	Provide ample clearances, insulation, air circulation. Check heating apparatus before leaving it unattended.
Combustion sparks	Rubbish-burning, foundry cupolas, furnaces and fireboxes, and process equipment.	Use incinerators of approved design. Provide spark arresters on stacks. Operate equipment carefully.
Overheated materials	Abnormal process temperatures. Materials in driers. Overheating of flammable liquids.	Have careful supervision and competent operators, supplemented by well-maintained automatic temperature controls.
Static electricity	Dangerous in presence of flammable vapors. Occurs at spreading and coating rolls or where liquid flows from pipes.	Ground equipment. Use static eliminators. Humidify the atmosphere.

Adapted from *Factory Mutual Record.*

company engineer, to help you conduct inspections and to make recommendations for eliminating fire hazards. If company policy permits, you can ask the local fire department for help as well.

Fire inspections should be made with a critical eye. Every shortcoming should be listed. You should not hedge in listing certain hazards for fear that they might reflect poor practice. Omission of a pertinent detail might result in a fire later; such an outcome would reflect poor supervision.

The National Fire Protection Association (NFPA) Inspection Manual, a pocket-sized book, is a valuable reference for the beginner as well as the experienced inspector. It covers both common and special fire hazards, their elimination or safeguarding, and human safety in all types of properties.

Informing the Work Force

Periodic inspections are an important part of any fire protection program. Under a complete program, your responsibilities extend further. As you become acquainted with actual or potential fire hazards, and all possible physical corrections have been made, familiarize the personnel in your department with each hazard and explain how it relates to them individually.

FIRE PREVENTION CHECKLIST

ELECTRICAL EQUIPMENT

- [ ] No makeshift wiring
- [ ] Extension cords serviceable
- [ ] Motors and tools free of dirt and grease
- [ ] Lights clear of combustible materials
- [ ] Safest cleaning solvents used

- [ ] Fuse and control boxes clean and closed
- [ ] Circuits properly fused or otherwise protected
- [ ] Equipment approved for use in hazardous areas (if required)
- [ ] Ground connections clean and tight and have electrical continuity

FRICTION

- [ ] Machinery properly lubricated

- [ ] Machinery properly adjusted and/or aligned

SPECIAL FIRE-HAZARD MATERIALS

- [ ] Storage of special flammables isolated

- [ ] Nonmetal stock free of tramp metal

WELDING AND CUTTING

- [ ] Area surveyed for fire safety

- [ ] Combustibles removed or covered
- [ ] Permit issued

OPEN FLAMES

- [ ] Kept away from spray rooms and booths

- [ ] Portable torches clear of flammable surfaces
- [ ] No gas leaks

PORTABLE HEATERS

- [ ] Set up with ample horizontal and overhead clearances
- [ ] Secured against tipping or upset
- [ ] Combustibles removed or covered

- [ ] Safely mounted on noncombustible surface
- [ ] Not used as rubbish burners
- [ ] Use of steel drums prohibited

HOT SURFACES

- [ ] Hot pipes clear of combustible materials
- [ ] Ample clearance around boilers and furnaces

- [ ] Soldering irons kept off combustible surfaces
- [ ] Ashes in metal containers

SMOKING AND MATCHES

- [ ] ''No smoking'' and ''smoking'' areas clearly marked
- [ ] Butt containers available and serviceable

- [ ] No discarded smoking materials in prohibited areas

SPONTANEOUS IGNITION

- [ ] Flammable waste material in closed, metal containers
- [ ] Flammable waste material containers emptied frequently

- [ ] Piled material, cool, dry, and well ventilated
- [ ] Trash receptacles emptied daily

STATIC ELECTRICITY

- [ ] Flammable liquid dispensing vessels grounded or bonded
- [ ] Moving machinery grounded

- [ ] Proper humidity maintained

HOUSEKEEPING

- [ ] No accumulations of rubbish
- [ ] Safe storage of flammables
- [ ] Passageways clear of obstacles
- [ ] Automatic sprinklers unobstructed

- [ ] Premises free of unnecessary combustible materials
- [ ] No leaks or dripping of flammables and floor free of spills
- [ ] Fire doors unblocked and operating freely with fusible links intact

EXTINGUISHING EQUIPMENT

- [ ] Proper type
- [ ] In proper location
- [ ] Access unobstructed
- [ ] Clearly marked

- [ ] In working order
- [ ] Service date current
- [ ] Personnel trained in use of equipment

Figure 17–2. This sample checklist serves as a guide for the supervisor in drawing up an inspection list. It should be reviewed regularly to keep it up-to-date.

Call their attention to all the physical safeguards that have been provided to prevent injury and destruction by fire. You should stress to individuals the precautions necessary for their jobs—the safe practices that complement mechanical protection.

If individuals on the job understand the reason for precautions and the possible consequences if fire-safety rules are not followed, they are more likely to comply. Patient explanation and persistent enforcement, in every case, are important fire prevention duties of the supervisor.

CAUSES OF FIRES

Everyone should be alert for potential causes of fires, shown in Table 17–A and discussed below. Other fire prevention concerns are also noted in the following sections.

Electrical Equipment

Special fire hazards are presented by electrical motors, switches, lights, and other electrical equipment exposed to flammable vapors, dusts, gases, or fibers. The NFPA's code and standards pamphlets designate the standards governing a particular hazard and indicate the special protective equipment needed. *The National Electrical Code*, ANSI/NFPA 70–1990 (the NEC), gives the specifications for the equipment required. Substandard substitutions or replacements must not be made. (See Chapter 16, Electrical Safety, for more details.)

Haphazard wiring, poor connects, and temporary repairs must be brought up to standard. Fuses should be of the proper type and size. Circuit breakers should be checked to see that they have not been locked in the closed position (which results in overloading), and to see that moving parts do not stick.

Cleaning electrical equipment with solvents can be hazardous because many solvents are flammable and toxic. Keep in mind that a solvent may not present a fire hazard, but still be a health hazard. For example, carbon tetrachloride is nonflammable, but its vapors are extremely toxic. Before a solvent is used, therefore, determine both its toxic and flammable properties.

A satisfactory solvent is inhibited methyl chloroform, or a blend of Stoddard solvent and perchloroethylene. These solvents are commonly used in industry, because they are relatively non-flammable and have a fairly high Threshold Limit Value® with respect to toxicity. (Chapter 8, Industrial Hygiene, includes a discussion.)

Many of the cleaning solvents encountered in industry are mixtures of different chemicals. They are usually marketed under nondescriptive trade names or code numbers. Currently, there are no absolutely safe cleaning solvents. Before any commercially available solvent is used, you must know its chemical composition. Without such knowledge, the hazards cannot be evaluated, nor the required safety controls be used (see Chapter 8, Industrial Hygiene.)

Friction

Friction generated by overhead transmission bearings and shafting—where dust and lint accumulate in locations such as grain elevators and woodworking plants—are frequent sources of ignition. Bearings, for example, can overheat and ignite dust in the surrounding area. Keep bearings lubricated so they do not run hot, and remove accumulations of combustible dust as part of a rigid housekeeping routine. Pressure lubrication fittings should be kept in place, and oil holes of bearings should be kept covered to prevent combustible dust and grit from entering the bearings and causing overheating.

Flammable Liquids

Flammable liquids are not really a cause of fire, although they are often referred to as such. More correctly, they are contributing factors to fires. A spark or minor source of ignition, which might otherwise be harmless, can start a fire or touch off explosive forces when flammable vapors, evaporated from liquids and then mixed with air, are present. A liquid is termed flammable if the liquid emits enough vapors to burn at normal temperatures (less than 100 F). Otherwise, the liquid would be termed a combustible liquid.

Almost all industrial plants use flammable liquids. It is the responsibility of the workers to see that safe practices are followed in storage, handling, and dispensing of such liquids. Because all flammable liquids are volatile, they are continually giving off invisible vapors. Use the following guidelines to teach workers safe handling of these materials:

- Flammable liquids should be stored in, and dispensed from, approved safety containers equipped with vapor-tight, self-closing caps or covers.
- Flammable liquids should be used only in rooms or areas having adequate and, if possible, positive ventilation. If the solvent hazard is especially high, solvents should be used only in places having local exhaust ventilation.
- When highly volatile and dangerous liquids are being used, a warning placard or sign should be placed near the operation, notifying personnel and giving warning that all open flames are hazardous and must be kept away.
- Vapors of flammable liquids are normally heavier than most air and will seek the floor or other lowest possible level, where they may not be easily detected. This mandates adequate ventilation and elimination of ignition sources.
- Wherever flammable liquids are used, it is essential that ignition sources—open flames and sparks—be eliminated or alternative preventive measures taken.

Depending on factors that determine how flammable liquids are stored and used, the following safeguards may be required:

- An approved flammable liquids storage vault or room
- Special explosion-proof fixtures and equipment
- Automatic suppression system
- Explosive-relief devices and panels
- Self-closing faucets and safety vents for drums
- Flammable liquids storage cabinets
- Safety cans
- Special ventilating equipment

Specific guidelines to help determine which procedures and equipment need to be implemented at a particular site can be found in the NFPA guidelines, OSHA 29 *CFR* 1910.106. Fire department inspectors can also be of invaluable assistance in determining the minimal requirements of the local code.

Flammable Gases

Most flammable gases are stored in high-pressure cylinders or in bulk tanks. The gases are then piped to the user's location. Cylinder storage and gas usage areas are normally well planned for fire protection. The storage and usage areas will normally include fire-resistive separations, automatic sprinklers, special ventilation, explosion-relieving vents, separation of gases from other materials, and separation of incompatible gases.

Explosive Dusts

Finely divided materials not only tend to burn more readily than do bulk materials, but also can create explosive atmospheres if suspended in air. Industries that generate explosive dusts normally have planned programs to limit the amount of dust accumulations and the number of ignition sources and have equipment to limit the effects of the destructive forces developed in dust explosions.

The programs also include other measures, such as specific cleaning procedures. Ignition sources are limited through special electrical equipment, no-smoking policies, welding and cutting policies, and open flame limitations. Areas expected to have dust often have explosion-relief panels, and potentially, explosion-suppressive systems.

Plastics

Over the past two decades, general plastic usage has increased dramatically. Plastic containers are used for combustible, flammable, and nonflammable materials. At times, plastic containers are stored empty or within corrugated cartons. Noncontainer use of plastics has also increased.

Regardless of storage configurations, plastics tend to burn hotter and faster and create more smoke than cellulosic materials such as wood or paper. Standard sprinkler systems may not control the high-challenge fires that plastics create. Supervisors of industrial plastic processes should be aware of the increased hazard this material represents.

Ordinary Combustibles

Paper and wood products are often referred to as ordinary combustibles. Rack storage and solid pile storage of these materials provide conditions that promote fire growth. Materials that are stacked above one another provide good flue spaces while blocking sprinkler water patterns, thus acting to spread a fire. Sprinkler systems are designed to overcome such fire growth.

A second category of ordinary combustibles that pose rapid fire-growth hazards is wooden pallets. Most sprinkler systems will not control fires in wooden pallets stacked over 6 ft tall. If the sprinkler density has been increased, a maximum 8-ft tall stack may be possible. In either case, the supervisor is usually faced with the problem of limiting pallet heights, while meeting production and space limitations. Extra pallets are often stored outside or in separate buildings to reduce exposure to serious fires.

Detailed information is contained in NSC Data Sheet 532, *Flammable and Combustible Liquids in Small Containers*. NFPA 30, *Flammable and Combustible Liquids*, published by the National Fire Protection Association, is also an excellent source of detailed information on this subject.

OTHER HAZARDOUS MATERIALS

There are many other types of materials that must be kept isolated to prevent fire. For example, some chemicals, such as sodium and potassium, decompose violently in the presence of water, evolve hydrogen, and ignite spontaneously. Yellow phosphorus may also ignite spontaneously on exposure to air. Other combinations, too numerous to mention here, may react with the evolution of heat and produce fire or explosion—in some cases, without air or oxygen being present. These materials must be handled in a special manner. NFPA 49, *Hazardous Chemicals Data*, lists about 100 such items and provides information on unusual shipping containers, fire hazards, life hazards, storage, firefighting phases, and additional data, where applicable. Check the NFPA for other publications relating to hazardous materials.

Some materials, principally the ethers, may become unstable during long periods of storage and eventually may explode. In such cases, using the oldest stock first is both good fire safety and sound housekeeping. Part of your job is seeing that this principle is followed.

When materials are used or stored in your department, you should find out their hazards—whether they explode when heated, react with water, heat spontaneously, yield hazardous decomposition products, or otherwise react in combination with other materials. Check the Material Safety Data Sheet (MSDS) for recommended extinguishing methods and reactivity data.

Obtain detailed information on all potentially hazardous materials and devise appropriate fire safety measures. The company's fire insurance carrier can be asked to help.

Each type of industry tends to have unique fire protection requirements. The guidelines for protecting such items as computer facilities, ovens, furnaces, boilers, and other specific equipment are covered by various NFPA design standards.

Welding and Cutting

Welding and cutting operations ideally should be conducted in a separate, well-ventilated room with a fire-resistant floor. This safety measure is not, of course, always practical.

If welding and cutting must be carried on in other locations, these operations must not be performed until (1) the areas have been surveyed for fire safety by experts who know the hazards; (2) the necessary precautions to prevent fires have been taken; and (3) a permit has been issued. This permit must not be stretched to cover an area, item, or time not originally specified—no matter how small the job may seem, or how little time may be required to do it. (For more information on hot work permit programs, see NSC's *Accident Prevention Manual for Business & Industry, Engineering & Technology* volume.)

If welding must be done over wood floors, they should be swept clean, wetted down, and then covered with fire-retardant blankets, metal, or other noncombustible covering. Pieces of hot metal and sparks must be kept from falling through floor openings onto combustible materials.

Sheet metal or flame-resistant canvas or curtains should be used around welding operations to keep sparks from reaching combustible materials. Welding or cutting should not be permitted in or near rooms containing flammable or combustible liquids, vapors, or dusts. Nor should these operations be done in or near closed tanks that contain—or have contained—flammable liquids, until the tanks have been thoroughly drained and purged, and tested free from flammable gases or vapors.

No welding or cutting should be done on a surface until combustible coverings or deposits have been removed. It is important that combustible dusts or vapors not be created during welding. Be sure to provide fire-extinguishing equipment at each welding or cutting location. Where extra-hazardous conditions cannot be eliminated completely or protected by isolation, consider installing the back-up protection of a fire hose.

Station a watcher at the welding operation who can prevent stray sparks or slag from starting fires, or who can immediately extinguish fires that do start

despite all the precautions taken. The area should be under fire surveillance for at least one-half hour after welding or cutting has been completed. Many fires are not detected until a few minutes after they ignite.

Open Flames

No open flames should be allowed in or near spray rooms or spray booths. Occasionally, workers may need to do indoor spray-painting or spray-cleaning outside of a standard spray room or booth. In such cases, provide adequate ventilation and eliminate all possible ignition sources, such as spark-producing devices and open flames.

Gasoline, kerosene, or alcohol torches should be placed so their flames are at least 18 in. (46 cm) away from wood surfaces. They should not be used in the presence of dusts or vapors, near flammable or combustible liquids, paper, excelsior, or similar material. Torches should never be left unattended while they are burning.

Portable Heaters

Gasoline furnaces, portable heaters, and salamanders always present a serious fire hazard. Discourage workers from using them. Upright models of fuel oil salamanders have been involved in numerous accidents. Replace these hazards with other types of low-profile heaters.

Fuel used in portable heaters should be restricted to liquefied petroleum gas, coal, coke, fuel oil, or kerosene. The area must be well ventilated, since the products of combustion contain carbon monoxide.

All these heating devices require special attention with respect to clearances and mounting. A clearance of 2 ft (0.6 m) horizontally and 6 ft (1.8 m) vertically should be maintained between a heater and any combustible material. If coal or coke is used, the heater should be supported on legs 6 in. (15 cm) high or on 4 in. (10 cm) of tile blocks and set on a noncombustible surface.

Combustible material overhead should be removed or shielded by noncombustible insulating board or sheet material with an air space between it and the combustible material. A natural-draft hood and flue of noncombustible material should be installed.

As a fire precaution, each unit must be carefully watched. Heaters should be shut down and allowed to cool off before being refueled. Coal and coke salamanders should not be moved until the fire is out.

All portable heating devices must be equipped with suitable handles for safe and easy carrying. Make sure they are secured or protected against workers accidentally upsetting them.

The most serious fire hazard occurs when heaters are improvised from old steel drums or empty paint containers, with scrap wood, tar paper, or other waste used as fuel. Residue or vapors can spread the fire. Let workers know that this practice is prohibited.

Hot Surfaces

If possible, smoke pipes from heating appliances should not pass through ceilings or floors. If a smoke pipe must be run through a combustible wall, provide a galvanized double-thimble with clearance equal to the diameter of the pipe and ventilated on both sides of the wall, for protection.

Soldering irons must not be placed directly upon wood benches or other combustibles. Special insulated rests, which will prevent dangerous heat transfer, can be used instead.

Smoking and Matches

With increasing local, state, and federal regulation, and with increasing public awareness of the hazards of smoking, management usually has very specific policies about cigarette/cigar/pipe smoking in the workplace. Smoking and no-smoking areas must be clearly defined and marked with conspicuous signs. Reasons for these restrictions must be explained to employees, and rigid enforcement must be maintained at all times.

Fire-safe, metal containers and safety ashtrays should be provided in any areas where smoking is permitted. Periodically check "no-smoking" areas, especially when they include stairways and other out-of-the-way places, for evidence of discarded smoking materials.

Spontaneous Combustion

Spontaneous heating is a chemical action in which the oxidation (burning) of a fuel produces a slow generation of heat. When adjacent materials provide sufficient insulating properties, the heating process can continue until spontaneous ignition occurs. Conditions leading to spontaneous ignition exist where there is sufficient air for oxidation but not enough ventilation to carry away the heat generated by the oxidation. Any factor that accelerates oxidation while other conditions remain constant obviously increases the likelihood of ignition.

Materials like unslaked lime and sodium chlorate are susceptible to spontaneous ignition, especially when wet. Such chemicals should be kept cool and dry, away from combustible material. Rags and waste saturated with linseed oil, paint, or petroleum products often cause fires because no provision is made for the generated heat to escape. By keeping such refuse in air-tight metal containers with self-closing covers (Figure 17–3), you limit the oxygen supply,

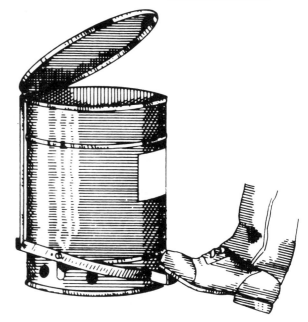

Figure 17–3. Check containers for oily waste to make sure the lid closes snugly against the top.

and a fire will quickly extinguish itself. These containers should be emptied daily.

Sound precautions against spontaneous ignition include total exclusion of air or good ventilation. The former is practical with small quantities of material through the use of air-tight containers. Ventilation can best be assured by storing material in small piles or by turning over a large pile at regular intervals.

To determine the progress of spontaneous heating, monitor the temperature within the interior of a mass of material in various locations. Exterior temperatures are not likely to provide a good index.

Static Electricity

Sparks due to static electricity may be a hazard wherever flammable vapors, gases, or combustible dusts exist. Precautions against static electricity are required in such areas. Static charges result from friction between small particles or from the contact and separation of two substances, one or both of which are nonconductive.

Static charges can be produced in many ways—for example, by the flow of gasoline through a nonconductive hose, by passing dry and powdered materials down a non-conductive chute, or through the friction of machine parts.

Although it is impossible to prevent generation of static electricity under the above circumstances, the hazard of static sparks can be avoided by preventing buildup of static charges. One or more of the following preventive methods can be used:

- Grounding
- Bonding
- Maintaining a specific level of humidity
- Ionizing the atmosphere

A combination of these methods may be advisable in some instances where the accumulation of static charges presents a severe hazard. NFPA 77, *Recommended Practice on Static Electricity*, gives additional information and guidance on this subject.

Grounding is accomplished by mechanically connecting a conductive machine or vessel (in which the generation of static may be a hazard) to ground by means of a low-resistance conductor. Another method is to make the entire floor and structure of the building conductive so that all equipment in contact with it will be grounded. If you use the first method, be sure to check the continuity of the ground circuit. Connections must be clean and the conductor unbroken (Figure 17–4). With the second method, make certain that the floor is free of wax, oil, or other insulating films.

As people learn from walking across a rug and then touching a doorknob or other conductive object, the human body can also carry an electric charge. Workers can ground themselves by wearing conductive

shoes with a floor of conductive material. The conductive parts of such shoes should be made of non-ferrous metal. As an added measure of safety, the conductive flooring may be made of spark-resistant metal. Ferrous contacts increase the risk of friction sparks.

When humidity is low, the hazard of static electricity is greatest. When humidity is high, the moisture content of the air serves as a conductor to drain off static charges as they are formed. When humidity is added to prevent accumulation of static charges, make sure that an effective relative humidity—usually 60 to 70%—is maintained. However, the minimum humidity required for safety may vary over a considerable range under different conditions. In fact, under some conditions the static charge cannot be controlled by humidifying the air. Consult with engineering authorities to determine the best approach.

When air is ionized, it has sufficient conductivity to prevent static accumulation. Ionization is produced by electrical discharges, radiation from radioactive substances, or gas flames. Only an electrostatic neutralizer designed for use in hazardous locations should be used. Otherwise, the neutralizer may itself be a source of ignition and touch off flammable vapor or dust. Neutralizers must be kept in good condition.

Figure 17–4. Grounds and bonds should be constructed of bare, flexible wire so that broken wires are not concealed. Wires must be attached securely.

EFFECTIVE HOUSEKEEPING FOR FIRE SAFETY

Good housekeeping is another important part of an effective fire protection program. As supervisor, you must maintain a positive attitude with the crew and enforce housekeeping rules at all times. Each person should be held personally responsible for preventing the accumulation of unnecessary combustible materials in the work area. Workers should be held accountable for their areas at the end of their shift. Here are the precautions to take:

- Combustible materials should be present in work areas only in quantities required for the job. They should be removed to a designated, safe storage area at the end of each work day.
- Quick-burning and flammable materials should be stored only in designated locations. Such locations always should be away from ignition sources and have special fire-extinguishing provisions. Covered metal receptacles are best.
- Vessels or pipes containing flammable liquids or gases must be airtight. Any spills should be cleaned up immediately.
- Workers should be careful not to contaminate their clothing with flammable liquids. If contamination does occur, these individuals must be required to change their clothing before continuing to work.
- Passageways and fire doors should be unobstructed. Stairwell doors must never be propped open, and material should never be stored in stairwells.
- Material must never block automatic sprinklers [within 18 in. (46 cm) of sprinkler head] or be piled around fire extinguisher locations or sprinkler and standpipe controls.

To obtain proper distribution of water, a minimum of 18 in. (46 cm) of clear space is required below sprinkler deflectors. However, clearance of 24 to 36 in. (60 to 90 cm) is recommended. If there are no sprinklers, clearance of 3 ft (0.9 m) between piled material and the ceiling is required to permit use of hose streams. Double these distances when stock is piled more than 15 ft (4.5 m) high.

Be sure to check applicable codes, especially *Life Safety Code*, ANSI/NFPA 101–1994.

FIRE PREVENTION INSPECTIONS

Fire prevention inspection needs will vary from facility to facility and department to department. However, all fire prevention inspections have four basic goals:

1. Minimize the size of fires by ensuring that combustible and flammable material storage is controlled.
2. Control ignition sources to reduce the possibility of fire.
3. Make sure that fire protection equipment is operational.
4. Make sure that personnel exit facilities are maintained.

You will normally have some responsibility for fire prevention inspections as part of your daily, weekly, and monthly inspection activities. Fire prevention inspections generally are conducted at the supervisory level as part of routine safety inspection activities.

The best way to ensure that proper items are inspected is to prepare a department checklist. The checklist should be as specific as possible in defining what is, and is not, permitted. This checklist can then be used to audit the department and can be given to employees to make sure that levels of housekeeping and other fire prevention items are defined clearly to all parties. Sample inspection checklists are included at the end of this section.

In order to prepare a checklist, the information in this chapter be used, along with information from local or corporate safety coordinators, fire department inspectors, and insurance inspectors. Once the checklist has been prepared, the inspection frequency should be defined. Deficient items should be noted and the corrective action taken as soon as possible. Follow-up is also necessary at a predetermined interval.

Fire Protection Equipment

One of the most important audit items will be to determine that the fixed fire protection equipment is on. The checklist items should include the fire protection water valves in your department which should be locked or supervised open. Other fire protection systems are normally controlled by electrical panels, which can be audited to ensure that the systems are on and that no trouble signals are showing. Automatic sprinkler patterns can be audited by making sure that there are at least 18 in. between products and the sprinkler deflectors.

Fire extinguishers should be tagged and marked with the date of the last inspection. Gages should show charge, and seals should be in place. The audit form should ensure that the extinguishers are hung in designated locations and that access is not blocked. The nozzles should be free and clear.

Fire Protection Systems

Fires can be extinguished by exhausting the fuel source, by manual fire fighting, or by fixed fire-fighting systems. Fire extinguishment typically occurs by lowering the temperature through dousing the fire with water, reducing the oxygen available through carbon dioxide or foam systems, or by interrupting the chemical chain reaction by using a halon or dry chemical system.

All systems and system components should be listed by recognized testing agencies according to their fire-fighting purposes. In addition, all systems should be inspected and/or tested annually or as required by applicable standards or manufacturers' instructions. Automatic sprinklers are available in wet-pipe systems, dry-pipe systems for piping subject to freezing, deluge systems for high-challenge fires, or pre-action systems for locations subject to water damage.

Halon, carbon dioxide, and foam systems typically are actuated by fire detection systems. These systems generally are heat, smoke, or flame detectors. Detection systems also are used often to reduce the threat to life and property by providing early warning to initiate manual fire fighting or evacuation of the workplace.

ALARMS, EQUIPMENT, AND EVACUATION

A fully effective fire protection program, of course, includes fire extinguishment. Whenever a fire breaks out, take immediate action:

1. Sound the fire alarm right away—regardless of the size of the fire.
2. Attempt to extinguish or control the fire with appropriate fire-extinguishing equipment.

Fire Alarms

Facilities in areas where municipal fire departments are available usually have an alarm box close to the plant entrance or located in one of the buildings. Others may have auxiliary alarm boxes, connected to the municipal fire alarm system, at various points in the facility. Another system is a direct connection to the nearest fire station that may register water flow alarms in the sprinkler system, be activated by fire detectors, or be set off manually. In some cases, the telephone may be the only means for signaling a fire alarm.

Whatever the alarm system used, all employees should be carefully instructed in how, when, and where to report a fire. These three steps are extremely important. Many fires get out of control simply because someone did not know how, when, or where to give the alarm.

Extinguishers

As the supervisor, you need to know the four classes of fires that might break out in your area. Before employees can combat beginning fires effectively, they must be familiar with and understand these classes of fires. Briefly, they are:

Class A—Fires in ordinary combustible materials, such as wood, paper, cloth, rubber, and many plastics, where the quenching and cooling effects of water or of solutions containing large percentages of water are of prime importance.

Class B—Fires in flammable liquids, greases, oils, tars, oil-base paints, lacquers, and similar materials, where smothering or exclusion of air and interrupting the chemical chain reaction are most effective. This class also includes flammable gases.

Class C—Fires in or near live electrical equipment, where the use of a nonconductive extinguishing agent is of first importance. The material that is burning is, however, either Class A or Class B in nature.

Class D—Fires that occur in combustible metals, such as magnesium, lithium, and sodium. Special extinguishing agents and techniques are needed for fires of this type.

Each of the fire extinguishers in the department or at a job location should have on it a plate identifying the class of fire for which it is intended (Figures 17–5a and 17–5b), its operating instructions, and its servicing instructions. The data plate should have the identifying symbol/name of a recognized testing facility to indicate that the unit has been listed or approved. Equipment that does not bear an approval label should be brought to the attention of management. Only listed or approved equipment should be furnished. See "Fire Protection Equipment List" (Underwriters Laboratories Inc., Northbrook, Ill. 60062). Also, "Approved Equipment for Industrial Fire Protection" (Factory Mutual Research Corp., Norwood, Mass. 02062).

Figure 17–5a. Fire extinguishers should have a plate identifying the class of fire they can put out, the operating instructions, and servicing instructions. This one uses the triangle-square-circle-star symbols. Note the emergency and rescue alarms next to the fire extinguisher (protected to prevent accidental tripping).

The information on these data plates can help you teach employees how to operate the extinguishing equipment. See to it that every worker knows the important details about each of the fire-extinguishing agents provided in the particular job area or shop (Table 17–B). Fire extinguisher manufacturers and distributors can be contacted for additional training materials, and local fire departments may be willing to provide training assistance. Information on selection, installation, use, inspection, and maintenance of extinguishers can be obtained from the National Fire Protection Association. NFPA 10, *Portable Fire Extinguishers*, is an authorative source of information and should be made available to all supervisors for reference use.

Normally, the location and installation of fire safety equipment at strategic places are the responsibility of higher management. However, you should know the location of units in your area, and recommend relocating them or obtaining additional units if such changes will afford more adequate protection.

Check to see that extinguishers are not blocked and that signs indicating their location are conspicuous (Figure 17–6). Marking the area of the floor directly under a fire extinguisher is an excellent way to keep employees from placing obstructions in front of the equipment.

The location of extinguishers can be identified by painting the support of the extinguisher with standard fire-protection red. Fire protection equipment itself, such as sprinkler system piping, is sometimes painted or marked in red. Make sure that each worker knows the location of the nearest unit and knows the

Figure 17–5b. Picture-symbol labels used by the National Association of Fire Equipment Distributors show the class of fire for which an extinguisher can be used. The symbols shown here would be on a Class A extinguisher (for extinguishing fires in trash, wood, or paper). The left symbol is blue. Because this extinguisher is not meant for use on Class B or C fires, those two picture-symbols are in black with a diagonal red line through them. A Class A/B extinguisher would have the first two symbols in blue and the third in black with a red diagonal. For use on a Class B/C, the last two would be in blue; on a Class A/B/C, all three would be in blue.

Table 17–B. Fire Extinguisher Selection Chart

	CLASS A						CLASS A/B	CLASS B/C			
	WATER TYPES		MULTIPURPOSE DRY CHEMICAL		AFFF FOAM	HALON 1211	AFFF FOAM	CARBON DIOXIDE	DRY CHEMICAL TYPES		HALON 1211
	STORED PRESSURE*	PUMP TANK*	STORED PRESSURE	CARTRIDGE OPERATED	STORED PRESSURE*	STORED PRESSURE	STORED PRESSURE*	SELF EXPELLING	STORED PRESSURE	CARTRDIGE OPERATED	STORED PRESSURE
SIZES AVAILABLE	2½ gal	2½ and 5-gal	2½-30 lb ALSO (Wheeled 150-350 lb)	5-30 lb ALSO (Wheeled 50-350 lb)	2½ gal	9 to 22 lb	2½ gal	5-20 lb ALSO Wheeled 50-100 lb	2½-30 lb ALSO Wheeled 150-350 lb	4-30 lb ALSO Wheeled 50-350 lb	2 to 22 lb
HORIZONTAL RANGE (APPROX.)	30 to 40 ft	30 to 40 ft	10-15 ft (Wheeled-15-45 ft)	10-20 ft (Wheeled-15-45 ft)	20-25 ft	14 to 16 ft	20 to 25 ft	3-8 ft (Wheeled-10 ft)	10-15 ft (Wheeled-15-45 ft)	10-20 ft (Wheeled-15-45 ft)	10 to 16 ft
DISCHARGE TIME (APPROX.)	1 min.	1 to 2 min	8-25 s (Wheeled-30-60 s)	8-25 seconds (Wheeled-20-60 s)	50 seconds	10 to 18 seconds	50 seconds	8-15 seconds (Wheeled-8-30 s)	8-25 seconds (Wheeled-30-60 s)	8-25 seconds (Wheeled-20-60 s)	8 to 18 seconds

	CLASS/A/B/C			CLASS D
	MULTIPURPOSE DRY CHEMICAL		HALON 1211	DRY POWDER
	STORED PRESSURE	CARTRIDGE OPERATED	STORED PRESSURE	CARTRIDGE OPERATED
SIZES AVAILABLE	2½-30 lb ALSO Wheeled 150-350 lb	5-30 lb ALSO Wheeled 50-350 lb	9 to 22 lb	30 lb ALSO Wheeled 150-350 lb
HORIZONTAL RANGE (APPROX.)	10-15 ft (Wheeled-15-45 ft)	10-20 ft (Wheeled-14-45 ft)	14-16 ft	5 ft (Wheeled-15 ft)
DISCHARGE TIME (APPROX.)	8-25 seconds (Wheeled-30-60 s)	8-25 seconds (Wheeled-20-60 s)	10 to 18 seconds	20 seconds (Wheeled-150 lb to 70 s, 350 lb 1¾ min.)

*Must be protected from freezing

Courtesy of the National Association of Fire Equipment Distributors

importance of keeping areas around extinguisher units clear.

It is a good idea to schedule fire drills, during which workers use extinguishers appropriate for their particular work areas (Figure 17–7). The company's safety professional or fire chief, the insurance company's safety engineer, or the local fire department representative will assist in conducting such drills.

Every organization should have a specific program for periodic inspection and servicing of portable fire-extinguishing equipment (see *CFR* 1910.157(e)(2)). Fire extinguishers must be visually inspected monthly. This routine work is usually outside the scope of the department supervisor and/or maintenance personnel. However, you can do some things to assist in this program. For example, during regular department inspections, you should double-check each data card to determine when each extinguisher was last inspected and serviced. Management should be notified when inspections have been missed or are outdated. If the extinguisher seal is broken, its pressure gage reads below normal, or any other unsatisfactory condition is observed, report these conditions immediately. Defective extinguishers may not work.

Fire extinguishers must meet the following requirements:

- Be kept fully charged and in their designated places.
- Be located along normal paths of travel where practical.
- Not be obstructed or obscured from view.
- Not be mounted higher than 5 ft (1.5 m) (to the top of the extinguisher) if they weigh 40 lb (18 kg) or less. If heavier than 40 lb, extinguishers must not be mounted higher than 3½ ft (1 m). There shall be a clearance of at least 4 in. (10 cm) between the bottom of the extinguisher and the floor.
- Be inspected by management or a designated employee, at least monthly, to make sure that they are in their designated places, have not been tampered with or actuated, and do not have corrosion or other impairment.
- Be examined at least yearly and/or recharged or repaired to ensure operability and safety. A tag must be attached to show the maintenance or recharge date, and the signature or initials of the person performing the service.
- Be hydrostatically tested. Extinguisher servicing agencies should be contacted to perform this service at appropriate intervals.
- Be selected on the basis of type of hazard, degree of hazard, and area to be protected.

Figure 17–6. Painting the floor under a fire extinguisher reminds employees not to obstruct the equipment.

- Be placed so that the maximum travel (walking) distances between extinguishers, unless there are extremely hazardous conditions, do not exceed 75 ft (23 m) for Class A extinguishers or 50 ft (15 m) where Class B extinguishers are used for hazardous-area protection. The travel distance requirement does not apply when Class B extinguishers are used for spot-hazard protection.

Fire Brigades

Many plants and construction job sites have organized fire brigades. However, they should be well enough acquainted with the form and activities of a fire brigade.

In some fire brigades, you, as supervisor, or a delegated foreman will be named a brigade chief or company captain. Whether or not you are chosen for this job, you should know each worker's fire-brigade assignment. On this basis, you can organize the work in your department so that brigade members may attend training and drills as designated by their brigade chiefs (Figure 17–8).

Regardless of your role in the fire-brigade organization, you should be familiar with the location and

operation of the following items in and adjacent to your department—standpipes and valves, sprinkler system valves, electric switches for fans and lights, fire-alarm boxes and telephones, fire doors, and emergency power for equipment and/or lighting. In particular, you must make sure that enclosed stairway doors are kept closed. Such doors are provided to keep smoke, combustion gases, and heat out of exit passageways and stairs, as well as help prevent the rapid spread of fire. Doors should be equipped with self-closing devices and should not be propped open. In cases where the exitway is used frequently, doors may be held open with fire-actuated detectors that will automatically close the door should fire break out.

To become familiar with the items listed above, accompany the fire inspector on inspection tours to learn how the items operate. Whether or not you will operate this equipment will depend upon the particular situation and specific policy and procedures in your plant. In some instances, if you know the system thoroughly, you may be the best one to use fire safety equipment. However, it can be extremely dangerous to let anyone without proper knowledge operate the equipment. Activating it at the wrong time or using it incorrectly might seriously endanger people's lives in case of fire.

Follow-Up for Fire Safety

Frequent inspections of the area, correction of hazardous conditions, and instruction of the workers in fire prevention and extinguishment measures are still not enough to ensure a fire-safe work area. Make sure you instill fire-safe attitudes in your personnel.

Make fire prevention and extinguishment the subject of frequent safety talks between workers. Also, discuss these issues in safety meetings so that workers will be fully aware that fire prevention is a vital part of the overall departmental safety program.

Special Fire Protection Problems

Construction. When construction is going on at an existing plant site, fire protection problems are compounded. You and the construction supervisor should work together in determining the hazards in each field of operations and in identifying the limits of your firefighting capabilities. On the basis of this knowledge, you should take steps to upgrade fire protection equipment and practices.

Construction activities may require measures such as using temporary wiring or portable heaters and gasoline engines that must be refueled on the spot. As a result, manufacturing processes, ordinarily not hazardous, may become so under these conditions.

Figure 17–7. These employees are being trained in the most effective ways to use extinguishers.

Figure 17–8. Fire brigade members undergo training in a simulated environment requiring the use of self-contained breathing apparatus. (Courtesy Survivair Division of U.S. Divers.)

See that employees are aware of additional ignition sources created by the construction operations. The construction supervisor should inform workers of additional fire hazard exposures present from plant activity.

When the construction work involves plumbing, the water supply available for fire extinguishment may be decreased or need to be shut off. To offset this possible shortage of water, obtain additional auxiliary fire-extinguishing equipment such as temporary hose lines or water tank trucks.

Sprinkler System Shutdown. Whenever a sprinkler system must be shut down, special precautions must be taken to ensure maximum fire safety. Check to see that all preliminary work is completed. Set out extra firefighting equipment, and notify firefighting units which areas will be without sprinkler protection. One procedure requires that a bright red tag be attached to the sprinkler valve when it is closed (Figure 17–9) and that the insurance carrier be notified. The work is completed as quickly as possible, at which time the tag is removed and the insurance company notified that service is restored. To make certain that the valve is left wide open, it is tested and then locked in the open position. All sprinkler valves should be checked regularly to make sure they have not been tampered with, locked, or shut.

Whenever the sprinkler system must be shut down for alterations or repairs, you and your maintenance people should do the work before or after normal working hours. If work must be done during the most hazardous times, special precautions may have to be taken, such as having hose lines laid and furnishing extra fire patrols.

Some plant and construction supervisors make a practice of attending each other's safety meetings. They benefit from exchange of their respective points of view and learn about each other's problems. This joint approach to common exposures is recommended.

Radioactive Materials. Use of radioactive materials for various manufacturing processes has become more common. Their presence requires special understanding and control in event of fire or other emergency.

Figure 17–9. Sprinkler system valves should be locked in the open position (*top*). When a sprinkler system must be shut down, a safe procedure is to do preliminary work first, then set out extra firefighting equipment, inform firefighting units which areas will be without sprinkler protection, notify the insurance carrier, and tag the closed sprinkler valve (*right*). When the work is completed, remove the tag, test and lock the valve in the fully open position, and notify firefighting units and the insurance carrier that service is restored. All sprinkler valves should be checked regularly to ensure they have not been unlocked, tampered with, or shut.

In view of the potential contamination hazards to employees, supervisors in the areas or departments having radioactive materials should be thoroughly familiar with the procedures to be followed in case of fire or other emergency, and should rigidly enforce compliance with those procedures. Access to the contaminated area following the emergency should be prohibited until authorized personnel have recovered the sources of radiation and have determined that the contamination is below the safe levels.

Valuable information on this problem can be secured from the Nuclear Regulatory Commission, 2120 L Street, N.W., Washington, D.C. 20555.

Evacuation

Prevention of fire is the primary objective of any fire protection program. Nevertheless, each program must include provisions to ensure the safety of employees in the event of a fire. Security personnel must be included in the overall plan (Figure 17–10).

Most facilities will have a plan to respond to emergencies. The plan may be as simple as an informal evacuation plan, or as complicated as a fully written emergency response and disaster recovery plan for various natural and man-made emergencies. Normally, your responsibilities during an emergency are for the people and department under your authority.

You should have access to the written emergency plan for periodic review and instruction of the workforce. You should develop a clear personal plan for complying with the facility's emergency plan and responding to evacuation, spills, fires, firefighting, natural disasters, and utility failures. Make sure you know proper procedures for reporting emergencies, and the meaning of alarm signals. You should actively participate in routine emergency drills.

In some cases, you or a member of your workforce will serve on the emergency response team and will have specific duties outside the department during an emergency. You should plan how to respond to an emergency with those members missing. Workers should be trained to respond in specific assigned duties,

Figure 17–10. Include security personnel in your fire protection program planning. This console is part of a comprehensive pro-prietary system capable of detecting fire, burglary, and any unwanted fluctuations in industrial process conditions. (Courtesy ADT—American District Telegraph Company.)

such as shutdown of equipment, accounting for all employees, assembling in assigned places, taking charge of specific responsibilities, sounding the all-clear, etc.

From time to time, the local fire department will tour your facility. The company can benefit by your active participation in these tours. The more the fire fighters know about flammable and combustible materials in the facility, the quicker the fire department can respond to minimize fire losses.

You can do much to prepare for the safety of fellow employees in the event of a serious fire. As a matter of routine, you should make sure that each man and woman knows the evacuation alarms, both primary and alternate exit and escape routes, what to do, and where to be during and after an evacuation.

You should make sure that each person knows that he or she, upon being alerted, must proceed at a fast walking pace—not a run—to an assigned exit or, if it is blocked, to the nearest clear one. Day in and day out, you should emphasize that exit routes and fire doors must be kept unobstructed.

Such essential steps in emergency procedures will best be instilled by periodic fire drills. Plan and conduct practice evacuations at regular intervals. You should also do everything possible to integrate fire emergency training with plant-wide drills and the company's overall evacuation plan. When it is impossible to hold drills, give oral instructions regarding evacuation procedures, and distribute printed information describing these policies to all employees.

FIRE PROTECTION EDUCATION

Fire protection education for the workforce typically involves the following:

- Emergency response
- Fire fighting
- Home fire safety

Emergency response training usually will be necessary one or more times per year. The workforce will need to know how to turn in and how to respond to alarms. The specific procedures should be rehearsed.

Workers also should practice use of portable fire extinguishers and hoses at least yearly if they are expected to use the equipment. A yearly review of the fixed fire protection systems and manual operation of those systems is usually necessary. If any of the employees travels, hotel fire safety is an appropriate topic to address. Fire Prevention Week in October is normally a convenient time to review these items. Safety meetings also provide a convenient forum. Brochures and written materials are available from the National Fire Protection Association in Quincy, Massachusetts. OSHA's Fire Prevention Plan (29 *CFR* 1910.38) can provide additional guidelines.

Fire Prevention Week is also a good time to address home fire safety. The proper use of smoking materials, smoke detector usage, and home fire drills are appropriate topics. The beginning of the heating season is a good time to introduce concepts of proper use of wood-burning equipment. In December, review the hazards and protective measures regarding holiday decorations and natural Christmas trees.

PROTECTIVE INSURANCE REQUIREMENTS

The property (fire) insurance carrier generally conducts inspection tours of the facility on a routine basis. These tours are intended to identify items that can help to reduce loss in a fire or minimize ignition sources. The supervisors of an area will often accompany the insurance representative on the tour to understand the insurance concerns. The representative also can be used as a resource to help solve fire protection problems.

The insurance carrier will require specific programs that you should understand. No-smoking requirements are one program, and the cutting and welding permit system is another. The locking or supervision of water supply valves, and testing and maintenance of fire protection equipment are other typical required programs.

Another common insurance requirement involves those times when fire protection systems are removed from service. The insurance carrier often will require a call and a specific tagging system. The carrier then tracks the time when the system is out of service to ensure that the system is returned to service. The carrier may require alternative fire protection such as standby charged hoses during those periods. The notification of the fire department is also a good procedure.

SUMMARY OF KEY POINTS

This chapter covered the following key points:

- The primary requirements for a successful fire prevention program include continuous training in fire-safe work practices, regular inspections of work areas, and close supervision of employee job performance. Ultimate responsibility for fire safety rests with the supervisor.

- Fire chemistry can be illustrated by the tetrahedron model in which elevated temperature, oxygen, fuel, and chain reaction must occur simultaneously for a fire to continue burning. Fires can be extinguished by one or more of the following: removing or reducing oxygen supply, removing fuel sources, reducing or cooling heat sources, and interrupting the chain reaction.

- The main causes of fires include faulty electrical equipment, friction, flammable liquids, flammable gases, explosive dusts, plastics, and ordinary combustibles such as paper or wood. Supervisors should see that electrical equipment is properly grounded, insulated, and maintained, and that all flammable or combustible materials are properly stored, ventilated, and contained.

- Other causes of fires include operations such as welding and cutting, open flames, portable heaters, hot surfaces, smoking and lighted matches, spontaneous combustion, and static electricity. Stringent safe work practices and housekeeping measures can reduce the possibility of fire from these causes.

- Supervisors must establish good housekeeping rules and guidelines and hold each employee accountable for following these procedures. These rules include preventing the accumulation of combustible materials, changing clothing contaminated with flammable liquids, keeping passageways and fire doors unobstructed, and making sure sprinkler systems are not blocked by stacked material.

- Supervisors should make regular inspections to identify fire hazards in their areas and take action to resolve them. Workers should be continuously informed of these hazards and the safeguards required to prevent fires from occurring. A fire prevention inspection has four goals: (1) minimize the size of fires by controlling combustible and flammable materials, (2) controlling ignition sources, (3) making sure fire protection equipment operates, and (4) ensuring that personnel exit facilities are maintained.

- Supervisors should make sure that employees know how to operate fire alarms and firefighting equipment, and understand evacuation procedures. Employees should know how, when, and where to report a fire by activating fire alarms. They should also know Class A, B, C, and D fires, and which fire-extinguishing equipment is appropriate for each class of fire. Firefighting and extinguishing equipment should have this information clearly printed on the data plate.

- Such equipment must be inspected periodically and repaired or replaced if found defective. Supervisors must follow up on inspections and instruction sessions to make sure employees are complying with safety procedures.
- In some facilities, companies organize fire brigades. Supervisors must know how the fire brigade operates and the specific responsibilities of each brigade member. Supervisors should also know the location and operation of all fire safety equipment and exit facilities.
- In some cases, construction onsite, sprinkler system shutdowns, or the presence of radioactive materials present special fire safety problems. Supervisors must make sure special fire prevention measures are taken to avoid accidents and injuries.

- Evacuation procedures should be communicated to every employee in the event of a serious fire or other emergency. Workers should know evacuation alarm signals, where to exit the facility, and how to secure their work areas before leaving. Supervisors should hold drills to practice evacuation procedures.
- Fire protection education for employees involves responding to emergencies, fighting fires, and instituting home fire safety procedures. Employees should practice using fire-extinguishing equipment and understand how to fireproof their homes.
- Insurance companies generally inspect facilities on a routine basis. Supervisors can accompany these inspectors on their tours and use them as a resource to help solve fire protection problems.

ACCIDENT CASE STUDIES

The following case studies are provided for classroom instructors and for supervisors conducting on-the-job training. The instructor should present only the case and allow the students to solve the problem through discussion and questioning. The background information, possible solutions, and summary are provided for the instructor's guidance.

CASE I

A maintenance man was working on a metal stair platform and wanted to use a ¼-inch electric drill. The drill had a three-wire power cord. An extension cord running from the source of power also was a three-wire cord, but was not long enough. In order to connect the drill to the extension cord, the man obtained another short extension cord from the tool crib.

As the man started to drill, he was electrocuted.

What caused the accident? What could have been done to prevent this accident? Similar accidents in the future?

Guide and Background Information for Case I

Explain to the group, if asked, that the drill and cord were in proper working order, no shorts. Also, the extension from the power source was okay. However, the extension cord from the tool crib had been improperly wired. The grounding lead had been connected to the "hot" terminal, so that the frame of the drill was energized. When the man grasped the grounded metal stair platform, he was electrocuted.

Once this fact is brought out, have the group pursue the question of preventing another such accident. In the state in which this accident occurred, three workmen had been electrocuted because of improperly wired electrical connections. (This report was from a State Industrial Commission.)

Following are other pertinent facts:

- The maintenance man was not standing on damp or wet ground. The area was dry.
- The extension cord was made up by a tool crib attendant.
- There were no rules in effect concerning procedure for maintenance and repair of equipment.
- The cord had not been tested.

Possible Solutions for Case I

1. All extension cords and electrical equipment should be immediately tested by a qualified electrician before allowing further use of such equipment.
2. A strict rule should be put into effect that only qualified persons be allowed to repair equipment.
3. Electric tools and equipment should be inspected on a regular schedule to make certain they are properly wired, grounded, and in proper working order. The equipment should also be marked to indicate to the user it had been inspected. The date of inspection should be shown.
4. All electric tools must be grounded regardless of use or location.
5. Make certain all workers understand they are not to make repairs to tools or equipment.
6. If tool crib attendants are to maintain equipment, they should be properly trained.

Summary

Summarize the discussion by pointing out that the unsafe condition must be eliminated—in this case, the improperly wired extension cord. The only way this can be done is to test all extensions and make certain there are no other defective ones. Once all the electrical equipment has been checked out, some procedures must be set up to make certain the equipment is kept in good condition.

Since there are also unsafe procedures involved, provisions must be made to help prevent a recurrence. In this case the tool crib attendant did not follow safe procedures and the maintenance man did not make certain the extension was okay. Safe procedures should be enacted and enforced. Stress the importance of eliminating all possible causes.

CASE II

A freight handler attempted to sharpen the point of a bale hook on a grinding wheel. The foreman of the department saw him, but assumed that "anybody can use a little bench grinder." The freight handler caught the point of his hook between the tool rest and the wheel. The wheel broke and a large piece of it sturck him in the face. He was permanently disabled by the injury.

What could have been done to prevent this accident? Similar future accidents?

Guide and Background Information for Case II

Explain to the group, if asked, that this bench grinder was properly guarded and mounted. Also, the wheel was properly tested before mounting.

Following are other pertinent facts:

- The tool rest was properly set $1/4$-inch from the wheel.
- Eye protection was available, but obviously not used (goggles).
- The eye shield was evidently not in place.
- There was no regulation regarding unauthorized use of equipment.
- The freight handler was not from this foreman's department.
- A grinding wheel of this type is proper for sharpening bale hooks.

Possible Solutions for Case II

1. It should be obvious to the group that the primary cause of this accident was one or more unsafe procedures. The foreman made a mistake by "assuming" the freight handler knew how to use a bench grinder. The freight handler committed several unsafe procedures. First, he did not use the eye protection that was available. Second, he must have put the hook in a position that let the point slip betwen the tool rest and the wheel.
2. While you may get such comments from the group as "Fire the foreman," the solution should focus on

a rigidly enforced rule, "No Unauthorized Use of Equipment."
3. The following procedures can also be set up:
   a. A policy regarding the sharpening of tools, such as returning tools to the tool crib for sharpening or replacement.
   b. A program of training if the freight handlers are to do the sharpening.
   c. A lockout device interlocked with the starting switch to prevent the wheel from operating when the eye shield is not in the proper position.

Summary

Point out to the group that the two cases, Case I and Case II, bring out the importance of unsafe conditions and unsafe procedures as the causes of accidents. Stress the importance of searching for all possible causes.

Point out that in Case II there were a number of places where the accident prevention program needed tightening in order to prevent similar accidents in the future.

Emphasize the importance of checking procedures and conditions in advance to prevent accidents of this type. Mention the importance of a job safety analysis as a tool for preventing accidents.

CASE III

A tool truck driver, making his routine crib stops, picked up a crib attendant. The attendant had requested a ride to another part of the plant, since he knew the driver would be going in that direction.

The truck driver deviated from his aisle route and angled through a cleared but darkened area. This area was being prepared for new machinery installation, and at that time of night was not fully lighted.

The driver, who was sitting on the right side of the cab, suddenly noticed that the truck was headed for a steel building column. Before he could warn the driver, the left front corner of the truck struck and glanced off the column.

The impact threw the driver against the column and about 15 ft (4.6 m) away from the truck. The truck continued for approximately 50 ft (15.3 m) before the rider could get behind the wheel to apply the brakes and bring the truck to a stop.

The driver suffered a skull fracture, concussion, and severe injuries to the left arm and chest. He was taken immediately to the local hospital, where he died from a blood clot about three weeks later.

What could have been done to prevent this accident? Similar accidents in the future?

Guide and Background Information for Case III

Explain to the group, if asked, that these trucks are not designed to carry passengers. In order for a rider to sit on the seat, the driver must move over, which puts him in an awkward position.

Following are other pertinent facts:

- The aisles were not marked in this particular area.
- The area was not "roped off" and there were no signs of any type to indicate equipment was being installed.
- The truck was not equipped with a seat belt.
- The accident happened on the second shift (about 10 p.m.).
- The machinery installation had been going on over an extended period of time.
- There were no rules concerning riders.
- The driver was experienced.

Possible Solutions for Case III

1. The driver's failure to stay within the main aisle was an important factor in the accident. Properly marked aisles might have prevented the driver from taking a short cut.
2. When there are properly marked aisles, truck drivers should be instructed and trained in proper procedures.
3. Better illumination might also have prevented the accident. Since the installation had been going on for some time, the area should have been properly illuminated.
4. The area could have been roped off or marked.
5. A rigidly enforced rule against "no riders" should have been instituted.
6. Installation of seat belts in equipment of this type is a possibility, but because of the nature of the work, not generally done.

Summary

Stress to the group that the accident was caused by a combination of factors:

- The unsafe conditions were poor lighting, unmarked aisles, and lack of signs.
- The unsafe procedures were the driver's "short cutting," and picking up the rider. The lack of a rule against riders did not exonerate the driver, since he had to make room for the rider and had to be aware that he was not in the best position to control the truck. The rider also must have been aware of the situation when he moved into the seat.

CASE IV

A truck driver and a millwright were in the process of moving a pump, weighing approximately 2,000 lbs, that was mounted on a skid. They were using a mobile crane, and as the operator was backing up with the crane boom elevated at about a 30-degree angle, the rear wheels of the crane rose off the ground. To compensate for this overbalanced condition, the operator, while proceeding backwards, raised the crane boom to an approximate 60-degree angle. His co-worker at the time was observing the load on the crane boom and did not notice an overhead 440-volt line.

The crane boom became entangled in the overhead 440-volt wires, located approximately 18 ft (5.4) above ground. When the operator observed this, he jumped off the operating platform and was not injured. The insulation on the 440-volt line was not cut, so the crane boom did not become energized.

What could have been done to prevent this accident? Similar accident in the future?

Guide and Background Information for Case IV

Explain to the group, if the question comes up, that this man was not injured. Had the insulation been cut, this man could have been electrocuted.

Following are other pertinent facts:

- The crane was of proper size to handle the load.
- It was not possible to have the electric current turned off.
- The crane was designed to operate both forward the backward.
- Both employees were experienced and had worked together before. Both worked for the same company.
- The normal procedure in this company was for a man to walk at the head of the mast in order to guide the movement of the crane. In this case, the millwright should have taken over the job.

Possible Solutions for Case IV

1. The millwright or another person should have walked at the head of the mast and made certain all was clear.
2. The job was poorly planned. When the rear wheels of the crane raised off the ground at the 30-degree angle, this should have been the signal to stop the operation and do some checking. The pump should have been placed on dollies and rolled under the wires, since it was quite obvious that with only 18 ft (5.4 m) of clearance and the boom raised to 60 degrees, there was going to be contact.

Summary

Better job procedures should be instituted and definite procedures established and followed in all operations of this type. A thorough inspection should be made of the area and the route to be followed before any attempt is made to move loads of this nature.

CASE V

A warehouse employee was in the act of removing a part from the top of a storage bin. To reach it, she climbed on a wooden container instead of getting a ladder available for this purpose.

While standing on the edge of the container, it began to tip. The employee, trying to check her fall, grasped for the top of the bin with her left hand. As she did so, the ring she wore on her ring finger became caught on a bolt end, damaging the finger so badly that it had to be amputated.

What could have been done to prevent this accident? Similar accidents?

Guide and Background Information for Case V

If the question is raised, there were no rules in this company prohibiting the wearing of jewelry at work.

Following are other pertinent facts:

- The bins contained parts that were not in great demand.
- There was only one ladder in the storage area, which was large and contained three aisles.
- The ladder, at the time of the accident, was not in the aisle in which the employee was injured.

Possible Solutions for Case V

1. A rule prohibiting the wearing of jewelry at work could be instituted. Better training on the hazards of wearing rings would help.
2. Availability of a ladder in each aisle would make one always handy and thus more likely to be used.
3. Better housekeeping would help, since the wooden container the employee stood on actually was not in its proper place.
4. The use of posters also would help with the problem.

Summary

Point out the hazards of wearing rings at work. Stress the need for education on the subject, since it may be difficult to enforce a rule against wearing rings because of certain religious beliefs.

CASE VI

An employee asked another employee to exchange a trim knife for him at the tool crib. The man placed the knife in the side pocket of his coveralls and started for the crib. On the way, he stopped to remove some stock

from a shelf 8 ft (2.4 m) from the floor. He was unable to reach the stock from the floor, so he used the bottom shelf on the rack as a platform to stand on. As he bent his right leg to step up on the shelf, the sharp point of the knife was forced up and penetrated his body approximately one inch. A deep laceration was the only injury.

What could have been done to prevent his accident? Similar accidents in the future?

Guide and Background Information for Case VI

If asked, explain there were no holders or tool bags provided for carrying tools of this type, such as punches, awls, screw drivers, etc. Also, it was common practice for workers to exchange tools for each other.

Following are other important facts that may help:

- The trim knife was not one of the tools used in the injured employee's own work. He did, however, use other similar tools.
- The injured employee started out to obtain stock, but was asked to exchange the trim knife on his way.
- There was a small stepladder nearby that the injured employee could have used.
- There were no rules prohibiting any employee from exchanging tools, even if the attendant at the tool crib knew it was for another person.

Possible Solutions for Case VI

1. Proper belt holders should be provided to carry tools, since these employees used sharp-pointed tools frequently.
2. A rule prohibiting employees from exchanging tools for each other could be put in effect.
3. Better training in the handling of tools would help.
4. More emphasis on the hazards of using shelves as a stepladder also would help prevent accidents.

Summary

Stress the number of accidents resulting from improperly handled tools. Emphasize the importance of a sound training program on the proper use and handling of sharp and pointed tools. Stress also the need for rigid rules for carrying tools of this type.

CASE VII

A worker was assigned to clean out sludge in the large degreaser he had been operating. His supervisor advised him to use gloves and respiratory protection before entering the degreasing tank and told him to make sure he had a co-worker watch him while he worked. An hour later, the supervisor noted the co-worker eating lunch. He asked the co-worker where the degreaser operator was. The co-worker responded that the operator had told him to go on to lunch, that he would be finished in the degreaser in five minutes. The supervisor and co-worker went immediately to check on the operator, whom they saw face down in the degreasing tank. The supervisor hurriedly descended into the tank to rescue the worker, but had to leave without retrieving him when he immediately began to feel light-headed. The co-worker called the emergency squad who, equipped with self-contained breathing apparatus, retrieved the unconscious operator. The operator survived, but spent several days in the hospital recuperating and has experienced some memory loss and problems in coordination.

What could have been done to prevent this accident? Similar situations in the future?

Guide and Background Information for Case VII

Explain that this is a typical confined space accident, in which the rescuer is as likely (or more likely) to be injured as the initial victim. In over half the recorded cases, the unprepared rescuer has also been a fatality. Other pertinent facts include:

- The degreaser operator had not received training in proper use of respiratory protection or confined space entry procedures.
- The operator was able to work in the pit for at least an hour before being overcome because:
- His organic vapor respirator did provide protection against the initial concentration of solvent vapors in the tank.
- The tank was not an oxygen deficient atmosphere initially (although under other circumstances, it might have been one), but became one later.
- The worker stirred and shoveled up the sludge. Solvent vapors were released from air pockets in the solid material, so the concentration of vapors increased in the tank, and began to displace oxygen. The respirator no longer protected him because the absorbent material had become saturated with the solvent, and more important, the respirator could not supply oxygen (which had been depleted by the end of the hour). When trying to rescue the worker, the supervisor began to experience the effects of a lack of oxygen.
- The cloth gloves the worker wore may have added to his solvent exposure, because they may have allowed the solvent to be absorbed through the skin.

Possible Solutions for Case VII

1. The job indicated poor or not planning. The operator, co-worker, and supervisor should have received training on the hazards and proper procedures for entry into confined spaces. These

include:
   a. Use of a permit entry system
   b. Pre-entry sampling of the atmosphere for flammable or toxic gases and vapors
   c. Use of air-supplied or self-contained breathing apparatus where dangerous chemical exposures could occur, as well as impervious clothing
   d. Use of lifelines and belts to aid in retrieval, if necessary
   e. Presence of an attendant throughout the duration of the work

2. The supervisor should have completed a confined space entry permit and communicated to the operator the equipment and precautions that were necessary, and assured that these were used.

3. The operator should have received training in the proper use of respiratory protection and protective clothing, and the distinction between air-purifying and air-supplying respirators.

4. Hazard communication training should have been provided to these workers to acquaint them with the toxic properties of the chemicals they worked with.

Summary

Better training should be provided and procedures instituted to prevent future incidents of this type. The hazards of the confined space should be assessed and understood before anyone attempts an entry.

CASE VIII

A worker in your department has taken an excessive number of sick days. You ask the company nurse what the problem might be. She says the worker has come in to ask for aspirin almost every day she has been at work for the last two months.

You go to observe the worker at her operation. Her job, which is in the rework area, involves using pliers to remove wires from a part before it is repaired. You ask the worker to demonstrate how she does her job. She shows you that she grasps the part in her left hand, and twists it as she uses the pliers to grasp and pull the wire out with her right hand.

You note that she does not use the stool provided. The worker says that it is uncomfortable; because she is short, her legs dangle. She explains that she has been absent due to pain in her back and numbness and tingling in her hands.

Guide and Background Informationfor Case VIII

Explain to the group, if asked, that there are at least two obvious ergonomic problems at the workstation: dangling feet and repetitive, forceful grasping combined with twisting of the wrist. The latter is probably responsible for the worker's back pain. Other pertinent facts include:

- Her doctor told her that her hand problem was due to hormonal changes caused by menopause and that her back pain was likely caused by stress. Her doctor had not seen her work area; however, he or she told the worker to take asperin to releive the symptoms.
- She told you she did not want to complain about her discomfort because she was afraid of losing her job.

Possible Solutions for Case VIII

1. Institute an ergonomics program that could include:
   –Observation of work operations with employee involvement and input
   –Medical surveillance to determine if and where workers are being injured
   –Management commitment to making positive ergonomic changes where necessary
   –Development of an ergonomically trained team of managers, engineers, medical personnel, supervisors, workers, and others

2. For this particular injured worker, the supervisor should observe which actions require forceful repetitive movement of the wrists. It is apparent that the worker had to twist both her left and right wrists while applying force to the pliers with her right hand. These repetitive motions have probably contributed to what appears to be carpal tunnel syndrome. Some solutions might be:
   –Automating the operation
   –Using a grip and clamp to hold the part, and using ergonomically designed pliers that prevent the wrist from bending
   –Reducing the work rate
   –Rotating the worker to other jobs that require use of different muscle groups

Summary

Institute an ergonomic program. Evaluate this and all similar operations to determine if other workers are experiencing similar problems. Investigate and establish specific solutions for these operations and obtain employee feedback on how well the solutions are working.

GLOSSARY

Abrasive blasting. A process for cleaning surfaces by means of such materials as sand, alumina, or steel grit in a stream of high-pressure air.

Accident. An unplanned event, not necessarily injurious or damaging to property, interrupting the activity in process.

Accident causes. Hazards and those factors (management oversight, system limitation/failure, etc.) that, individually or in combination, directly cause accidents.

Acclimatization. The process of becoming accustomed to new conditions (i.e., heat, cold).

Accuracy (instrument). Quite often used incorrectly as precision (see Precision). Accuracy refers to the agreement between the true value and a reading or observation obtained from an instrument or a technique.

ACGIH. American Conference of Governmental Industrial Hygienists. ACGIH develops and publishes recommended occupational exposure limits for hundreds of chemical substances and physical agents. (See TLV.)

Acoustic, Acoustical. Containing, producing, arising from, actuated by, related to, or associated with sound.

Acoustic trauma. Hearing loss caused by sudden loud noise in one ear, or by sudden blow to head. In most cases, hearing loss is temporary, although there may be some permanent loss.

Acrylic. A family of synthetic resins made by polymerizing esters of acrylic acids.

Action level. Term used by OSHA and NIOSH to express the level of toxicant that requires medical surveillance, usually one half the PEL.

Acuity. The sense that pertains to the sensitivity of receptors used in hearing or vision.

Acute effect. An adverse effect on a human or animal body, with several symptoms d eveloping rapidly and coming quickly to a crisis.

Acute exposure. Symptoms usually appear within 24 hours of an exposure; the exposure is usually brief and highly concentrated.

Acute toxicity. The adverse (acute) effects resulting from a single dose of or exposure to a substance.

Adhesion. The ability of one substance to stick to another. There are two types of adhesion: mechanical, which depends on the penetration of the surface, and molecular or polar adhesion, in which adhesion to a smooth surface is obtained because of polar groups such as carboxyl groups.

Administrative controls. Methods of controlling employee exposures by job rotation, work assignment, or time periods away from the hazard.

Adsorption. The condensation of gases, liquids, or dissolved substances on the surfaces of solids.

Aerosol. Liquid droplets or solid particles dispersed in air; are of fine enough particle size (0.01 to 100 micrometers) to remain so dispersed for a period of time.

Agency or agent. The principal object, such as a tool, machine, or equipment, involved in an accident; is usually the object inflicting injury or property damage.

Air. The mixture of gases that surrounds the earth; its major components are as follows: 78.08 percent nitrogen, 20.95 percent oxygen, 0.03 percent carbon dioxide, and 0.93 percent argon. Water vapor (humidity) varies. (See Standard air.)

Air cleaner. A device designed to remove atmospheric airborne impurities, such as dusts, gases, vapors, fumes, and smokes.

Air conditioning. The process of treating air so as to control simultaneously its temperature, humidity, cleanliness, and distribution to meet requirements of the conditioned space.

Air filter. An air-cleaning device to remove light

particulate matter from normal atmospheric air.

Air hammer. A percussion-type pneumatic tool, fitted with a handle at one end of the shank and a tool chuck at the other, into which a variety of tools may be inserted.

Air-line respirator. A respirator that is connected to a compressed breathing air source by a hose.

Air monitoring. The sampling for and measuring of pollutants in the atmosphere.

Air-purifying respirator. Respirator that uses filters or sorbents to remove harmful substances from the air.

Air-regulating valve. An adjustable valve used to regulate airflow to the facepiece, helmet, or hood of an air-line respirator.

Air-supplied respirator. Respirator that provides a supply of breathable air from a clean source outside of the contaminated work area.

Algorithm. A precisely stated procedure or set of instructions that can be applied stepwise to solve a problem.

Aliphatic. (Derived from Greek word for oil.) Pertaining to an open-chain carbon compound. Usually applied to petroleum products derived from a paraffin base and having a straight or branched chain and saturated or unsaturated molecular structure. Substances such as methane and ethane are typical aliphatic hydrocarbons.

Alkali. Any chemical substance that forms soluble soaps with fatty acids. Alkalis are also referred to as bases. They may cause severe burns to the skin. Alkalis turn litmus paper blue and have pH values from 8 to 14.

Allergic reaction. An abnormal physiological response to chemical or physical stimuli by a sensitive person.

Allergy. An abnormal response of a hypersensitive person to chemical and physical stimuli. Allergic manifestations of major importance occur in about 10 percent of the population.

Alloy. A mixture of metals.

Alpha particle (alpha-ray, alpha-radiation). A small, electrically charged particle of very high velocity thrown off by many radioactive materials, including uranium and radium. It is made up of two neutrons and two protons. Its electric charge is positive.

Ambient noise. The all-encompassing noise associated with a given environment, being usually a composite of sounds from many sources.

Ampere. The standard unit for measuring the strength of an electrical current.

Anemometer. A device to measure air velocity.

Anesthetic. A chemical that causes a total or partial loss of sensation. Overexposure to anesthetics can cause impaired judgment, dizziness, drowsiness,

headache, unconsciousness, and even death. Examples include alcohol, paint remover, and degreasers.

Anneal. To treat by heat with subsequent cooling for drawing the temper of metals, that is, to soften and render them less brittle. (See Temper.)

ANSI. The American National Standards Institute is a voluntary membership organization (run with private funding) that develops consensus standards nationally for a wide variety of devices and procedures.

Antidote. A remedy to relieve, prevent, or counteract the effects of a poison.

Antioxidant. A compound that retards deterioration by oxidation. Antioxidants for human food and animal feeds, sometimes referred to as freshness preservers, retard rancidity of fats and lessen loss of fat-soluble vitamins (A, D, E, K). Antioxidants also are added to rubber, motor lubricants, and other materials to inhibit deterioration.

Appearance. A description of a substance at normal room temperature and normal atmospheric conditions. Appearance includes the color, size, and consistency of a material.

Approved. Tested and (listed) as satisfactory meeting predetermined requirements by an authority having jurisdiction.

Arc welding. One form of electrical welding using either uncoated or coated rods.

Arc-welding electrode. A component of the welding circuit through which current is conducted between the electrode holder and the arc.

Article. A manufactured item formed to a specific shape or design which has an end-use function dependent in whole, or in part, upon its shape or use during end use and which does not release or otherwise result in exposure to a hazardous chemical under normal conditions of use.

Asbestos. A hydrated magnesium silicate in fibrous form.

Asphyxia. Suffocation from lack of oxygen. Chemical asphyxia is produced by a substance such as carbon monoxide that combines with hemoglobin to reduce the blood's capacity to transport oxygen. Simple asphyxia is the result of exposure to a substance such as methane, that displaces oxygen.

Asphyxiant. A vapor or gas that can cause unconsciousness or death by suffocation (lack of oxygen).

Assistant Secretary. The Assistant Secretary of Labor for Occupational Safety and Health, U.S. Department of Labor, or a designee.

ASTM. American Society for Testing and Materials; voluntary membership organization with members from broad spectrum of individuals, agencies, and industries concerned with materials. As the world's largest source of voluntary consensus standards for materials, products, systems, and services, ASTM is a resource for sampling and testing methods, health and safety aspects of materials, safe performance guidelines, and effects of physical and biological agents and chemicals.

Atmosphere-supply respirator. A respirator that provides breathing air from a source independent of the surrounding atmosphere. There are two types: air-line and self-contained breathing apparatus.

Atmospheric pressure. The pressure exerted in all directions by the atmosphere. At sea level, mean atmospheric pressure is 29.92 inches Hg, 14.7 psi, or 407 inches w.g.

Atmospheric tank. A storage tank that has been designed to operate at pressures from atmopheric through 0.5 psig (3.5 kPa).

Atomic energy. Energy released in nuclear reactions. Of particular interest is the energy released when a neutron splits an atom's nucleus into smaller pieces (fission) or when two nuclei are joined together under millions of degrees of heat (fusion). "Atomic energy" is really a popular misnomer. It is more correctly called "nuclear energy."

Atomic power. See Atomic energy.

Atomic waste. The radioactive ash produced by the splitting of uranium (nuclear) fuel, as in a nuclear reactor. It may include products made radioactive in such a device.

Audible range. The frequency range over which normal ears hear approximately 20 Hz through 20,000 Hz. Above the range of 20,000 Hz, the term "ultrasonic" is used. Below 20 Hz, the term "subsonic" is used.

Audiogram. A record of loss of hearing level measured at several different frequencies—usually 500 to 6,000 Hz. The audiogram may be presented graphically or numerically. Hearing level is shown as a function of frequency.

Audiometer. A signal generator or instrument for measuring objectively the sensitivity of hearing in decibels.

Autoignition temperature. The temperature to which a closed, or nearly closed, container must be heated in order that a flammable liquid, when introduced into the container, will ignite spontaneously or burn.

Babbitt. An alloy of tin, antimony, copper, and lead used as a bearing metal.

Background noise. Noise coming from sources other than the particular noise source being monitored. (See Background radiation.)

Background radiation. The radiation coming from sources other than the radioactive material to be measured. This "background" is primarily due to cosmic rays, which constantly bombard the earth from outer space.

Bacteria. Microscopic organisms living in soil, water, organic matter, or the bodies of plants and animals characterized by lack of a distinct nucleus and lack of ability to photosynthesize. Singular: Bacterium.

Bag house. Many different trade meanings. Term commonly used for the housing containing bag filters for recovery of fumes of arsenic, lead, sulfur, etc., from flues of smelters.

Balancing by dampers. Method of designing local exhaust system ducts using adjustable dampers to distribute airflow after installation.

Balancing by static pressure. Method of designing local exhaust system ducts by selecting the duct diameters that generate the static pressure to distribute airflow without dampers.

Ball mill. A grinding device using balls usually of steel or stone in a rotating container.

Base. A compound that reacts with an acid to form a salt. It is another term for alkali. It turns litmus paper blue.

Bauxite. Impure mixture of aluminum oxides and hydroxides; the principal source of aluminum.

Beat elbow. Bursitis of the elbow; occurs from use of heavy vibrating tools.

Beat knee. Bursitis of the knee joints due to friction or vibration; common in mining.

Beehive kiln. A kiln shaped like a large beehive; used usually for calcining ceramics.

Benign. Not malignant. A benign tumor is one that does not metastasize or invade tissue. Benign tumors may still be lethal, due to pressure on vital organs.

Beta particle (beta-radiation). A small, electrically charged particle thrown off by many radioactive materials. It is identical with the electron. Beta particles emerge from radioactive material at high speeds.

Billet. A piece of semifinished iron or steel, nearly square in section, made by rolling and cutting an ingot.

Binder. The nonvolatile portion of a coating vehicle which is the film-forming ingredient used to bind the paint pigment particles together.

Biodegradable. Capable of being broken down into innocuous products by the action of living things.

Biohazard. A combination of the words "biological hazard." Organisms or products of organisms that present a risk to humans.

Biohazard area. Any area (a complete operating complex, a single facility, a room within a facility, etc.) in which work has been or is being performed with biohazardous agents or materials.

Biohazard control. Any set of equipment and procedures utilized to prevent or minimize the exposure of humans and their environment to biohazardous agents or materials.

Biomechanics. The study of the human body as a system operating under two sets of law: the laws of Newtonian mechanics and the biological laws of life.

Black light. Ultraviolet (UV) light radiation between 3,000 and 4,000 angstroms (0.3-0.4 micrometers).

Black liquor. A liquor composed of alkaline and organic matter resulting from digestion of wood pulp and cooking acid during the manufacture of paper.

Bleaching bath. Chemical solution used to bleach colors from a garment preparatory to dyeing it, a solution of chlorine or sodium hypochlorite commonly being used.

Blind spot. Normal defect in visual field due to position at which optic nerve enters the eye.

Boiling point. The temperature at which a liquid changes to a vapor state, expressed in degrees.

BOM. Bureau of Mines of the U.S. Department of Interior. BOM began approving air-breathing apparatus in 1918, later added all types of respirators. BOM's respirator testing/approval activities have been discontinued. BOM-approved Type 14F gas masks are still acceptable; all other BOM approvals have expired or have been replaced by NIOSH approvals.

Bonding. The interconnecting of two objects by means of a clamp and bare wire. Its purpose is to equalize the electrical potential between objects. (See Grounding.)

Brake horsepower. The horsepower required to drive a unit; it includes the energy losses in the unit and can be determined only by actual test. It does not include driver losses between motor and unit.

Brass. An alloy of copper and zinc; may also contain a small portion of lead.

Brattice. A partition constructed in underground passageways to control ventilation in mines.

Braze. To solder with any alloy that is relatively infusible.

Brazing furnace. Used for heating metals to be joined by brazing. Requires a high temperature.

Breathing tube. A tube through which air or oxygen flows to the facepiece, helmet, or hood.

Breathing zone. Imaginary globe of two-foot radius surrounding the head.

Broach. A cutting tool for cutting nonround holes.

Bubble tube. A device used to calibrate air-sampling pumps.

Buffer. Any substance in a fluid that tends to resist the change in pH when acid or alkali is added.

Bulk density. Mass of powdered or granulated solid material per unit of volume.

Bulk plant. That portion of a property where flammable or combustible liquids are received by tank vessel, pipelines, tank car, or tank vehicle and are sorted or blended in bulk for the purpose of distributing such liquids by tank vessel, pipeline, tank car, tank vehicle, or container.

Burns. Result from the application of too much heat to the skin. First-degree burns show redness of the unbroken skin; second-degree, skin blisters and some breaking of the skin; third-degree, skin blisters and destruction of the skin and underlying tissues, which can include charring and blackening.

cc. Cubic centimeter; a volume measurement in the metric system, equal in capacity to one milliliter (ml)—approximately twenty (20) drops. There are 16.4 cc in one cubic inch.

Calcination. The heat treatment of solid material to bring about thermal decomposition, to lose moisture or other volatile material, or to be oxidized or reduced.

Calender. An assembly of rollers for producing a desired finish on paper, rubber, artificial leather, plastics, or other sheet material.

Calking. The process or material used to fill seams of boats, cracks in tile, etc.

Capture velocity. Air velocity at any point in front of the hood necessary to overcome opposing air currents and to capture contaminated air by causing it to flow into the exhaust hood.

Carbon black. Essentially a pure carbon, best known as common soot. Commercial carbon black is produced by making soot under controlled conditions. It is sometimes called channel black, furnace black, acetylene black, or thermal black.

Carbon monoxide. A colorless, odorless, toxic gas produced by any process that involves the incomplete combustion of carbon-containing substances.

Carcinogen. A substance or agent that can cause a growth of abnormal tissue or tumors in humans or animals.

Carcinogenic. Cancer-producing.

Carding. The process of combing or untangling wool, cotton, etc.

Carpal tunnel. A passage in the wrist through which the median nerve and many tendons pass to the hand from the forearm.

Carpal tunnel syndrome. A common affliction caused by compression of the median nerve in the carpal tunnel. Often associated with tingling, pain, or numbness in the thumb and first three fingers.

CAS number. Identifies a particular chemical by the Chemical Abstract Service, a service of the American Chemical Society that indexes and compiles abstracts of worldwide chemical literature called "Chemical Abstracts."

Case-hardening. A process of surface-hardening metals by raising the carbon or nitrogen content of the outer surface.

Cask (or coffin). A thick-walled container (usually lead) used for tranporting radioactive materials.

Casting. The pouring of a molten material into a mold and permitting it to solidify to the desired shape.

Catalyst. A substance that changes the speed of a chemical reaction but undergoes no permanent change itself. In respirator use, a substance that converts a toxic gas (or vapor) into a less toxic gas (or vapor). Usually catalysts greatly increase the reaction rate, as in conversion of petroleum to gasoline by cracking. In paint manufacture, catalysts, which hasten the film-forming, generally become part of the final product. In most uses, however, they do not, and can often be used over again.

Cathode. The negative electrode.

Catwalk. A narrow suspended footway usually for inspection or maintenance purposes.

Caustic. Something that strongly irritates, burns, corrodes, or destroys living tissue. (See Alkali.)

Ceiling limit (C). A maximum airborne concentration of a toxic substance in the work environment. ACGIH terminology. (See TLV.)

Cellulose ($C_6H_{10}OH_5$). A carbohydrate that makes up the structural material of vegetable tissues and fibers.

Celsius. The Celsius temperature scale is a designation of the scale previously known as the centigrade scale.

Cement, portland. Portland cement commonly consists of hydraulic calcium silicates to which the addition of certain materials in limited amounts is permitted. Ordinarily, the mixture consists of calcareous materials such as limestone, chalk, shells, marl, clay, shale, blast furnace slag, etc. In some specifications, iron ore and limestone are added. The mixture is fused by calcining at temperatures usually up to 1,500 C.

Central nervous system. The brain and spinal cord. These organs supervise and coordinate the activity of the entire nervous system. Sensory impulses are transmitted into the central nervous system, and motor impulses are transmitted out.

Centrifuge. An apparatus that uses centrifugal force to separate or remove particulate matter suspended in a liquid.

Ceramic. A term applied to pottery, brick, and the tile products molded from clay and subsequently calcined.

CFR. *Code of Federal Regulations*. A collection of the regulations that have been promulgated under U.S. law.

Chemical. Any element, chemical compound, or mixture of elements and/or compounds.

Chemical burns. Generally similar to those caused by heat. After emergency first aid, their treatment is the same as that for thermal burns.

Chemical cartridge. A chemical cartridge is used with a respirator for removal of low concentrations of specific vapors and gases.

Chemical cartridge respirator. A respirator that uses various chemical substances to purify inhaled air of certain gases and vapors.

Chemical engineering. That branch of engineering concerned with the development and application of manufacturing processes in which chemical or certain physical changes of materials are involved.

Chemical manufacturer. An employer with a workplace where chemicals are produced for use or distribution.

Chemical name. The scientific designation of a chemical in accordance with the nomenclature system developed by the International Union of Pure and Applied Chemistry (IUPAC) or the Chemical Abstracts Service (CAS), or a name that will clearly identify the chemical for the purpose of conducting a hazard evaluation.

Chemical reaction. A change in the arrangement of atoms or molecules to yield substances of different composition and properties. Common types of reaction are combination, decomposition, double decomposition, replacement, and double replacement.

CHEMTREC. Chemical Transportation Emergency Center.

Chronic. Persistent, prolonged, repeated.

Chronic exposure. Symptoms are usually delayed and cumulative, and result from repeated exposure to low-level concentrations of hazardous chemicals.

Chronic toxicity. Adverse (chronic) effects resulting from repeated doses of or exposures to a substance over a relatively prolonged period of time.

Circuit. A complete path over which electrical current may flow.

Circuit breaker. A device that automatically interrupts the flow of an electrical current when the current becomes excessive.

Clays. A great variety of aluminum-silicate-bearing rocks, plastic when wet, hard when dry. Used in pottery, stoneware, tile, bricks, cements, fillers, and abrasives. Kaolin is one type of clay. Some clay deposits may include appreciable quartz. Commercial grades of clays may contain up to 20 percent quartz.

Clean Air Act. Federal law enacted to regulate/reduce air pollution. Administered by EPA.

Clean Water Act. Federal law enacted to regulate/reduce water pollution. Administered by EPA.

Coagulation. Formation of a clot or gelatinous mass.

Coated welding rods. The coatings of welding rods vary. For the welding of iron and most steel, the rods contain manganese, titanium, and a silicate.

Code of Federal Regulations (CFR). The rules

promulgated under U.S. law and published in the *Federal Register* and actually in force at the end of a calendar year are incorporated in this code.

Coffin. A thick-walled container (usually lead) used for transporting radioactive materials; also called cask.

Cohesion. Molecular forces of attraction between particles of like composition.

Colloid mill. A machine that grinds materials into very fine state of suspension, often simultaneously placing this in suspension in a liquid.

Combustible. Able to catch on fire and burn.

Combustible liquids. Combustible liquids are those having a flash point at or above 37.8 C (100 F).

Comfort ventilation. Airflow intended to maintain comfort of room occupants (heat, humidity, and odor).

Comfort zone. The average range of effective temperatures over which the majority (50 percent or more) of adults feel comfortable.

Common name. Any designation or identification such as code name, code number, trade name, brand name, or generic name used to identify a chemical other than by its chemical name.

Communicable. Refers to a disease whose causative agent is readily transferred from one person to another.

Compaction. The consolidation of solid particles between rolls, or by tamp, piston, screw, or other means of applying mechanical pressure.

Compound. A substance composed of two or more elements joined according to the laws of chemical combination. Each compound has its own characteristic properties different from those of its constituent elements.

Compressed gas. (1) Any gas or mixture of gases having, in a container, an absolute pressure exceeding 40 psi at 70 degrees F (21.2 degrees C); or (2) a gas or mixture of gases having, in a container, an absolute pressure exceeding 104 psi at 130 degrees F (54.4 degrees C) regardless of the pressure at 70 degrees F (21.2 degrees C); or (3) a liquid having a vapor pressure exceeding 40 psi at 100 degrees F (37.8 degrees C) as determined by ASTM D-323-72.

Concentration. The amount of a given substance in a stated unit of measure. Common methods of stating concentration are percent by weight or by volume, weight per unit volume, normality, etc.

Condensate. The liquid resulting from the process of condensation.

Condensation. Act or process of reducing from one form to another denser form, such as steam to water.

Conditions to avoid. Conditions encountered during handling or storage that could cause a substance to become unstable.

Conductive hearing loss. Type of hearing loss; not caused by noise exposure, but due to any disorder in the middle or external ear that prevents sound from reaching the inner ear.

Confined space. Any area that has limited openings for entry and exit that would make escape difficult in an emergency, has a lack of ventilation, contains known and potential hazards, and is not intended nor designated for continuous human occupancy. (See "Permit Required Confined Space," 29 *CFR* 1910.146(b)(23) proposed.)

Contact dermatitis. Dermatitis caused by contact with a substance—gaseous, liquid, or solid. May be due to primary irritation or an allergy.

Container. Any bag, barrel, drum, bottle, box, can, cylinder, reaction vessel, storage tank, or the like that contains a hazardous chemical. For the purposes of the HCS, engines, fuel tanks, and other operating systems in a vehicle are not considered to be containers.

Controlled area. A specified area in which exposure of personnel to radiation or radioactive material is controlled and which is under the supervision of a person who has knowledge of the appropriate radiation protection practices, including pertinent regulations, and who has responsibility for applying them.

Convection. The motions in fluids resulting from differences in density and the action of gravity.

Coolants. Coolants are transfer agents used in a flow system to convey heat from its source.

Corrective lens. A lens ground to the wearer's individual prescription.

Corrosion. Physical change, usually deterioration or destruction, brought about through chemical or electrochemical action as contrasted with erosion caused by mechanical action.

Corrosive (COR). As defined by DOT, a corrosive material is a liquid or solid that causes visible destruction or irreversible alterations in human skin tissue at the site of contact or—in the case of leakage from its packaging—a liquid that has a severe corrosion rate on steel. Two common examples are caustic soda and sulfuric acid.

Cosmic rays. High-energy rays of great penetrating power that bombard the earth from outer space.

Cottrell precipitator. A device for dust collection using high-voltage electrodes.

Counter. A device for counting. (See Geiger counter and Scintillation counter.)

Covered electrode. A composite filler metal electrode consisting of a core of bare electrode or metal-cored electrode to which a covering (sufficient to provide a slag layer on the weld metal) has been applied; the covering may contain materials providing such functions as shielding from the atmosphere, deoxidation, and arc stabilization and can serve as a source of metallic additions to the weld.

CPR. Cardiopulmonary resuscitation.

CPSC. Consumer Products Safety Commission; federal agency with responsibility for regulating hazardous materials when they appear in consumer goods. For CPSC purposes, hazards are defined in the Hazardous Substances Act and the Poison Prevention Packaging Act of 1970.

Cristobalite. A crystalline form of free silica, extremely hard and inert chemically; very resistant to heat. Quartz in refractory bricks and amorphous silica in diatomaceous earth are altered to cristobalite when exposed to high temperatures (calcined).

Critical pressure. The pressure under which a substance may exist as a gas in equilibrium with a liquid at the critical temperature.

Critical temperature. The temperature above which a gas cannot be liquefied by pressure alone.

Crucible. A heat-resistant, barrel-shaped pot used to hold metal during melting in a furnace.

Crude petroleum. Hydrocarbon mixtures that have a flash point below 65.6 C (150 F), and that have not been processed in a refinery.

Cry-, cryo- (prefix). Very cold.

Cryogenics. The field of science dealing with the behavior of matter at very low temperatures.

Cubic centimeter (cc). A volumetric measurement that is also equal to one milliliter (mL.)

Cubic meter (m³). A measure of volume in the metric system.

Cutting fluids (oils). The cutting fluids used in industry today are usually an oil or an oil-water emulsion used to cool and lubricate a cutting tool. Cutting oils are usually light or heavy petroleum fractions.

CW laser. Continuous wave laser. (See Laser.)

Cyanide (as CN). Cyanides inhibit tissue oxidation upon inhalation or ingestion and cause death.

Cyanosis. Blue appearance of the skin, especially on the face and extremities, indicating a lack of sufficient oxygen in the arterial blood.

Cyclone separator. A dust-collecting device that has the ability to separate particles by size. Typically used to collect respirable dust samples.

Damage risk criterion. The suggested base line of noise tolerance which, if not exceeded, should result in no hearing loss due to noise. A damage risk criterion may include in its statement a specification of such factors as time of exposure, noise level and frequency, amount of hearing loss that is considered significant, percentage of the population to be protected, and method of measuring the noise.

Damp. A harmful gas or mixture of gases occurring in coal mining.

Dampers. Adjustable sources of airflow resistance used to distribute airflow in a ventilation system.

Dangerous to life or health, immediately (IDLH). Used to describe very hazardous atmospheres where employee exposure can cause serious injury or death within a short time or serious delayed effects.

dBA. Sound level in decibels read on the A-scale of a sound level meter. The A scale discriminates against very low frequencies (as does the human ear) and is, therefore, better for measuring general sound levels. (See also Decibel.)

Decibel (dB). A unit used to express sound power level (Lw). Sound power is the total acoustic output of a sound source in watts (W). By definition, sound power level, in decibels, is: Lw = 10 log W divided by W o, where W is the sound power of the source and W o is the reference sound power.

Decomposition. Breakdown of a material or substance (by heat, chemical reaction, electrolysis, decay, or other process) into parts or elements or simpler compounds.

Decontaminate. To make safe by eliminating poisonous or otherwise harmful substances, such as noxious chemicals or radioactive material.

Density. The mass (weight) per unit volume of a substance. For example, lead is much more dense than aluminum.

Depressant. A substance that reduces a functional activity or an instinctive desire of the body, such as appetite.

Dermal toxicity. Adverse effects resulting from skin exposure to a substance. Ordinarily used to denote effects in experimental animals.

Dermatitis. Inflammation of the skin.

Dermatosis. A broader term than dermatitis; it includes any cutaneous abnormality, encompassing folliculitis, acne, pigmentary changes, and nodules and tumors.

Designated representative. Any individual or organization to whom an employee gives written authorization to exercise said employee's rights under the standard. A recognized or certified collective bargaining agent is treated automatically as a designated representative without regard to written employee authorization.

Diaphragm. (1) The musculomembranous partition separating the abdominal and thoracic cavities; (2) any separating membrane or structure; (3) a disk with one or more openings in it, or with an adjustable opening, mounted in relation to a lens, by which part of a light may be excluded from an area.

Diatomaceous earth. A soft, gritty, amorphous silica composed of minute siliceous skeletons of small aquatic plants. Used in filtration and decolorization of liquids, insulation, filler in dynamite, wax, textiles, plastics, paint, and rubber. Calcined and flux-calcined diatomaceous earth contains appreciable amounts of cristobalite, and dust levels should be controlled the same as for cristobalite.

Die. A hard metal or plastic form used to shape material to a particular contour or section.

Differential pressure. The difference in static pressure between two locations.

Diffusion rate. A measure of the tendency of one gas or vapor to disperse into or mix with another gas or vapor.

Dike. A barrier constructed to control or confine substances and prevent their movement.

Diluent. A liquid that is blended with a mixture to reduce concentration of the active agents.

Dilution. The process of increasing the proportion of solvent or diluent (liquid) to solute or particulate matter (solid).

Dilution ventilation. Airflow designed to dilute contaminants to acceptable levels. Also referred to as general ventilation or exhaust. (See General exhaust.)

Director. The Director, National Institute for Occupational Safety and Health, U.S. Department of Health and Human Services, or a designee.

Direct-reading instrumentation. Those instruments that give an immediate indication of the concentration of matter.

Disinfectant. An agent that frees from infection by killing the vegetative cells of microorganisms.

Dispersion. The general term describing systems consisting of particulate matter suspended in air or other fluid; also, the mixing and dilution of contaminant in the ambient environment.

Distal. Away from the central axis of the body.

Distillery. A plant or that portion of a plant where flammable or combustible liquids produced by fermentation are concentrated, and where the concentrated products may also be mixed, stored, or packaged.

Distributor. A business, other than a chemical manufacturer or importer, that supplies hazardous chemicals to other distributors or employees.

DOL. U.S. Department of Labor; includes the Occupational Safety and Health Administration (OSHA) and Mine Safety and Health Administration (MSHA).

Dose. A term used (1) to express the amount of a chemical or of ionizing radiation energy absorbed in a unit volume or an organ or individual. Dose rate is the dose delivered per unit of time. (See also Roentgen, Rad, Rem.) (2) Used to express amount of exposure to a chemical substance.

Dose equivalent, maximum permissible (MPD). The largest equivalent received within a specified period which is permitted by a regulatory agency or other authoritative group on the assumption that receipt of such dose equivalent creates no appreciable somatic or genetic injury. Different levels of MPD may be set for different groups within a population. (By popular usage, "dose, maximum permissible," is an accepted synonym.)

Dosimeter (dose meter). An instrument used to determine the full-shift exposure a person has received to a physical hazard.

DOT hazard class. DOT requires that hazardous materials offered for shipment be labeled with the proper DOT hazard class. These classes include corrosive, flammable liquid, organic peroxide, ORM-E, poison B, etc. The DOT hazard class may not adequately describe all the hazard properties of the material.

Drier. Any catalytic material that, when added to a drying oil, accelerates drying or hardening of the film.

Drop forge. To forge between dies by a drop hammer or drop press.

Droplet. A liquid particle suspended in a gas. The liquid particle is generally of such size and density that it settles rapidly and remains airborne for an appreciable length of time only in a turbulent atmosphere.

Dross. The scum that forms on the surface of molten metals, largely oxides and impurities.

Dry chemical. A powdered fire-extinguishing agent usually composed of sodium bicarbonate, monoammonium phosphate, potassium bicarbonate, etc.

Duct. A conduit used for conveying air at low pressures.

Dust collector. An air-cleaning device to remove heavy particulate loadings from exhaust systems before discharge to outdoors; usual range is loadings of 0.003 gr/cu ft (0.007 mg/m^3) and higher.

Dusts. Solid particles generated by handling, crushing, grinding, rapid impact, detonation, and decrepitation of organic or inorganic materials, such as rock, ore, metal, coal, wood, and grain. Dusts do not tend to flocculate except under electrostatic forces; they do not diffuse in air but settle under the influence of gravity.

Dynometer. Apparatus for measuring force or work output external to a subject. Often used to compare external output with associated physiological phenomena to assess physiological work efficiency.

EAP. Employee Assistance Program.

Ear. The entire hearing apparatus, consisting of three parts: external ear, the middle ear or tympanic cavity, and the inner ear or labyrinth. Sometimes the pinna is called the ear.

Effective temperature. An arbitrary index that combines into a single value the effect of temperature, humidity, and air movement on the sensation of warmth and cold on the human body.

Effective temperature index. An empirically determined index of the degree of warmth perceived on exposure to different combinations of temperature, humidity, and air movement. The determination of effective temperature requires simultaneous determinations of dry bulb and wet bulb temperatures.

Effluent. Generally something that flows out or forth, like an outflow of a sewer, storage tank, canal, or other channel.

Electrical current. The flow of electricity measured in amperes or milliamperes.

Electrical precipitator. A device that removes particles from an airstream by charging the particles and collecting the charged particles on a suitable surface.

Electrolysis. The process of conduction of an electric current by means of a chemical solution.

Electromagnetic radiation. The propagation of varying electric and magnetic fields through space at the speed of light, exhibiting the characteristics of wave motion.

Electroplate. To cover with a metal coating (plate) by means of electrolysis.

Element. Solid, liquid, or gaseous matter that cannot be further decomposed into simpler substances by chemical means.

Elutriator. A device used to separate particles according to mass and aerodynamic size by maintaining a laminar flow system at a rate which permits the particles of greatest mass to settle rapidly while the smaller particles are kept airborne by the resistance force of the flowing air for longer times and distances. The various times and distances of deposit may be used to determine representative fractions of particle mass and size.

Emery. Aluminum oxide, natural and synthetic abrasive.

Emission factor. Statistical average of the amount of a specific pollutant emitted from each type of polluting source in relation to a unit quantity of material handled, processed, or burned.

Emission inventory. A list of primary air pollutants emitted into a given community's atmosphere in amounts per day, by type of source.

Emission standards. The maximum amount of pollutant permitted to be discharged from a single polluting source.

Employee. A worker who may be exposed to hazardous chemicals under normal operating conditions or in foreseeable emergencies. However, not included in this definition are workers such as office workers or bank tellers who encounter hazardous chemicals only in nonroutine, isolated instances.

Employer. A person engaged in a business where chemicals are either used, distributed, or produced for use or distribution, including a contractor or subcontractor.

Emulsifier or emulsifying agent. A chemical that holds one insoluble liquid in suspension in another. Casein, for example, is a natural emulsifier in milk, keeping butterfat droplets dispersed.

Emulsion. A suspension, each in the other, of two or more unlike liquids that usually will not dissolve in each other.

Enamel. A paintlike oily substance that produces a glossy finish to a surface to which it is applied. Often contains various synthetic resins. It is lead free. In contrast is the ceramic enamel, that is, porcelain enamel, which contains lead.

Engineering controls. Methods of controlling employee exposures by modifying the source or reducing the quantity of hazards.

Entrance loss. The loss in static pressure of a fluid that flows from an area into and through a hood or duct opening. The loss in static pressure is due to friction and turbulence resulting from the increased gas velocity and configuration of the entrance area.

Entry loss. Loss in pressure caused by air flowing into a duct or hood.

Environmental toxicity. Information obtained as a result of conducting environmental testing designed to study the effects of various substances on aquatic and plant life.

Enzymes. Delicate chemical substances, mostly proteins, that enter into and bring about chemical reactions in living organisms.

EPA. U. S. Environmental Protection Agency.

EPA number. The number assigned to chemicals regulated by the Environmental Protection Agency.

Ergonomics. A multidisciplinary activity dealing with interactions between people and their total working environment plus stresses related to such environmental elements as atmosphere, light, and sound as well as all tools and equipment of the workplace.

Esters. Organic compounds that may be made by interaction between an alcohol and an acid, and by other means. Esters are nonionic compounds, including solvents and natural fats.

Etch. To cut or eat away material with acid or other corrosive substance.

Evaporation. The process by which a liquid is changed into the vapor state.

Evaporation rate. The ratio of the time required to evaporate a measured volume of a liquid to the time required to evaporate the same volume of a reference liquid (ethyl ether) under ideal test conditions. The higher the ratio, the slower the evaporation rate. FAST evaporation rate greater than 3.0. Examples: Methyl Ethyl Ketone (MEK) = 3.8, Acetone = 5.6, Hexane = 8.3. MEDIUM evaporation rate 0.8 to 3.0. Examples: 190 proof (95%) Ethyl Alcohol = 1.4, VM&P Naphtha = 1.4, MIBK = 1.6.SLOW evaporation rate less than 0.8. Examples: Xylene = 0.6, Isobutyl Alcohol = 0.6, Normal Butyl Alcohol = 0.4, Water = 0.3, Mineral Spirits = 0.1.

Exhalation valve. A device that allows exhaled air to leave a respirator and prevents outside air from entering through the valve.

Exhaust ventilation. The removal of air (usually by mechanical means) from any space. The flow of air between two points is due to the occurrence of a pressure difference between the two points. This pressure difference will cause air to flow from the high pressure to the low pressure zone.

Explosive limit. See Flammable limit.

Explosive. A reaction that causes a sudden, almost instantaneous release of pressure, gas, and heat.

Exposure. Contact with a chemical, biological, or physical hazard.

Extinguishing medium. The fire-fighting substance to be used to control a material in the event of a fire. It is usually named by its generic name, such as fog, foam, water, etc.

Extrusion. The forcing of raw material through a die or a form in either a heated or cold state, and in a solid state or in partial solution.

Eyepiece. Gastight, transparent window(s) in a full facepiece through which the wearer may see.

Eye protection. Recommended safety glasses, chemical splash goggles, face shields, etc. to be utilized when handling a hazardous material.

Face velocity. Average air velocity into the exhaust system measured at the opening into the hood or booth.

Facepiece. That portion of a respirator that covers the wearer's nose and mouth in a half-mask facepiece, or nose, mouth, and eyes in a full facepiece.

Facing. In foundry work, the final touch-up work of the mold surface to come in contact with metal is called the facing operation and the fine powdered material used is called the facing.

Fainting. Technically called syncope, a temporary loss of consciousness as a result of a diminished supply of blood to the brain.

Fallout. Dust particles that contain radioactive fission products resulting from a nuclear explosion. The wind can carry fallout particles many miles.

Fan static pressure. The pressure added to a system by a fan. It equals the sum of pressure losses in the system minus the velocity pressure in the air at the fan inlet.

Farmer's lung. Fungus infection and ensuing hypersensitivity from grain dust.

Fatal accident. An accident that results in one or more deaths within one year.

FDA. The U.S. Food and Drug Administration; under the provisions of the federal Food, Drug and Cosmetic Act, the FDA establishes requirements for the labeling of foods and drugs to protect consumers from misbranded, unwholesome, ineffective, and hazardous products. FDA also regulates materials used in contact with food and the conditions under which such materials are approved.

Federal Register. Publication of U.S. government documents officially promulgated under the law, documents whose validity depends upon such publication. (See *Code of Federal Regulations.*)

Fertilizer. Plant food usually sold in mixed formula containing basic plant nutrients: compounds of nitrogen, potassium, phosphorus, sulfur, and sometimes other minerals.

Fever. A condition in which the body temperature is above its regular or normal level.

FIFRA. Federal Insecticide, Fungicide, and Rodenticide Act; regulations administered by EPA under this Act require that certain useful poisons, such as chemical pesticides, sold to the public contain labels that carry health hazard warnings to protect users.

Film badge. A piece of masked photographic film worn by nuclear workers. Because it is darkened by nuclear radiation, radiation exposure can be checked by inspection of the film.

Filter. (1) A device for separating components of a signal on the basis of its frequency. It allows components in one or more frequency bands to pass relatively unattenuated, and it attenuates greatly components in other frequency bands. (2) A fibrous medium used in respirators to remove solid or liquid particles from the airstream entering the respirator. (3) A sheet of material that is interposed between patient and the source of x-rays to absorb a selective part of the x-rays. (4) A fibrous or membrane medium used to collect air samples of dust, fume, or mist.

Filter, HEPA. High-efficiency particulate air filter that is at least 99.97 percent efficient in removing thermally generated monodisperse dioctylphthalate smoke particles with a diameter of 0.3 < GM >.

Firebrick. A special clay that is capable of resisting high temperatures without melting or crumbling.

Fire damp. In mining, the accumulation of an explosive gas, chiefly methane gas. Miners refer to all dangerous underground gases as "damps."

Fire point. The lowest temperature at which a material can evolve vapors to support continuous combustion.

First aid. Emergency measures to be taken before regular medical help can be obtained.

Fission. The splitting of an atomic nucleus into two parts accompanied by the release of a large amount of radioactivity and heat.

Flame propagation. See Propagation of flame.

Flammable. A chemical that falls into one of the following categories: Aerosol flammable; gas flammable (includes forming a flammable mixture with air at a 13% concentration of the chemical, or less, and forming a flammable range, when mixed with air, that is wider than 12% by volume, regardless of the lower limit); liquid flammable; and solid flammable.

Flammable limits. Flammables have a minimum concentration below which propagation of flame does not occur on contact with a source of ignition. This is known as the lower flammable explosive limit (LEL). There is also a maximum concentration of vapor or gas in air above which propagation of flame does not occur. This is known as the upper flammable explosive limit (UEL). These units are expressed in percent of gas or vapor in air by volume.

Flammable liquid. Any liquid having a flash point below 37.8 C (100 F).

Flammable range. The difference between the lower and upper flammable limits, expressed in terms of percentage of vapor or gas in air by volume; is also often referred to as the "explosive range."

Flange. In an exhaust system, a rim or edge added to a hood to reduce the quantity of air entering the hood from behind the hood.

Flashback. Occurs when flame from a torch burns back into the tip, the torch, or the hose.

Flash blindness. Temporary visual disturbance resulting from viewing an intense light source.

Flash ignition. See Flash point.

Flash point. The lowest temperature at which a liquid gives off enough vapor to form an ignitable mixture with air and produce a flame when a source of ignition is present.

Flask. n foundry work, the assembly of the cope and the drag constitutes the flask. It is the wooden or iron frame containing sand into which molten metal is poured. Some flasks may have three or four parts.

Flocculation. The process of forming a very fluffy mass of material held together by weak forces of adhesion.

Flotation. A method of ore concentration in which the mineral is caused to float due to chemical frothing agents while the impurities sink.

Flow coefficient. A correction factor used for figuring volume flow rate of a fluid through an orifice. This factor includes the effects of contraction and turbulence loss (covered by the coefficient of discharge), plus the compressibility effect, and the effect of an upstream velocity other than zero. Since the latter two effects are negligible in many instances, the flow coefficient is often equal to the coefficient of discharge.

Flow meter. An instrument for measuring the rate of flow of a fluid.

Fluid. A substance tending to flow or conform to the outline of its container. It may be liquid, vapor, gas, or solid (like raw rubber).

Fluorescence. Emission of light from a crystal, after the absorption of energy.

Fluorescent screen. A screen coated with a fluorescent substance so that it emits light when irradiated with x-rays.

Flux. Usually refers to a substance used to clean surfaces and promote fusion in soldering. However, fluxes of various chemical nature are used in the smelting of ores, in the ceramic industry, in assaying silver and gold ores, and in other endeavors. The most common fluxes are silica, various silicates, lime, sodium, and potassium carbonate, and litharge and red lead in the ceramic industry. (See Galvanizing.)

Fly ash. Finely divided particles of ash entrained in flue gases arising from the combustion of fuel.

Foot-candle. A unit of illumination. The illumination at a point on a surface which is one foot from, and perpendicular to, a uniform point source of one candle.

Foot-pound. A unit of work equal to the amount of energy needed to raise one pound one foot upward.

Foot-pounds of torque. A measurement of the physiological stress exerted upon any joint during the performance of a task. The product of the force exerted and the distance from the point of application to the point of stress. Physiologically, torque that does not produce motion nonetheless causes work stress, the severity of which depends on the duration and magnitude of the torque. In lifting an object or holding it elevated, torque is exerted and applied to the lumbar vertebrae.

Force. That which changes the state of rest or motion in matter.

Foreseeable emergency. Any potential occurrence such as, but not limited to, equipment failure, rupture of containers, or failure of control equipment that could result in an uncontrolled release of a hazardous chemical into the workplace.

Frequency (in hertz or Hz). Rate at which pressure oscillations are produced. One hertz is equivalent to one cycle per second.

Friable. Particles or compounds in a crumbled or pulverized state.

Friction factor. A factor used in calculating loss of pressure due to friction of a liquid, solid, or gas flowing through a pipe or duct.

Friction loss. The pressure loss due to friction.

Fume. Airborne particulate formed by the evaporation

of solid materials, e.g., metal fume emitted during welding. Usually less than one micron in diameter.

Fuse. A wire or strip of easily melted metal, usually set in a plug, placed in an electrical circuit as a safeguard. If the current becomes too strong, the metal melts, thus breaking the circuit.

Fusion. The joining of atomic nuclei to form a heavier nucleus, accomplished under conditions of extreme heat (millions of degrees). If two nuclei of light atoms fuse, the fusion is accompanied by the release of a great deal of energy. The energy of the sun is believed to be derived from the fusion of hydrogen atoms to form helium. In welding, the melting together of filler metal and base metal (substrate), or of base metal only, which results in coalescense.

Gage pressure. Pressure measured with respect to atmospheric pressure.

Galvanizing. An old but still used method of providing a protective coating for metals by dipping them in a bath of molten zinc.

Gamma-rays (gamma radiation). The most penetrating of all radiation. Gamma-rays are very high-energy x-rays.

Gangue. In mining or quarrying, useless chipped rock.

Gas. A state of matter in which the material has very low density and viscosity; can expand and contract greatly in response to changes in temperature and pressure; easily diffuses; neither a solid nor liquid.

Gas metal arc-welding (GMAW). An arc-welding process that produces coalescense of metals by heating them with an arc between a continuous filler metal (consumable) electrode and the work; shielding is obtained entirely from an externally supplied gas or gas mixture; some methods of this process are called MIG or CO_2 welding.

Gas tungsten arc-welding (GTAW). An arc-welding process that produces coalescence of metals by heating them with an arc between a tungsten (nonconsumable) electrode and the work; shielding is obtained from a gas or gas mixture. Pressure may or may not be used and filler metal may or may not be used. (This process has sometimes been called TIG welding.)

Gate. A groove in a mold to act as a passage for molten metal.

Geiger counter. A gas-filled electrical device that counts the presence of atomic particles or rays by detecting the ions produced. (Sometimes called a "Geiger-Mueller" counter.)

General exhaust. A system for exhausting air containing contaminants from a general work area; usually accomplished via dilution.

General ventilation. System of ventilation consisting of either natural or mechanically induced fresh air movements to mix with the dilute contaminants

in the workroom air.

Generic name. A nonproprietary name for a material.

Genetic effects. Mutations or other changes that are produced by irradiation of the germ plasm.

Genetically significant dose (GSD). The dose which, if received by every member of the population, would be expected to produce the same total genetic injury to the population as do the actual doses received by various individuals.

Glove box. A sealed enclosure in which all handling of items inside the box is carried out through long impervious gloves sealed to ports in the walls of the enclosure.

Gob. "Gob pile" is waste mineral material such as from coal mines sometimes containing sufficient coal that gob fires may arise from spontaneous combustion.

Grab sample. A sample that is taken within a very short time period.

Gram (g). A metric unit of weight; one ounce equals 28.4 grams.

Gravity, specific. The ratio of the mass of a unit volume of a substance to the mass of the same volume of a standard substance at a standard temperature. Water at 4 C (39.2 F) is the standard substance usually referred to. For gases, dry air, at the same temperature and pressure as the gas, is often taken as the standard substance.

Gravity, standard. A gravitational force that will produce an acceleration equal to 9.8 m/sec^2 or 32.17 ft/sec^2. The actual force of gravity varies slightly with altitude and latitude. The standard was arbitrarily established as that at sea level and 45-degree latitude.

Gray iron. The same as cast iron and, in general, any iron containing high carbon.

Ground-fault circuit interrupter (GFCI). A device that measures the amount of current flowing to and from an electrical source. When a difference is sensed, indicating a leakage of current that could cause an injury, the device very quickly breaks the circuit.

Grounding. The procedure used to carry an electrical charge to ground through a conductive path. (See Bonding.)

Grooving. Designing a tool with grooves on the handle to accommodate the fingers of the user. A bad practice because of the great variation in the size of workers' hands. Grooving interferes with sensory feedback. Intense pain may be caused by the grooves to an arthritic hand.

Half-life, radioactive. For a single radioactive decay process, the time required for the activity to decrease to half its value by that process.

Halogenated hydrocarbon. A chemical material that has carbon plus one or more of the halogen elements: chlorine, fluorine, bromine, or iodine.

Hammer mill. A machine for reducing the size of stone or other bulk material by means of hammers usually placed on a rotating axle inside a steel cylinder.

Hand protection. Specific type of gloves or other hand protection required to prevent harmful exposure to hazardous materials.

Hardness. A relative term to describe the penetrating quality of radiation. The higher the energy of the radiation, the more penetrating (harder) is the radiation.

Hazard. An unsafe condition which, if left uncontrolled, may contribute to an accident.

Hazardous chemical. Any chemical that is a physical or health hazard.

Hazard warning. Any words, pictures, symbols, or combination thereof appearing on a label or other appropriate form of warning that conveys the hazard(s) of the chemical(s) in the container(s).

Hazardous decomposition products. Any hazardous materials that may be produced in dangerous amounts if the material reacts with other agents, burns, or is exposed to other processes, such as welding. Examples of hazardous decomposition products formed when certain materials are heated include carbon monoxide and carbon dioxide.

Hazardous material. Any substance or compound that has the capability of producing adverse effects on the health and safety of humans.

Heading. n mining, a horizontal passage or drift of a tunnel, also the end of a drift or gallery. In tanning, a layer of ground bark over the tanning liquor.

Health hazard. A chemical for which there is statistically significant evidence based on at least one study conducted in accordance with established scientific principles that acute or chronic health effects may occur in exposed employees.

Hearing conservation. The prevention or minimizing of noise-induced deafness through the use of hearing protection devices; also, the control of noise through engineering methods, annual audiometric tests, and employee training.

Hearing level. The deviation in decibels of an individual's hearing threshold from the zero reference of the audiometer.

Heat stress. Relative amount of thermal strain from the environment.

Heat stress index. Index which combines the environmental heat and metabolic heat of the body into an expression of stress in terms of requirement for evaporation of sweat.

Heatstroke. A serious disorder resulting from exposure to excess heat. It results from sweat suppression and increased storage of body heat. Symptoms include hot, dry skin, high temperature, mental confusion, convulsions, and coma. Heatstroke is fatal if not treated promptly.

Heat treatment. Any of several processes of metal modification such as annealing.

Helmet. A device that shields the eyes, face, neck, and other parts of the head.

HEPA filter. See Filter, HEPA.

Hertz. The frequency measured in cycles per second. 1 cps = 1 Hz.

High-frequency loss. Refers to a hearing deficit starting with 2,000 Hz and higher.

Homogenizer. A machine that forces liquids under high pressure through a perforated shield against a hard surface to blend or emulsify the mixture.

Hood. (1) Enclosure, part of a local exhaust system; (2) a device that completely covers the head, neck, and portions of the shoulders.

Horsepower. A unit of power, equivalent to 33,000 foot-pounds per minute (746 W). (See Brake horsepower.)

Hot. In addition to meaning "having a relatively high temperature," this is a colloquial term meaning "highly radioactive."

Human-equipment interface. Area of physical or perceptual contact between man and equipment. The design characteristics of the human-equipment interface determine the quality of information. Poorly designed interfaces may lead to excessive fatigue or localized trauma, e.g., calluses.

Humidify. To add water vapor to the atmosphere; to add water vapor or moisture to any material.

Humidity. (1) Absolute humidity is the weight of water vapor per unit volume, pounds per cubic foot or grams per cubic centimeter. (2) Relative humidity is the ratio of the actual partial vapor pressure of the water vapor in a space to the saturation pressure of pure water at the same temperature.

Hydration. The process of converting raw material into pulp by prolonged beating in water; to combine with water or the elements of water.

Hydrocarbons. Organic compounds composed solely of carbon and hydrogen.

ICC. Interstate Commerce Commission.

Identity. Any chemical or common name indicated on the MSDS for the chemical. The identity used shall permit cross-references to be made among the required list of hazardous chemicals, the label, and the MSDS.

IDLH. Immediately dangerous to life or health.

Ignitable. Capable of being set afire.

Immediate use. The hazardous chemical will be used by, or under the control of, only the person who transfers it from a labeled container and only within the work shift in which it was transferred.

Immiscible. Not miscible. Any liquid that will not mix with another liquid, in which case it forms two separate layers or exhibits cloudiness or turbidity.

Impaction. The forcible contact of particles of matter; a term often used synonymously with impingement, but generally reserved for the case where particles are contacting a dry surface.

Impervious. A material that does not allow another substance to pass through or penetrate it.

Impingement. As used in air sampling, impingement refers to a process for the collection of particulate matter in which particles containing gas are directed against a wetted glass plate and the particles are retained by the liquid.

Importer. The first business, with employees and within the Customs Territory of the United States, that receives hazardous chemicals produced in other countries for the purpose of supplying them to distributors or employees within the United States.

Inches of mercury column. A unit used in measuring pressure. One inch of mercury column equals a pressure of 1.66 kPa (0.491 lb per sq in.).

Inches of water column. A unit used in measuring pressure. One inch of water column equals a pressure of 0.25 kPa (0.036 lb per sq in.).

Incidence rate (as defined by OSHA). The number of injuries and/or illnesses or lost workdays per 100 full-time employees per year or 200,000 hours of exposure.

Incompatible. Materials that could cause dangerous reactions from direct contact with one another.

Incubation. Holding cultures of microorganisms under conditions favorable to their growth.

Induration. Heat hardening that may involve little more than thermal dehydration.

Industrial hygiene. The science (or art) devoted to the recognition, evaluation, and control of those environmental factors or stresses (i.e., chemical, physical, biological, and ergonomic) that may cause sickness, impaired health, or significant discomfort to employees or residents of the community.

Inert gas. A gas that does not normally combine chemically with another substance.

Inert gas, welding. An electric welding operation utilizing an inert gas such as helium to flush away the air to prevent oxidation of the metal being welded.

Infrared radiation. Electromagnetic energy with wavelengths from 770 nm to 12,000 nm.

Ingestion. (1) The process of taking substances into the stomach, as food, drink, medicine, etc. (2) With regard to certain cells, the act of engulfing or taking up bacteria and other foreign matter.

Ingot. A block of iron or steel cast in a mold for ease in handling before processing.

Inhalation. The breathing in of a substance in the form of a gas, vapor, fume, mist, or dust.

Inhalation valve. A device that allows respirable air to enter the facepiece and prevents exhaled air from leaving the facepiece through the intake opening.

Injury. Damage or harm to the body as the result of violence, infection, or other source.

Inorganic. Term used to designate compounds that generally do not contain carbon. Source: matter, other than vegetable or animal. Examples: sulfuric acid and salt. Exceptions are carbon monoxide, carbon dioxide.

Insoluble. Incapable of being dissolved.

Intermediate. A chemical formed as a "middle-step" in a series of chemical reactions, especially in the manufacture of organic dyes and pigments. In many cases, it may be isolated and used to form a variety of desired products. In other cases, the intermediate may be unstable or used up at once.

Intoxication. Means either drunkenness or poisoning.

Inverse square law. The propagation of energy through space is inversely proportional to the square of the distance it must travel. An object 3 m (9.8 ft) away from an energy source receives $1/9$ as much energy as an object 1 m (3.3 ft) away.

Ionizing radiation. Refers to (1) electrically charged or neutral particles, or (2) electromagnetic radiation that will interact with gases, liquids, or solids to produce ions. There are five major types: alpha, beta, x (or x-ray), gamma, and neutrons.

Irradiation. The exposure of something to radiation.

Irritant. A substance that produces an irritating effect when it contacts skin, eyes, nose, or respiratory system.

Isotope. One of two or more atomic species of an element differing in atomic weight but having the same atomic number. Each contains the same number of protons but a different number of neutrons.

Jigs and fixtures. Often used interchangeably; precisely, a jig holds work in position and guides the tools acting on the work, while a fixture holds but does not guide.

Kaolin. A type of clay composed of mixed silicates and used for refractories, ceramics, tile, and stoneware. In some deposits, free silica may be present as an impurity.

kg. Kilogram; a metric unit of weight, about 2.2 U.S. pounds. (Also see mg.)

Kilogram (kg). A unit of weight in the metric system equal to 2.2 lb.

Kinetic energy. Energy due to motion. (See Work.)

L. Liter; a metric unit of capacity. A liter is about the same as one quart (0.946 L).

Label. Any written, printed, or graphic material that is displayed on or affixed to containers of hazardous chemicals.

Lacquer. A collodial dispersion or solution of nitro-cellulose, or similar film-forming compounds, resins, and plasticizers in solvents and diluents used as a protective and decorative coating for various surfaces.

Lapping. The operation of polishing or sanding surfaces such as metal or glass to a precise dimension.

Laser. The acronym for Light Amplification by Stimulated Emission of Radiation. The gas laser is a type in which laser action takes place in a gas medium, usually a continous wave (CW) laser, which has a continuous output—with an off time of less than 1 percent of the pulse duration time.

Laser light region. A portion of the electromagnetic spectrum that includes ultraviolet, visible, and infrared light.

Lathe. A machine tool used to perform cutting operations on wood or metal by the rotation of the workpiece.

Latex. Original meaning: milky extract from rubber tree, containing about 35 percent rubber hydrocarbon; the remainder is water, proteins, and sugars. This word also is applied to water emulsions of synthetic rubbers or resins. In emulsion paints, the film-forming resin is in the form of latex.

LC. Lethal concentration; a concentration of a substance being tested that will kill a test animal.

LD. Lethal dose; a concentration of a substance being tested that will kill a test animal.

Lead poisoning. Poisoning produced when lead compounds are swallowed or inhaled. Inorganic lead compounds commonly cause symptoms of lead colic and lead anemia. Organic lead compounds can attack the nervous system.

Leakage radiation. Radiation emanating from diagnostic equipment in addition to the useful beam; also, radiation produced when the exposure switch or timer is not activated.

LEL. See Lower explosive limit.

Lethal. Capable of causing death.

LFL. Lower flammable limit. See LEL.

Liquefied petroleum gas. A compressed or liquefied gas usually composed of propane, some butane, and lesser quantities of other light hydrocarbons and impurities; obtained as a by-product in petroleum refining. Used chiefly as a fuel and in chemical synthesis.

Liquid. A state of matter in which the substance is a formless fluid that flows in accord with a law of gravity.

Liter (L). A metric measure of capacity—one quart equals 0.9 L.

Local exhaust. A system for capturing and exhausting contaminants from the air at the point where the contaminants are produced.

Local exhaust ventilation. A ventilation system that captures and removes contaminants at the point they are being produced before they escape into the workroom air.

Loudness. The intensive attribute of an auditory sensation, in terms of which sounds may be ordered on a scale extending from soft to loud. Loudness depends primarily upon the sound pressure of the stimulus, but it also depends upon the frequency and wave form of the stimulus.

Low-pressure tank. A storage tank designed to operate at pressures at more than 0.5 psig but not more than 15 psig (3.5 to 103 kPa).

Lower explosive limit (LEL). The lower limit of flammability of a gas or vapor at ordinary ambient temperatures expressed in percent of the gas or vapor in air by volume.

LP-gas. See Liquefied petroleum gas.

Lumen. The flux on one square foot of a sphere, one foot in radius, with a light source of one candle at the center that radiates uniformly in all directions.

M. Meter; a unit of length in the metric system. One meter is about 39 inches.

M³. Cubic meter; a metric measure of volume, about 35.3 cubic feet or 1.3 cubic yards.

Makeup air. Clean, tempered outdoor air supplied to a workspace to replace air removed by exhaust ventilation or some industrial process.

Manometer. Instrument for measuring pressure; essentially a U-tube partially filled with a liquid (usually water, mercury, or a light oil), so constructed that the amount of displacement of the liquid indicates the pressure being exerted on the instrument.

Maser. Microwave Amplification by Stimulated Emission of Radiation. When used in the term "optical maser," it is often interpreted as molecular amplification by stimulated emission of radiation.

Material Safety Data Sheet (MSDS). Written or printed material concerning a hazardous chemical that is prepared in accordance with paragraph (g) of the Hazard Communication Standard (HCS).

Maximum permissible concentration (MPC). The concentrations set by the National Committee on Radiation Protection (NCRP). They are recommended maximum average concentrations of radionuclides to which a worker may be exposed, assuming that he or she works 8 hours a day, 5 days a week, 50 weeks a year.

Maximum permissible dose (MPD). Currently, a permissible dose is defined as the dose of ionizing radiation that, in the light of present knowledge, is not expected to cause appreciable bodily injury to a person at any time during his or her lifetime.

Maximum use concentration (MUC). The product of the protection factor of the respiratory protection equipment and the permissible exposure limit (PEL).

Mechanical filter respirator. A respirator used to protect against airborne particulate matter like dusts, mists, metal fume, and smoke. Mechanical filter respirators do not provide protection against gases, vapors, or oxygen-deficient atmospheres.

Mechanical ventilation. A powered device, such as a motor-driven fan or vacuum hose attachment, for exhausting contaminants from a workplace, vessel, or enclosure.

Mega. One million—for example, 1 megacurie = 1 million curies.

Melt. In glass industry, the total batch of ingredients that may be introduced into a pot or furnace.

Melting point. The temperature at which a solid substance changes to a liquid state.

Membrane filter. A filter medium made from various polymeric materials such as cellulose, polyethylene, or tetrapolyethylene. Membrane filters usually exhibit narrow ranges of effective pore diameters and, therefore, are very useful in collecting and sizing microscopic and submicroscopic particles and in sterlizing liquids.

Meter. See M.

Mev. Million electron volts.

mg. Milligram; a metric unit of weight. There are 1,000 milligrams in one gram (g) of a substance. One gram is equivalent to almost $4/100$ of an ounce.

mg/kg. Milligrams per kilogram; an expression of toxicological dose.

mg/m³. Milligrams per cubic meter; a unit used to express concentrations of dusts, gases, fumes, or mists in air.

Mica. A large group of silicates of varying composition, but similar in physical properties. All have excellent cleavage and can be split into very thin sheets. Used in electrical insulation.

Microphone. An electroacoustic transducer that responds to sound waves and delivers essentially equivalent electric waves.

Milliampere. $1/1000$ of an ampere.

Milligram (mg). A unit of weight in the metric system. One thousand milligrams equal one gram. (See mg.)

Milligrams per cubic meter (mg/m³). Unit used to measure air concentrations of dusts, gases, mists, and fumes. (See mg/m³.)

Milliliter (mL). A metric unit used to measure volume. One milliliter equals one cubic centimeter.

Millimeters of mercury (mmHg). The unit of pressure equal to the pressure exerted by a column of liquid mercury one millimeter high at a standard temperature. (See mmHg.)

Millwright. A mechanic engaged in the erection and maintenance of machinery.

Mineral pitch. Tar from petroleum or coal in distinction to wood tar.

Mineral spirits. A petroleum fraction with boiling range between 149 and 230 C (300 and 400 F).

Mist. Suspended liquid droplets generated by condensation from the gaseous to the liquid state or by breaking up a liquid into a dispersed state, such as by splashing, foaming, or atomizing.

Mixture. A combination of two or more substances that may be separated by mechanical means. The components may not be uniformly dispersed. (See also Solution.)

mL. Milliliter; a metric unit of capacity, equal in volume to one cubic centimeter (cc), or about $1/16$ of a cubic inch. There are 1,000 milliliters in one liter (L).

mmHg. Millimeters (mm) of mercury (Hg); a unit of measurement for low pressures or partial vacuums.

Mold. (1) A growth of fungi forming a furry patch, as on stale bread or cheese. (2) A hollow form or matrix into which molten material is poured to produce a cast.

Monaural hearing. Refers to hearing with one ear only.

MPE. Maximum permissible exposure.

MPL. May be either maximum permissible level, or limit, or dose. It refers to the tolerable dose rate of humans exposed to nuclear radiation.

MSDS. Material Safety Data Sheet.

MSHA. The Mine Safety and Health Administration of the U.S. Department of Labor; federal agency with safety and health regulatory and enforcement authority for the mining industry established by the Mine Safety and Health Act.

Mutation. A transformation of the gene that may alter characteristics of the offspring.

Naphthas. Hydrocarbons of the petroleum type that contain substantial portions of paraffins and naphthalenes.

Narcosis. A state of stupor, unconsciousness, or arrested activity produced by the influence of narcotics or other chemicals.

Natural gas. A combustible gas composed largely of methane and other hydrocarbons with variable amounts of nitrogen and noncombustible gases.

Natural radioactivity. The radioactive background or, more properly, the radioactivity that is associated with the heavy naturally occurring elements.

Natural ventilation. Air movement caused by wind, temperature difference, or other nonmechanical factors.

Nausea. An unpleasant sensation and that often precedes vomiting.

Neurotoxin. Chemicals that produce their primary effect on the nervous system.

Neutral wire. Wire carrying electrical current back to the source, thus completing a circuit.

Neutralize. To eliminate potential hazards by inactivating strong acids, caustics, and oxidizers. For example, acids can be neutralized by adding an appropriate amount of caustic substance.

NFPA. The National Fire Protection Association; a voluntary organization whose aim is to promote and improve fire protection and prevention. The NFPA publishes 16 volumes of codes known as the National Fire Codes.

NIOSH. The National Institute for Occupational Safety and Health; a federal agency that conducts research on health and safety concerns, tests and certifies respirators, and trains occupational health and safety professionals.

Noise. Any unwanted sound.

Noise-induced hearing loss. The terminology used to refer to the slowly progressive inner ear hearing loss that results from exposure to continuous noise over a long period of time as contrasted to acoustic trauma or physical injury to the ear.

Nonferrous metal. Metal (such as nickel, brass, or bronze) that does not include any appreciable amount of iron.

Nonflammable. Not easily ignited, or if ignited, not burning rapidly.

Nonionizing radiation. Electromagnetic radiation that does not cause ionization. Includes ultraviolet, laser, infrared, microwave, and radiofrequency radiation.

Nonsparking tools. Tools made from beryllium—or copper or aluminum—bronze that greatly reduce the possibility of igniting dusts, gases, or flammable vapors. Although these tools may emit some sparks when striking metal, the sparks have a low heat content and are not likely to ignite most flammable liquids.

Nonvolatile matter. The portion of a material that does not evaporate at ordinary temperatures.

NRC. (1) National Response Center; a notification center in the Coast Guard Building in Washington, DC, with a toll-free telephone number (1-800-424-8802) that must be called when significant oil or chemical spills or other environmentally related accidents occur; (2) Nuclear Regulatory Commission of the U.S. Department of Energy.

Nuclear energy. The energy released in a nuclear reaction, such as fission or fusion; mistakenly called "atomic energy."

Nuclear reaction. Result of the bombardment of a nucleus with atomic or subatomic particles or very high energy radiation.

Nuclear reactor. A machine for producing a controlled chain reaction in fissionable material. It is the heart of nuclear power plants where it serves as a source of heat.

Nuisance dust. Has long history of little adverse effect on the lungs and does not produce significant organic disease or toxic effect when exposures are kept under reasonable control.

Odor. That property of a substance that affects the sense of smell.

Ohm. The unit of electrical resistance.

Ohm's Law. The steady current through certain electrical circuits is directly proportional to the applied electromotive force (voltage).

Oil dermatitis. Blackheads and acne caused by oils and waxes that plug the hair follicles and sweat ducts.

Olefins. A class of unsaturated hydrocarbons characterized by relatively great chemical activity. Obtained from petroleum and natural gas. Examples: butene, ethylene, and propylene. Generalized formula: C_nH_{2n}.

Olfactory. Relating to the sense of smell. The olfactory organ in the nasal cavity is the sensing element that detects odors and transmits information to the brain through the olfactory nerves.

Ophthalmologist. A physician who specializes in the structure, function, and diseases of the eye.

Oral. Used in or taken into the body through the mouth.

Oral toxicity. Adverse effects resulting from taking a substance into the body via the mouth. Ordinarily used to denote effects in experimental animals.

Organic. Term used to designate chemicals that contain carbon. To date nearly one million organic compounds have been synthesized or isolated. Many occur in nature; others are produced by chemical synthesis. (See also Inorganic.)

Organic peroxide. An organic compound that contains the bivalent -O-O- structure and which may be considered to be a structural derivative of hydrogen peroxide, where one or both of the hydrogen atoms have been replaced by an organic radical.

Orifice. The opening that serves as an entrance and/or outlet of a body cavity or organ, especially the opening of a canal or a passage.

Orifice meter. A flowmeter, employing as the measure of flow rate the difference between the pressures measured on the upstream and downstream sides of a restriction within a pipe or duct.

Oscillation. The variation, usually with time, of the magnitude of a quantity with respect to a specified reference when the magnitude is alternately greater and smaller than the reference.

OSHA. Occupational Safety and Health Administration; federal agency in the U.S. Department of Labor with safety and health regulatory and enforcement authority for most U.S. industry and business.

Osmosis. The passage of fluid through a semipermeable membrane as a result of osmotic pressure.

Overexposure. Exposure to a hazardous material beyond the allowable exposure levels.

Oxidation. Process of combining oxygen with some other substance; technically, a chemical change in which an atom loses one or more electrons whether or not oxygen is involved. Opposite of reduction.

Oxidizer. A chemical, other then a blasting agent or explosive, that initiates or promotes combustion in other materials, thereby causing a fire either by itself or through the release of oxygen or other gases.

Oxygen deficiency. An atmosphere having less than the percentage of oxygen found in normal air.

Paraffins, Paraffin series. Those straight- or branched-chain hydrocarbon components of crude oil and natural gas whose molecules are saturated (i.e., carbon atoms attached to each other by single bonds) and therefore very stable. Examples: methane and ethane. Generalized formula: C_nH_{2n+2}.

Particle. A small discrete mass of solid or liquid matter.

Particulate. A particle of solid or liquid matter.

Particulate matter. A suspension of fine solid or liquid particles in air, such as dust, fog, fume, mist, smoke, or sprays. Particulate matter suspended in air is commonly known as an aerosol.

PEL. Permissible exposure limit; the legally enforced exposure limit for a substance established by OSHA regulatory authority. The PEL indicates the permissible concentration of air contaminants to which nearly all workers may be repeatedly exposed 8 hours a day, 40 hours a week, over a working lifetime (30 years) without adverse health effects.

Pelleting. In various industries, a material in powder form that may be made into pellets or briquettes for convenience. The pellet is a distinctly small briquette. (See Pelletizing.)

Pelletizing. Refers primarily to extrusion by pellet mills. The word "pellet," however, carries its lay meaning in that it is also applied to other small extrusions and to some balled products. Pellets are generally regarded as being larger than grains but smaller than briquettes.

Percent impairment of hearing. Percent hearing loss; an estimate of a person's ability to hear correctly. It is usually based, by means of an arbitrary rule, on the pure tone audiogram. The specific rule for calculating this quantity from the audiogram varies from state to state according to rule or law.

Percent volatile. The percentage of a liquid or solid (by volume) that will evaporate at an ambient temperature of 70 F (unless some other temperature is stated). Examples: butane, gasoline, and paint

thinner (mineral spirits) are 100% volatile; their individual evaporation rates vary, but over a period of time each will evaporate completely.

Permanent disability (or permanent impairment). Any degree of permanent, nonfatal injury. It includes any injury that results in the partial loss, to complete loss of use, of any part of the body, or any permanent impairment of functions of the body or a part thereof.

Permissible exposure limit (PEL). An exposure limit that is published and enforced by U.S. OSHA as a legal standard.

Personal protective equipment. Devices worn by workers to protect against hazards in the environment.

Pesticides. General term for that group of chemicals used to control or kill such pests as rats, insects, fungi, bacteria, weeds, etc., that prey on man or agricultural products. Among these are insecticides, herbicides, fungicides, rodenticides, miticides, fumigants, and repellents.

Petrochemical. A term applied to chemical substances produced from petroleum products and natural gas.

pH. Means used to express the degree of acidity or alkalinity of a solution with neutrality indicated at seven.

Phosphors. Materials capable of absorbing energy from suitable sources, such as visible light, cathode rays, or ultraviolet radiation, and then emitting a portion of the energy in the ultraviolet, visible, or infrared region of the electromagnetic spectrum. In short, they are fluorescent or luminescent.

Physical hazard. A chemical for which there is scientifically valid evidence that it is a combustible liquid, a compressed gas, explosive, flammable, an organic peroxide, an oxidizer, pyrophoric, unstable, or water reactive.

Pig. (1) A container (usually lead) used to ship or store radioactive materials. The thick walls protect the person handling the container from radiation. (2) In metal refining, a small ingot from the casting of blast furnace metal.

Pigment. A finely divided, insoluble substance that imparts color to the material to which it is added.

Pilot plant. Small-scale operation preliminary to a major enterprise. Common in the chemical industry.

Pink noise. Noise that has been weighted, especially at the low end of the spectrum, so that the energy per band (usually octave band) is about constant over the spectrum.

Pitot tube. A device consisting of two concentric tubes, one serving to measure the total or impact pressure existing in an airstream, the other to measure the static pressure only. When the annular space between the tubes and the interior of the center tube are connected across a pressure-measuring device, the pressure difference automatically nullifies the static pressure, and the velocity pressure alone is registered.

Plasma arc welding (PAW). An arc welding process that produces coalescence of metals by heating them with a constricted arc between an electrode and the workpiece (transferred arc) or the electrode and the constricting nozzle (non-transferred arc). Shielding is obtained by the hot, ionized gas issuing from the orifice, which may be supplemented by an auxiliary source of shielding gas. Shielding gas can be an inert gas or a mixture of gases. Pressure may or may not be used, and filler metal may or may not be supplied.

Plastic. Officially defined as any one of a large group of materials that contains as an essential ingredient an organic substance of large molecular weight. Two basic types: thermosetting (irreversibly rigid) and thermoplastic (reversibly rigid). Before compounding and processing, plastics often are referred to as (synthetic) resins. Final form may be as film, sheet, solid, or foam; flexible or rigid.

Plenum chamber. An air compartment connected to one or more ducts or connected to a slot in a hood used for air distribution.

Plutonium. A heavy element that undergoes fission under the impact of neutrons. It is a useful fuel in nuclear reactors. Plutonium cannot be found in nature, but can be produced and "burned" in reactors.

Point source. A source of radiation whose dimensions are small enough compared with distance between source and receptor for those dimensions to be neglected in calculations.

Poison. A material introduced into a reactor core to absorb neutrons. Any substance that, when taken into the body, is injurious to health.

Poison, Class A. A DOT hazard class for extremely dangerous poisons, that is, poisonous gases or liquids of such nature that a very small amount of the gas, or vapor of the liquid, mixed with air is dangerous to life. Some examples: phosgene, cyanogen, hydrocyanic acid, nitrogen peroxide.

Poison, Class B. A DOT hazard class for liquid, solid, paste, or semisolid substances—other than Class A poisons or irritating materials—that are known (or presumed on the basis of animal tests) to be so toxic to man as to afford a hazard to health during transportation. Some examples: arsenic, beryllium chloride, cyanide, mercuric oxide.

Pollution. Contamination of soil, water, or atmosphere beyond that which is natural.

Polymer. A high molecular weight material formed by the joining together of many simple molecules (monomers). There may be hundreds or even thousands of the original molecules linked end to end and often crosslinked. Rubber and cellulose are naturally occurring polymers. Most resins are chemically produced polymers.

Portal. Place of entrance.

Portland cement. See Cement, portland.

Positive displacement pump. Any type of air mover pump in which leakage is negligible, so that the pump delivers a constant volume of fluid, building up to any pressure necessary to deliver that volume (unless, of course, the motor stalls or the pump breaks).

Potential energy. Energy due to position of one body with respect to another or to the relative parts of the same body.

Power. Time rate at which work is done; units are the watt (one joule per second) and the horsepower (33,000 foot-pounds per minute). One horsepower equals 746 watts.

PPE, personal protective equipment. Includes items such as gloves, goggles, respirators, and protective clothing.

ppm. Parts per million; part of air by volume of vapor or gas or other contaminant.

Precision. The degree of agreement of repeated measurements of the same property, expressed in terms of dispersion of test results about the mean result obtained by repetitive testing of a homogeneous sample under specified conditions.

Pressure. Force applied to, or distributed over a surface; measured as force per unit area. (See Atmospheric pressure, Gage pressure, Standard temperature and pressure, Static pressure, and Velocity pressure.)

Pressure vessel. A storage tank or vessel designed to operate at pressures greater than 15 psig (103 kPa).

PRF laser. A pulsed recurrence frequency laser, which is a pulsed-typed laser with properties similar to a CW laser if the frequency is very high.

Probe. A tube used for sampling or for measuring pressures at a distance from the actual collection or measuring apparatus. It is commonly used for reaching inside stacks or ducts.

Produce. To manufacture, process, formulate, or repackage a chemical.

Propagation of flame. The spread of flame through the entire volume of a flammable vapor-air mixture from a single source of ignition.

Protection factor (PF). With respiratory protective equipment, the ratio of the ambient airborne concentration of the contaminant to the concentration inside the facepiece.

Protective atmosphere. A gas envelope surrounding the part to be brazed, welded, or thermal sprayed, with the gas composition controlled with respect to chemical composition, dew point, pressure, flow rate, etc.

Protective coating. A thin layer of metal or organic material, such as paint, applied to a surface primarily to protect it from oxidation, weathering, and corrosion.

psi. Pounds per square inch. For technical accuracy, pressure must be expressed as psig (pounds per square inch gauge) or psia (pounds per square inch absolute); that is, gauge pressure plus sea level atmospheric pressure, or psig plus about 14.7 pounds per square inch. (See also mmHg.)

Pulsed laser. A class of laser characterized by operation in a pulsed mode, i.e., emission occurs in one or more flashes of short duration (pulse length).

Pumice. A natural silicate, such as volcanic ash or lava. Used as an abrasive.

Push-pull hood. A hood consisting of an air supply system on one side of the contaminant source blowing across the source and into an exhaust hood on the other side.

Pyrethrum. A pesticide obtained from the dried, powdered flowers of the plant of the same name; mixed with petroleum distillates, it is used as an insecticide.

Pyrophoric. A chemical that will ignite spontaneously in air at a temperature of 130 F (54.4 C) or below.

Q fever. Disease caused by rickettsial organism that infects meat and livestock handlers; similar but not identical to tick fever.

Q-switched laser (also known as Q-spoiled). A pulsed laser capable of extremely high peak powers for very short durations (pulse length of several nanoseconds).

Quality. A term used to describe the penetrating power of x-rays or gamma-rays.

Quartz. Vitreous, hard, chemically resistant, free silica, the most common form in nature. The main constituent in sandstone, igneous rocks, and common sands.

Quenching. A heat-treating operation in which metal raised to the desired temperature is quickly cooled by immersion in an oil bath.

rad. Radiation absorbed dose.

Radiation (nuclear). The emission of atomic particles or electromagnetic radiation from the nucleus of an atom.

Radiation (thermal). The transmission of energy by means of electromagnetic waves longer than visible light. Radiant energy of any wavelength may, when absorbed, become thermal energy and result in the increase in the temperature of the absorbing body.

Radiation protection guide (RPG). The radiation dose that should not be exceeded without careful consideration of the reasons for doing so; every effort should be made to encourage the maintenance of radiation doses as far below this guide as practical.

Radiation source. An apparatus or a material emitting or capable of emitting ionizing radiation.

Radiator. That which is capable of emitting energy in wave form.

Radioactive. The property of an isotope or element that is characterized by spontaneous decay to emit radiation.

Random noise. A sound or electrical wave whose instantaneous amplitudes occur as a function of time, according to a normal (Gaussian) distribution curve. Random noise is an oscillation whose instantaneous magnitude is not specified for any given instant of time.

Rated line voltage. The range of potentials in volts of an electrical supply line.

Reaction. A chemical transformation or change; the interaction of two or more substances to form new substances.

Reactive. See Unstable.

Reactivity. A description of the tendency of a substance to undergo chemical reaction with the release of energy. Undesirable effects—such as pressure buildup, temperature increase, formation of noxious, toxic, or corrosive by-products—may occur because of the reactivity of a substance to heating, burning, direct contact with other materials, or other conditions in use or in storage.

Reagent. Any substance used in a chemical reaction to produce, measure, examine, or detect another substance.

Recoil energy. The emitted energy that is shared by the reaction products when a nucleus undergoes a nuclear reaction, such as fission or radioactive decay.

Reducing agent. In a reduction reaction (which always occurs simultaneously with an oxidation reaction), the reducing agent is the chemical or substance that (1) combines with oxygen or (2) loses electrons to the reaction.

Refractory. A material that is especially resistant to the action of heat, such as fireclay, magnetite, graphite, and silica; used for lining furnaces, etc.

Regimen. A regulation of the mode of living, diet, sleep, exercise, etc., for a hygienic or therapeutic purpose; sometimes mistakenly called regime.

Relative humidity. See Humidity.

Reliability. The degree to which an instrument, component, or system retains its performance characteristics over a period of time.

Rem. Roentgen equivalent man, a dose unit that equals the dose in rads multiplied by the appropriate value of RBE for the particular radiation.

Resistance. (1) In electricity, any condition that

retards current or the flow of electrons; it is measured in ohms. (2) Opposition to the flow of air, as through a canister, cartridge, particulate filter, or orifice. (3) A property of conductors, depending on their dimensions, material, and temperature, that determines the current produced by a given difference in electrical potential.

Respirable size particulates. Particles in the size range that permits them to penetrate deep into the lungs upon inhalation.

Respirator. A device to protect the wearer from inhalation of harmful contaminants.

Respiratory protection. Devices that will protect the wearer's respiratory system from overexposure by inhalation of airborne contaminants.

Respiratory system. Consists of (in descending order)—the nose, mouth, nasal passages, nasal pharynx, pharynx, larynx, trachea, bronchi, bronchioles, air sacs (alveoli) of the lungs, and muscles of respiration.

Responsible party. Someone who can provide additional information on the hazardous chemical and appropriate emergency procedures, if necessary.

Reverbatory furnace. A furnace in which heat is supplied by burning of fuel in a space between the charge and a low roof.

Riser. In metal casting, a channel in a mold to permit escape of gases.

Roentgen (R). A unit of radioactive dose, or exposure was called a roentgen (pronounced rentgen). (See rad.)

Route of entry. The path by which chemicals can enter the body; primarily inhalation, ingestion, and skin absorption.

Rosin. Specifically applies to the resin of the pine tree and chiefly derives from the manufacture of turpentine. Widely used in the manufacture of soap, flux.

Rotameter. A flow meter, consisting of a precision-bored, tapered, transparent tube with a solid float inside.

Rotary kiln. Any of several types of kilns used to heat material, such as in the portland cement industry.

Rouge. A finely powdered form of iron oxide used as a polishing agent.

Safety. The control of recognized hazards to attain an acceptable level of risk.

Safety can. An approved container, of not more than 19 L (5 gal) capacity, having a spring-closing lid and spout cover, and so designed that it will safely relieve internal pressure when subjected to fire exposure.

Safety program. Activities designed to assist employees in the recognition, understanding, and control of hazards in the workplace.

Salamander. A small furnace, usually cylindrical in shape, without grates, used for heating.

Salt. A product of the reaction between an acid and a base.

Sampling. The withdrawal or isolation of a fractional part of a whole.

Sandblasting. A process for cleaning metal castings and other surfaces with sand by a high-pressure airstream.

Sandhog. Any worker doing tunneling work requiring atmospheric pressure control.

Sanitize. To reduce the microbial flora in or on articles, such as eating utensils, to levels judged safe by public health authorities.

SCBA. See Self-contained breathing apparatus.

Sealed source. A radioactive source sealed in a container or having a bonded cover, where the container or cover has sufficient mechanical strength to prevent contact with and dispersion of the radioactive material under the conditions of use and wear for which it was designed.

Self-contained breathing apparatus (SCBA). A respiratory protection device that consists of a supply or a means of obtaining respirable air, oxygen, or oxygen-generating material carried by the wearer.

Self-ignition. See Auto-ignition temperature.

Sensible. Capable of being perceived by the sense organs.

Sensitivity. The minimum amount of contaminant that can repeatedly be detected by an instrument.

Sensitization. The process of rendering an individual sensitive to the action of a chemical.

Sensitizer. A substance which, on first exposure, causes little or no reaction in man or test animals but which, on repeated exposure, may cause a marked response not necessarily limited to the contact site.

Sensory feedback. Use of external signals perceived by sense organs (e.g., eye, ear) to indicate quality or level of performance of an event triggered by voluntary action.

Shakeout. In the foundry industry, the separation of the solid, but still not cold, casting from its molding sand.

Shale. Many meanings in industry, but in geology a common fossil rock formed from clay, mud, or silt somewhat stratified but without characteristic cleavage.

Shale oil. Some shale is bituminous and on distillation yields a tarry oil.

Shield, shielding. Interposed material (like a wall) that protects workers from harmful radiations released by radioactive materials.

Shielded metal arc welding (SMAW). An arc-welding process that produces coalescence of metals by heating them with an arc between a covered metal electrode and the work. Shielding is obtained from decomposition of the electrode covering. Pressure is not used and filler metal is obtained from the electrode.

Shock. Primarily the rapid fall in blood pressure following an injury, operation, or the administration of anesthesia.

Shootblasting. A process for cleaning of metal castings or other surfaces by small steel shot in a high-pressure airstream. This process is a substitute for sandblasting to avoid silicosis.

Short-term exposure limit (STEL). ACGIH-recommended exposure limit. Maximum concentration to which workers can be exposed for a short period of time (15 minutes) for only four times throughout the day with at least one hour between exposures.

Silica gel. A regenerative absorbent consisting of the amorphous silica manufactured by the action of HCl on sodium silicate. Hard, glossy, quartzlike in appearance. Used in dehydrating and in drying and as a catalyst carrier.

Silicates. Compounds of silicon, oxygen, and one or more metals with or without hydrogen. Silicate dusts cause nonspecific dust reactions, but generally do not interfere with pulmonary function or result in disability.

Silicon. A nonmetallic element being, next to oxygen, the chief elementary constitutent of the earth's crust.

Silicones. Unique group of compounds made by molecular combination of the element silicon or certain of its compounds with organic chemicals. Produced in a variety of forms, including silicone fluids, resins, and rubber. Silicones have special properties, such as water repellency, wide temperature resistance, high durability, and great dielectric strength.

Silver solder. A solder of varying components but usually containing an appreciable amount of cadmium.

Sintering. Process of making coherent powder of earthy substances by heating, but without melting.

Skin absorption. Ability of some hazardous chemicals to pass directly through the skin and enter the bloodstream.

Skin dose. A special instance of tissue dose, referring to the dose immediately on the surface of the skin.

Skin sensitizer. See Sensitizer.

Skin toxicity. See Dermal toxicity.

Slag. The dross of flux and impurities that rise to the surface of molten metal during melting and refining.

Sludge. In general, any muddy or slushy mass.

Slurry. A thick, creamy liquid resulting from the mixing and grinding of limestone, clay, and other raw materials with water.

Smelting. One step in the procurement of metals from ore—hence, to reduce, to refine, to flux, or to scorify.

Smog. Irritating hazard resulting from the sun's effect on certain pollutants in the air.

Smoke. An air suspension (aerosol) of particles, originating from combustion or sublimation.

Soaking pit. A device in steel manufacturing in which ingots with still molten interiors stand in a heated, upright chamber until solidification is complete.

Soap. Ordinarily a metal salt of a fatty acid, usually sodium stearate, sodium oleate, sodium palmitate, or some combination of these.

Soapstone. Complex silicate of varied composition similar to some talcs with wide industrial application such as in rubber manufacture.

Solder. A material used for joining metal surfaces together by filling a joint or covering a junction.

Solid-state laser. A type of laser that utilizes a solid crystal such as ruby or glass. This type is most commonly used in impulse lasers.

Solubility in water. A term expressing the percentage of a material (by weight) that will dissolve in water at ambient temperature. Solubility information can be useful in determining spill cleanup methods and fire-extinguishing agents and methods for diluting a material. Terms used to express solubility are:

Negligible	Less than 0.1 percent
Slight	0.1 to 1.0 percent
Moderate	1 to 10 percent
Appreciable	More than 10 percent
Complete	

Solution. Mixture in which the components lose their identities and are uniformly dispersed. All solutions consist of a solvent (water or other fluid) and the substance dissolved, called the "solute." A true solution is homogeneous such as salt in water.

Solvent. A substance that dissolves another substance.

Soot. Agglomerations of particles of carbon impregnated with "tar," formed in the incomplete combustion of carbonaceous material.

Sorbent(s). (1) A material that removes toxic gases and vapors from air inhaled through a canister or cartridge. (2) Material used to collect gases and vapors during air-sampling. (3) Nonreactive materials used to clean up chemical spills. Examples: clay and vermiculite.

Sound. An oscillation in pressure, stress, particle displacement, particle velocity, etc., which is propagated in an elastic material, in a medium with internal forces (e.g., elastic, viscous), or the superposition of such propagated oscillations.

Sound absorption. The change of sound energy into some other form, usually heat, as it passes through a medium or strikes a surface.

Sound analyzer. A device for measuring the band pressure level or pressure-spectrum level of a sound as a function of frequency.

Sound level. A weighted sound pressure level, obtained by the use of metering characteristics and the weighting A, B, or C specified in ANSI S1.4.

Sound-level meter and octave-band analyzer. Instruments for measuring sound pressure levels in decibels referenced to 0.0002 microbar.

Sound pressure level (SPL). The level, in decibels, of a sound is 20 times the logarithm to the base 10 of the ratio of the pressure of the sound to the reference pressure. The reference pressure must be explicitly stated.

Sour gas. Slang for either natural gas or a gasoline contaminated with odor-causing sulfur compounds. In natural gas, the contaminant is usually hydrogen sulfide; in gasolines, usually mercaptans.

Source. Any substance that emits radiation. Usually refers to a piece of radioactive material conveniently packaged for scientific or industrial use.

Spasm. Tightening or contraction of any set(s) of muscles.

Special fire-fighting procedures. Special procedures and/or personal protective equipment that is necessary when a particular substance is involved in a fire.

Specific chemical identity. The chemical name, Chemical Abstracts Service (CAS) Registry Number, or any other information that reveals the precise chemical designation of a substance.

Specific gravity. The weight of a material compared to the weight or an equal volume of water; an expression of the density (or heaviness) of the material. (See Gravity, specific.)

Specific weight. The weight per unit volume of a substance, same as density.

Spectrophotometer. An instrument used for comparing the relative intensities of the corresponding colors produced by chemical reaction.

Spectrum. The distribution in frequency of the magnitudes (and sometimes phases) of the components of a wave. Spectrum also is used to signify a continuous range of frequencies, usually wide in extent, within which waves have some specified common characteristics. Also, the pattern of red-to-violet light observed when a beam of sunlight passes through a prism and then projects upon a surface.

Speech interference level (SIL). The speech interference level of a noise is the average, in decibels, of the sound pressure levels of the noise in the three octave bands of frequency 600-1,200, 1,200-2,400, and 2,400-4,800 Hz.

Speech perception test. A measurement of hearing acuity by the administration of a carefully controlled list of words. The identification of correct responses is evaluated in terms of norms established by the average performance of normal listeners.

Spontaneously combustible. A material that ignites as a result of retained heat from processing, or which will oxidize to generate heat and ignite, or which absorbs moisture to generate heat and ignite.

Spot welding. One form of electrical-resistance welding in which the current and pressure are restricted to the spots of metal surfaces directly in contact.

Spray coating painting. The result of the application of a spray in painting as a substitute for brush painting or dipping.

Stability. An expression of the ability of a material to remain unchanged. For MSDS purposes, a material is stable if it remains in the same form under expected and reasonable conditions of storage or use. Conditions which may cause instability (dangerous change) are stated—i.e., temperatures above 150 F, shock from dropping, etc.

Stain. A dye used to color microorganisms as an aid to visual inspection.

Stamping. Many different usages in industry, but a common one is the crushing of ores by pulverizing.

Standard air. Air at standard temperature and pressure. The most common values are 21.1 C (70 F) and 101.3 kPa (29.92 in. Hg).

Standard conditions. In industrial ventilation, 21.1 C (70 F), 50 percent relative humidity, and 101.3 kPa (29.92 in. of mercury) atmosphere pressure.

Standard Industrial Classification (SIC) Code. A classification system for places of employment according to major type of activity by U.S. Government.

Standard man. A theoretical physically fit man of standard (average) height, weight dimensions, and other parameters (blood composition, percentage of water, mass of salivery glands, to name a few), used in studies of how heat or ionizing radiation affects humans.

Standard temperature and pressure. See Standard air.

Static pressure. The potential pressure exerted in all directions by a fluid at rest.

STEL. Short-term exposure limit; ACGIH terminology. See TLV.

Sterile. Free of living microorganisms.

Sterility. Inability to reproduce.

Sterilization. The process of making sterile; the killing of all forms of life.

Stink damp. In mining, hydrogen sulfide.

Stress. A physical, chemical, or emotional factor that causes bodily or mental tension and may be a factor in disease causation or fatigue.

Stressor. Any agent or thing causing a condition of stress.

Strip mine. A mine in which coal or ore is extracted from the earth's surface after removal of overlayers of soil, clay, and rock.

Supplied-air respirators. Air-line respirators or self-contained breathing apparatus.

Supplied-air suit. A one- or two-piece suit that is impermeable to most particulate and gaseous contaminants and is provided with an adequate supply of respirable air.

Surface-active agent or surfactant. Any of a group of compounds added to a liquid to modify surface of interfacial tension. In synthetic detergents, which is the best known use of surface-active agents, reduction of interfacial tension provides cleansing action.

Surface coating. Term used to include paint, lacquer, varnish, and other chemical compositions used for protecting and/or decorating surfaces. See Protective coating.

Suspect carcinogen. A material that is believed to be capable of causing cancer but for which there is limited scientific evidence.

Sweating. (1) Visible perspiration; (2) the process of uniting metal parts by heating solder so that it runs between the parts.

Sweetening. The process by which petroleum products are improved in odor by chemically changing certain sulfur compounds of objectionable odor into compounds having little or no odor.

Swing grinder. A large power-driven grinding wheel mounted on a counterbalanced swivel-supported arm guided by two handles.

Synthesis. The reaction or series of reactions by which a complex compound is obtained from simpler compounds or elements.

Synthetic. (From Greek work *synthetikos*—that which is put together.) "Man-made'synthetic' should not be thought of as a substitute for the natural," states *Encyclopedia of the Chemical Process Industries;* it adds, "Synthetic chemicals are frequently more pure and uniform than those obtained naturally." Example: synthetic indigo.

Synthetic detergents. Chemically tailored cleaning agents soluble in water or other solvents. Originally developed as soap substitutes. Because they do not form insoluble precipitates, they are especially valuable in hard water. They may be composed of surface-active agents alone, but generally are combinations of surface-active agents and other substances, such as complex phosphates, to enhance detergency.

Synthetic rubber. Artificially made polymer with rubberlike properties. Various types have varying composition and properties. Major types designated as S-type, butyl, neoprene (chloroprense polymers), and N-type. Several synthetics duplicate the chemical structure of natural rubber.

Systemic toxicity. Adverse effects caused by a substance that affects the body in a general rather than local manner.

Tailings. In mining or metal recovery processes, the gangue rock residue from which all or most of the metal has been extracted.

Talc. A hydrous magnesium silicate used in ceramics, cosmetics, paint, and pharmaceuticals, and as filler in soap, putty, and plaster.

Tall oil. (Name derived from Swedish word tallolja; material first investigated in Sweden—not synonymous with U.S. pine oil). Natural mixture of rosin acids, fatty acids, sterols, high-molecular weight alcohols, and other materials derived primarily from waste liquors of sulfate wood pulp manufacture. Dark brown, viscous, oily liquid often called liquid rosin.

Tar. A loose term embracing wood, coal, or petroleum exudations. In general, represents complex mixture of chemicals of top fractional distillation systems.

Tar crude. Organic raw material derived from distillation of coal tar and used for chemicals.

Tare. A deduction of weight, made in allowance for the weight of a container or medium. The initial weight of a filter, for example.

Temper. To relieve the internal stresses in metal or glass and to increase ductility by heating the material to a point below its critical temperature and cooling slowly.

Tempering. The process of heating or cooling make-up air to the proper temperature.

Temporary threshold shift (TTS). The hearing loss suffered as the result of noise exposure, all or part of which is recovered during an arbitrary period of time when one is removed from the noise.

Temporary total disability. An injury that does not result in death or permanent disability, but which renders the injured person unable to perform regular duties or activities on one or more calendar days after the day of injury. (This is a definition established by OSHA.)

Tendon. Fibrous component of a "muscle." It frequently attaches at the area of application of tensile force. When its cross section is small, stresses in the tendon are high, particularly because the total force of many muscle fibers is applied at the single terminal tendon. (See Tenosynovitis.)

Tenosynovitis. Inflammation of the connective tissue sheath of a tendon.

Teratogen. A substance or agent to which exposure of a pregnant female can result in malformations in the fetus. An example is the drug thalidomide.

Terminal velocity. The terminal rate of fall of a particle through a fluid as induced by gravity or other external force; the rate at which frictional drag balances the accelerating force (or the external force).

Thermal pollution. Discharge of heat into bodies of water to the point that increased warmth activates all sewage, depletes the oxygen the water needs to cleanse itself, and eventually destroys some of the fish and other organisms in the water.

Thermonuclear reaction. A fusion reaction, that is, a reaction in which two light nuclei combine to form a heavier atom, releasing a large amount of energy. This is believed to be the sun's source of energy. It is called thermonuclear because it occurs only at a very high temperature.

Thermoplastic. Capable of being repeatedly softened by heat.

Thermosetting. Capable of undergoing a chemical change from a soft to a hardened substance when heated.

Thermosetting plastics. Those that are heat-set in their final processing to a permanently hard state. Examples: phenolics, ureas, and melamines.

Thinner. A liquid used to increase the fluidity of paints, varnishes, and shellac.

Threshold. The level where the first effects occur; also the point at which a person just begins to notice a tone is becoming audible.

Threshold limit value (TLV). A guideline for exposure developed by ACGIH. Refers to airborne concentrations of substances and represents conditions under which it is believed that nearly all workers may be repeatedly exposed day after day without adverse effect. Some workers may be affected at, or below, the TLV limits.

Threshold limit value ceiling (TLV-C). The ceiling level of the exposure that should never be exceeded. This value has been established as the maximum level to be used in computing the TWA and STEL limits. The ceiling value is under the IDLH limit for a given substance.

Threshold limit value short-term exposure limit (TLV-STEL): An exposure level considered safe to work in for short periods of time (15 minutes), four times a day maximum, with at least 60 minutes between exposures. No irritation or other adverse effects should be experienced. This level is higher than the TLV-TWA for a given substance.

Threshold limit value time-weighted average (TLV-TWA). The time-weighted average concentration of a substance for a normal 8-hour workday and a 40-hour workweek to which nearly all workers may be repeatedly exposed, day after day,

without adverse effect.

Time-weighted average concentration (Refers to concentrations of airborne toxic materials that have been weighted for a certain time duration, usually 8 hours.

Tinning. Any work with tin, such as tin roofing; but in particular in soldering, the primary coating with solder of the two surfaces to be united.

Tolerance. (1) The ability of a living organism to resist the usually anticipated stress. (2) Also, the limits of permissible inaccuracy in the fabrication of an article above or below its design specifications.

Tolerance dose. See Maximum permissible concentration and MPL.

Toxic substance. Any substance that can cause acute or chronic injury to the human body, or that is suspected of being able to cause diseases or injury under some conditions.

Toxicity. The sum of adverse effects resulting from exposure to a material, generally by the mouth, skin, or respiratory tract.

Toxin. A poisonous substance that is derived from an organism.

Trade name. The commercial name or trademark by which something is known.

Trade secret. Any confidential formula pattern, process, device, information or compilation of information (including chemical name or other unique chemical identifier) that is used in an employer's business, and that gives the employer an opportunity to obtain an advantage over competitors who do not know or use it.

Trauma. An injury or wound brought about by an outside force.

TSCA. Toxic Substances Control Act; federal environmental legislation, administered by EPA, for regulating the manufacture, handling, and use of materials classified as "toxic substances."

Tumbling. An industrial process, such as in founding, in which small castings are cleaned by friction in a rotating drum (tumbling mill, tumbling barrel), which may contain sand, sawdust, stone, etc.

Turbid. Cloudy.

Turning vanes. Curved pieces added to elbows or fan inlet boxes to direct air and so reduce turbulence losses.

See TLV-UEL. See Upper explosive limit. (Also see Lower explosive limit.)

Ultraviolet. Those wavelengths of the electromagnetic spectrum that are shorter than those of visible light and longer than x-rays, 10 cm to 100 cm wavelength.

UN number. A registry number assigned to dangerous commonly carried goods by the United Nations Committee of Experts on the Transport of Dangerous Goods. The UN number is required in shipping documentation and on packaging as part of the DOT regulations for shipping hazardous materials.

Unsafe condition. That part of the work environment that contributed or could have contributed to an accident.

Unstable. A chemical that, in the pure state, or as produced or transported, will vigorously polymerize, decompose, condense, or become self-reactive under conditions of shock, pressure, or temperature. Such chemicals are also referred to as reactive.

Upper explosive limit (UEL). The highest concentration (expressed in percent vapor or gas in the air by volume) of a substance that will burn or explode when an ignition source is present.

USC. *United States Code*, the official compilation of federal statutes.

USDA. U.S. Department of Agriculture.

Use. To package, handle, react, or transfer a substance.

Valve. A device that controls the direction of air or fluid flow or the rate and pressure at which air or fluid is delivered, or both.

Vapor. The gaseous form of substances that are normally in the solid or liquid state (at room temperature and pressure).

Vat dyes. Water insoluble, complex coal tar dyes that can be chemically reduced in a heated solution to a soluble form that will impregnate fibers. Subsequent oxidation then produces insoluble color dyestuffs that are remarkably fast to washing, light, and chemicals.

Vector. (1) Term applied to an insect or any living carrier that transports a pathogenic microorganism from the sick to the well, inoculating the latter; the organism may or may not pass through any developmental cycle. (2) Anything (e.g., velocity, mechanical force, electromotive force) having magnitude, direction, and sense that can be represented by a straight line of appropriate length and direction.

Velocity. A vector that specifies the time rate of change or displacement with respect to a reference.

Velocity pressure. The kinetic pressure in the direction of flow necessary to cause a fluid at rest to flow at a given velocity. When added to static pressure, it gives total pressure.

Velometer. A device for measuring air velocity.

Ventilation. Circulating fresh air to replace contaminated air.

Ventilation, dilution. Airflow designed to dilute contaminants to acceptable levels.

Ventilation, mechanical. Air movement caused by a fan or other air-moving device.

Ventilation, natural. Air movement caused by wind, temperature difference, or other nonmechanical factors.

Vibration. An oscillation motion about an equilibrium position produced by a disturbing force.

Vinyl. A general term applied to a class of resins such as polyvinyl chloride, acetate, butyral, etc.

Viscose rayon. The type of rayon produced from the reaction of carbon disulfide with cellulose and the hardening of the resulting viscous fluid by passing it through dilute sulfuric acid, causing the evolution of hydrogen sulfide gas.

Viscosity. The tendency of a fluid to resist internal flow without regard to its density.

Visual acuity. Ability of the eye to sharply perceive the shape of objects in the direct line of vision.

Volatile. The percentage of a liquid or solid (by volume) that will evaporate at an ambient temperature of 70 F (unless some other temperature is stated). Examples: butane, gasoline, and paint thinner (mineral spirits) are 100% volatile; their individual evaporation rates vary, but over a period of time each will evaporate completely.

Volt. The practical unit of electromotive force or difference in potential between two points in an electrical field. Just as pressure in a water pipe causes water to flow, electrical pressure—or voltage—pushes the current of electrons through wires to receptacles, such as light fixtures and outlets.

Volume flow rate. The quantity (measured in units of volume) of a fluid flowing per unit of time, as cubic feet per minute, gallons per hour, or cubic meters per second.

Vulcanization. Process of combining rubber (natural, synthetic, or latex) with sulfur and accelerators in presence of zinc oxide under heat and usually pressure in order to change the material permanently from a thermoplastic to a thermosetting composition, or from a plastic to an elastic condition. Strength, elasticity, and abrasion resistance also are improved.

Water column. A unit used in measuring pressure. (See also Inches of water column.)

Water curtain or waterfall booth. In spray painting, the water running down a wall into which excess paint spray is drawn or blown by fans; the water carries the paint downward to a collecting point.

Waterproofing agents. These usually are formulations of three distinct materials: (1) a coating material, (2) a solvent, and (3) a plasticizer. Among the materials used in waterproofing are cellulose esters and ether, polyvinyl chloride resins or acetates, and variations of vinyl chloride-vinylidine chloride polymers.

Water-reactive. A chemical that reacts with water to release a gas that is either flammable or presents a health hazard.

Watt (w). A unit of electrical power, equal to one joule per second.

Weight. The force with which a body is attracted toward the earth.

Weld. A localized coalescense of metals or nonmetals produced either by heating the materials to suitable temperatures, with or without the application of pressure, or by the application of pressure alone, and with or without the use of filler material.

Welding. The several types of welding are electric arc welding, oxyacetylene welding, spot welding, and inert or shielded gas welding utilizing helium or argon. The hazards involved in welding stem from (1) the fumes from the weld metal such as lead or cadmium metal, or (2) the gases created by the process, or (3) the fumes or gases arising from the flux.

Welding rod. A rod or heavy wire that is melted and fused into metals in arc welding.

White damp. In mining, carbon monoxide.

White noise. A noise whose spectrum density (or spectrum level) is substantially independent of frequency over a specified range.

Work. When a force acts against a resistance to produce motion in a body, the force is said to do work. Work is measured by the product of the force acting and the distance moved through against a resistance. The units of measurement are the erg (the joule is 1×10^7 ergs) and the footpound.

Work area. A room or defined space in a workplace where hazardous chemicals are produced or used and where employees are present.

Workers' Compensation. Insurance plan that compensates injured workers or their survivors.

Work-hardening. The property of a metal to become harder and more brittle on being "worked," that is, bent repeatedly or drawn.

Work hours. The total number of hours worked by all employees.

Work injuries (including occupational illnesses). Those that arise out of and in the course of gainful employment regardless of where the accident occurs. Excluded are work injuries to private household workers and injuries occurring in connection with farm chores, which are classified as home injuries.

Workplace. An establishment, job site, or project at one geographical location containing one or more work areas.

Work strain. The natural physiological response reaction of the body to the application of work stress. The locus of the reaction may often be remote from the point of application of work stress. Work strain is not necessarily traumatic, but may appear as trauma when excessive, either directly or cumulatively, and must be considered by the industrial engineer in equipment and task design. Thus, a moderate increase of heart rate is nontraumatic work strain resulting from physical work strain if caused by undue work stress on the wrists.

Work stress. Biomechanically, any external force acting on the body during the performance of a task. It always produces work strain. Application of work stress to the human body is the inevitable consequence of performance of any task, and is, therefore, only synonymous with "stressful work conditions" when excessive. Work stress analysis is an integral part of task design.

Workers. All persons gainfully employed, including owners, managers, other paid employees, the self-employed, and unpaid family workers, but excluding private household workers.

CONVERSION OF UNITS

ALL PHYSICAL UNITS OF MEASUREMENT can be reduced to three basic dimensions—mass, length, and time. Not only does reducing units to these basic dimensions simplify the solution of problems, but standardization of units makes comparison between operations (and between operations and standards) easier.

For example, air flows are usually measured in liters per minute, cubic meters per second, or cubic feet per minute. The total volume of air sampled can be easily converted to cubic meters or cubic feet. In another situation, the results of atmospheric pollution studies and stack sampling surveys are often reported as grains per cubic foot, grams per cubic foot, or pounds per cubic foot. The degree of contamination is usually reported as parts of contaminant per million parts of air.

If physical measurements are made or reported in different units, they must be converted to the standard units if any comparisons are to be meaningful.

To save time and space in reporting data, many units have standard abbreviations. Because the metric system (SI) is becoming more frequently used, conversion factors are given for the standard units of measurement.

FUNDAMENTAL UNITS

Conversion factors for various measurement units are listed in the tables in this section. To use a table to find the numerical value of the quantity desired, locate the unit to be converted in the first column. Then multiply this value by the number appearing at the intersection of the row and the column containing the desired unit. The answer will be the numerical value in the desired unit.

Various English systems and metric system units are given for the reader's convenience. The standard system of measurement, however, is the International System of Units (SI). The official conversion factors and an explanation of the system are given to 6- or 7-place accuracy in ASTM Standard E 380-93.

Briefly, the SI System being adopted throughout the world is a modern version of the MKSA (meter, kilogram, second, ampere) system. Its details are published and controlled by an international treaty organization, the International Bureau of Weights and Measures (BIPM), set up by the Metre Convention signed in Paris, France, on May 20, 1875. The United States and Canada are member states of this Convention, as implemented by the Metric Conversion Act of 1975 (Public Law 94-168).

HELPFUL ORGANIZATIONS

The following four groups in the U.S. and Canada are deeply involved in planning and implementing metric conversion:

American National Metric Council
4330 East-West Highway, #1117
Bethesda, MD 20814-4408

Canadian Metric Association
P.O. Box 35
Fonthill, ON L0S 1E0, Canada

National Institute of Standards and Technology
U.S. Dept. of Commerce
Metric Program, Bldg. 820
Gaithersburg, MD 20899

U.S. Metric Association
10245 Andasol Avenue
Northridge, CA 91325-1504

Conversion of Units

FAHRENHEIT-CELSIUS CONVERSION TABLE

Fahrenheit-Celsius Conversion.—A simple way to convert a Fahrenheit temperature reading into a Celsius temperature reading or vice versa is to enter the accompanying table in the center or boldface column of figures. These figures refer to the temperature in either Fahrenheit or Celsius degrees. If it is desired to convert from Fahrenheit to Celsius degrees, consider the center column as a table of Fahrenheit temperatures and read the corresponding Celsius temperature in the column at the left. If it is desired to convert from Celsius to Fahrenheit degrees, consider the center column as a table of Celsius values, and read the corresponding Fahrenheit temperature on the right.

To convert from "degrees Fahrenheit" to "degrees Celsius" (formerly called "degrees centigrade"), use the formula:

$$t_c = \frac{(t_f - 32)}{1.8} \text{ or } \frac{5}{9}(t_f - 32)$$

Conversely,
$$t_f = 1.8\, t_c + 32 \text{ or } \frac{9}{5}\, t_c + 32$$

Example, convert the boiling point of water in F to C:

$$212 \ \text{F} - 32 = 180$$

$$\frac{5}{9}(180) = 100 \ \text{C}$$

Fahrenheit—Celsius Conversion Table

Deg C		Deg F	Deg C		Deg F	Deg C		Deg F	Deg C		Deg F
−273	**−459.4**	. . .	−129	**−200**	−328	−13.9	**7**	44.6	1.1	**34**	93.2
−268	**−450**	. . .	−123	**−190**	−310	−13.3	**8**	46.4	1.7	**35**	95.0
−262	**−440**	. . .	−118	**−180**	−292	−12.8	**9**	48.2	2.2	**36**	96.8
−257	**−430**	. . .	−112	**−170**	−274	−12.2	**10**	50.0	2.7	**37**	98.6
−251	**−420**	. . .	−107	**−160**	−256	−11.7	**11**	51.8	3.3	**38**	100.4
−246	**−410**	. . .	−101	**−150**	−238	−11.1	**12**	53.6	3.9	**39**	102.2
−240	**−400**	. . .	− 96	**−140**	−220	−10.6	**13**	55.4	4.4	**40**	104.0
−234	**−390**	. . .	− 90	**−130**	−202	−10.0	**14**	57.2	5.0	**41**	105.8
−229	**−380**	. . .	− 84	**−120**	−184	− 9.4	**15**	59.0	5.6	**42**	107.6
−223	**−370**	. . .	− 79	**−110**	−166	− 8.9	**16**	60.8	6.1	**43**	109.4
−218	**−360**	. . .	− 73	**−100**	−148	− 8.3	**17**	62.6	6.7	**44**	111.2
−212	**−350**	. . .	− 68	**− 90**	−130	− 7.8	**18**	64.4	7.2	**45**	113.0
−207	**−340**	. . .	− 62	**− 80**	−112	− 7.2	**19**	66.2	7.8	**46**	114.8
−201	**−330**	. . .	− 57	**− 70**	− 94	− 6.7	**20**	68.0	8.3	**47**	116.6
−196	**−320**	. . .	− 51	**− 60**	− 76	− 6.1	**21**	69.8	8.9	**48**	118.4
−190	**−310**	. . .	− 46	**− 50**	− 58	− 5.6	**22**	71.6	9.4	**49**	120.2
−184	**−300**	. . .	− 40	**− 40**	− 40	− 5.0	**23**	73.4	10.0	**50**	122.0
−179	**−290**	. . .	− 34	**− 30**	− 22	− 4.4	**24**	75.2	10.6	**51**	123.8
−173	**−280**	. . .	− 29	**− 20**	− 4	− 3.9	**25**	77.0	11.1	**52**	125.6
−169	**−273**	−459.4	− 23	**− 10**	14	− 3.3	**26**	78.8	11.7	**53**	127.4
−168	**−270**	−454	−17.8	**0**	32−	− 2.8	**27**	80.6	12.2	**54**	129.2
−162	**−260**	−436	−17.2	**1**	33.8	− 2.2	**28**	82.4	12.8	**55**	131.0
−157	**−250**	−418	−16.7	**2**	35.6	− 1.7	**29**	84.2	13.3	**56**	132.8
−151	**−240**	−400	−16.1	**3**	37.4	− 1.1	**30**	86.0	13.9	**57**	134.6
−146	**−230**	−382	−15.6	**4**	39.2	− 0.6	**31**	87.8	14.4	**58**	136.4
−140	**−220**	−364	−15.0	**5**	41.0	0−	**32**	89.6	15.0	**59**	138.2
−134	**−210**	−346	−14.4	**6**	42.8	0.6	**33**	91.4	15.6	**60**	140.0

Deg C		Deg F	Deg C		Deg F	Deg C		Deg F	Deg C		Deg F
16.1	61	141.8	50.0	122	251.6	83.9	183	361.4	276.7	530	986
16.7	62	143.6	50.6	123	253.4	84.4	184	363.2	282.2	540	1004
17.2	63	145.4	51.1	124	255.2	85.0	185	365.0	287.8	550	1022
17.8	64	147.2	51.7	125	257.0	85.6	186	366.8	293.3	560	1040
18.3	65	149.0	52.2	126	258.8	86.1	187	368.6	298.9	570	1058
18.9	66	150.8	52.8	127	260.6	86.7	188	370.4	304.4	580	1076
19.4	67	152.6	53.3	128	262.4	87.2	189	372.2	310.0	590	1094
20.0	68	154.4	53.9	129	264.2	87.8	190	374.0	315.6	600	1112
20.6	69	156.2	54.4	130	266.0	88.3	191	375.8	321.1	610	1130
21.1	70	158.0	55.0	131	267.8	88.9	192	377.6	326.7	620	1148
21.7	71	159.8	55.6	132	269.6	89.4	193	379.4	332.2	630	1166
22.2	72	161.6	56.1	133	271.4	90.0	194	381.2	337.8	640	1184
22.8	73	163.4	56.7	134	273.2	90.6	195	383.0	343.3	650	1202
23.3	74	165.2	57.2	135	275.0	91.1	196	384.8	348.9	660	1220
23.9	75	167.0	57.8	136	276.8	91.7	197	386.6	354.4	670	1238
24.4	76	168.8	58.3	137	278.6	92.2	198	388.4	360.0	680	1256
25.0	77	170.6	58.9	138	280.4	92.8	199	390.2	365.6	690	1274
25.6	78	172.4	59.4	139	282.2	93.3	200	392.0	371.1	700	1292
26.1	79	174.2	60.0	140	284.0	93.9	201	393.8	376.7	710	1310
26.7	80	176.0	60.6	141	285.8	94.4	202	395.6	382.2	720	1328
27.2	81	177.8	61.1	142	287.6	95.0	203	397.4	387.8	730	1346
27.8	82	179.6	61.7	143	289.4	95.6	204	399.2	393.3	740	1364
28.3	83	181.4	62.2	144	291.2	96.1	205	401.0	398.9	750	1382
28.9	84	183.2	62.8	145	293.0	96.7	206	402.8	404.4	760	1400
29.4	85	185.0	63.3	146	294.8	97.2	207	404.6	410.0	770	1418
30.0	86	186.8	63.9	147	296.6	97.8	208	406.4	415.6	780	1436
30.6	87	188.6	64.4	148	298.4	98.3	209	408.2	421.1	790	1454
31.1	88	190.4	65.0	149	300.2	98.9	210	410.0	426.7	800	1472
31.7	89	192.2	65.6	150	302.0	99.4	211	411.8	432.2	810	1490
32.2	90	194.0	66.1	151	303.8	100.0	212	413.6	437.8	820	1508
32.8	91	195.8	66.7	152	305.6	104.4	220	428.0	443.3	830	1526
33.3	92	197.6	67.2	153	307.4	110.0	230	446.0	448.9	840	1544
33.9	93	199.4	67.8	154	309.2	115.6	240	464.0	454.4	850	1562
34.4	94	201.2	68.3	155	311.0	121.1	250	482.0	460.0	860	1580
35.0	95	203.0	68.9	156	312.8	126.7	260	500.0	465.6	870	1598
35.6	96	204.8	69.4	157	314.6	132.2	270	518.0	471.1	880	1616
36.1	97	206.6	70.0	158	316.4	137.8	280	536.0	476.7	890	1634
36.7	98	208.4	70.6	159	318.2	143.3	290	554.0	482.2	900	1652
37.2	99	210.2	71.1	160	320.0	148.9	300	572.0	487.8	910	1670
37.8	100	212.0	71.7	161	321.8	154.4	310	590.0	493.3	920	1688
38.3	101	213.8	72.2	162	323.6	160.0	320	608.0	498.9	930	1706
38.9	102	215.6	72.8	163	325.4	165.6	330	626.0	504.4	940	1724
39.4	103	217.4	73.3	164	327.2	171.1	340	644.0	510.0	950	1742
40.0	104	219.2	73.9	165	329.0	176.7	350	662.0	515.6	960	1760
40.6	105	221.0	74.4	166	330.8	182.2	360	680.0	521.1	970	1778
41.1	106	222.8	75.0	167	332.6	187.8	370	698.0	526.7	980	1796
41.7	107	224.6	75.6	168	334.4	193.3	380	716.0	532.2	990	1814
42.2	108	226.4	76.1	169	336.2	198.9	390	734.0	537.8	1000	1832
42.8	109	228.2	76.7	170	338.0	204.4	400	752.0	565.6	1050	1922
43.3	110	230.0	77.2	171	339.8	210	410	770.0	593.3	1100	2012
43.9	111	231.8	71.8	172	341.6	215.6	420	788	621.1	1150	2102
44.4	112	233.6	78.3	173	343.4	221.1	430	806	648.9	1200	2192
45.0	113	235.4	78.9	174	345.2	226.7	440	824	676.7	1250	2282
45.6	114	237.2	79.4	175	347.0	232.2	450	842	704.4	1300	2372
46.1	115	239.0	80.0	176	348.8	237.8	460	860	732.2	1350	2462
46.7	116	240.8	80.6	177	350.6	243.3	470	878	760.0	1400	2552
47.2	117	242.6	81.1	178	352.4	248.9	480	896	787.8	1450	2642
47.8	118	244.4	81.7	179	354.2	254.4	490	914	815.6	1500	2732
48.3	119	246.2	82.2	180	356.0	260.0	500	932	1093.9	2000	3632
48.9	120	248.0	82.8	181	357.8	265.6	510	950	1648.9	3000	5432
49.4	121	249.8	83.3	182	359.6	271.1	520	968	2760.0	5000	9032

From Machinery's Handbook, *18th ed. (The Industrial Press)*

Above 1000 in the center column, the table increases in increments of 50. To convert 1462 degrees F to Celsius, for instance, add to the Celsius equivalent of 1400 degrees F ⅝ths of 62 or 34 degrees, which equals 794 C.

LENGTH

To Obtain → Multiply Number of by↓	meter (m)	centimeter (cm)	millimeter (mm)	micron (μ) or micrometer (μm)	angstrom unit, (A)	inch (in.)	foot (ft)
meter	1	100	1000	10^6	10^{10}	39.37	3.28
centimeter	0.01	1	10	10^4	10^8	0.394	0.0328
millimeter	0.001	0.1	1	10^3	10^7	0.0394	0.00328
micron	10^{-6}	10^{-4}	10^{-3}	1	10^4	3.94×10^{-5}	3.28×10^{-6}
angstrom	10^{-10}	10^{-8}	10^{-7}	10^{-4}	1	3.94×10^{-9}	3.28×10^{-10}
inch	0.0254	2.540	25.40	2.54×10^4	2.54×10^8	1	0.0833
foot	0.305	30.48	304.8	304,800	3.048×10^9	12	1

AREA

To Obtain → Multiply Number of By↓	square meter (m²)	square inch (sq in.)	square foot (sq ft)	square centimeter (cm²)	square millimeter (mm²)
square meter	1	1,550	10.76	10,000	10^6
square inch	6.452×10^{-3}	1	6.94×10^{-3}	6.452	645.2
square foot	0.0929	144	1	929.0	92,903
square centimeter	0.0001	0.155	0.001	1	100
square millimeter	10^{-6}	0.00155	0.00001	0.01	1

DENSITY

To Obtain → Multiply Number of By↓	gm/cm³	lb/cu ft	lb/gal
gram/cubic centimeter	1	62.43	8.345
pound/cubic foot	0.01602	1	0.1337
pound/gallon (U.S.)	0.1198	7.481	1

1 grain/cu ft = 2.28 mg/m³

FORCE

To Obtain → Multiply Number of By ↘	dyne	newton (N)	kilogram-force	pound-force (lbf)
dyne	1	1.0×10^{-5}	1.02×10^{4}	2.248×10^{4}
newton	1.0×10^{5}	1	0.1020	0.2248
kilogram-force	9.807×10^{-5}	9.807	1	2.205
pound-force	4.448×10^{-5}	4.448	0.4536	1

MASS

To Obtain → Multiply Number of By ↘	gram (gm)	kilogram (kg)	grains (gr)	ounce (avoir) (oz)	pound (avoir) (lb)
gram	1	0.001	15.432	0.03527	0.00220
kilogram	1,000	1	15,432	35.27	2.205
grain	0.0648	6.480×10^{-5}	1	2.286×10^{-3}	1.429×10^{-4}
ounce	28.35	0.02835	437.5	1	0.0625
pound	453.59	0.4536	7,000	16	1

VOLUME

To Obtain → Multiply Number of By ↘	cu ft	gallon (U.S. liquid)	liters	cm³	m³
cubic foot	1	7.481	28.32	28,320	0.0283
gallon (U.S. liquid)	0.1337	1	3.785	3,785	3.79×10^{-3}
liter	0.03531	0.2642	1	1,000	1×10^{-3}
cubic centimeters	3.531×10^{-5}	2.64×10^{-4}	0.001	1	10^{-6}
cubic meters	35.31	264.2	1,000	10^{6}	1

VELOCITY

To Obtain ⟶ Multiply Number of By	cm/s	m/s	km/hr	ft/s	ft/min	mph
centimeter/second	1	0.01	0.036	0.0328	1.968	0.02237
meter/second	100	1	3.6	3.281	196.85	2.237
kilometer/hour	27.78	0.2778	1	0.9113	54.68	0.6214
foot/second	30.48	0.3048	18.29	1	60	0.6818
foot/minute	0.5080	0.00508	0.0183	0.0166	1	0.01136
mile per hour	44.70	0.4470	1.609	1.467	88	1

PRESSURE

To Obtain ⟶ Multiply Number of By	lb/sq in. (psi)	atm	in. (Hg) 32 F 0 C	mm (Hg) 32 F 0 C	kPa (kN/m²)	ft (H₂O) 60 F 15 C	in. (H₂O)	lb/sq ft
pound/square inch	1	0.068	2.036	51.71	6.895	2.309	27.71	144
atmospheres	14.696	1	29.92	760.0	101.32	33.93	407.2	2,116
inch (Hg)	0.4912	0.033	1	25.40	3.386	1.134	13.61	70.73
millimeter (Hg)	0.01934	0.0013	0.039	1	0.1333	0.04464	0.5357	2.785
kilopascals	0.1450	9.87×10^{-3}	0.2953	7.502	1	0.3460*	4.019	20.89
foot (H₂0)(15 C)	0.4332	0.0294	0.8819	22.40	2.989*	1	12.00	62.37
inch (H₂0)	0.03609	0.0024	0.073	1.867	0.2488	0.0833	1	5.197
pound/square foot	0.0069	4.72×10^{-4}	0.014	0.359	0.04788	0.016	0.193	1

*at 4C

HEAT, ENERGY, OR WORK

To Obtain → Multiply Number of By ↓	joule	ft-lb	kwh	hp-hour	kcal	cal	Btu
joules	1	0.737	2.773×10^{-7}	3.725×10^{-7}	2.39×10^{-4}	0.2390	9.478×10^{-4}
foot-pound	1.356	1	3.766×10^{-7}	5.05×10^{-7}	3.24×10^{-4}	0.3241	1.285×10^{-3}
kilowatt-hour	3.6×10^6	2.66×10^6	1	1.341	860.57	860,565	3,412
hp-hour	2.68×10^6	1.98×10^6	0.7455	1	641.62	641,615	2,545
kilocalorie	4,184	3,086	1.162×10^{-3}	1.558×10^{-3}	1	1,000	3.9657
calorie	4.184	3.086	1.162×10^{-6}	1.558×10^{-6}	0.001	1	.00397
British thermal unit	1,055	778.16	2.930×10^{-4}	3.93×10^{-4}	0.252	252	1

BIBLIOGRAPHY

American Conference of Governmental Industrial Hygienists, 1330 Kemper Meadow Drive, Cincinnati, OH 45240. "Threshold Limit Values for Chemical Substances and Physical Agents in the Workroom Environment." (Issued annually.)

American Industrial Hygiene Association, 345 White Pond Drive, Akron, OH 44320. *Industrial Ventilation—A Manual of Recommended Practice.* (Latest edition.)

American National Red Cross, 17th and D Streets N.W., Washington, DC, 20006. *First Aid Textbook.* (Latest edition.)

American National Standards Institute, 11 West 42nd Street, New York, NY 10036. "Catalog of American National Standards." (Issued annually.)

American Society for Training and Development (ASTD), 1630 Duke Street, P.O. Box 1443, Alexandria, VA 22313. *Literature Index.* (Quarterly.)

American Welding Society (AWS), 550 LeJeune Road N.W., P.O. Box 351040, Miami, FL 33135. *Welding Handbook.*

Bird, FE, Jr. *Management Guide to Loss Control.* Loganville, GA: Institute Publications ILCI, 1974.

Bird, FE, Jr., and RG Loftus. *Loss Control Management.* Loganville, GA: Institute Publications ILCI, 1976.

Blanchard, K, et al. *The One Minute Manager Gets Fit.* New York: Nightingale-Conant, 1989.

Currance, PL and Bronstein, AC. *Emergency Care For Hazardous Materials Exposure No. 2.* St. Louis: Mosby, 1994.

"Care and Operating Instructions for Various Electric Tools." Milwaukee Electric Tool Corp., 13171 West Lisbon Road, Brookfield, WI 53005.

Chemical Manufacturers Association, 1300 Wilson Blvd., Arlington, VA 22209. Chemical Data Sheets.

CHEMTREC: Chemical Transportation Emergency Center, Washington, DC. Access via telephone: (800) 424–9300 or (202) 483–7616.

Computer Data Bases (some purchased and some accessed by phone)
OHMTADS: Accessed through Chemical Information System, Inc. (800) 247–8737.
CHRIS: Chemical Hazard Response Information System Accessed through Chemical Information System.
CAMEO: Computer-Aided Management of Emergency Operations.

DeCristoforo, RJ. *Complete Book of Stationary Power Tool Techniques.* New York: Sterling, 1988.

Factory Mutual Engineering Corporation, 1151 Boston-Providence Turnpike, Norwood, MA 02062. *Factory Mutual System Approval Guide.* (Annual.) *Handbook of Property Conservation. Loss Prevention Data Books.* (Periodic.)

Fallon, WK. Ed. *AMA Management Handbook.* 2nd ed. New York: AMACOM, 1983.

Ferry, TS. *Modern Accident Investigation and Analysis.* 2nd ed. New York: John Wiley & Sons, 1988.

Firenze, RJ. *The Process of Hazard Control.* Dubuque, IA: Kendall/Hunt Publishing Company, 1978.

Grandjean, E. *Fitting the Task to the Man: An Ergonomics Approach.* 4th ed. London: Taylor and Francis, 1988.

Hammer, W. *Occupational Safety Management and Engineering*, 4th ed. Englewood Cliffs, NJ: Prentice-Hall, Inc., 1989.

Hawley's Condensed Chemical Dictionary, 12th ed., New York: Reinhold Publishing Corporation, 1993.

Heinrich, HW, D Petersen, and N Roos. *Industrial Accident Prevention: A Safety Management Approach.* 5th rev. ed. New York: McGraw-Hill Book Company, 1980.

Human Factors Section, Eastman Kodak Laboratory.

Ergonomics Design for People at Work, vol. 1. New York: VanNostrand Reinhold, 1983.

Illuminating Engineering Society, 345 East 47th Street, New York: 10017. *IES Lighting Handbook (The Standard Lighting Guide). Practice for Industrial Lighting* (ANSI/IES RP7).

International Labor Office, 1750 New York Avenue, Suite 330, Washington, DC 20006. *Encyclopedia of Occupational Health and Safety.*

Johnson, WG. *MORT Safety Assurance Systems.* New York: Marcel Dekker, 1980.

Kroemer, KHE. Coupling the hand with the handle. *Human Factors* 28 (1986): 337–39.

Engineering Anthropometry, in Salvendy, G., ed. *Handbook of Human Factors/Ergonomics.* New York: John Wiley & Sons, 1987, pp. 154–68.

Office Ergonomics: Work Station Dimensions. In Alexander, DC, and Pulat, BM, eds. *Industrial Ergonomics.* Norcross, GA: Institute of Industrial Engineers, 1985, pp. 187–201.

Testing individual capability to lift material: Repeatability of a dynamic test compared with static testing. *Journal of Safety Research* 16(1985): 1–7.

VDT Workstation Design. In Helander, MG, ed. *Handbook of Human Computer Interaction.* New York: Elsevier Publications, 1989.

Mosby, C. V. International Agency for Research on Cancer (IARC) Monographs.

National Association of Suggestion Systems (NASS), 230 North Michigan Avenue, Suite 1200, Chicago, IL 60601. *NASS Horizons.* (Quarterly.)

National Fire Protection Association. Batterymarch Park, Quincy, MA 02269. *Fire Protection Handbook. Fire Protection Guide on Hazardous Materials. Flammable and Combustible Liquids Code Handbook. Industrial Fire Hazards Handbook. Inspection Manual. Life Safety Code Handbook. National Electrical Code Handbook.* "Standards and Recommended Practices." (Catalog available.)

National Instutite of Occupational Safety and Health (NIOSH), Public Health Service, Centers for Disease Control, Department of Health and Human Service, 1600 Clifton Road, Atlanta, GA 30333. *Certified Personal Protective Equipment.* "Criteria Documents." *The Industrial Environment: Its Evaluation and Control. Machine Guarding—Assessment of Need.* "Publications Catalog" *Safety Program Practices in High Versus Low Accident Rate Companies. Toxic Substances List.* "Welding Safely."

National Safety Council, 1121 Spring Lake Drive, Itasca, IL 60143-3201. *Accident Facts* (Annually.) *Accident Investigation. Accident Prevention Manual for Business & Industry* (3 volumes). *Chemicals, the Press and the Public. Electrical Inspection Illustrated. Family Safety and Health.* (Quarterly.) "5 Minute Safety Talks." (Series.) *Forging Safety Manual. Fundamentals of Industrial Hygiene. Safeguarding Concepts Illustrated. Occupational Safety & Health Data Sheets.* (Index available.) *Out in Front: Effective Supervision in the Workplace. Power Press Safety Manual. Safety and Health.* (Monthly.)

Ottoboni, MA. *The Dose Makes the Poison: A Plain Language Guide to Toxicology.* 2nd ed. New York: Van Nostrand Reinhold, 1991.

Petersen, DC. *Techniques of Safety Management*, 3rd ed. Goshen, NY: Aloray, Inc., 1989.

Prevent Blindness America, 500 East Remington Road, Schaumburg, IL 60173. Eyesight in Industry.

Sax, NI and RJ Lewis. D*angerous Properties of Industrial Materials*, 9th ed., 3 vols. New York: Van Nostrand Reinhold, 1993.

Simonds, RH, and JV Grimaldi. *Safety Management*, 5th ed. Homewood, IL: Richard D. Irwin, Inc., 1989.

Steil, L, et al. *Listening: It Can Change Your Life.* New York: John Wiley & Sons, 1984.

TOXLINE: Toxicology Information On-Line, MEDLARS Management Section, Specialized Information Systems, National Library of Medicine, 8600 Rockville Pike, Bethesda, MD 20909, (301) 496–6193, (800) 638–8480.

Underwriters Laboratories Inc., 333 Pfingsten Road, Northbrook, IL 60062. "Product Directories."

U.S. Department of Commerce, NOAA Hazardous Materials Response Branch, 7600 Sandy Point Way NE, Seattle, WA 98115, (206) 526–6317Can also be purchased through the National Safety Council.

U.S. Department of Labor, Occupational Safety and Health Administration, 200 Constitution Avenue, Washington, DC 20210. *An Illustrated Guide to Electrical Safety*, Pub. 3073. *A Brief Guide to Recordkeeping Requirements for Occupational Injuries and Illnesses.* O.M.B. No. 1220–0029. June 1986.

U.S. General Services Administration, National Archives and Records Administration, Office of the Federal Register, Washington, DC 20408. *Code of Federal Regulations*: Title 10—"Energy." Title 29—"Labor." Title 40—"Protection of the Environment." *CFR*, 1910, General Industry Safety and Health. *CFR*, 49, Transportation, Parts 100–199. Standards OSHA 2206.

Note: Government publications are available from the Superintendent of Documents, U.S. Government Printing Office, Washington, DC 20402.

Wolvin, AD, and C Coakley. *Listening*, 4th ed. Dubuque, IA: Brown and Benchmark, 1992.

INDEX